D0982962

Production Operations

Well Completions, Workover, and Stimulation
Volume 1

Third Edition

Third Edition

Volume

1

Production Operations

Well Completions, Workover, and Stimulation

Thomas O. Allen
and
Alan P. Roberts

Oil & Gas Consultants International, Inc.
Tulsa

4554 South Harvard Avenue
Tulsa, Oklahoma 74135

The ideas and techniques presented herein are correct to the best of our knowledge and ability. However, we do not assume responsibility for results obtained through their application

Library of Congress Catalog Card Number: 89-60701
International Standard Book Number: 0-930972-12-0
Production Operations Set (Volumes 1 and 2) ISBN: 0-930972-11-2
Printed in U.S.A.
First Edition–March, 1978
Second printing–April, 1979
Third printing–March, 1981
Fourth printing–January, 1982
Second Edition–May, 1982
Second printing–February, 1984
Third Edition–June, 1989

PREFACE

The third edition of Production Operations includes major revisions and updating of many chapter in Volumes 1 and 2. In Volume 2, a new chapter, "Use of PC's in Production Operations," recognizes the considerable impact of personal computer technology in increasing the efficiency of the individual engineer, explorationist, and manager. In addition to major revisions three appendices have been added to the Fracturing section in line with recent advances in determining rock mechanics parameters and in frac job design. A new appendix has also been added to the Formation Damage section describing use of SEM and EDX technology to improve our understanding of reservoir rock materials. Major revision in perforating include specific recommendations to offset the apparent long-term trend of shortcutting basic perforating requirements, resulting in poor perforating results worldwide, and correspondingly a loss of production and loss of recoverable oil and gas reserves. Major revisions in acidizing include the requirement of determining chemical analysis of specific formation rock prior to acidizing new fracture acidizing techniques, improved control of iron precipitants in both sweet and sour crude wells during acidizing and new techniques for sandstone acidizing. The trend toward cheaper more versatile non-ionic surfactants for use in acids is noted with specific recommendations for surfactant use.

The primary aim of OGCI's Production Operations I course and text books is to provide the best of proved field production technology for field operating engineers and supervisors. We have presented this course and sold Production Operations books to thousands of engineers and supervisors on a worldwide basis. We feel this course and these text books offer the best and least expensive opportunity currently available to the Petroleum Industry to increase producing profits and economically recoverable oil and gas reserves.

In an overview of the Industry, well completion, workover, and stimulation, and well operations may seem to play only a small part. The same is true even if we limit our view to the exploration and production phases of the Industry. From our vantage, however, the focal point of exploration and production is a successful well completion that obtains, and maintains, effective communication with the desired reservoir fluids.

A well completion is not merely a mechanical process of drilling a hole, setting casing, and perforating a hydrocarbon section. The importance of total reservoir description; the role of effective communication between the reservoir and the wellbore; the hazards of flow restrictions around the wellbore; the importance of knowing where fluids are and where they are moving to; and the rigors of excluding undesirable fluids all become more and more evident as we move deeper into the areas of enhanced methods of maximizing recovery of increasingly valuable hydrocarbon fluids.

In preparing Production Operations, Volume 1 and Volume 2, we have separated well completion and well operation technology into packages to permit detailing the more important facets. However, all applicable technology must come together to accomplish effective well completion and recompletion operations. The use of all available technology from many sources is required to solve specific problems.

Volumes 1 and 2 are the product of more than twenty three years of conducting training programs throughout the world for Industry groups, including engineers, managers, geologists, technicians, foremen, service company personnel, and others.

The question is often asked, "What's new in well completion technology?" Our answer must be that new technology per se is not the real issue in considering improvement in producing operations. "The key to optimizing oil and gas recovery and profits is the effective application of proved technology." This has been the theme of our Production Operations courses since our first effort in 1966, and is the theme of these two books on Production Operations. A primary objective of our technical training has been to assist operating groups reduce the length of time required for "proved profit-making techniques" to become routine field practice.

The business of well completion is continually changing. The learning process continues, technology improves, and just as important, the rules of the game change with the times and with the area. In many areas, effective and economic recovery of hydrocarbons from more and more marginal reservoirs is the name of the game. In other areas where costs are tremendous due to the complications of deep wells, offshore activities, or geographic location, high production rates, which are needed to provide sufficient return on the incomprehensible investment required, provide the winning combination.

Response to the first two editions of Production Operations, Volumes 1 and 2, reflects Industry acceptance of our efforts. We anticipate that the much improved third edition will be even more valuable for production operating personnel. The widespread awareness of the need to update petroleum personnel at all levels in the application of proved technology provides OGCI with the incentive to invest time and money in providing new and improved training courses and technical books. To meet this need, OGCI is offering additional courses each year and is continuing the development of a series of technical books for the Petroleum Industry. Finally, OGCI Software has become well established in the development of Petroleum Industry software to improve the efficiency of the industry professional.

T. O. Allen
Alan P. Roberts

Tulsa, Oklahoma
June, 1989

ACKNOWLEDGMENTS

Authors receive many different kinds of assistance from many different people. We wish to acknowledge a number of these contributions by name and also give our thanks to many others, as well.

We owe a special debt of gratitude to the various oilwell service companies. Without their help, there would be no oil industry; nor would there be a book on Production Operations. We particularly want to acknowledge extensive help and counsel from Baker Hughes, Dowell Schlumberger, Western Atlas Integrated Technology, Halliburton Company, and Schlumberger.

For assistance on the first edition, special recognition is due Dr. Scott P. Ewing (deceased), who prepared the original writeup of the chapter on Corrosion (Vol. 2), and to Wallace J. Frank, who contributed a great deal to the first edition.

C. Robert Fast, OGCI, and C. C. McCune, Chevron Research, reviewed the chapter on Acidizing (Vol. 2) and made valuable suggestions. Dr. D. A. Busch, Dr. P. A. Dickey, Dr. G. M. Friedman, and Dr. Glenn Visher, OGCI, rendered valuable assistance in the preparation of the chapter on Geology (Vol. 1).

C.P. Coppel, Chevron Research, reviewed the chapter on Surfactants (Vol. 2), Norman Clark (deceased) and Dr. Charles Smith made valuable technical contributions to the Reservoir Engineering chapter (Vol. 1). Ray Leibach, OGCI, assisted in the preparation of the chapter on Sand Control (Vol. 2), G. W. Tracy, OGCI, reviewed the section on production testing, Well Testing chapter, Vol. 1. John E. Eckel, OGCI, contributed to the chapter on Downhole Production Equipment.

For work and suggestions on the second edition, Wayne Hower made valuable contributions in the revision of several chapters. We also wish to acknowledge the assistance of C. Robert Fast, Raymond E. Leibach, and Carl E. Montgomery, who had many helpful suggestions for the revision of the Hydraulic fracturing chapter. Also, Raymond E. Leibach made a significant contribution on the revision of "Tubing Strings, Packers, Subsurface Control Equipment." And we wish to thank Dwight Smith for his assistance in updating and revising the Primary Cementing chapter.

For the third edition the contributions of several people and Industry groups on the fracturing and acidizing sections was significant. Particularly included are C. Robert Fast, Calvin Saunders, A. R. Hendrickson, and Loyd Jones, OGCI associates; Jerry Hinkle and Curtis Crowe, Dowell Schlumberger, Jim McGowen, Ed Stahl, and Gene Broaddus, Halliburton Services; Chris Parks, Kelco Oil Field Group; Harry McLeod, Conoco; and William Daniel, Daniel-Price Exploration. Charles George, Halliburton Services, made helpful suggestions on updating the cementing sections. Jim Erdle, OGCI Software, prepared the new section on use of PC's in production operations.

Our thanks and appreciation go, too, to the various operating companies that have participated in our Production Operations course sessions. They have helped hone and refine many generations of lecture notes into this two volume textbook.

We appreciate the assistance of Gerald L. Farrar, publications editor of the first edition, Patricia Duyfhuizen, publications editor of the second edition, and Verma Hughes, publications editor of the third edition of the books.

Finally, we acknowledge the valuable contribution of Jewell O. Hough, who over a twelve year period prepared dozens of revisions, leading up to the publication of these books.

—The Authors

Contents Vol. 1

Contents Vol. 2

Chapter 1 Geologic Considerations In Producing Operations

INTRODUCTION

Geologic studies have provided data for the finding, development, and operation of oil and gas reservoirs during the more than 125 years of oil and gas field operation.

Engineers, as well as geologists, have employed isopachs, structure maps, isobaric maps, core and log information, production tests, and other data as a guide in decisions relative to the development and operation of oil and gas fields. These tools, however, often proved to be inadequate for correct extrapolation or prediction of the vertical and horizontal continuity of porosity and permeability beyond the wellbore. Transient pressure tests provide additional clues, but only represent average values of both vertical and lateral variations in reservoir properties of porous zones in communication with the wellbore.

Various types of well tests, particularly "pulse" tests, are useful in determining continuity of permeability between wells. Pressure drawdown tests may help determine reservoir limits. However, the results of all types of transient pressure tests are questionable in layered zones, unless dynamic tests with tools such as through tubing flowmeters indicate flow from *all* permeable layered zones open to the wellbore.

Research personnel and earth scientist/engineer teams during the past forty years have focused attention on the importance of more comprehensive analysis of the reservoir geology, particularly the interrelation of fluid flow in the reservoir with depositional environment and post-depositional history. The more general use of this approach in reservoir evaluation has brought about a better understanding of the quality, permeability distribution, and continuity of specific reservoirs and their contained fluids. The result has been increased oil recovery for a number of reservoirs from the 30–40% range to 50–83%.

The aim of this chapter is to highlight geologic technology useful in solving production problems and to present case studies to illustrate the value of improved reservoir description for both sandstone and carbonate reservoirs. Total geologic description of the reservoir is integrated with reservoir fluid flow data, reservoir production history and reservoir performance predictions to provide information required to plan effective depletion of the reservoir.

Harris[18] in his treatise on "The Role of Geology in Reservoir Simulation" outlined geologic controls on reservoir properties and pointed out the significance of these controls on reservoir performance. Figure 1-1 (from Harris) shows combined geologic and reservoir engineering data input required for good reservoir description.

Some of the functions of the geologist in this type of study may be to:

—Select core samples for both geologic and reservoir studies
—Identify the depositional environment and source rocks
—Develop a depositional model modified by post-depositional changes
—Construct a structure map
—Develop cross sections or other representations to show changes in rock properties throughout the reservoir
—Develop porosity and permeability trends including both horizontal and vertical barriers to fluid flow—particularly those trends predictable from depositional and post depositional history of the reservoir

Figure 1-2 illustrates typical data obtained from combined studies of geologic and reservoir engineering aspects.

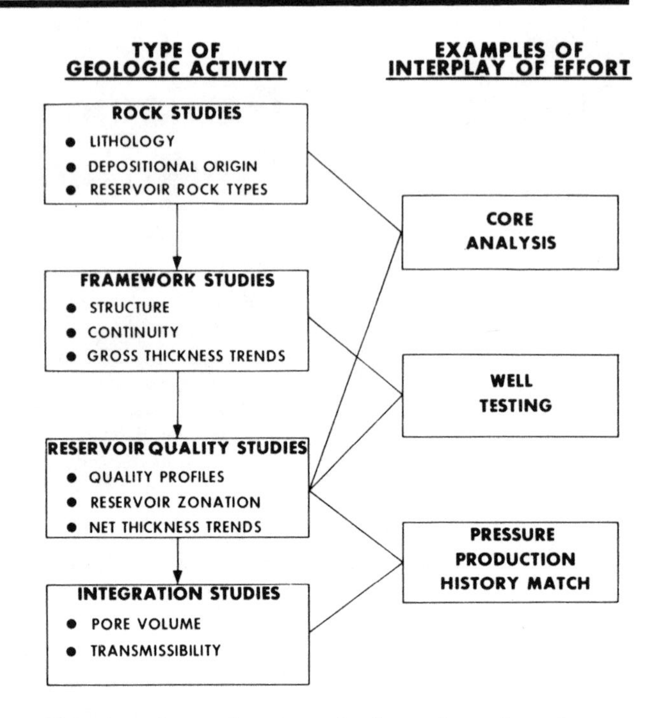

FIG. *1-1—Interrelated geologic and engineering activities in reservoir description.[18] Permission to publish by The Society of Petroleum Engineers.*

The Habitat of Oil and Gas

Most oil and gas reservoirs are found in sandstones or carbonates. There are very limited occurrences in shale, volcanic rock, and fractured basement rock (basalt). Comparing the significance of sandstone and carbonate reservoirs, sandstones are more abundant,

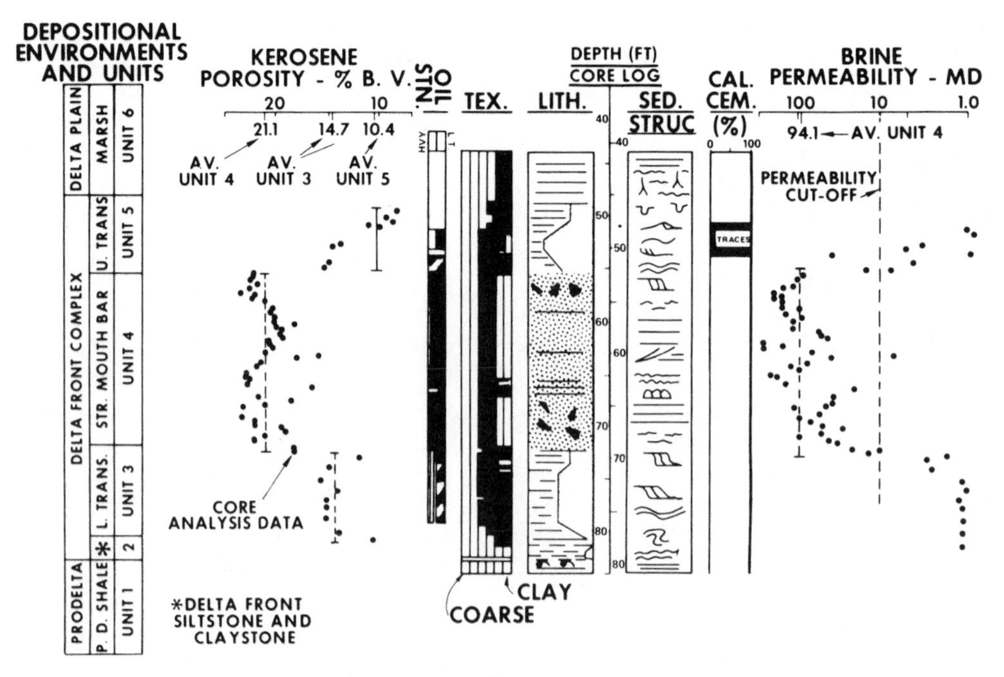

FIG. *1-2—Example reservoir quality profile.[18] Permission to publish by The Society of Petroleum Engineers.*

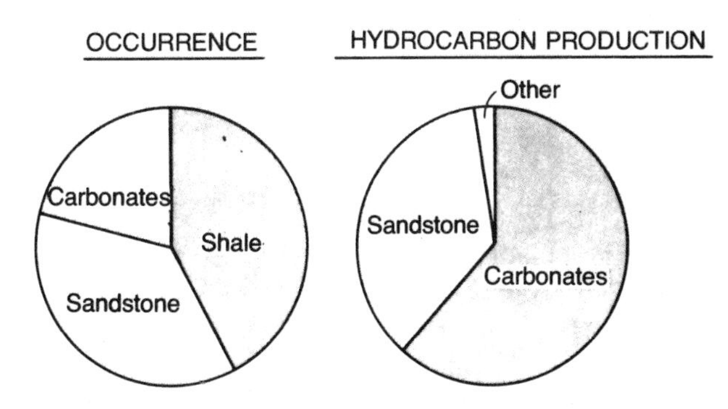

FIG. *1-3—Significance of sandstone and carbonates.*[15]

yet limestones are more important as reservoirs for hydrocarbons, as shown in Figure 1-3. The higher percentage of hydrocarbons in carbonates is greatly influenced by the numerous large reservoirs in the Middle East.

Traps for Oil and Gas Accumulation

The three prerequisites to a commercial accumulation of oil and/or gas are:

1. Source rock
2. Porous and permeable container rock
3. Impermeable caprock or seal

A trap may be structural, stratigraphic, or a combination of the two. That portion of the trap in which oil or gas is stored in nature is usually referred to as a petroleum reservoir.

Figure 1-4 illustrates an oil accumulation in an anticlinal structure where depositional strata have been bent upward. This very common type of structure accounts for many large accumulations of oil and gas.

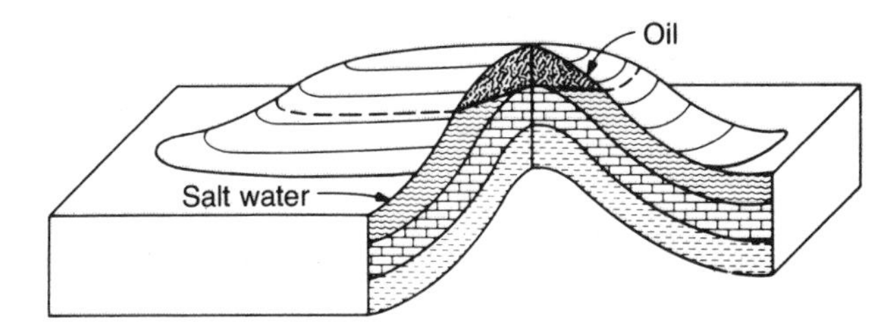

FIG. *1-4—Oil accumulation in an anticlinal structure.* Elements of Petroleum Reservoirs. *Permission to publish by The Society of Petroleum Engineers.*

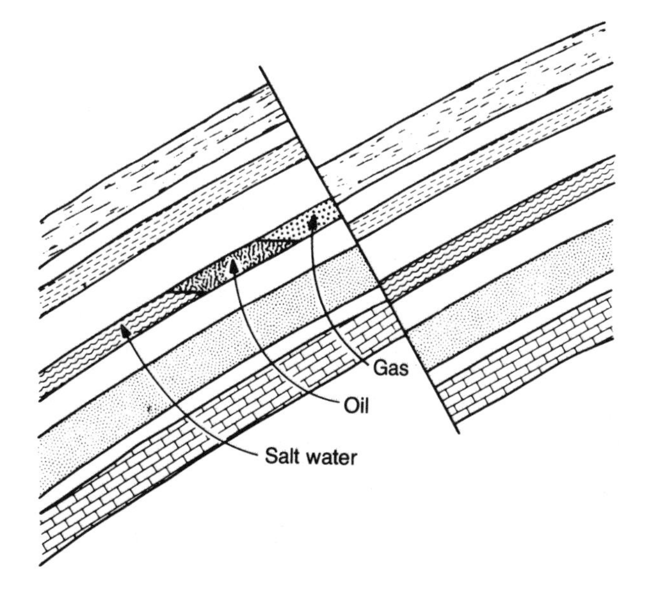

FIG. *1-5—Structural trap resulting from faulting.* Elements of Petroleum Reservoirs. *Permission to publish by The Society of Petroleum Engineers.*

Figure 1-5 is a structural trap. Oil and gas is trapped against impermeable beds as a result of faulting. Many oil and gas accumulations are associated with growth faults formed in ancient river deltas. These "normal" or growth faults were caused by rapid sedimentation on top of salt beds or undercompacted marine clays.

Figure 1-6 represents an oil accumulation under an unconformity where the upward movement of oil in

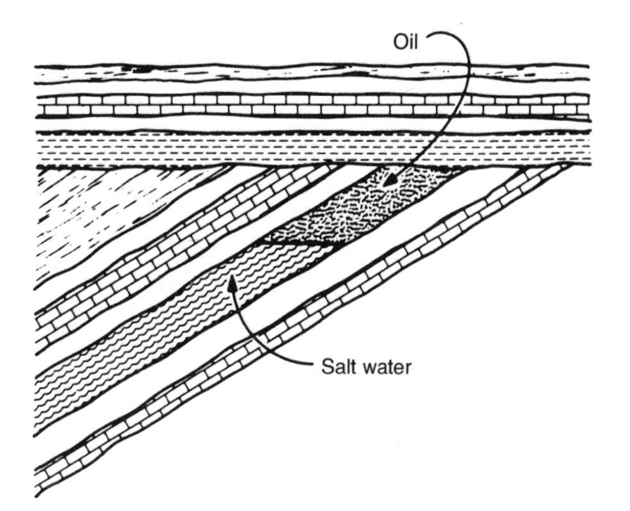

FIG. *1-6—Oil accumulation under an unconformity.* Elements of Petroleum Reservoirs. *Permission to publish by The Society of Petroleum Engineers.*

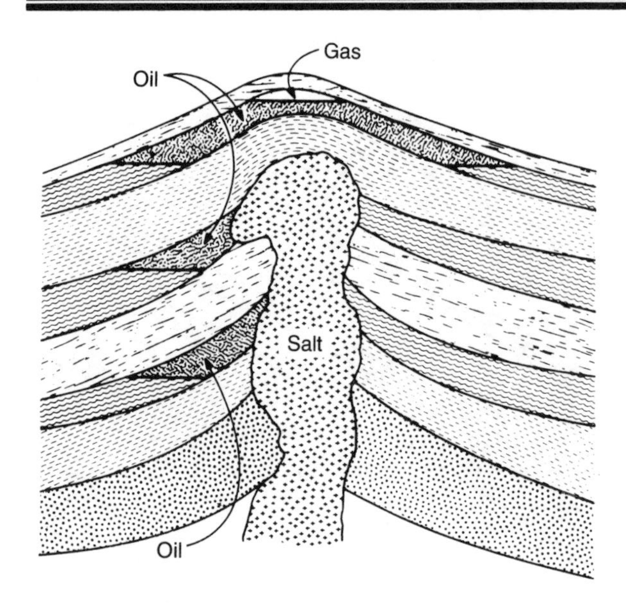

FIG. *1-7—Oil accumulation in the vicinity of a piercement-type salt dome.* Elements of Petroleum Reservoirs. *Permission to publish by The Society of Petroleum Engineers.*

a permeable zone has been blocked by impermeable deposits laid down on the weathered surface of lower beds. The East Texas field, which represents the largest known oil reservoir in the contiguous forty-eight states of the U.S., was formed in this manner.

Figure 1-7 represents an oil accumulation on top and along the sides of a salt dome. Many oil and gas reservoirs along the Gulf Coast of the U.S. are associated with both deep-seated and piercement-type salt domes. In this situation, riverborne sediments were rapidly deposited in the Gulf of Mexico on top of thick beds of plastic clays or salt. With continued deposition the added weight on the shale or salt caused it to compress, resulting in growth faults above these plastic beds. In some areas, the plastic salt flowed upward thousands of feet through faults or fractures, forming huge salt plugs or domes.

As the salt plug moved upward, much like a massive pile driver, many radial faults were created around the salt dome. Flow of oil or gas in an upward direction along depositional strata is often blocked by the salt domes. Radial faults caused the hydrocarbon accumulation around a salt dome to be broken into many reservoirs.

Figure 1-8 shows an oil accumulation in lenses within a sandstone bar, representing one type of stratigraphic trap. Oil and gas are found in many different types of stratigraphic traps in sandstones and carbonates.

Figure 1-9 shows a cross-section through a limestone reef with an oil accumulation in the top. Limestone reefs may be an accumulation of shells, detritus, or coral. It has been erroneously assumed by some engineers and others that most reefs are homogeneous with continuous horizontal and vertical permeability.

No reservoir should be considered to have continuous vertical permeability from top to bottom unless this is proved by vertical permeability measurements of the full diameter cores throughout the entire permeable section.

Fractures and Joints in Reservoir Rocks

Mechanical characteristics of reservoir rocks can vary considerably. Some may be plastic or semiplastic, and bend or deform without fracturing. Others are hard and brittle, and fracture or break rather than bend. Reservoir rocks may be fractured by either tensional or compressive forces. Shales found at depths to about 15,000 ft in the Gulf Coast of Texas and Louisiana are examples of plastic rocks.

Most dolomites have a high strength with little plasticity and frequently fracture rather than bend when subjected to high stress. Fracture systems, such as those found in the huge oil fields of the Middle East, may result from regional stress. Fractures provide major flow channels in these reservoirs whereas matrix permeability of the carbonates varies considerably.

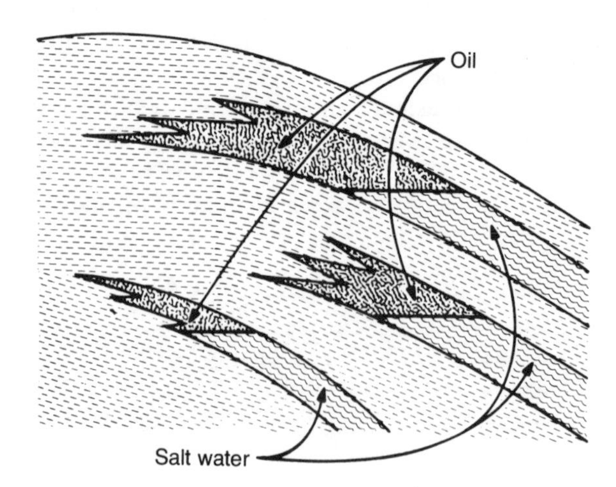

FIG. *1-8—Oil accumulation in sand lenses of the sand bar type.* Elements of Petroleum Reservoirs. *Permission to publish by The Society of Petroleum Engineers.*

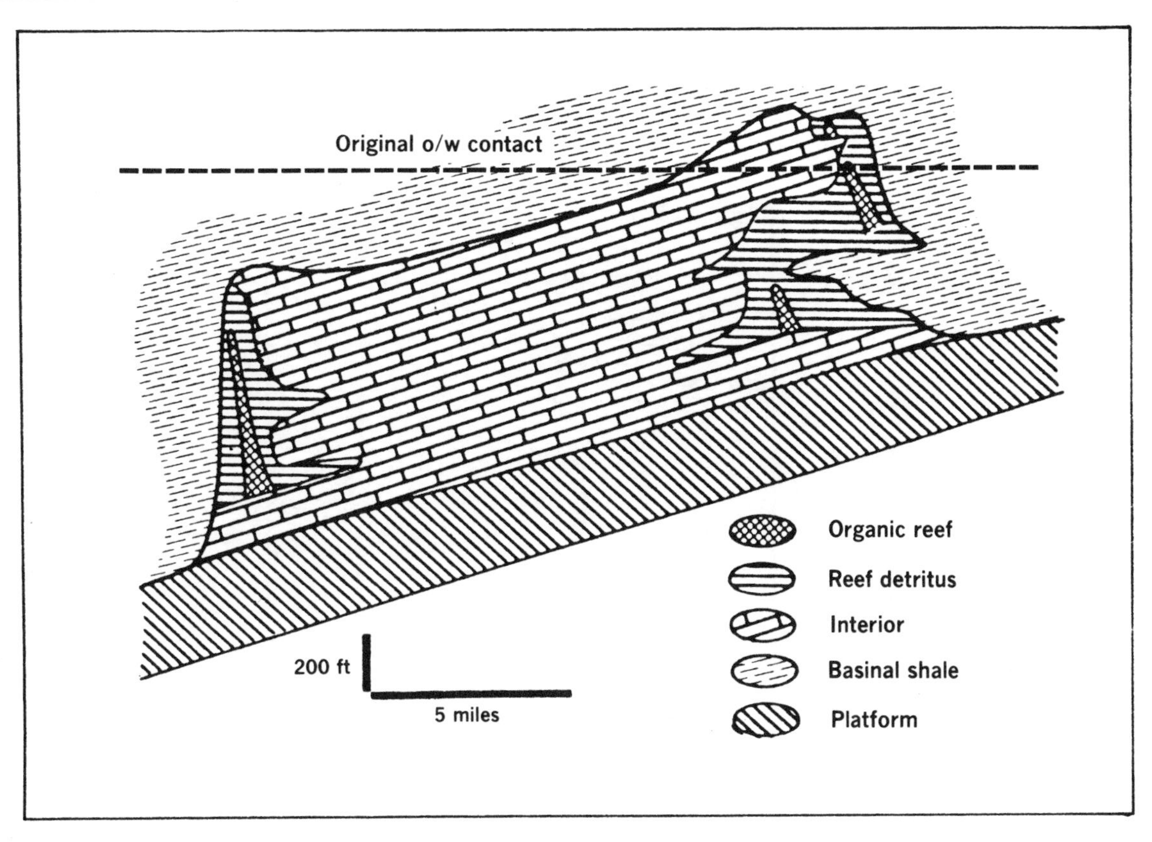

FIG. *1-9—Reef with oil accumulation in top.*[20] *Permission to publish by The Society of Petroleum Engineers.*

Faults associated with salt domes in the Texas and Louisiana Gulf Coast are representative of fracturing caused by local rather than regional stress, although regional stress may contribute to total fracturing and faults around salt domes. These faults are usually closed and seldom provide flow paths along the fracture face.

Joints may be defined as closed, vertical or near vertical, smooth straight fractures. They occur in parallel sets with spacings ranging from a few centimeters to several meters apart. The direction of fracture planes may be constant over large areas. Although joints are usually closed at depth, they may be propped open during hydraulic fracturing, or etched during fracture acidizing to form flow channels. If joints are forced open by hydraulic pressure during drilling, cementing, acidizing or well killing operations, the result may be partial loss of circulating fluid or complete loss of returns under more extreme cases. It is common for joints to be held open during waterflood operations, resulting in loss of sweep efficiency. If reservoir pressure is low relative to well depth, a high percentage of injected fluids may enter joints.

Joints are found in most oil and gas producing reservoirs throughout the world; however, they occur most frequently in high compressive strength rock, especially in carbonates. Parallel joints may be continuous for many miles, and joint systems may cover hundreds of square miles. Linears or joints can sometimes be observed on the surface, and their direction may or may not be the same as joints in the subsurface. Joints are often seen in cores as vertical or near-vertical fractures; however, widely spaced joints may not always be penetrated by a core bit. The Spraberry trend in the Permian Basin near Midland, Texas and the Marmaton trend in North Texas and Western Oklahoma are examples of extensively jointed oil-bearing reservoirs.

SANDSTONE RESERVOIRS

Most oil and gas production from sandstone is derived from deposits originating from river-borne sediments. Environments and depositional models for

ENVIRONMENTS					DEPOSITIONAL MODELS
CONTINENTAL	ALLUVIAL (FLUVIAL)	ALLUVIAL FANS (APEX, MIDDLE & BASE OF FAN)	STREAM FLOWS	CHANNELS	ALLUVIAL FAN (APEX OF FAN; BRAIDED CHANNELS (WASHES) AND ABANDONED CHANNELS; MIDDLE OF FAN; EDGE OF MOUNTAINS; PAVEMENT; BASE OF FAN; SALT PAN)
				SHEETFLOODS	
				"SIEVE DEPOSITS"	
			VISCOUS FLOWS	DEBRIS FLOWS	
				MUDFLOWS	
		BRAIDED STREAMS		CHANNELS (VARYING SIZES)	BRAIDED STREAM (LONGITUDINAL BARS; TRANSVERSE BARS)
				BARS: LONGITUDINAL	
				BARS: TRANSVERSE	
		MEANDERING STREAMS (ALLUVIAL VALLEY)	MEANDER BELTS	CHANNELS	MEANDERING STREAM (MEANDER BELT; FLOODBASIN; MEANDER BELT; FLOODBASIN; NATURAL LEVEE; POINT BAR; CHANNEL; CREVASSE; DIRECTION OF POINT BAR ACCRETION)
				NATURAL LEVEES	
				POINT BARS	
			FLOODBASINS	STREAMS, LAKES & SWAMPS	
	EOLIAN	DUNES	COASTAL DUNES	TYPES: TRANSVERSE, SEIF (LONGITUDINAL), BARCHAN, PARABOLIC, DOME-SHAPED	COASTAL DUNES (BEACH; PREVAILING WIND DIRECTION; OCEAN; OLDER DEPOSITS)
			DESERT DUNES		DESERT DUNES (BARCHANS & SAND SHEETS; SEIF DUNES)
			OTHER DUNES		

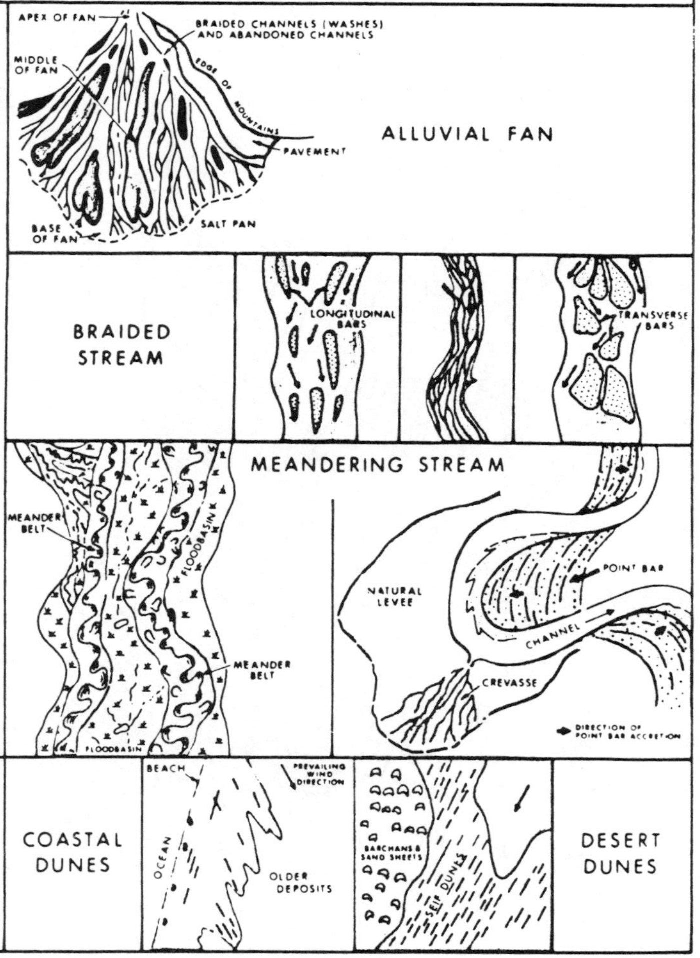

FIG. 1-10—*Alluvial (fluvial) and eolian environments and models of clastic sedimentation.*[22] *Permission to publish by AAPG.*

ENVIRONMENTS					DEPOSITIONAL MODELS
TRANSITIONAL	DELTAIC	UPPER DELTAIC PLAIN	MEANDER BELTS	CHANNELS	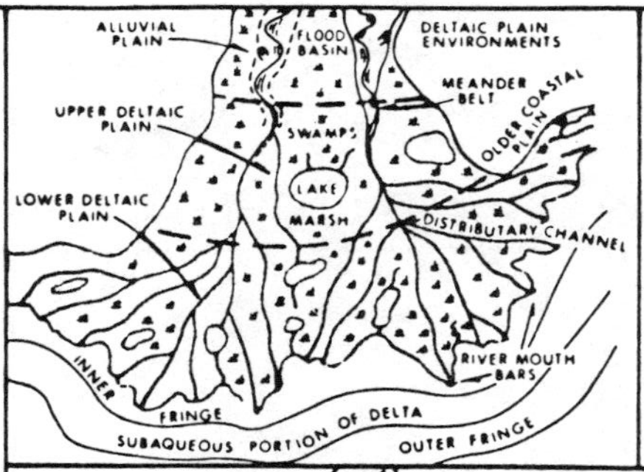 TYPES OF DELTAS: BIRDFOOT-LOBATE DELTA (ALLUVIAL PLAIN; FLOOD BASIN; DELTAIC PLAIN ENVIRONMENTS; MEANDER BELT; OLDER COASTAL PLAIN; UPPER DELTAIC PLAIN; SWAMPS; LAKE; MARSH; LOWER DELTAIC PLAIN; DISTRIBUTARY CHANNEL; RIVER MOUTH BARS; INNER FRINGE; SUBAQUEOUS PORTION OF DELTA; OUTER FRINGE)
				NATURAL LEVEES	
				POINT BARS	
			FLOODBASINS	STREAMS, LAKES & SWAMPS	
		LOWER DELTAIC PLAIN	DISTRIBUTARY CHANNELS	CHANNELS	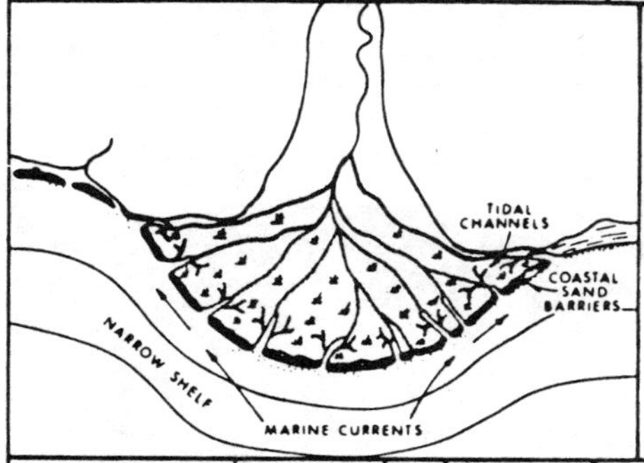 CUSPATE-ARCUATE DELTA (TIDAL CHANNELS; COASTAL SAND BARRIERS; NARROW SHELF; MARINE CURRENTS)
				NATURAL LEVEES	
			INTER-DISTRIBUTARY AREAS	MARSH, LAKES, TIDAL CHANNELS & TIDAL FLATS	
		FRINGE	DELTA FRONT: INNER	RIVER-MOUTH BARS BEACHES & BEACH RIDGES TIDAL FLATS	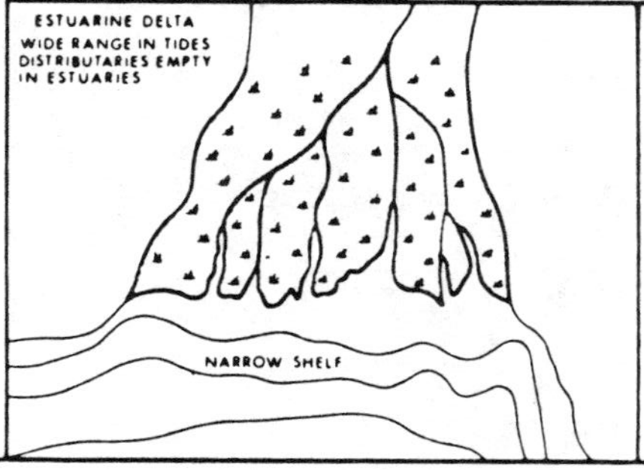 ESTUARINE DELTA (ESTUARINE DELTA WIDE RANGE IN TIDES DISTRIBUTARIES EMPTY IN ESTUARIES; NARROW SHELF)
			DELTA FRONT: OUTER		
		DISTAL			

FIG. 1-11—*Deltaic environments and models of clastic sedimentation.*[22] *Permission to publish by AAPG.*

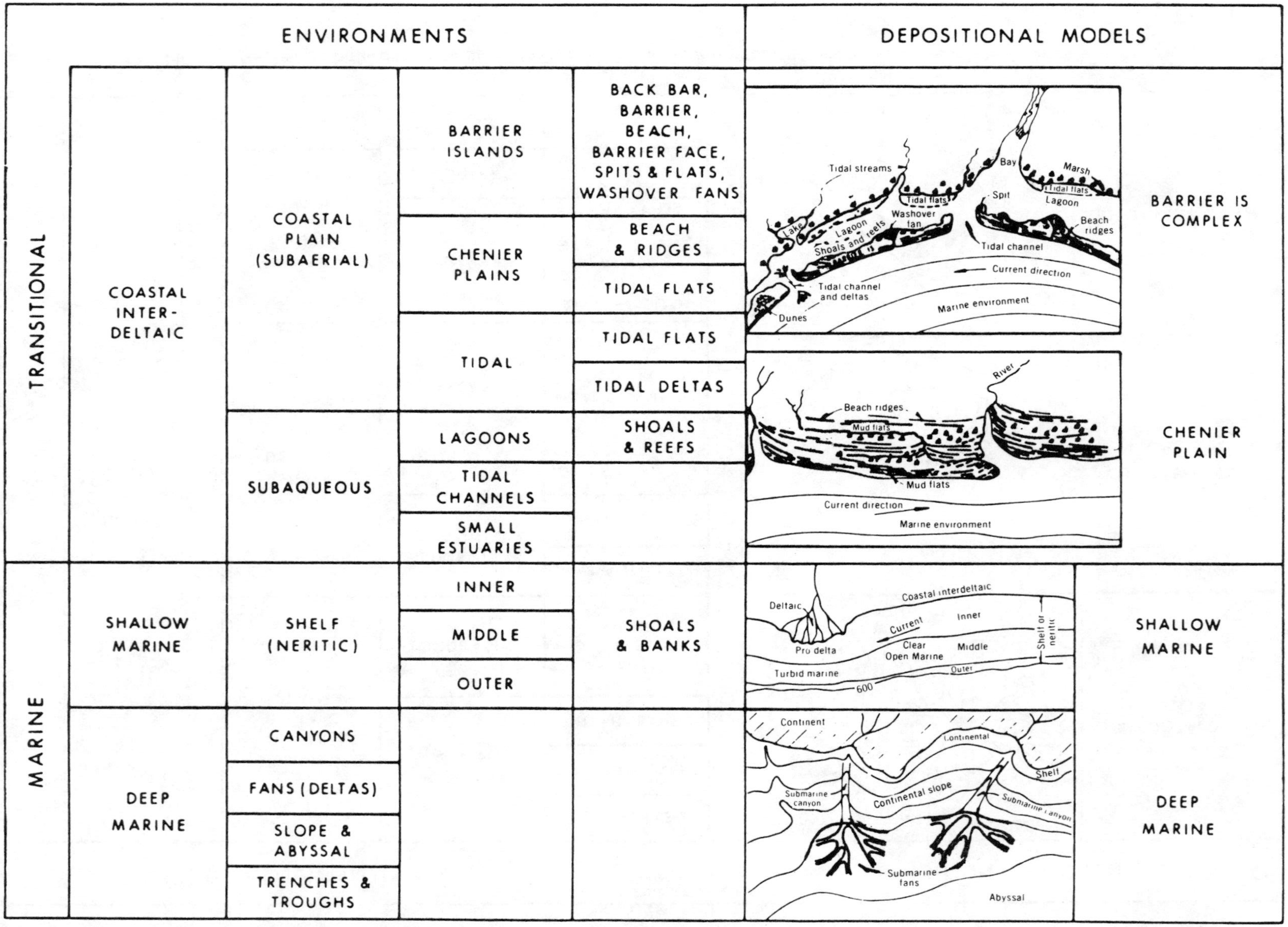

	ENVIRONMENTS				DEPOSITIONAL MODELS
TRANSITIONAL	COASTAL INTER-DELTAIC	COASTAL PLAIN (SUBAERIAL)	BARRIER ISLANDS	BACK BAR, BARRIER, BEACH, BARRIER FACE, SPITS & FLATS, WASHOVER FANS	BARRIER IS COMPLEX
			CHENIER PLAINS	BEACH & RIDGES	
				TIDAL FLATS	
			TIDAL	TIDAL FLATS	CHENIER PLAIN
				TIDAL DELTAS	
		SUBAQUEOUS	LAGOONS	SHOALS & REEFS	
			TIDAL CHANNELS		
			SMALL ESTUARIES		
MARINE	SHALLOW MARINE	SHELF (NERITIC)	INNER	SHOALS & BANKS	SHALLOW MARINE
			MIDDLE		
			OUTER		
	DEEP MARINE	CANYONS			DEEP MARINE
		FANS (DELTAS)			
		SLOPE & ABYSSAL			
		TRENCHES & TROUGHS			

FIG. 1-12—*Coastal-interdeltaic and marine environments and models of clastic sedimentation.*[22] *Permission to publish by AAPG.*

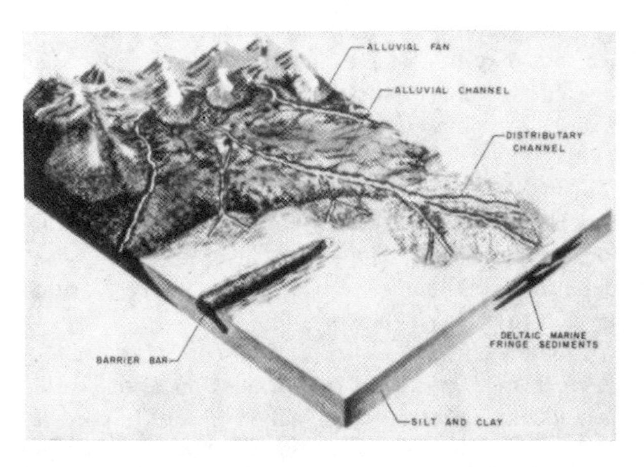

FIG. *1-13—Types of sandstone deposits, Elk City field, Oklahoma.[28] Permission to publish by The Society of Petroleum Engineers.*

river-borne sediments forming continental, transitional, and marine deposits are shown in Figures 1-10, 1-11, and 1-12 (from Geometry of Sandstone Reservoir Bodies by Rufus J. LeBlanc, AAPG Memoir 18[22]). Although LeBlanc illustrates many types of sandstones, a majority of sandstone is deposited in river deltas either as channel deposits, river mouth bars, or shallow marine bars. Examples of deposits having river-borne sediments as their source material are the hundreds of reservoirs laid down by ancestral rivers during the Tertiary period, including those related to the Mississippi River, and the Niger River.

The Latrobe River deposited sediments in Bass Straits fields off Australia, and the Orinoco River accounts for the prolific sandstone deposits off Trinidad. The McKenzie delta reservoirs in Canada, the Prudhoe Bay field in Alaska, Burgan field in Kuwait, and the many reservoirs in Indonesia are primarily river-borne deposits.

Representative types of deposits found in a delta complex are shown in Figure 1-13, which depict the alluvial channels, distributary channels, deltaic bar (marine fringe), and shallow marine barrier bar deposits. This model of deposits in the Elk City field, Oklahoma is illustrative of river-borne deposits found all over the world.

Figure 1-14 shows how the current in a river erodes the outside of a bend, and deposits sediment on the inside of a bend to form point bars, which are scattered all over the flood plain of a meandering river.

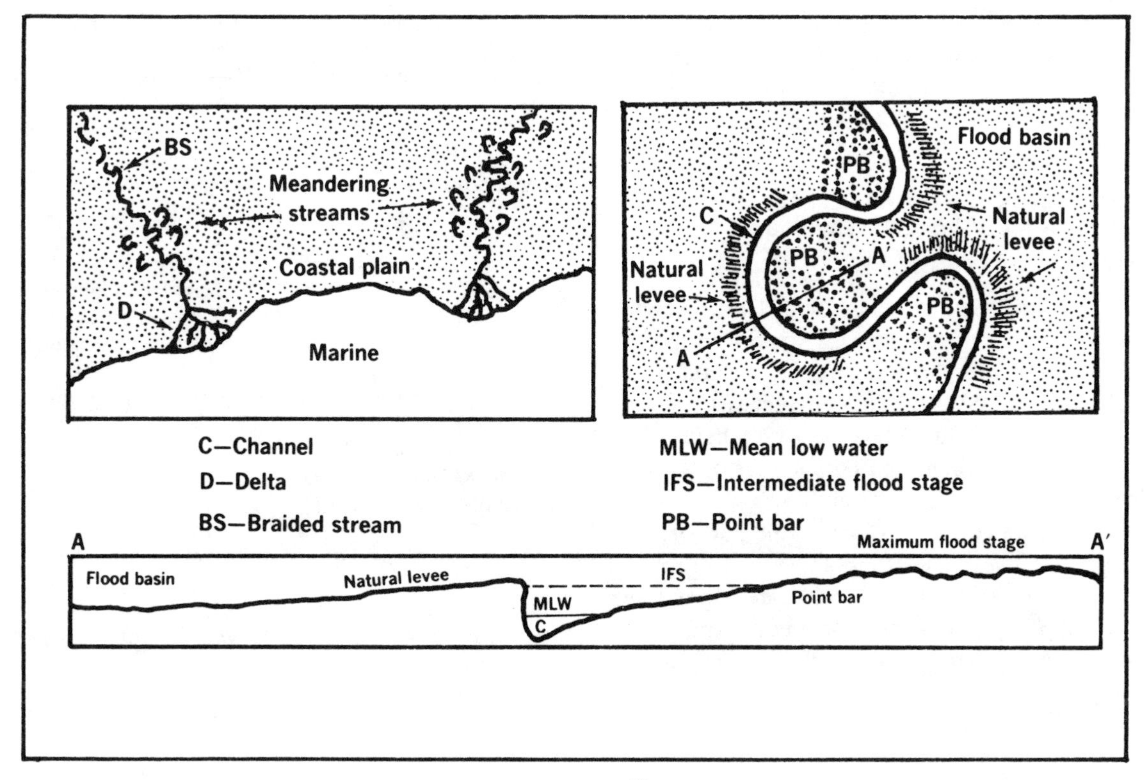

FIG. *1-14—Point bars formed in river bends.[22] Permission to publish by AAPG.*

Many significant oil and gas fields produce from point bars. Point bars are found all along a river from its source to the delta. The Niger delta is somewhat of an exception because a large number of producing reservoirs in this delta complex are point bars.

Oil and gas reservoirs are found frequently in sand bars deposited in braided streams, which are illustrated in Figure 1-10.

Deep marine deposits or turbidites, illustrated in Figure 1-12, can be deposited on the continental slope or may flow to the ocean floor. A number of very thick hydrocarbon reservoirs in California are examples of very prolific deep marine or turbidite deposition.

Significant wind-deposited dune sand reservoirs are relatively rare. A notable exception is the Rotliegendes formation of Permian age, extending from Germany through the Netherlands to southeastern England where gas production is obtained primarily from ancient sand dune deposits. The most significant reservoir producing from the Rotliegendes is the Groningen gas field in Holland, which contained the largest gas reserve in Europe.

Rotliegendes deposits under the North Sea generally exhibit a much lower permeability than in the Groningen field. The Leman field and others off southeastern England are representative of this lower permeability group with permeabilities averaging 1 md or less where pores in fine-grained sand have been filled primarily with secondary silica cement.

GEOLOGIC FACTORS AFFECTING RESERVOIR PROPERTIES IN SANDSTONE RESERVOIRS

Porosity and Permeability

Porosity is the pore volume divided by the bulk volume of the rock, and is usually expressed in percent. It provides the container for the accumulation of oil and gas, and gives the rock characteristic ability to absorb and hold fluids. The ease with which fluids move through inter-connected pore spaces of the reservoir rock is defined as permeability.

Compared with carbonates, sandstone porosity appears to be relatively consistent and easy to predict. Primary factors controlling porosity are grain size, sorting, texture (grain shape, roundness, and packing), type and manner of cementation of grains, and the amount and location of clay or other minerals associated with the sand. Layering, and intrusion of clay, carbonates, or other materials in the interstices of a sandstone deposit can appreciably lower porosity and permeability. Cementing material is usually silica, but may be calcite, clay or other minerals.

Table 1-1 provides a brief analysis of sandstone porosity. Porosities in sandstone reservoirs are usually in the 15–30% range; however, sandstones having much lower porosity are currently being economically produced, primarily through the use of hydraulic fracturing. Because of sedimentary layering in sandstone deposits, continuous vertical permeability is often limited to a few inches or a few feet.

It is evident that a 1-in. vertical core plug per foot of sand, frequently used in evaluating sandstone wells, is inadequate for the determination of vertical porosity and permeability of a reservoir. Continuity of vertical permeability is especially significant in studies to determine whether vertical movement of unwanted fluids within a producing oil or gas reservoir is due to con-

TABLE 1-1
Porosity in Sandstone Reservoirs[6]

Aspect	*Description*
Primary porosity in sediments	Commonly 25–40%
Ultimate porosity in rocks	Commonly 10–30%
Types of primary porosity	Almost exclusively interparticle.
Types of ultimate porosity	Primary interparticle, greatly modified by precipitation of authigenic clay minerals and silica.
Sizes of pores	Diameter and throat sizes closely related to sedimentary particle sizing and sorting.
Shape of pores	Initially dependent on particle shape, but greatly modified by secondary clay and silica.
Uniformity of size, shape, and distribution	Commonly fairly uniform
Influences of diagenesis (changes in rock since deposition)	Often large reduction of primary porosity by compaction and cementation.
Influence of fracturing	Generally not of major importance in reservoir properties.
Adequacy of core analysis for reservoir evaluation	Core plugs of 1-in. diameter commonly used for "matrix" porosity.
Permeability-porosity interrelations	Relatively consistent: commonly dependent on particle size and sorting.

ing, fingering, a poor cement job, or mechanical problems. Vertical permeability analysis of full diameter cores over the entire producing zone is required to determine whether coning is possible in the vicinity of a well. Unless it has been proved conclusively that a specific hydrocarbon-bearing zone has continuous vertical permeability, perforation of only part of the zone may result in low production and loss of economically recoverable oil or gas reserves.

Shale Break Prediction From a Study of Outcrops

Planning of well completions, including selection of the zones to be perforated, requires detailed data concerning continuity of both horizontal and vertical permeability and the location of barriers to movement of reservoir fluids. Some very thin shale breaks have been correlated for miles through log, core and outcrop studies of producing reservoirs.[16]

Zeito's sandstone study[33] of shallow marine, deltaic and channel origin provides valuable guidelines in this regard. Table 1-2 shows that shale breaks in shallow marine sandstone may be reliably predicted to have lateral continuity for considerable distances. Zeito reported that shale or silt layers deposited in channel or fluviatile sands usually converge on adjacent shales; thus, sands tend to be pinched out by shales or silt. Because of converging shales or silt layers, only 17% of the channel sand deposits in Zeito's study were continuous for over 500 ft., and only 34% were continuous for over 250 ft. Further Zeito reported:

—Shale breaks in shallow marine sandstone are generally parallel, and concentrated near the top and bottom of a sandstone unit.

—Shale breaks in channel and other onshore sandstone deposits tend to be converging, randomly distributed and unpredictable.

—Shale breaks in onshore deltaic sands tend to be more continuous than in channel sands in alluvial deposits and less continuous than in shallow marine deposits.

TABLE 1-2
Lateral Continuity of Shale Breaks in Marine Sandstone[33]

% of shale breaks	*Lateral extent, ft*	*Confidence*
96	>250	99%
89	>500	86%
83	>1,000	52%
80	>2,000	48%

Implications of Zeito's Outcrop Studies—In channel and on-shore deltaic deposits, the tendency for a shale stringer to converge upon an adjacent shale stringer has serious connotations in regard to trapping oil or gas. Converging shales can reduce recovery efficiency of waterfloods and enhanced recovery. Selection of perforation interval can be critical in channel and deltaic sands. Crossbedding tends to place a directional restriction on sweep efficiency of waterfloods, and enhanced recovery.

The lateral continuity of shale breaks and sandstone members in shallow marine deposits provides favorable depositional conditions for waterflooding and enhanced recovery operations.

Effect of Silt and Clay Content on Sandstone Permeability[19]

Figure 1-15 relates permeability to percent of silt and clay for several sandstone formations in the United States and Canada. In all nine reservoirs an increase in silt and clay decreased permeability. However, direct correlation between clay content and permeability is usually restricted to a single genetic sand unit as illustrated by comparing the Stevens and Bow Island sandstones. In the Stevens sand, 9% silt and clay reduced permeability to 1 md. In the Bow Island sand, permeability was about 1000 md with 9% clay

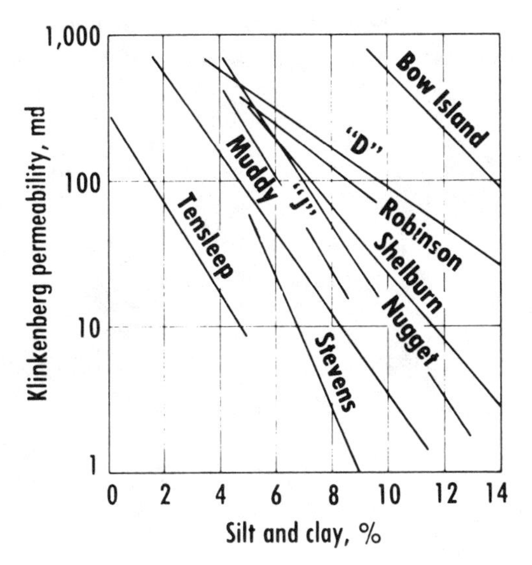

FIG. *1-15—Effect of silt and clay on sandstone permeability.[19] Permission to publish by The Petroleum Publishing Co.*

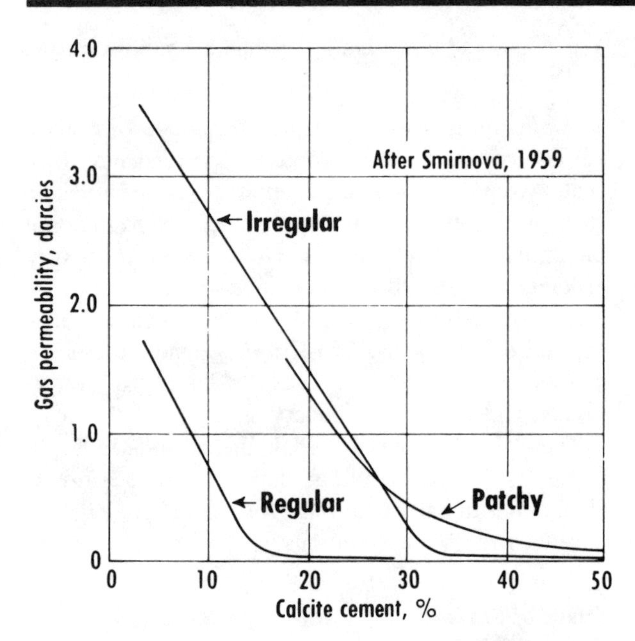

FIG. *1-16—How calcite cement alters permeability.*[19] *Permission to publish by The Petroleum Publishing Co.*

and silt. This wide difference in permeability of the two sandstones, each with 9% silt and clay, is due primarily to the much coarser texture of the Bow Island sand.

Effect of Mineral Cement in Sandstone Permeability[19]

Figure 1-16 shows the effect of calcite cement distribution on fine-grained quartz sandstones from the Ugersko suite of the Miocene in Russia. With uniform distribution of calcite within the reservoir rock, approximately 20% calcite reduced gas permeability to less than 50 md. If 20% calcite was irregularly distributed within the rock, this Miocene sand had a gas permeability of 1500 md.

Permeability Variations with Texture Changes

The sand grain size analysis, shown in Figure 1-17,[19] indicates a characteristic fluviatile sandstone reservoir having a fine-grained, ripple-laminated upper zone and a medium-grained, cross-stratified lower zone. The finer grained upper zone had lower water-oil relative permeability ratios at all water saturations.

Texture contrast causes some variation in permeability and relative permeability. A study by Hutchinson of a shallow marine sandstone showed that sand texture caused a maximum variation in permeability of 5:1. By comparison, the degree and type of cementation caused a 100:1 variation in permeability.

Relation of Permeability to Irreducible Water Saturation

Figure 1-18[19] presented by White shows the relationship between premeability and irreducible water saturation for a number of formations. In general, permeability to oil or gas decreases as irreducible water saturation increases. For a single genetic sand unit, the irreducible water relationship is sufficiently precise for calculation and prediction purposes.

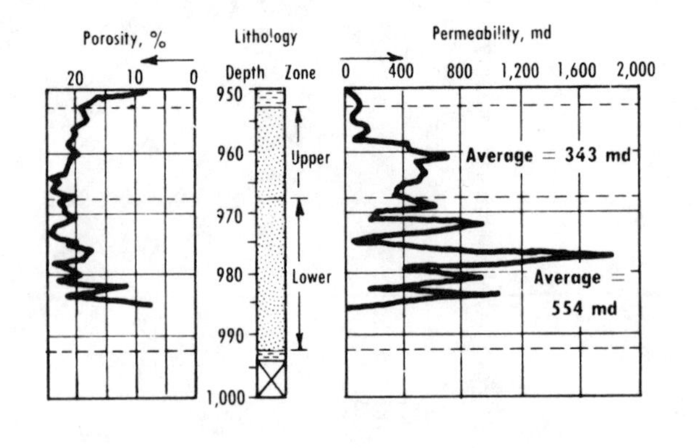

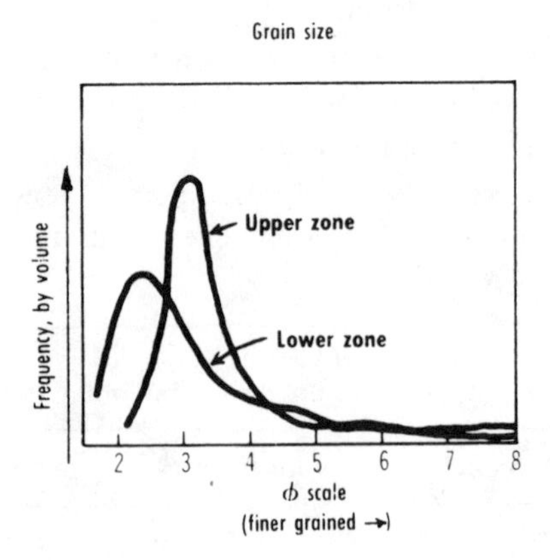

FIG. *1-17—Data from a layered Pennsylvanian sandstone reservoir.*[19] *Permission to publish by The Petroleum Publishing Co.*

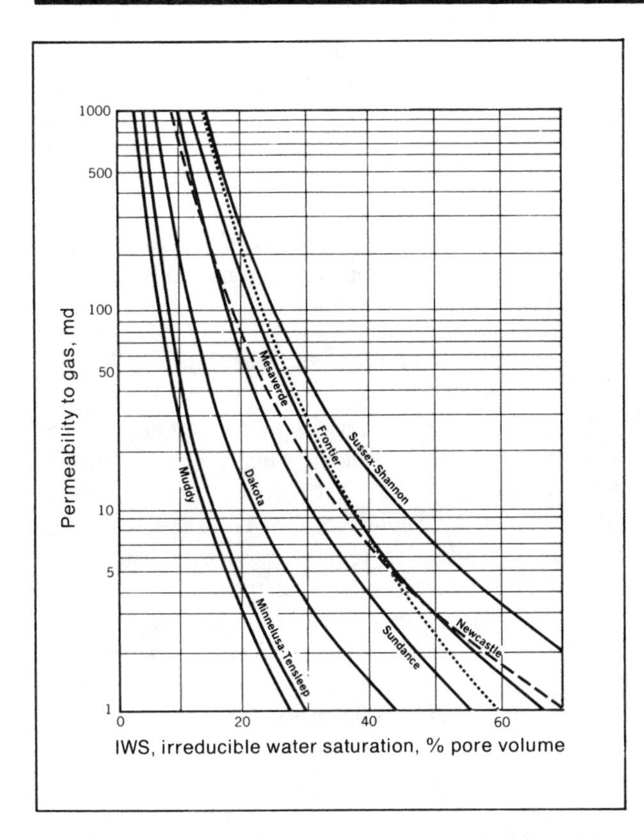

FIG. *1-18—How permeability and irreducible water saturation are related.*[19] *Permission to publish by The Petroleum Publishing Co.*

Identification of Channel or Fluviatile Sandstones

Hewitt's detailed studies[19] of a Pennsylvania sandstone reservoir provides a useful guide to help identify channel or fluviatile reservoirs. Core analysis and logs illustrated in Figure 1-19, along with production tests, structure maps, and isopach maps, are frequently the only available guides in the development and operation of oil and gas fields.

Usual interpretation of log and core data would provide the following information.

—Lithology log: sand from reservoir top at 885 ft to 942 ft. Neither the SP log or resistivity logs provide reliable data on continuity of vertical permeability and porosity in this sand-shale sequence.

—Core porosity: 15% to 25%, average 20%.

—Core horizontal permeability: lower two-thirds erratic, varying from 50 to 750 md. In upper one third, permeability gradually decreased upward to near zero at 885 ft.

These core and log data per se are insufficient to allow extrapolation or prediction of reservoir values beyond the wellbore in which they were measured. Transient pressure analysis could indicate the extent of the reservoir, but provides only an average of vertical and lateral variations in reservoir properties.

Better methods are required to detail the continuity of shale and sandstone members. Determination of

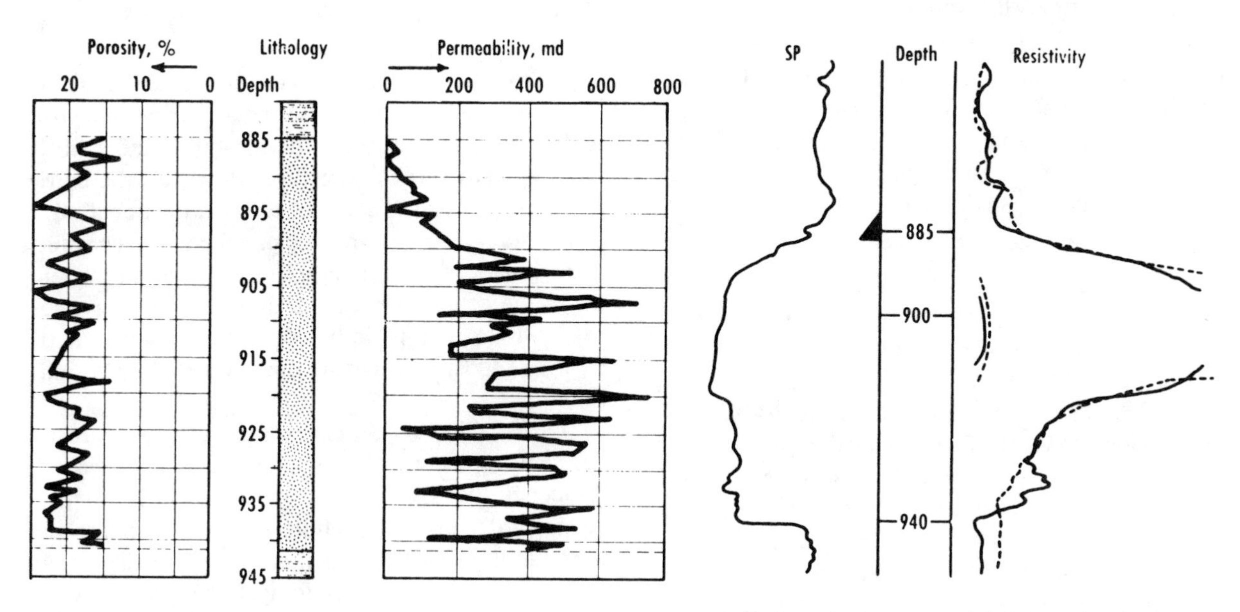

FIG. *1-19—Properties of a Pennsylvanian sandstone reservoir.*[19] *Permission to publish by the Petroleum Publishing Co.*

the depositional environment is a big step in this direction, as illustrated in Figure 1-19.

Examination by Hewitt of slabbed, continuous cores of the specific Pennsylvanian sandstone along with previously available data, revealed the following characteristics of this deposit.

—Gradational upper contact of sand body
—Sharp erosional basal contact
—High-angle cross stratification in the middle part of the reservoir
—Common slumping and distortion of the cross strata
—Trough-shaped ripples in the upper part of the reservoir
—Fragments of carbonized wood
—Few burrows by organisms in the overlying shales

These observations indicated the reservoir to be a *non-marine fluviatile* sandstone. Such characteristics are generally representative of any fluviatile or channel sandstone reservoir and should be useful to reservoir and production engineers involved with channel sandstone reservoir and should be useful to reservoir engineers, production engineers, and geologists involved with channel sand reservoirs:

—Trend of deposit: Long dimension parallel with paleoslope. Large angle to shoreline.
—Shape: Commonly lenticular and elongate, with length many times the width.
—Boundaries: Lower boundary—sharp and erosional. Upper boundary—gradational. Lateral boundaries—sharp lateral boundary if erosional depression completely filled with sand; gradual lateral boundary if sandstone body is small relative to size of a large alluvial plane.
—Texture: Grain size may decrease upward in reservoir.
—Structures: Trough-shaped ripples; high angle cross-stratification; slump common; burrows rare.
—Permeability: Decreases upward as grain size decreases. Maximum horizontal permeability is parallel to the direction of sediment transport, which usually parallels long dimension of sand body. This directional permeability results from (1) trough-shaped ripples, (2) high angle cross-stratification, and (3) the long axis of sand grains being oriented in direction of stream flow during transport and deposition.

The most significant single criterion for identifying a fluviatile sandstone is downward thickening, at the expense of the subjacent (eroded) strata. If a channel sandstone occupies a valley formed by a stream eroding into shale, the sandstone is usually biconvex in the subsurface due to differential compaction of the laterally adjacent shales.

Although no single criterion is diagnostic, various combinations of the above criteria usually are adequate for precise determination.

A Comparison of Channel and Bar Deposits in a Deltaic Environment[37]

Most oil and gas reservoirs in the deltaic environment are (1) channel sands, which include point bar deposits, and (2) deltaic bar (fringe) sands. Figure 1-20, presented by Sneider et al,[37] is an idealized permeability profile and log response to illustrate the dramatic differences between channel and deltaic bar (fringe) sands in a deltaic environment.

Deltaic Bar (Fringe) Sands—The center of a deltaic bar (fringe) deposit is at the mouth of a distributary channel. Permeability increases upward throughout the bar and is highest at the top. Also, permeability in the deltaic bar sands is highest near shore and decreases in a seaward direction. The entire deposit usually consists of a number of depositional sequences, illustrated in Figure 1-20(d,e,f), as a result of periodic floods. Each sequence is usually separated by clay or silt. However, there is a general upward increase in grain size and permeability for each sequence.

Shallow marine barrier bars, illustrated in Figure 1-13, are similar in nature to deltaic (fringe) deposits with a general increase in permeability and porosity in an upward vertical direction. However, the barrier bar deposit trend is parallel to the directional strike of the marine strata, which is parallel to the coastline. Barrier bars, being more affected by ocean currents and wave action, are cleaner and generally have less sand-silt-sand-silt periodic sequences than deltaic bar (fringe) deposits, which are more affected by periodic floods.

Deltaic Channel Sands—Permeability decreases vertically upward. Also, silt/clay layers increase in frequency, thickness, continuity, and aerial extent in an upward vertical direction. The log response and permeability profiles for deltaic channel sands, shown in Figure 1-20, also apply to upstream channel sands related to braided streams.

Figure 1-14 illustrates point bar (channel) sands that are scattered all over the flood plains of a mean-

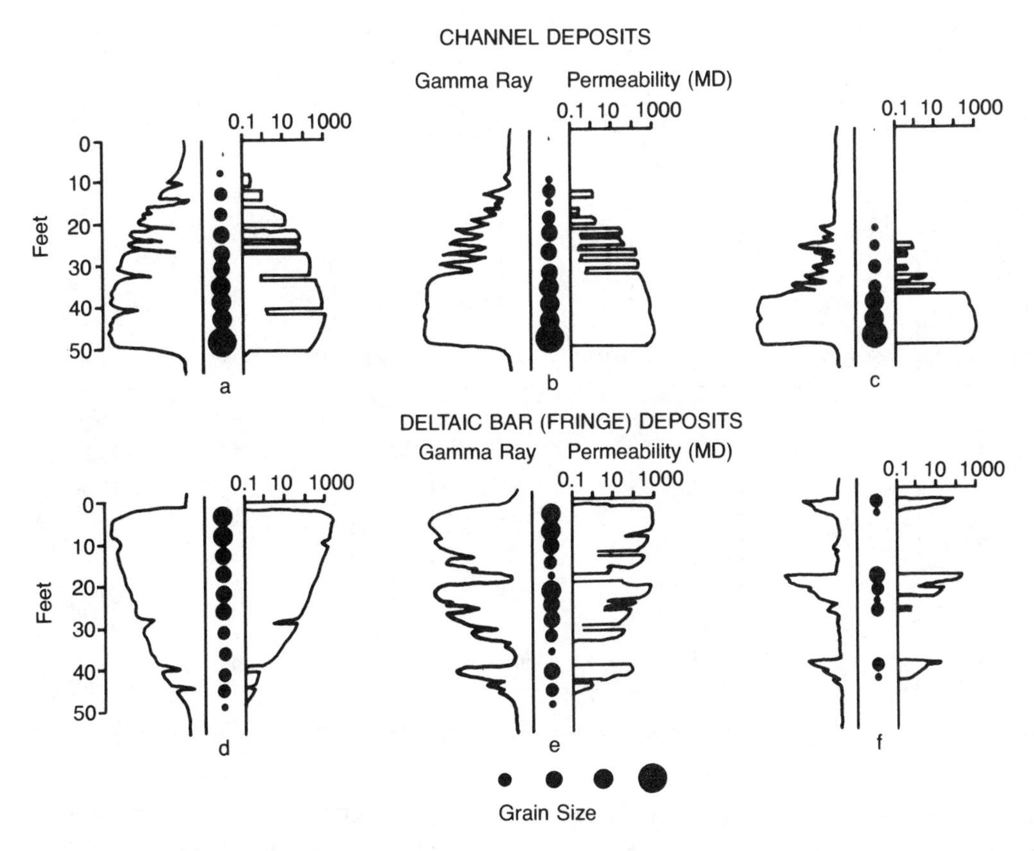

FIG. 1-20—*Idealized permeability profiles and log responses—deltaic deposits.*

dering river. The Niger delta contains an unusually large number of point bar deposits.

Summary of Deltaic Sand Deposits—Initial pore space in both channel and bar sands is controlled by grain size and sorting. For unconsolidated to moderately consolidated sand reservoirs, the general relationship between texture and pore space (porosity, pore size, and permeability) can be predicted. These relationships, illustrated in Figure 1-21,[37] apply to deltaic bar (fringe) deposits, and channel sands in the deltaic environment. They also apply generally to shallow marine barrier bars and upstream channel sands.

Geologic Control Summary for Sandstone Reservoirs[19]

Figure 1-22 outlines the relation of depositional environment, source material, and post depositional history to various sandstone reservoir parameters, such as volume limits of the sand deposit, water sensitivity, porosity, horizontal and vertical permeability, relative permeability, surface area of deposited material, irreducible water saturation, and homogeneity. This information, plus log studies, core analysis, various types of well tests, and reservoir pressure studies, provide most of the basic information required for reservoir description needed for planning, developing, operating and maintaining of wells and reservoirs.

APPLICATION OF GEOLOGIC CONCEPTS IN SPECIFIC SANDSTONE RESERVOIRS

Reservoir Description in the Niger Delta

Extensive work on reservoir description of the Niger Delta was reported by K. J. Weber[31] and by Weber and Daukoru,[32] of Shell BP Nigeria Ltd. These studies cover a twenty-year period and are being continued and expanded. Techniques used in these studies have increased recoverable reserves, decreased the number of dry holes, and increased the profitability of operations.

Commercial oil was discovered by Shell-BP in the Niger River Delta in 1955. By 1975, Niger delta pro-

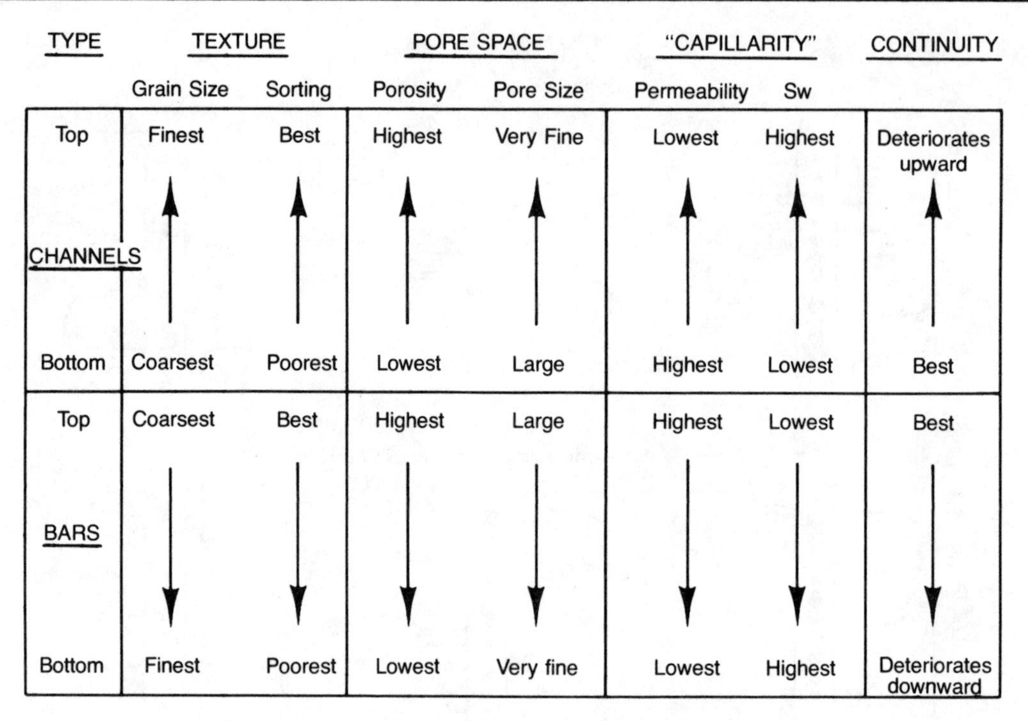

FIG. *1-21—Deltaic reservoir conceptual relationships.*

duction made Nigeria the seventh largest oil producer in the world for a period of time.

The Niger delta sequence starts with a thick deposit of under-compacted marine clays overlain by near shore deposits, in turn overlain by progradational continental sands. All of this is a composite of sediments deposited during repetitious cycles of prograding sedimentation. Basement faulting affects delta development and sediment thickness distribution. In the near-shore sequence, roll-over structures associated with growth faults trapped the hydrocarbons. Depositional environments of these reservoir sands strongly influence well completions, well productivity, and oil and gas recovery.

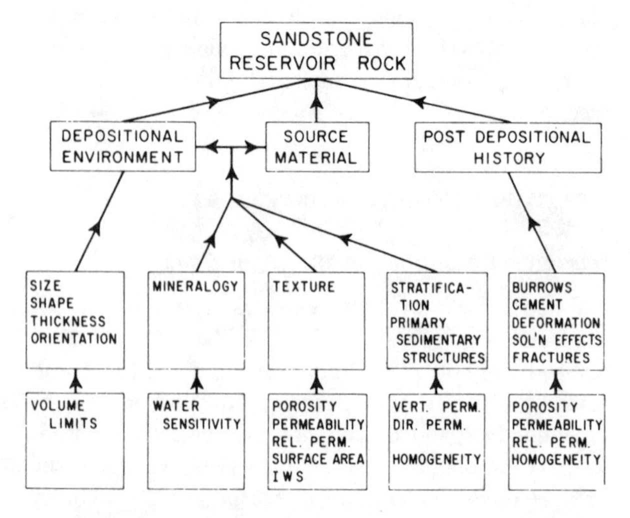

FIG. *1-22—Potential value of geologic studies of sandstones in reservoir evaluation.*[19] *Permission to publish by The Petroleum Publishing Co.*

Figure 1-23 illustrates the depositional environments and growth faulting in the Niger delta. Producible sand deposits are found in shallow marine barrier bars, river mouth bars, tidal channels, distributary channels, and point bars.

These different types of sand bodies can usually be distinguished by electric logs, gamma-ray logs, and by grain-size distribution obtained from conventional and sidewall cores. Permeability, porosity, and reservoir characteristics differ markedly from one type of deposit to another.

Barrier bar sands are deposited in a marginal marine environment on top of the finer grained barrier-foot deposits, as illustrated in Figure 1-23. Some general characteristics of barrier bars are that:

—Sand grain size and permeability increase from the bottom to the top of the bar;

—Clay "breaks" in barrier bars can be correlated

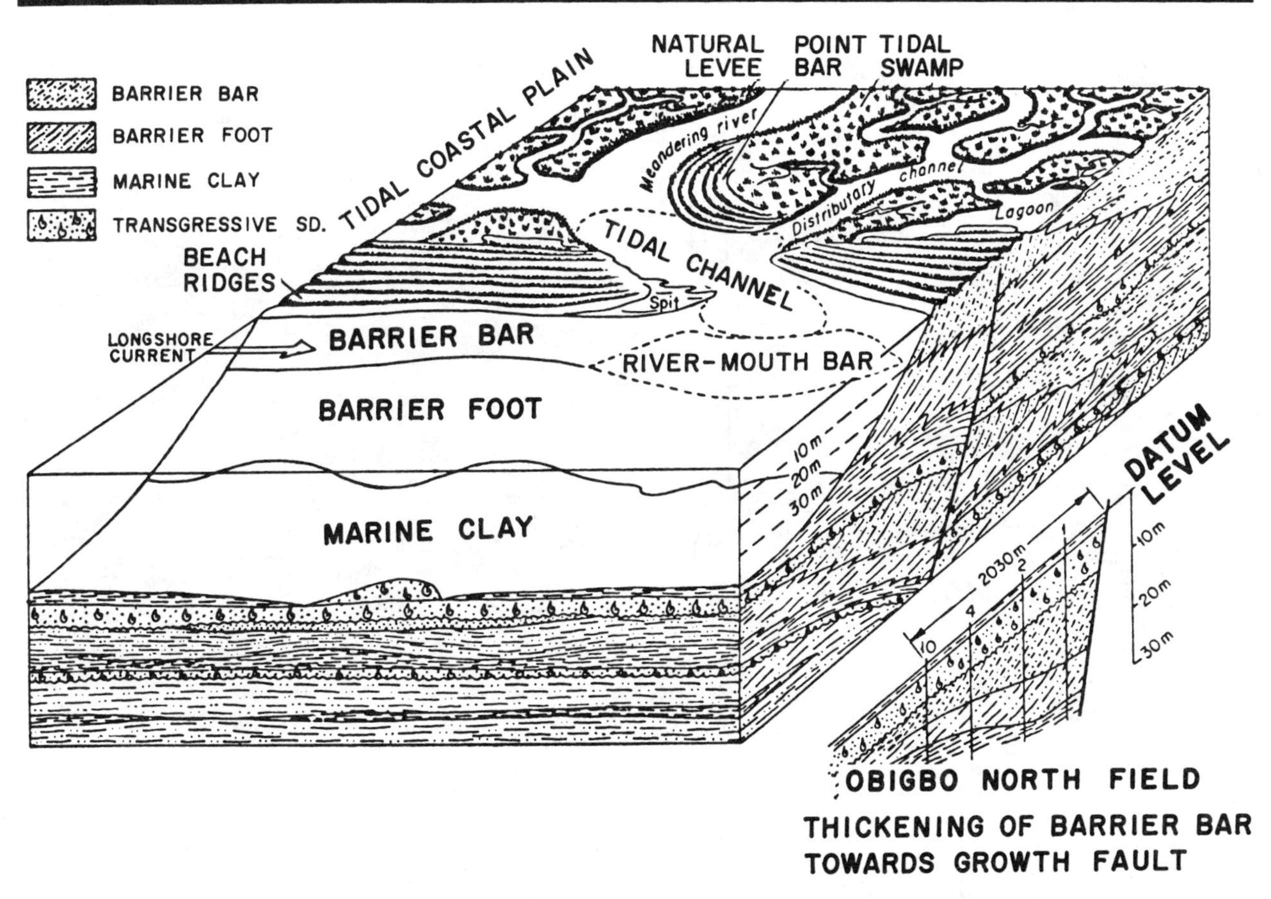

FIG. *1-23—Block diagram of typical Niger delta sediments.*[31]

over long distances. These clay "breaks" may limit or stop vertical flow of fluids;

—A series of barrier-bar sands may be deposited on top of each other with only a thin marine clay and/ or thin interval of barrier foot between clean bar sands;

—Barrier bars may be eroded in places by crosstrending river channels which, in turn, may be filled with reservoir sandstone.

Tidal channel fills often consist of a series of thin cross-bedded sequences fining upwards with a clay pebble or gravelly lag deposit at the base, and separated by thin clay beds. Grain size distribution is similar to fluviatile or point bar deposits. Clay breaks generally are difficult to correlate in tidal channel fills. Tidal channel locations are shown in Figure 1-23. River-mouth bars may be associated with tidal channels but are usually similar to barrier bars in depositional character.

Log correlation of various zones for part of the Egwa field is illustrated in Figure 1-24 with the various types of deposits shown for each well. A tidal channel is the topmost oil productive zone in wells 1 and 8. Well 9 penetrated back swamp and lagoon deposits and lacked permeable sands.

Fluviatile or Channel Deposits—These are formed by meandering rivers, and thus form point bars in a brickwork fashion covering much of the ancient river flood plain. River channel fills are shown below the tidal channel deposit in Wells 1 and 8, Figure 1-24. Fluviatile sediments are characterized by upward fining of grain-size distribution in the upper part of the fills, which are composed of laminated wavy-bedded clay and silty sand. However, the clayey part on top of the point bars frequently erodes, leaving coarse-grained sand on top.

Fluviatile sand bodies often have a sharp erosive base, which is easy to identify on SP and Gamma ray logs. Oval point bars are often bounded on three sides by clay plugs of the oxbow type lakes. Distributary channel fills are usually formed nearer the coast than

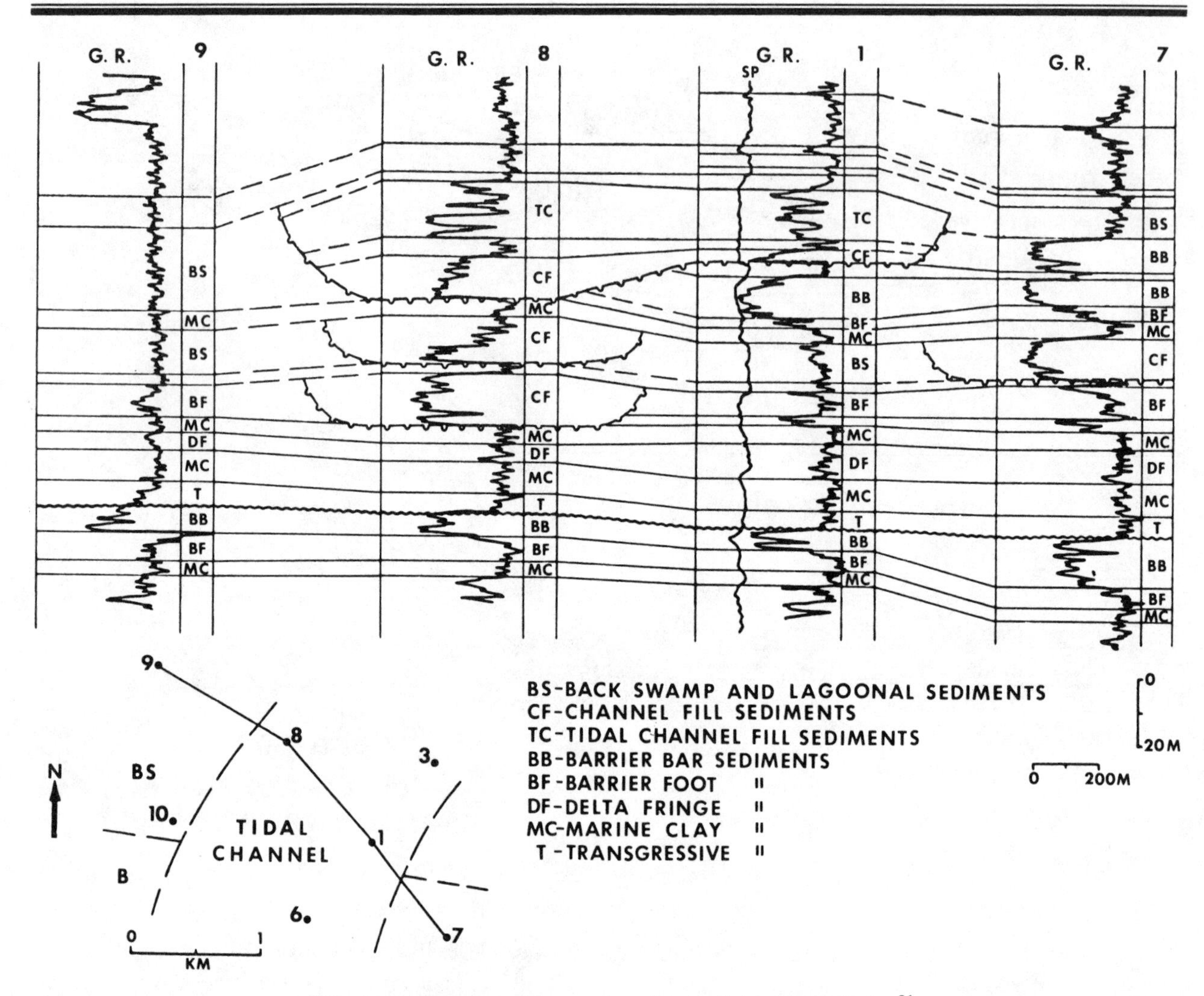

FIG. *1-24—Stratigraphic section, Egwa field, Niger Delta.*[31]

point bars. Distributary channels frequently erode their course through barrier bar sediments.

Growth Faults or Step Faults—Almost all of the hydrocarbon accumulations in the Niger delta are contained in roll-over structures associated with growth faults. These growth faults were caused by rapid sedimentation on top of undercompacted marine clays along the delta edge. Figure 1-25 illustrates roll-over structures, trapping hydrocarbons in the closures on the downthrown side of a boundary (growth) fault. Combinations of crestal and antithetic faults in Figures 1-25c and 1-25d produced collapsed graben-like structures with smaller and more numerous traps.

Practical Use of Sedimentological Information in Oil and Gas Field Development

Figure 1-26 illustrates the practical use of sedimentological information. Sedimentary conditions must be considered in making well locations and well completions if reservoirs are to be efficiently produced. In differentiating between formation damage and naturally occurring low permeability, knowledge of type of deposit may be critical.

For example, typically the completion in a barrier-bar sand may produce only a tenth as much oil per day as a channel-fill completion with the same length of interval. The higher productivity and permeability

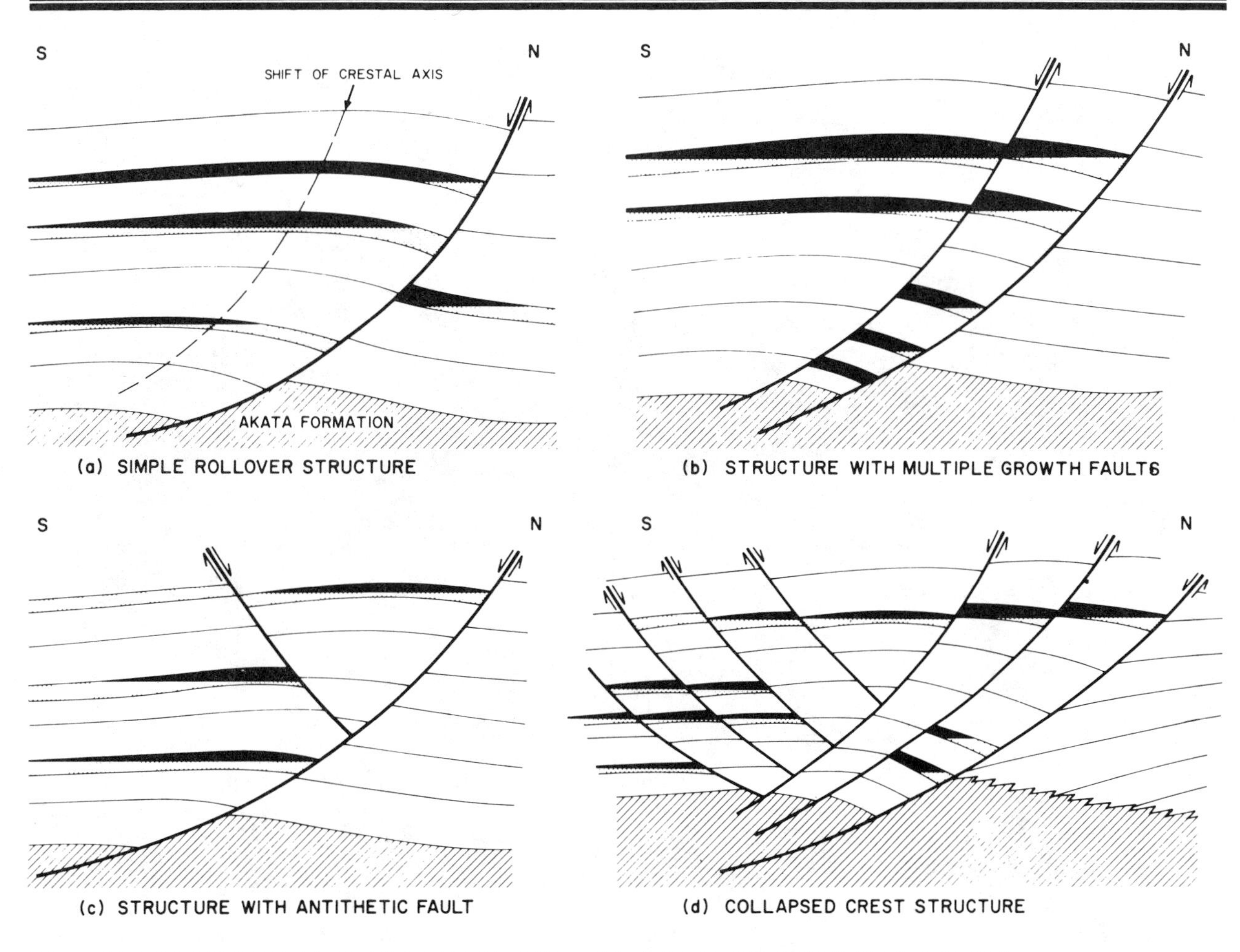

NOTE: ONLY A FEW RESERVOIR SANDS ARE SHOWN IN THE SCHEMATICAL SECTIONS AND THE SAND THICKNESS HAS BEEN ENLARGED

FIG. *1-25—Types of traps in the Niger delta.*[31]

of channel fills usually is due to larger grained sand deposited in the channels.

Following are some of the significant highlights of the Niger delta studies.

1. Continuity of clay breaks or other impermeable barriers can be predicted from depositional environment. This is very significant in predicting gas and water coning. The success of squeeze cementing in shutting off unwanted water or gas depends on knowledge of barriers to vertical movement of fluids. Correct location of vertical barriers to fluid flow can affect ultimate recovery.

2. Channel fills, point bars, and barrier bars have characteristic distributions of permeability both vertically and laterally. This can be critical in selecting the location of perforations to drain each sand member.

3. Sand grains in the Niger delta are generally large with many sands being quite unconsolidated. Identification of type of deposit is made from sidewall cores and conventional cores plus petrophysical data obtained from logs.

4. Channels often eroded across barrier bars. These sand-filled channels may have higher productivities than barrier bars because of larger sand grains; however, sand grains may be poorly sorted in channel sands.

5. Grain size analysis from conventional and side wall cores has been a most significant tool to identify the various types of deposits and also to select the high productivity intervals for completion.

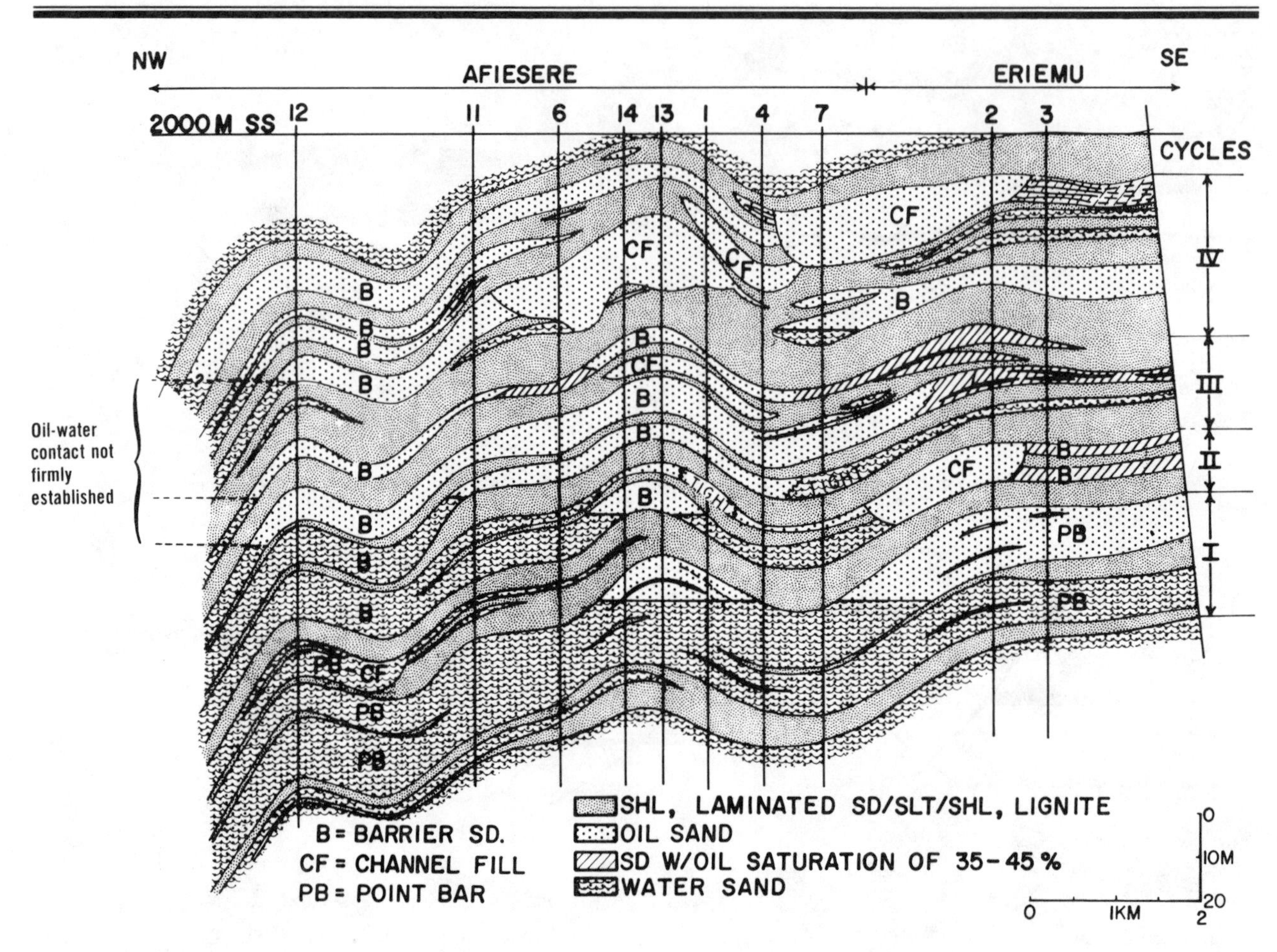

FIG. *1-26—Example of a complex hydrocarbon distribution in a series of four sedimentary cycles in Niger Delta.*[31]

Reservoir Description For Waterflood Planning—Elk City Field, Oklahoma[28]

A reservoir description study of the Elk City field was carried out by the Shell group as part of the planning phase for waterflooding. This study is an excellent example of the step by step process required for effective reservoir description. It includes many useful techniques and methods of presenting data.

Detailed knowledge of the distribution of porosity, permeability, and barriers influencing fluid flow must be integrated with reservoir data to select, plan, and implement any improved recovery project. Studies must adequately account for inhomogeneities or variations in reservoir and nonreservoir rock properties. Better understanding of sand body genesis—how it was deposited and the environment of deposition—make possible a more accurate reservoir description required for predicting recovery performance.

A total of 310 wells were drilled on 40-acre spacing in this 100-million bbl field, producing from a ±500-ft. thick zone of Pennsylvanian sandstones and conglomerates. The structure is anticline with eight reservoir zones between 8,800 and 11,000 ft.

Approach and Methods of Study—In this investigation, data were integrated from genetic sand studies, petrologic-petrophysical analysis, and rock-log calibrations to characterize and map porespace distribution and to help predict fluid-flow response of the waterflood. Thickness maps, based on sand genesis, of net permeable sand were prepared to represent floodable sand volume. Predictions were made from genetic facies maps of the spatial distribution of permeability and permeability barriers, as well as anticipated flood performance.

Interrelated studies covered are:

1. Lithology and petrophysical properties, with logs being calibrated with lithology in cored intervals

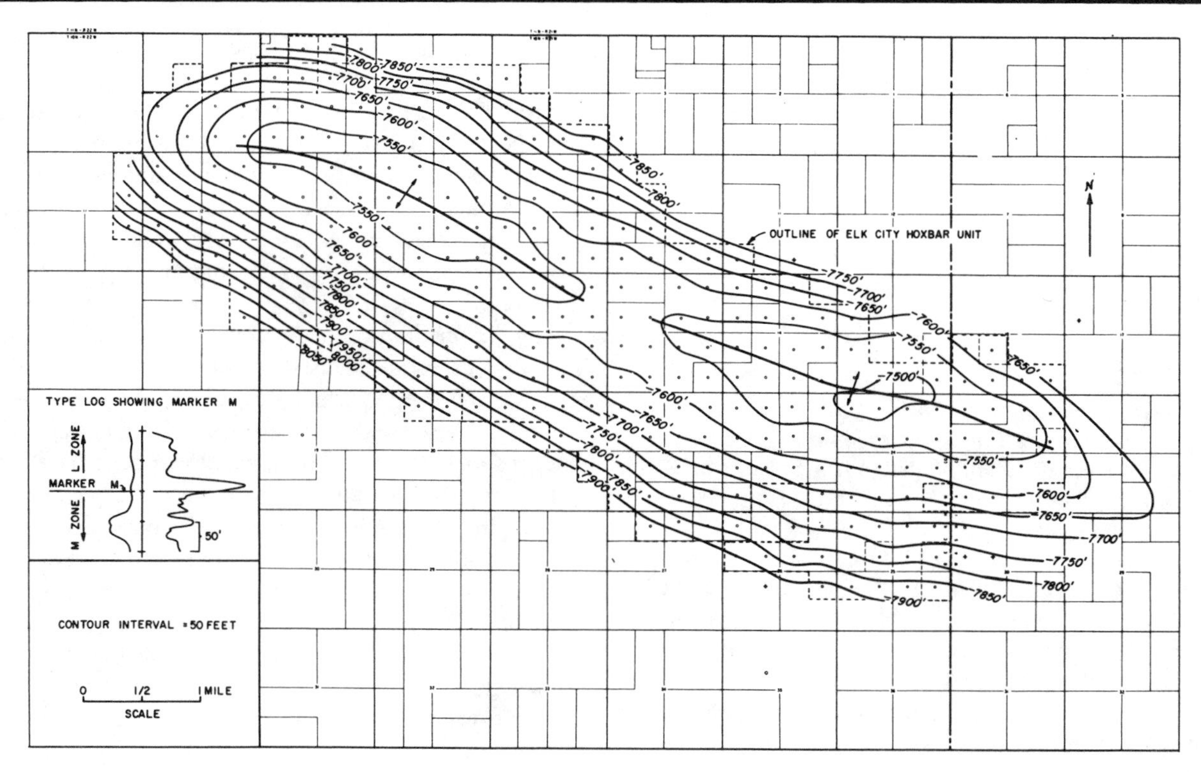

FIG. *1-27—Structure contour map on marker M, Elk City field.[28] Permission to publish by The Society of Petroleum Engineers.*

2. Interpretation of genesis of reservoir and reservoir rocks
3. The establishment of formation correlations
4. Mapping by subzones of net sand and sequences of rock types.

Most of the 310 wells were logged with the SP log, the 8- and 16-in. Normal Resistivity logs, the 24-ft. Lateral log, and the Microlog. Gamma ray-neutron logs were run on sixteen wells in place of the Microlog. In addition, over 1700 feet of cores from twenty-six wells were available representing 3 to 9 in. of section from each foot cored.

One of the starting points for this study was a structure map, Figure 1-27, on a limestone bed easily correlated across the field, and located between the primary productive zones.

For the lithologic study, mineralogy, grain size, sorting, and sedimentary structures were examined visually and microscopically. Pore sizes and geometry, and the type and amount of pore filling material were obtained from thin sections of cores. Grain size and sorting were determined either by sieve analysis or with a grain size-sorting comparator. Porosity and permeability were measured on a large number of cores and capillary pressure measurements were made on 35 cores.

Lithology and Petrophysical Properties and Relations—Reservoir rocks in this field are conglomerates and sandstones. Nonreservoir rocks are siltstones, shales, and limestone. Sand grain sizes range from very fine (0.062 mm diameter) to very coarse (2.0 mm). Most conglomerates range from granule (2.0 mm to 4.0 mm) to pebble size (4.0 to 64.00 mm). The finer grained rocks are the best sorted. With increase in median grain size, the sorting becomes progressively poorer.

Porosity, pore size, and permeability of any sandstone or conglomerate depend primarily on (1) grain size and sorting, and (2) the amount of cementation and compaction. Reservoir rocks in the Elk City field are all compacted. Cement or pore-fill material is less than 7% by volume. In these rocks, pore space correlates with grain size and sorting. Figure 1-28 shows a plot of grain size and sorting versus porosity. The fine grained rocks have highest porosities, and with an increase in grain size, sorting is poorer and porosity decreases.

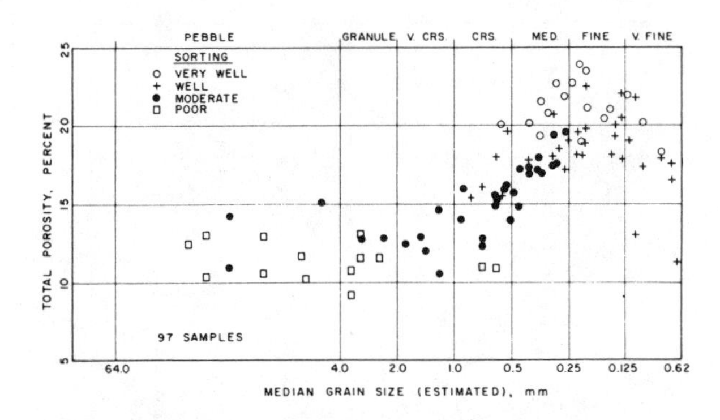

FIG. *1-28—Relationship among sorting, grain size, and porosity in Elk City sandstones and conglomerates.[28] Permission to publish by The Society of Petroleum Engineers.*

Pore size was estimated from pore-size measurements in thin sections, and from capillary pressure curves. Capillary pressure curves of representative reservoir rock types are shown in Figure 1-29. The fine-grained rocks have predominantly fine and very fine pores. With an increase in grain size, the average pore size and the number of medium and large pores increase.

From Figure 1-29, it is interesting to note the rather high permeability of 1638 md and the relatively low porosity of 12.5% in pebble conglomerate with well-sorted coarse sand matrix. On the other end of the grain-size scale, the very fine, very well sorted sandstone had a permeability of 1.6 md and a porosity of 20.0%.

Log Calibration—Logs were correlated with cores so that logs could be used to supplement core data. The SP and 8-in. Normal Resistivity curves and the Microlog were used to determine rock type and sequences of rock type as illustrated in Figure 1-30.

Impermeable siltstones, calcareous shales and limestones are distinguished from sandstones and conglomerates on the basis of SP development. For the sandstones and conglomerates, which exhibit moderate to well-developed SP, grain size is indicated by the 8-in. Normal Resistivity curve. The fine-grained rocks exhibit low resistivity. With an increase in grain size, the values of resistivity increase. Grain size determinations of the very silty, very shaly, or well-cemented rocks present a problem. The measured resistivity of these rocks is higher than clean or noncemented rocks of equivalent grain size. The Microlog opposite these rocks shows no separation or a characteristically "hashy" appearance.

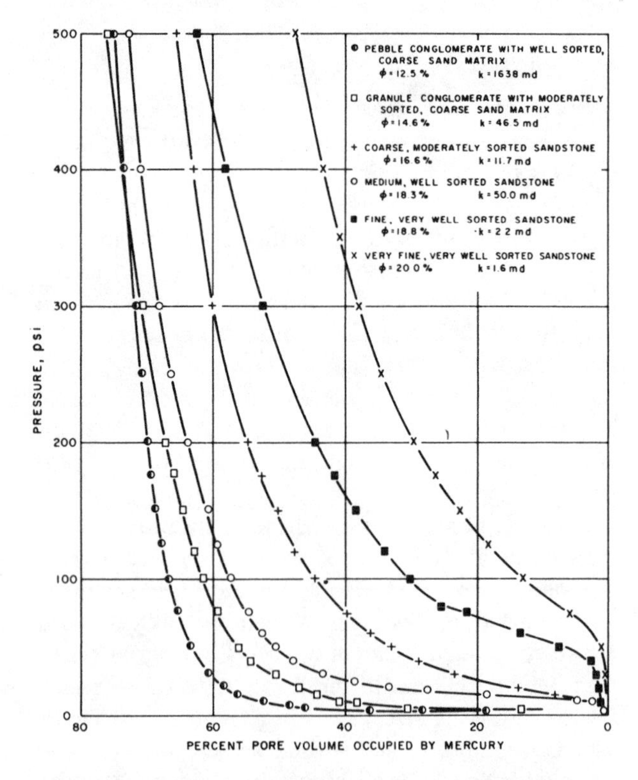

FIG. *1-29—Capillary-pressure curves of typical reservoir rock types.[28] Permission to publish by The Society of Petroleum Engineers.*

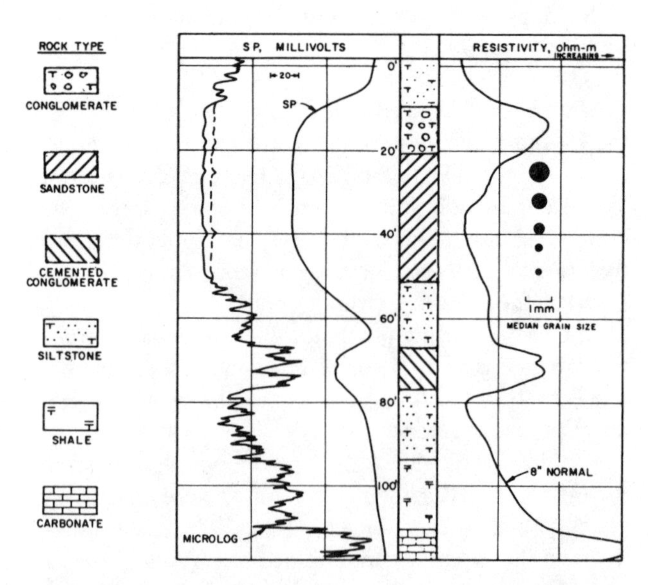

FIG. *1-30—General relations between rock type and log response.[28] Permission to publish by The Society of Petroleum Engineers.*

Correlation between grain size and log resistivity is apparent because of the relationship between rock types and pore space. The better sorted fine-grained rocks have the highest porosity and largest number of fine pores. This results in low values of formation factor and resistivity. Formation factor and resistivity increase as grain size increases, sorting becomes poorer, and porosity and the number of fine pores decrease. Hydrocarbon saturation has only a minor influence on resistivity values from the 8-in. Normal Resistivity curve because this curve seldom probes beyond the invaded zone adjacent to the well bore.

The distribution of porosity, permeability, and pore size follows in general the distribution of rock types.

Genesis of Sand Bodies—From core studies, including thin sections, the reservoirs consist principally of one or more of the following genetic types: barrier bar, alluvial channel, deltaic distributary channel, and deltaic marine fringe. Figure 1-13 illustrates the trend and distribution of the various deposits.

Barrier Bar Deposit—These deposits are composed of rocks ranging from siltstone to pebble conglomerate with progressive increase in grain size upward. A typical barrier bar sequence is illustrated by Figure 1-31. It may be noted that the resistivity gradually increases as grain size increases in an upward direction. Figure 1-32 shows that permeability follows the porosity trend and generally increases in an upward direction within the barrier bar.

The barrier bar deposits trend parallel with the depositional strike of the marine strata. Figure 1-33 shows the barrier bar deposits in relation to adjacent

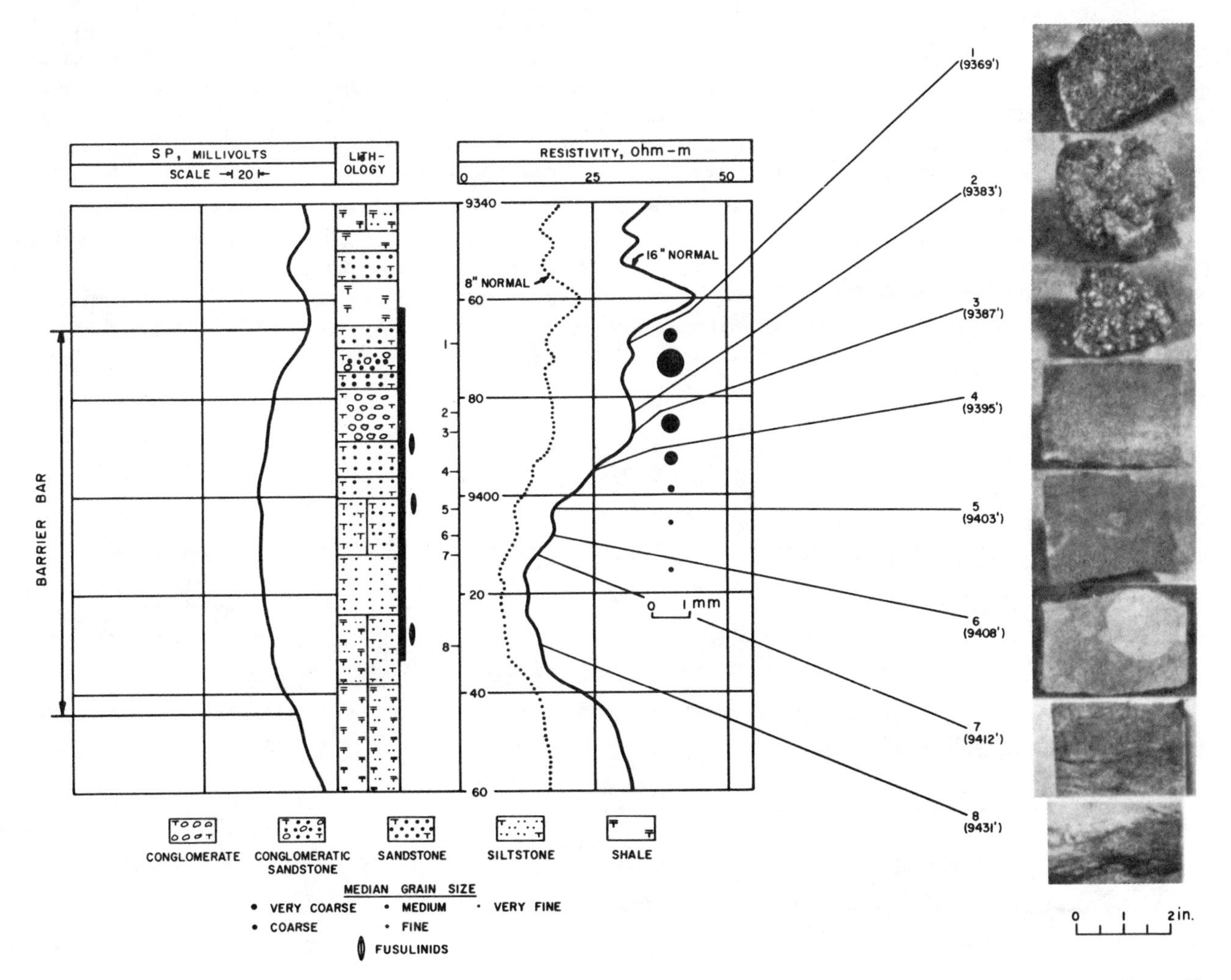

FIG. 1-31—*Vertical sequence of rock types and log response of a barrier bar deposit, Shell, G. Slatten No. 1.*[28] *Permission to publish by The Society of Petroleum Engineers.*

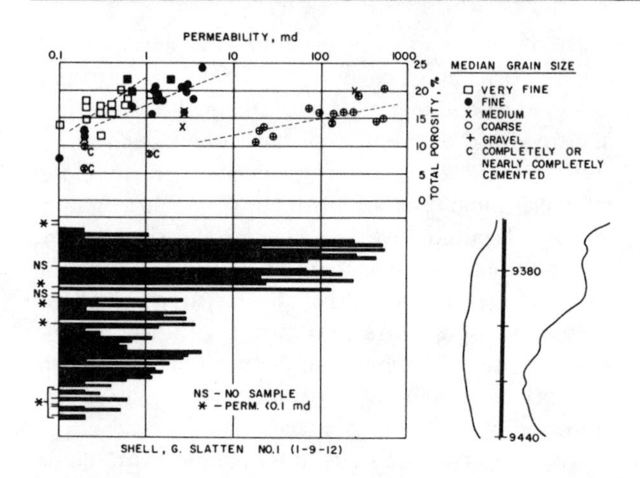

FIG. *1-32—Plot of porosity vs. permeability and vertical permeability profiles, Shell, G. Slatten No. 1.*[28] *Permission to publish by The Society of Petroleum Engineers.*

deltaic deposits and a lower marine limestone marker.

The coarsest-grained material is concentrated on the landward side and at the top of the deposits. The overall grain-size and sand thickness decreases in a seaward direction across the trend of the deposit. The permeability profile is essentially the same, parallel with the sand-body trend and decreasing uniformly toward the seaward side of the bar and is highest on the shore or back side of the deposit.

No shale breaks exist in the bar deposits except near the base. The lower and lateral boundaries are gradational with adjacent impermeable siltstones and shale.

Conglomerates, coarse sandstones, and some fine and medium-grained sandstones are massively-bedded. The fine- and medium-grained rocks have faintly developed, slightly inclined or horizontal bedding. The fine sandstones and siltstones are laminated and ripple bedded and show some reworking by marine organisms.

Deltaic Marine Fringe Deposits—Deltaic deposits in the Elk City field consist of either one or two stacked genetic units. Rock types range from shale to pebble conglomerate.

The vertical sequence of rock types in these deposits is characterized by (1) interbedding of sandstone with shale and siltstone, (2) a general upward decrease in the number of shale and siltstone beds and in the amount of interstitial clay- and clay-size particles in the sandstones, and (3) a general upward increase in grain size for each depositional sequence.

The deltaic marine fringe deposits are irregular in geometry and distribution. For each sequence in the delta buildup, permeability increases in an upward direction. The lower part of each sequence may be shaly or silty with low permeability. Each sequence is separated by impermeable or low permeability rocks.

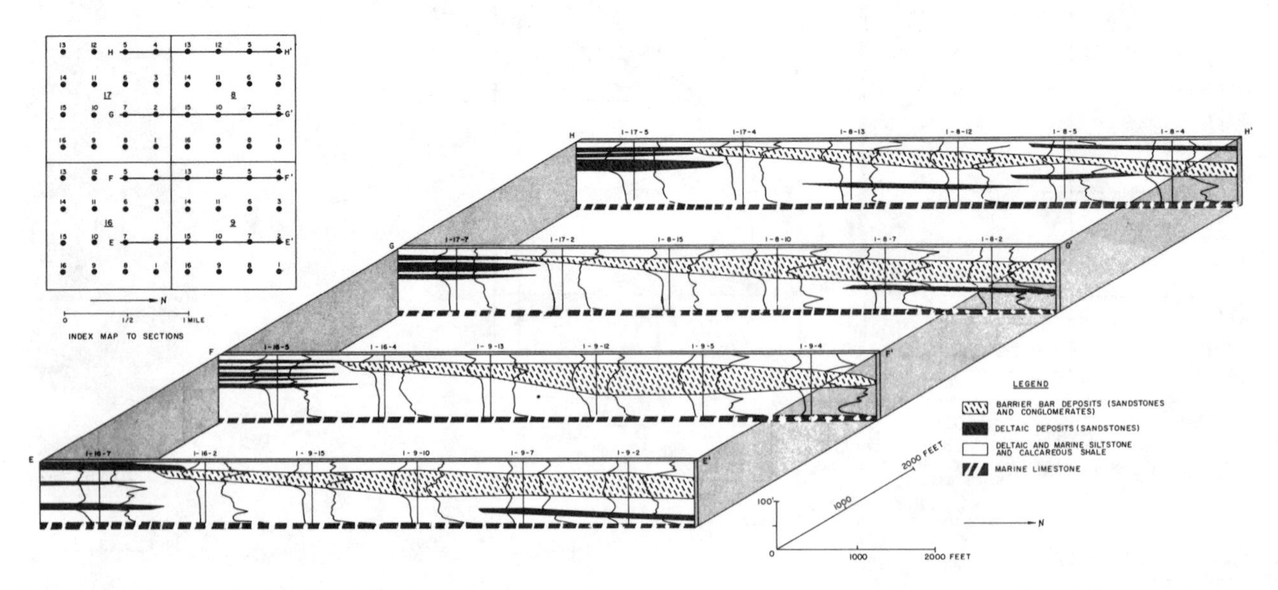

FIG. *1-33—Cross sections through a barrier bar deposit.*[28] *Permission to publish by The Society of Petroleum Engineers.*

Toward the lateral or basinward edges of the fringe deposits, the permeability gradually decreases as sand grades into impermeable silt or clay.

Permeability distribution in delta marine fringe deposits is similar to that found in barrier bars. However, the thin siltstone/shale beds that cap each cycle are widespread and are effective barriers to vertical flow.

Alluvial Channel Deposits—This type of channel deposit is predominantly coarse grained rock with a cyclical vertical variation in grain size due to being deposited by periodic floods from high gradient braided streams. Conglomerates and conglomeratic coarse-grained sandstones are the dominant rocks. Fine to very fine grained sandstones, siltstones, and silty shales are interbedded with coarser rocks but are not so abundant.

Typical log response of the SP, Microlog and 8-in Normal Resistivity curves for the channel deposits is shown in Figure 1-34.

The SP curve is well developed opposite the sandstones and conglomerates, with the SP being depressed slightly toward the shale baseline opposite silty sands. Opposite the permeable sandstones and conglomerates, the Microlog has positive separation. In the low permeability silty members, the Microlog is "hashy" and shows no positive separation, thus indicating small grain size.

The alluvial channel deposits are linear in trend and tend to be oriented perpendicular to the depositional

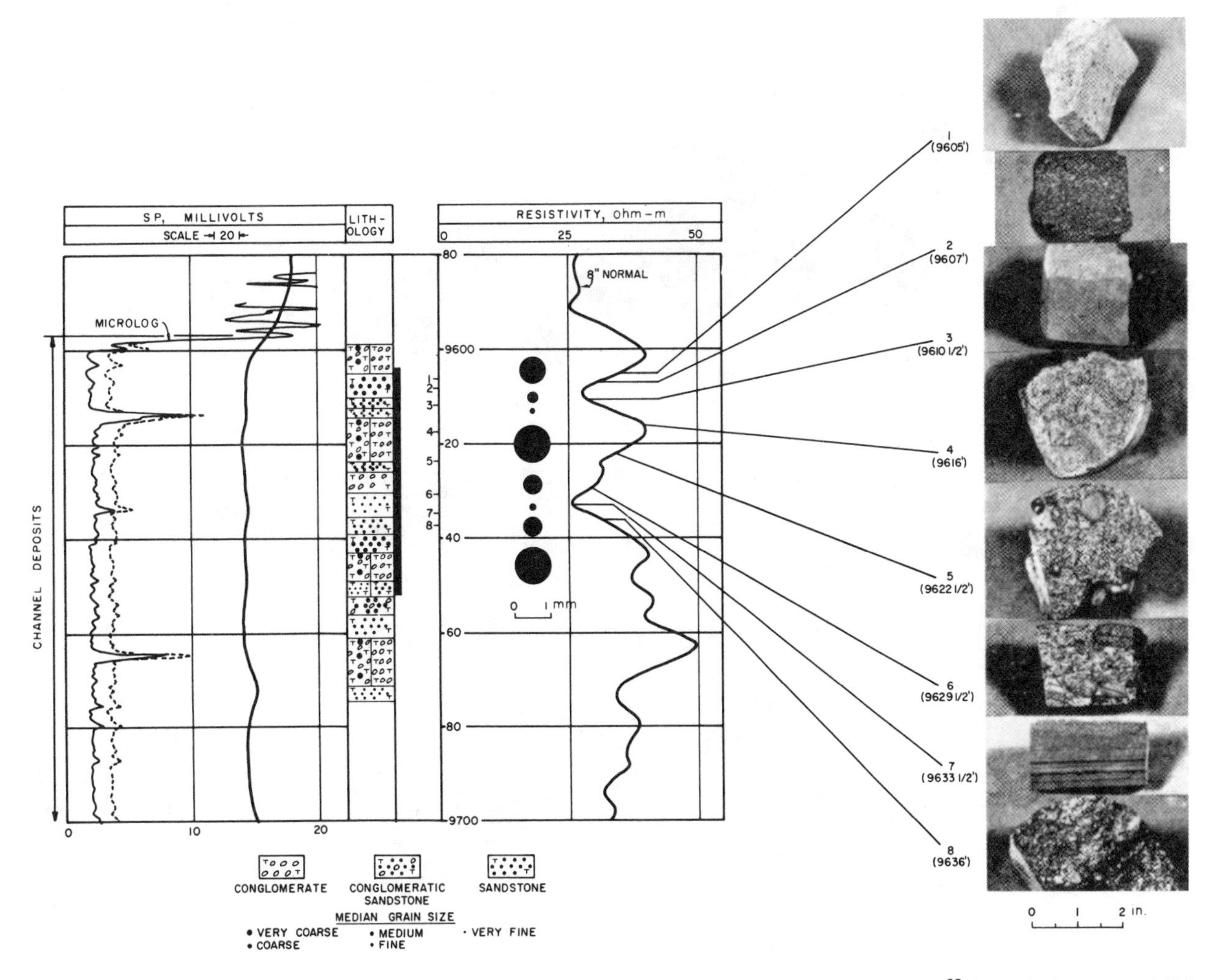

FIG. 1-34—Vertical sequence of rock types and log response, Shell, B. Pinkerton No. 1.[28] Permission to publish by The Society of Petroleum Engineers.

strikes of the marine strata. The distribution of deposits is controlled by size and shape of the stream-cut valley. Sand and gravel filled much of the valley, with silt and clay filling the remainder.

Rocks composing the channel sands have low porosity (10 to 15%), but high permeabilities (75 to 1500 md). Because of the relative frequency of depositional cycles, vertical and lateral permeabilities may be quite variable.

Distributary Channel Deposits—These deposits are similar to the alluvial channel deposits except that the rocks are not as coarse.

Genetic Sand-Body Identification and Delineation—Most sandstone deposits may be divided into individual genetic units deposited during a single cycle of sedimentation. For example, a genetic unit or cycle for a fluviatile sand may be the sand, siltstone, or shale layers deposited during a single flood cycle. Such a cycle can be correlated across a field unless interrupted or cut into by other deposits. Figure 1-35 shows the lateral distribution of genetic sand units for one zone within the reservoir. It may be noted that all types of sand found in the field are represented in this single zone.

Figure 1-36 is a sand thickness map of the L_3 Zone. This type of map aids in understanding the heterogeneity of the reservoir and well problems associated with both primary recovery, pressure maintenance, and enhanced recovery. A sand thickness map is particularly helpful in providing better communication between field personnel and those involved in technical studies.

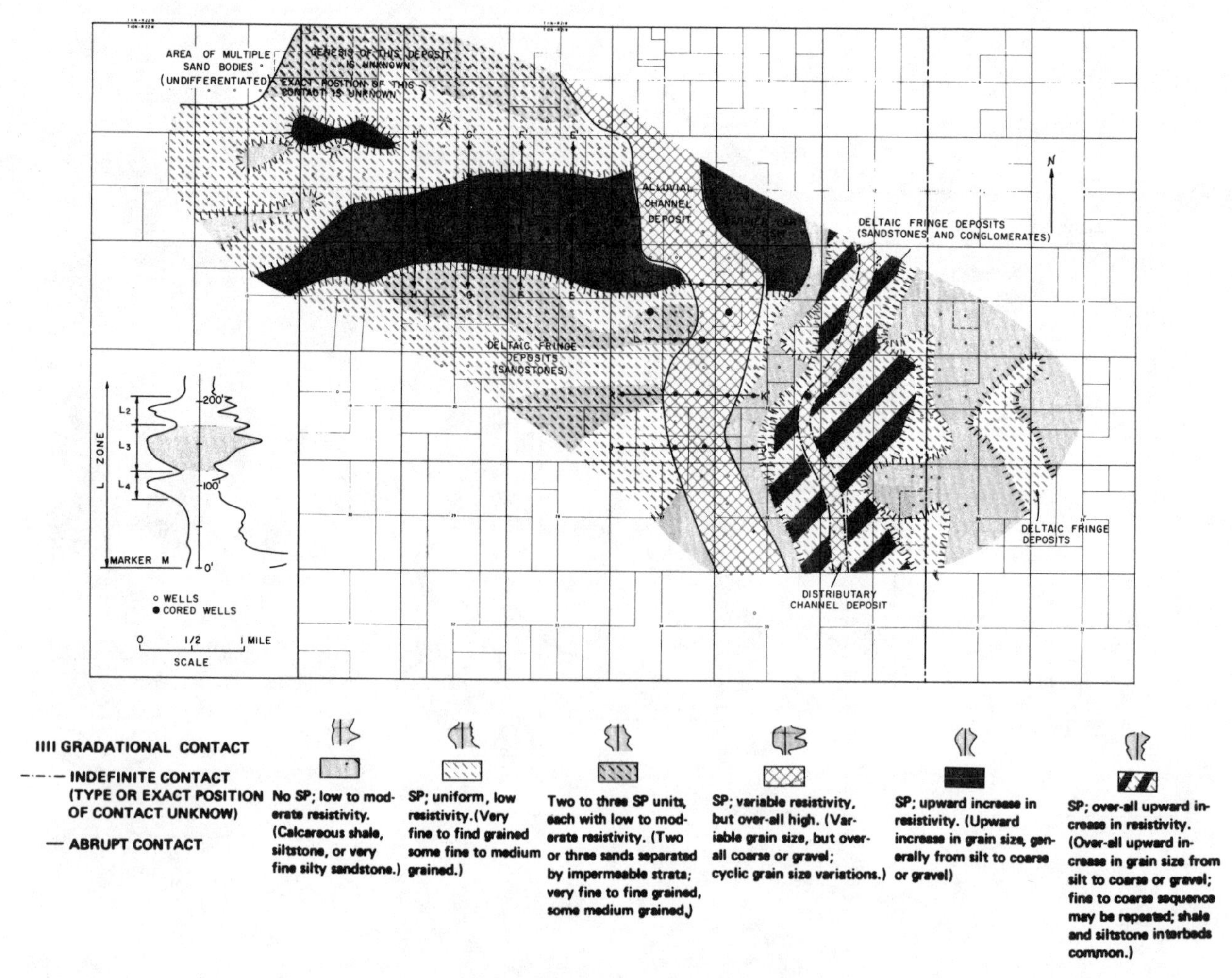

FIG. 1-35—*Distribution of sequence of rock types, L_3 subzone, showing the distribution of the genetic types of sand bodies.*[28] *Permission to publish by The Society of Petroleum Engineers.*

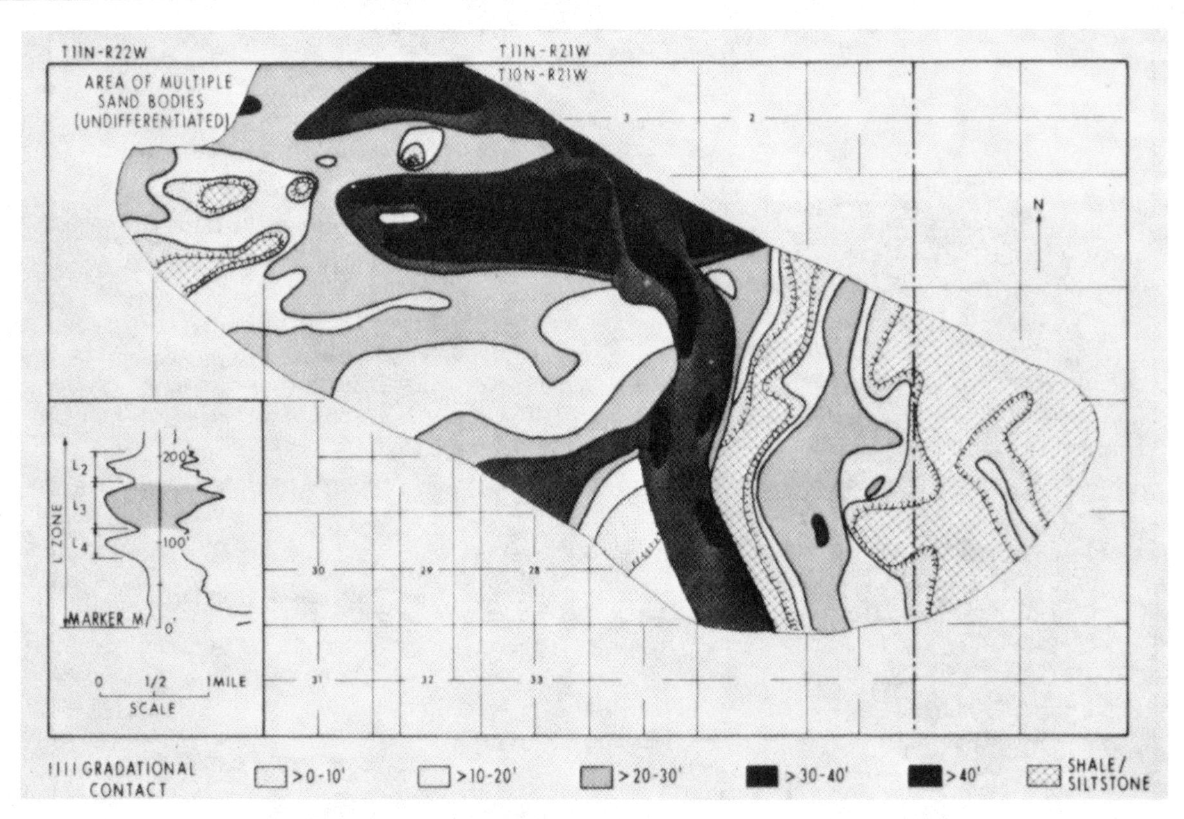

FIG. *1-36—Thickness map of net sand, L_3 subzone.*[28] *Permission to publish by The Society of Petroleum Engineers.*

Lithology and pore space in many types of sand are distributed in a systematic and predictable manner. The distribution and continuity of porosity, permeability and lithology can be predicted with reasonable certainty after identifying the genesis of sand members in a field.

From the identification and delineation of genetic sand bodies in the L_3 subzone, Figures 1-35 and 1-36, it can readily be seen how important this type of study can be from the standpoint of locating and completing wells for the waterflood project. There is no doubt that this type of study would have been quite useful in planning operations under primary recovery as well as improved recovery projects.

CARBONATE RESERVOIRS

Carbonate deposits are usually laid down as calcium carbonate ($CaCO_3$), primarily originating from a mixture of ground-up shells and excrement of marine organisms. Because of the heterogeneous nature of carbonates, a detailed description of any carbonate reservoir is required.

Limestones are composed of granular particles, a fine matrix, and cementing material. The grains may vary in size from a few microns to recognizable fragments of shells or corals. There are several different kinds of grains, the most important of which are (1) shell fragments, called ''bio''; (2) fragments of previously deposited limestone called ''intraclasts''; (3) small round pellets, the excreta of worms; and (4) ooliths spheres.

The matrix is usually calcareous mud containing clay-size calcite particles called micrite. The clear secondary calcite cement is called sparite. A rock consisting primarily of clear secondary calcite with intraclast grains is called ''intrasparite''.

A modification of Folk's classification of limestone is shown in Figure 1-37.

A rock consisting mainly of micrite (lime mud) with grains of broken shell fragments would be called ''biomicrite''. Biomicrite and pelmicrite are the most common limestone types.

In addition to these classifications, some limestones consist only of micrite; others consist of the remains of upstanding reef-building organisms.

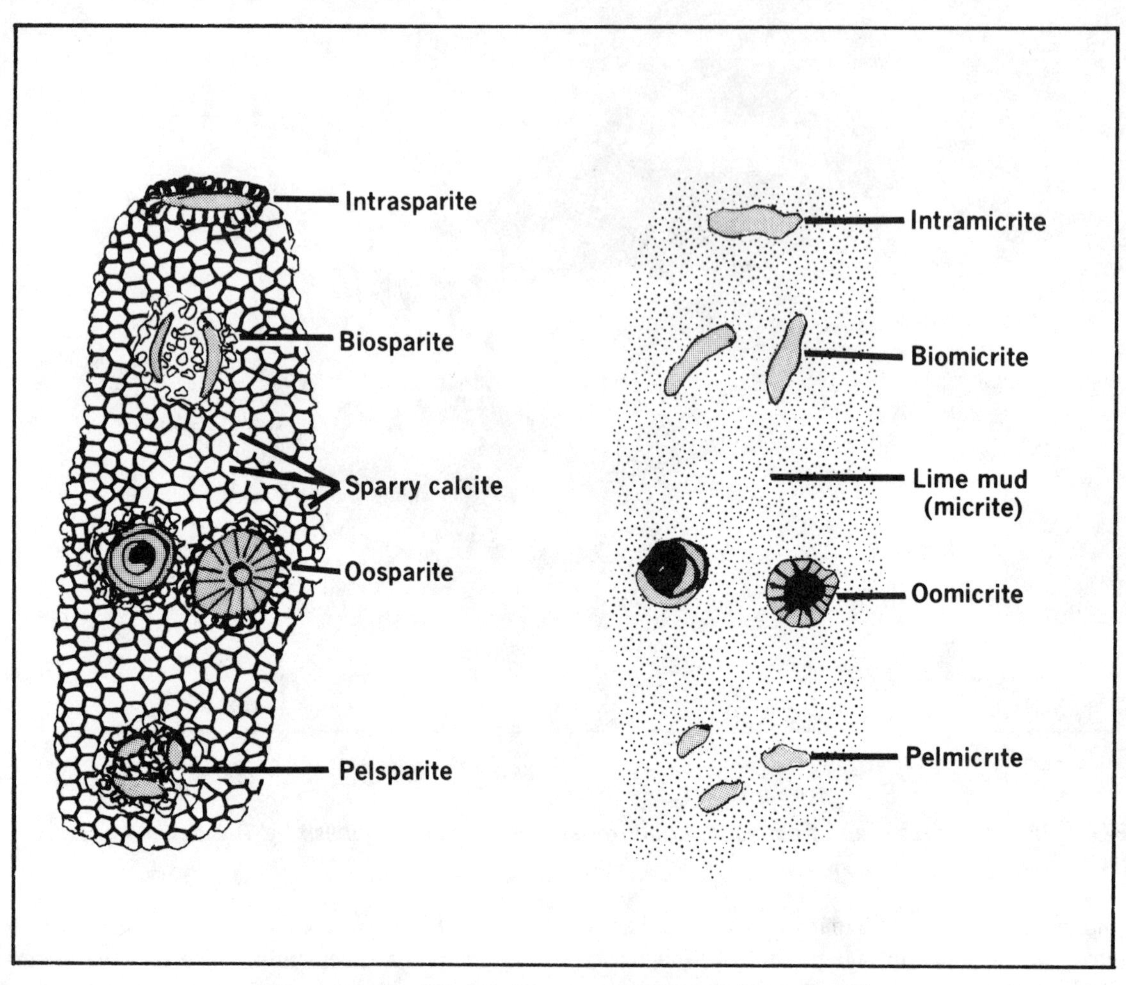

FIG. *1-37—Modified Folk's classification of limestone.*[15]

Pores in limestone may be either primary or secondary. The primary pores are those formed during deposition in environments where there were strong waves or currents to remove the fine muds. In such cases the limestones may consist mostly of grains with a minimum of micrite or sparite.

Secondary pores are formed by solution and reprecipitation of limestone after deposition. The most common types are "molds" and "channels". In the case of molds, certain grains are dissolved preferentially, leaving cavities. In the case of channels, the solutions moving through the rock dissolved out small tunnels. These are usually small in dimension (0.1 to 10 mm) and are lined with calcite crystals.

Dolomite, $CaMg(CO_3)_2$, is formed by recrystallizing micritic limestone. In the dolomitization process, water containing magnesium dissolves $CaCO_3$ and reprecipitates $CaMg(CO_3)_2$. Dolomitization may completely alter the texture of original limestone, and it may destroy such structures as bedding and fossils. The final texture is medium to coarsely crystalline, with a general massive overall structure.

Carbonates, being heterogeneous, may change from dense to porous within a few feet or even inches. It is very important to determine pore structure of carbonates. This is primarily controlled by the type of reservoir rock.

Carbonate Porosity

A few limestone reservoirs contain their original porosity. Chalks, oolitic limestone, and fragmented limestone may have high porosity, and many have some continuous porosity (see Table 1-3). A major problem in nearly all limestone is the lack of continuous porosity over a great distance. Fractures and joints often provide the best source of continuous porosity and permeability, unless fractures have been recemented with a secondary deposition.

Joints developed under a given stress pattern will be usually near vertical and parallel. If the reservoir rock is later subjected to a different stress pattern, a different set of joints may develop and may intersect previously formed joints.

TABLE 1-3
Porosity in Carbonate Rocks[6]

Aspect	*Significance*
Primary porosity in sediments	Commonly 40–70%.
Ultimate porosity in rocks	Commonly none or only small fraction of initial porosity; 5–15% common in reservoir facies.
Types of primary porosity	Interparticle commonly predominates, but intraparticle and other types are important.
Types of ultimate porosity	Widely varies because of post-depositional modifications.
Sizes of pores	Diameter and throat sizes commonly show little relation to sedimentary particle size or sorting.
Shape of pores	Greatly varied, ranges from strongly dependent "positive" or "negative" of particles to form completely independent of shapes of depositional or diagenetic components.
Uniformity of size, shape, and distribution	Variable, ranging from fairly uniform to extremely heterogeneous, even within body made up of single rock type.
Influence of diagenesis (change in rock since deposition)	Major; can create, obliterate, or completely modify porosity; cementation and solution important.
Influence of fracturing	Of major importance in reservoir properties if present.
Adequacy of core analysis for reservoir evaluation	Core plugs commonly inadequate; whole cores may be inadequate for large pores.
Permeability—porosity interrelations	Greatly varied; commonly independent of particle size and sorting.

The great variations in limestones are due primarily to the continuous change and modification which occurs as long as there is some porosity filled with water. When the water is at rest, solution and precipitation of carbonates may occur to develop larger crystals. If the water is circulating, recrystallization, solution, or precipitation of introduced materials will occur.

The original, primary porosity of most limestone reservoirs has been greatly altered by solution and reprecipitation performed by water moving through the rock. Table 1-3 provides general information on carbonate porosity.

Porosity Through Dolomitization—Porosity may be developed during the process of converting limestone ($CaCO_3$) to dolomite $CaMg(CO_3)_2$. Continuous porosity sufficient to produce oil or gas may depend on the degree of dolomitization in many carbonate reservoirs. Larger continuous flow channels or cavities are probably due to leaching out of limestone or calcitic fossils.

Porosity and pore size increase as complete dolomitization is approached. However, about 70% of the limestone must be dolomitized before appreciable porosity is formed. Saccharoidal or "grainy" dolomite usually is a good reservoir rock with well sorted intercrystalline porosity.

Types of Porosity—To properly develop and operate carbonate reservoirs, the basic type of porosity must be known. Because of the great variations in porosity in both the vertical and horizontal directions in a carbonate reservoir, detailed geologic and chemical description of reservoir rocks should be available from a representative number of wells penetrating the reservoir.

The seven most abundant types of carbonate porosity are:

1. *Interparticle Porosity*—Porosity between grains or particles of depositional origin is the dominant type of porosity in most carbonate sediments. Secondary interparticle porosity may be developed by selective dissolution of fine particles between larger particles.

2. *Moldic Porosity*—A mold is a pore formed by selective removal, normally by solution, of a rock constituent such as a shell or oolith.

3. *Fenestral Porosity*—Mud-supported and grain-supported fabrics, where the openings are larger than interparticle openings give this porosity.

4. *Intraparticle Porosity*—Internal chambers or other openings within individual or colonial skeletal

organisms are the most common of intraparticle pores.

5. *Intercrystal Porosity*—Porosity between crystals of equal size, as is found in porous dolomite, is of this type.

6. *Fracture Porosity*—This is formed by fracturing or breaking the reservoir rocks, normally by regional or localized tectonic stress.

7. *Vugular Porosity*—An opening in the rock large enough to be visible with the unaided eye. The opening does not conform in position, shape, or boundaries to particular fabric elements of the reservoir rock. Most vugs were created by dissolution. Vugs or fractures usually are required for dense carbonates to produce oil and gas. In such instances oil or gas in "pinpoint" porosity may "bleed" into the fractures.

Types of Carbonate Reservoir Traps—The single most important facies that serves as a trap for hydrocarbons is the reef. A reef is a stratigraphic trap, with some being enclosed in shale. A reef has drape and flank beds which commonly dip away from the reef. Reefs may occur in large complexes or may be scattered over an area as patch reefs. A spire-shaped reef, alone or cresting as a summit, is often called a pinnacle reef.

Figure 1-38 shows the relative location of three very common types of carbonate traps, namely, reef, backreef, and lagoon. The reef is normally deposited where waves are strong and original porosity may be high. Lagoon type carbonates are formed from reef debris which has become calcareous mud, with deposition in a low energy environment. Backreef deposits are formed in a slightly higher energy environment than lagoonal deposits.

Fractured carbonates with either high or low matrix permeability can contain huge quantities of oil and gas if fractures are open. Shells and other carbonate particles may be carried by long shore current and deposited in the same manner as sandstone barrier bars. Some of the better quality carbonate reservoirs are reefs, shoals, and carbonate banks.

APPLICATION OF GEOLOGIC CONCEPTS IN CARBONATE RESERVOIRS

Geologic concepts have obvious application in any study of a carbonate reservoir. The real question is how much and what type of geologic effort should be exerted to optimize economic recovery from the reservoir. The following four case histories show that detailed reservoir description was, or would have been, very beneficial.

In one case, improved well stimulation resulted. In two reservoirs, improved reservoir description showed the reasons for poor waterflood efficiency and indicated required remedial action. The fourth example involves a pinnacle reef, which was believed to be relatively homogeneous; however, horizontal barriers prevented vertical sweep of the reef with a miscible fluid, appreciably reducing ultimate recovery.

Variations in the San Andres Reservoirs Significant in Well Completions and Well Stimulation[2,3]

The San Andres in the Permian Basin of West Texas is an important carbonate deposit penetrated by more than one hundred thousand wells. One of the most significant parts of this study was the identification through core analyses and thin section studies of the basic types of carbonate present in the porous intervals of each well.

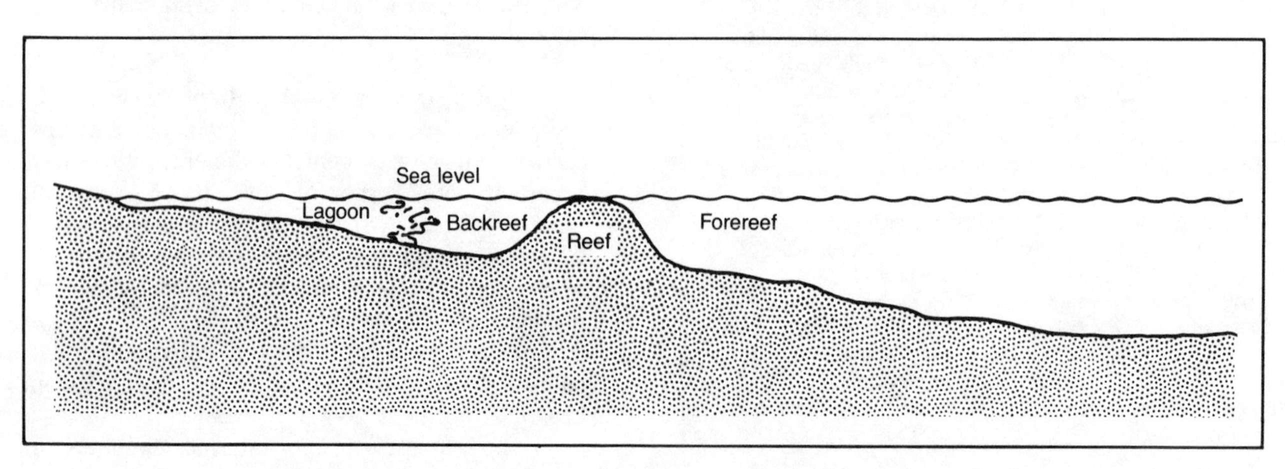

FIG. *1-38—Cross section of carbonate depositional environment.*[2]

The San Andres consists of three basic rock types: lagoon, backreef and reef, resulting from variations in the depositional environment during the Permian Period of the Paleozoic Era.

An important characteristic of the San Andres is its gradation from a full dolomitic section in the southern part of the Central Basin Platform to an almost entirely evaporitic section of anhydrite, gypsum, and rock salt with a thin dolomite bed in north central Texas, the Texas Panhandle, and western Oklahoma.

When the seas covered the Permian Basin, larger particles of coral and fossil fragments settled and formed reef-banks in the high energy zone away from the shoreline. These reef-banks later formed high-porosity, high permeability formations classified as reef type rock.

Shelfward from these banks, the quiet water environment allowed finer particles to deposit, which later resulted in the lower porosity backreef type rock. In the shallow water near the shoreline, even finer grained sedimentation eventually caused the formation of lagoon type rock.

Rock Type Description in San Andres—The *reef rock* of the San Andres formation has a massive, nonbedded appearance and may contain large nodules of anhydrite. It has a granular, sucrosic appearance with intergranular porosity. Permeabilities range from 4 to 32 md and porosities from 10 to 18%. Grain size in this rock is generally larger than 100 microns. Hydrochloric acid solubilities averaged about 85% for reef cores from the large Wasson field, but acid solubilities may be higher in some areas.

Permeabilities in the San Andres *backreef rock* usually range from less than 0.1 md to 3.0 md with porosities in the 4 to 10% range. The backreef has intergranular porosity in addition to porosity from shell fragments. Grain particle size ranges from 10 to 250 microns. Backreef cores average about 75% solubility in HCl.

The *lagoon rock type* is characterized by very fine, micritic grains, low intergranular porosity, poor permeability, and well developed bedding planes. Lagoon type rock usually has a particle size less than 10 microns with permeabilities less than 0.1 md. Due to its low porosity and permeability it is generally not considered commercially productive of hydrocarbons unless natural fractures or vugs are present. Vugs in lagoon type rock are formed by burrowing of marine life in the calcareous mud near marine channels.

Productivity Improvement in San Andres Wells—After identifying the rock types in specific depositional environments, laboratory and field studies can be made to plan well completion, workover, and well stimulation. Many failures of well stimulation probably have resulted from the lack of knowledge of the reservoir rock penetrated by the well being stimulated.

Figure 1-39 is an example of type of laboratory data obtained from backreef cores during a study involving stimulation of the San Andres reported in 1975[3].

Fracture flow efficiency tests were run, as illustrated in Figure 1-39, to determine the optimum acid treatment required for specific types of rock in each well. These etching tests indicated that a viscous preflush followed by 15% HCl was the most efficient acid fracturing treatment for San Andres backreef cores tested.

Table 1-4 shows stimulation results of all wells completed in essentially 100% backreef type rock[2] in this test program. As may be noted, hydraulic fracturing has been quite successful in San Andres backreef dolomite using water base frac fluids. Laboratory tests indicated the release of large quantities of fines when backreef cores were reacted with acid.

Results of acid fracturing using a viscous preflush followed by HCl, shown in Table 1-4, tends to verify the acid etching test results. Also the value of using a fines suspending agent in acidizing San Andres backreef is indicated in Table 1-4. Although more data is needed, field results suggest that acid fracturing may be competitive with hydraulic fracturing and propping, provided sufficient data is available on the

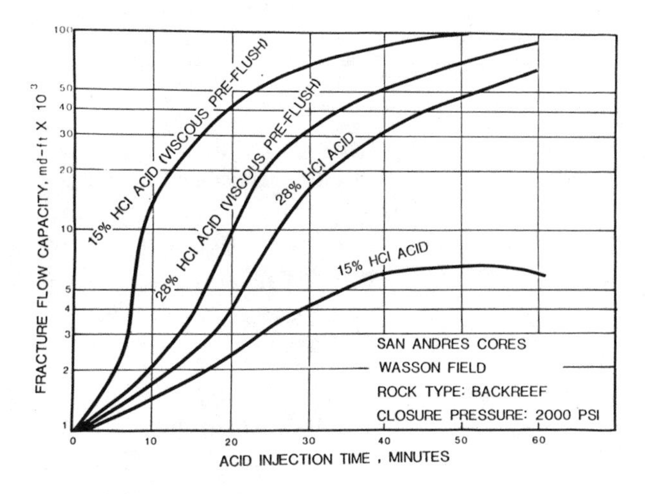

FIG. *1-39—Etched-fracture flow capacity vs. acid injection time on Backreef cores.[2] Permission to publish by Southwestern Petroleum Short Course.*

TABLE 1-4[2]
Stimulation Results in San Andres Backreef

Well No.	Fluid type	Rate bpm	Volume gal × 10^3	Sand, Sks	Production, bpd Before	After	Remarks
				Hydraulic Fracturing Jobs			
1	Viscous water gel	15	27.5	180 (20-40) 550 (10-20)	20 (oil) 0 (water)	45 (oil) 10 (water)	2.25 Fold increase
2	Gelled water with 25 lb F.L.A. per 1000 gallons	5	20.0	300 (20-40)	30 (oil) 9 (water)	52 (oil) 18 (water)	1.7 Fold increase six months after frac.
3	Gelled 1% KCl water	8	10.0	29.8 (20-40)	71 (oil) 5 (water)	102 (oil) 32 (water)	1.44 Fold increase one year after frac.
4	Gelled 1% KCl water	8	20.0	27.3 (20-40)	25 (oil) 4 (water)	156 (oil) 173 (water) (30 days)	6.24 Fold increase. (After one year 40 B/D oil and 80 B/D water.)
8	Gelled water	5.0	20.0	300 (20-40)	26 (oil) 6 (water)	113 (oil) 59 (water)	4.35 Fold
				Fracture Acidizing Jobs			
5	Viscous emulsion prepad followed by 20% HCl acid	4	10.5	None	90 (oil) 16 (water)	114 (oil) 31 (water)	1.27 Fold increase one year after acidizing.
6	Viscous emulsion prepad followed by 20% HCl with fines suspending agent	6.5	10.0	None	9.1 (oil) 3.9 (water)	35.1 (oil) 3.9 (water)	3.86 Fold increase
7	20% HCl with fines suspending agent	6.5	5.0	None	9.1 (oil) 3.9 (water)	35.1 (oil) 3.9 (water)	Volume small-type treatment questionable. (After 4 mo., 7.2 B/D oil and 10.8 B/D water.)
9	15% HCl with fines suspending agent	1.5	2.5	None	20 (oil) 200 (water)	33 (oil) 400 (water)	1.65 Fold increase after 5 months on old well.

type of rock, including etching results and the anticipated quantity of fines resulting from acid fracturing.

The results of these stimulation tests of backreef are an excellent example of the value of both geologic and acid etching studies in planning acid stimulation of carbonates.

Application of Carbonate Environmental Concepts to Well Completions in Secondary Recovery Projects[12]

A study of the Monahans Clearfork reservoir in west Texas, reported by Dowling, shows the value of comprehensive depositional environment studies to both primary and improved recovery projects in carbonate reservoirs.

The Monahans Clearfork reservoir was discovered in 1949. A pilot flood was initiated in 1959 and converted to full waterflood in 1961. By 1969, after injecting 23,700,000 barrels of water into the reservoir, some 20,000,000 barrels of water could not be accounted for. The waterflood was a failure up to that point.

As part of studies designed to plan remedial action, ten wells were drilled and cored. Analyses of slabbed cores and thin sections from these wells plus all available logs allowed reconstruction of the depositional environment. These studies pin-pointed factors limiting the productive quality of the reservoir, and led to locating additional oil reserves.

Since the porosity of productive dolomites is approximately equal to porosity of non-productive supratidal (formed above mean tide level) dolomites, these two environmental types could not be differentiated with porosity logs. In order to locate productive zones, some type of permeability measurement was required. Drill stem tests, production tests, flowmeters, tracer and temperature surveys, and other production logging techniques were used for this purpose.

Figure 1-40 is a typical cross-section showing net pay by depositional types. Net pay isopachs drawn for each zone and subzone aided in defining the limits, quality, and continuity of each reservoir. These isopachs also provided considerable information on the expected waterflood performance in each productive zone as well as suspected thief zones.

The original poor distribution of injection water is indicated by a 1968 injection profile on Well No. 45 showing 100% of the water going into the G-1 zone, and a 1967 profile on Well No. 65 showing 20% of the injected water going into the G-1 zone and 80% into the G-4 zone.

Use of permeability measurement tools to define pay zones rather than relying on log porosity measurements resulted in the location of additional oil which would not have been recovered. Mapping the vugular porous zones, formed by burrowing of marine organisms in calcareous mud near ancient marine channels, also indicated additional oil.

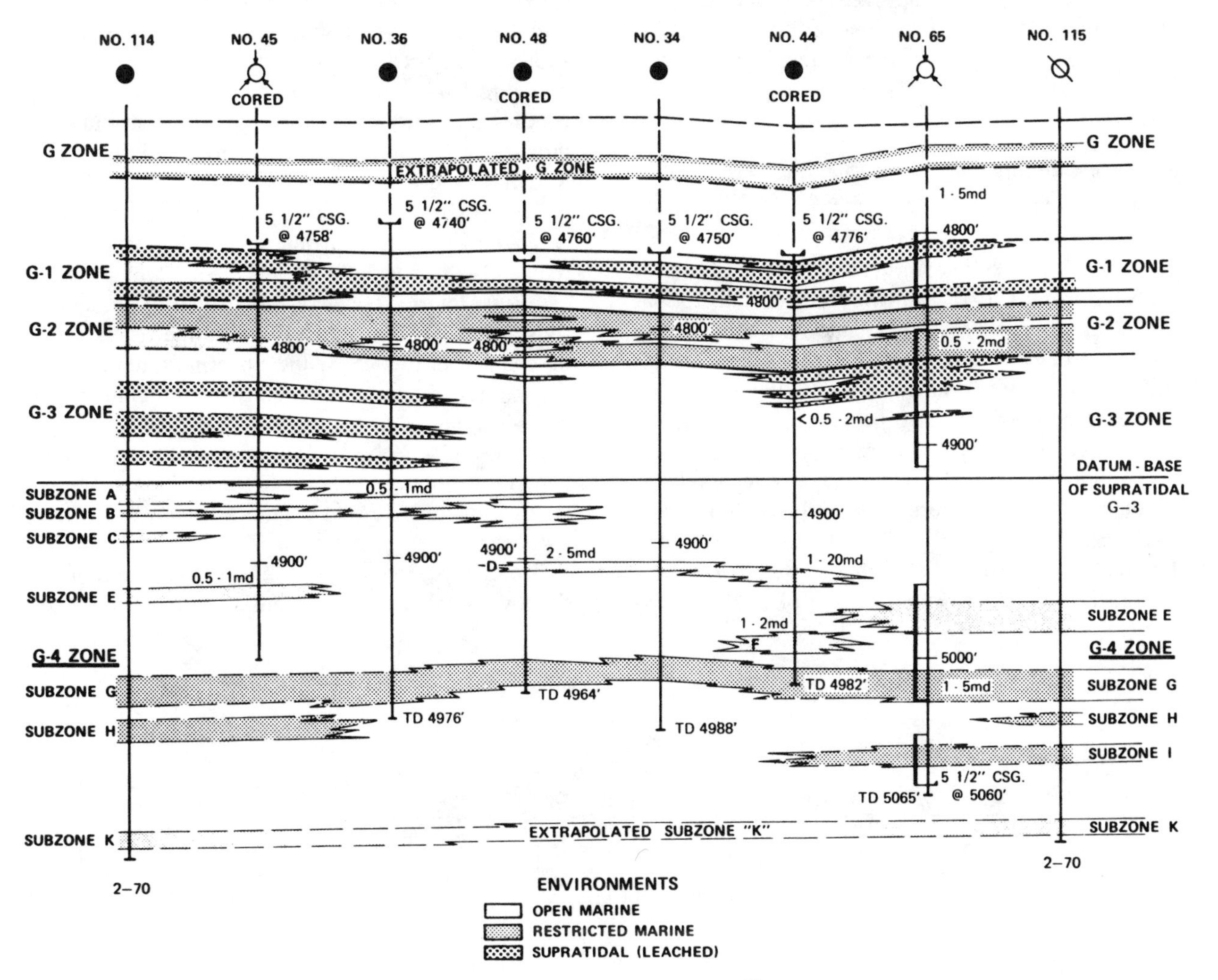

FIG. *1-40—West-East cross section of Monahans Clearfork reservoir.*[12] *Permission to publish by The Society of Petroleum Engineers.*

Prior to mapping these productive areas near ancient marine channels, zones of lagoon rock type were assumed to be nonproductive. Productive zones were located behind the casing in some wells. Other wells required deepening to penetrate indicated productive zones. New wells were drilled when required.

It is significant to note that all well workovers, deepening jobs, and new wells drilled as a result of this study paid out job costs and made a substantial profit.

The geologic reconstruction of the sedimentary environment and post depositional changes as an aid to reservoir definition proved to be effective in planning redevelopment of the reservoir. This study indicates that much more data is usually required than can be obtained from log analysis to plan well completions for both primary and improved recovery projects. This is especially true of carbonate reservoirs which undergo many changes in the character of porosity after deposition.

Geological-Engineering Team Changes Peripheral Flood to a Pattern Flood to Increase Recovery at Judy Creek[20]

The Judy Creek field in central Alberta, Canada, produces from a reef in the Beaverhill Lake formation. This reservoir has an organic reef rim with an interior of detrital carbonates. Porosity is best developed in the organic reef and reef detritus of the perimeter. Good porosity also occurs in a zone of reef detritus across the top of the reef.

The field was discovered in 1959; a waterflood was initiated in 1962 to arrest the rapidly declining reservoir pressure. By early 1973, it was apparent that the combined peripheral and bottom water flood was not satisfactory. A pressure gradient of 1,900 psi existed across the reservoir. Reservoir pressure in the vicinity of injection wells was about 1,000 psi above original pressure, and some internal areas of the pool were below bubble point. It was assumed that barriers within the reef were causing malfunction of the flood.

The first step in remedying the poor performance of the waterflood was to form a study team of reservoir geologists, stratigraphers, log analysts, computer specialists, reservoir engineers, and production engineers.

The study plan was divided into two parts. Phase I included reservoir description and fluid distribution. Phase II covered well workovers, artificial lift equipment, and infill drilling. Phase II also involved the predictive aspects of the study, employing multidimensional, multiphase models based on detailed description and fluid distribution compiled during the initial phase. A work sequence plan for the study is shown in Figure 1-41.

Figure 1-42 is an east-west cross-section developed by the study group showing porosity distribution of three facies-controlled porosity groups, designated as S-3, S-4, and S-5.

S-3 consists of a narrow, peripheral rim of organic reef with an interior lagoon of reef detritus and lime muds. S-4 has a thick organic reef buildup and an extensive interior facies of interbedded, tight and porous limestone. Immediately inside the rim is a zone of coarse detritus. S-5, the uppermost unit, has no organic reef framework. It consists predominantly of coarse organic detritus and lime sand.

A cross plot of horizontal permeability vs. percent porosity resulted in subdivision of the environmental facies into three reservoir families, shown in Figure 1-42.

Group I reservoirs occur in the organic-reef and shallow shoal facies and consist of reef framework

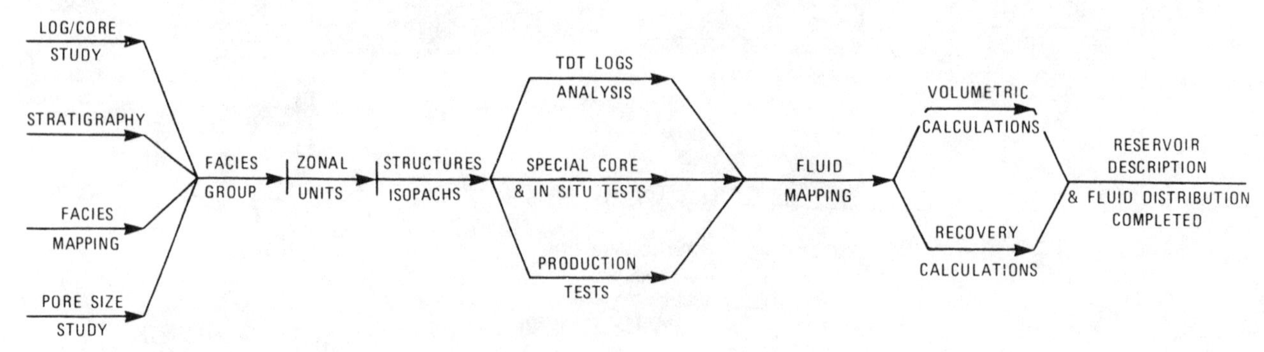

FIG. *1-41—Study sequence of Judy Creek field reservoir description and fluid dynamics.*[20] *Permission to publish by The Society of Petroleum Engineers.*

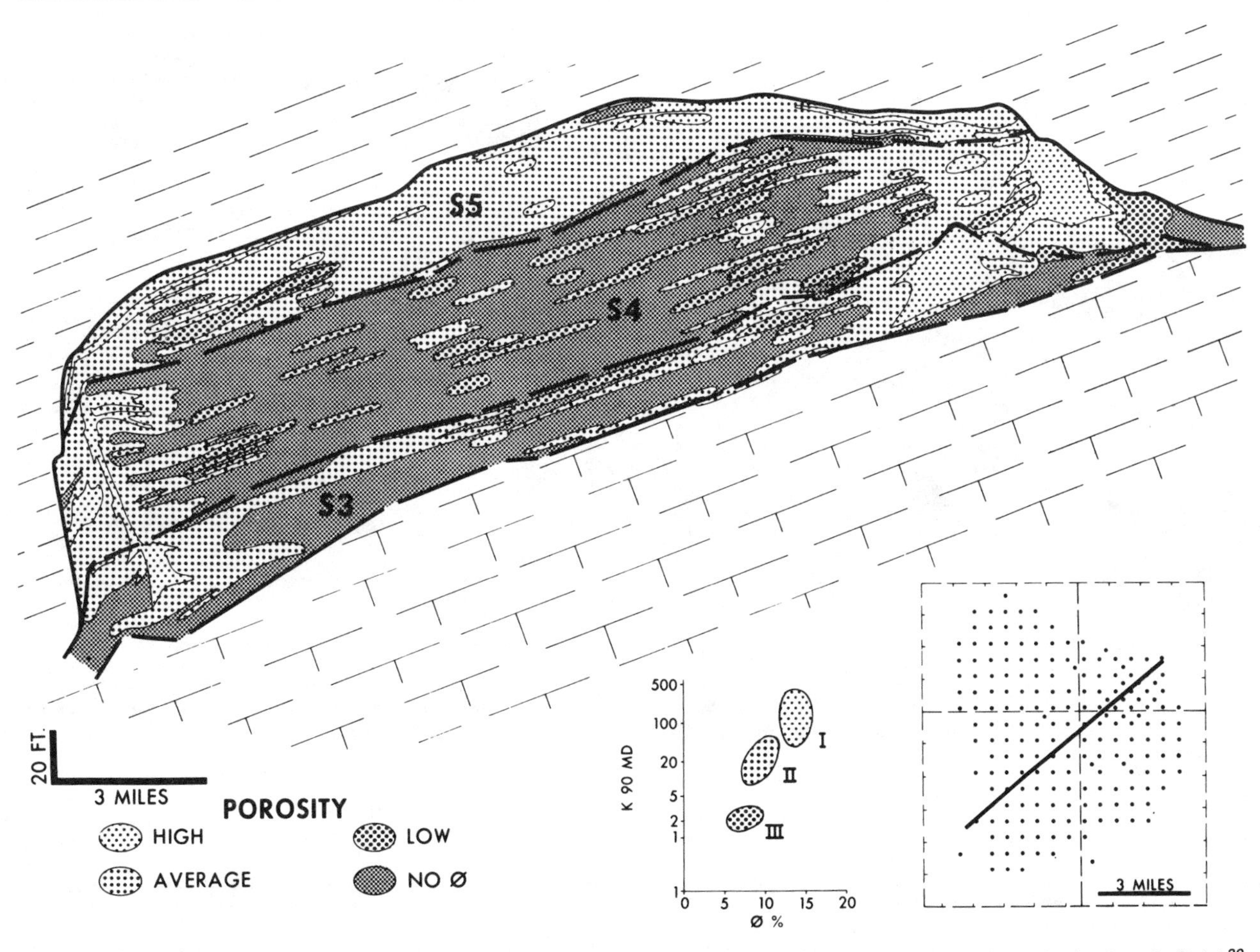

FIG. *1-42—Cross section showing distribution of three facies-controlled porosity groups, Judy Creek field.*[20] *Permission to publish by The Society of Petroleum Engineers.*

and associated reef detritus. Average porosity is 12.5% and average permeability is 170 md.

Group II reservoirs are found in shoals in the fore and backreef where reef detritus and algal laminate rocks were deposited. Porosity averages 9.5% and average permeability is 40 md.

Group III reservoirs occur in deeper water facies of the fore reef and in the interior lagoonal facies. Rock types consist of organic debris and pellets in a lime-mud matrix. Average porosity is 6.5% and average permeability is 3 md.

Pressure maintenance, initiated in 1962, concentrated water injection in the downdip periphery of the reef, principally into the S-3 and S-4 zones. Figure 1-43 shows the differential advance of water injected through March 1974. The water had advanced far updip in the S-3 unit and has fingered far updip on top of the S-5 zone which contains the highest continuous permeability. The discontinuous porosity in S-4 was being bypassed.

Figure 1-44 shows an isobaric map for March 1974 with the very poor pressure profile from bottom water and peripheral water injection. Also shown is the October 1975 isobaric map with the much improved pressure profile through the use of the centrally located pattern flood.

The improvement in the flood efficiency resulted from completing pattern injection wells in all porous intervals in S-3, S-4, and S-5 zones. All porous intervals in all producing wells within the pattern waterflood were perforated and high volume pumps were installed to place a pressure drawdown on all probable oil productive zones. In addition, all wells previously deemed to be flooded out with water were recompleted and reactivated.

Many of these reactivated wells produced more

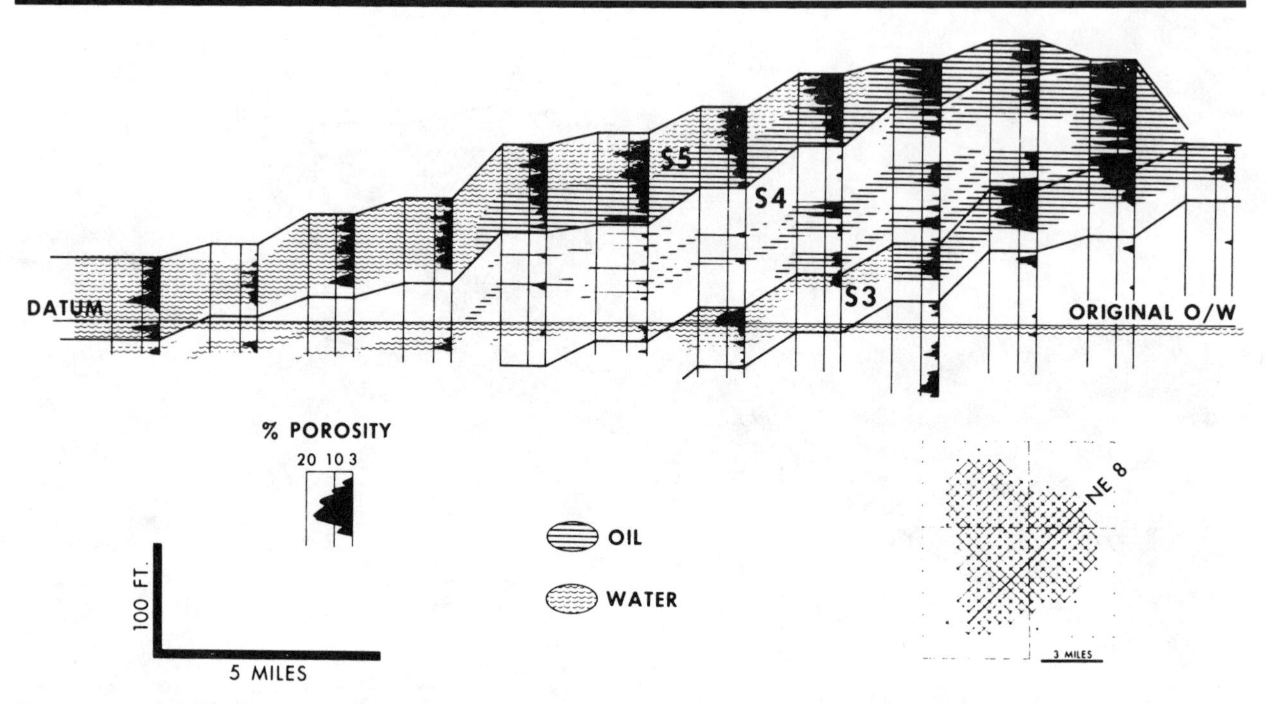

FIG. *1-43—Northeast-southwest structural cross section, Judy Creek, showing the differential advance of water injected downdip.*[20] *Permission to publish by The Society of Petroleum Engineers.*

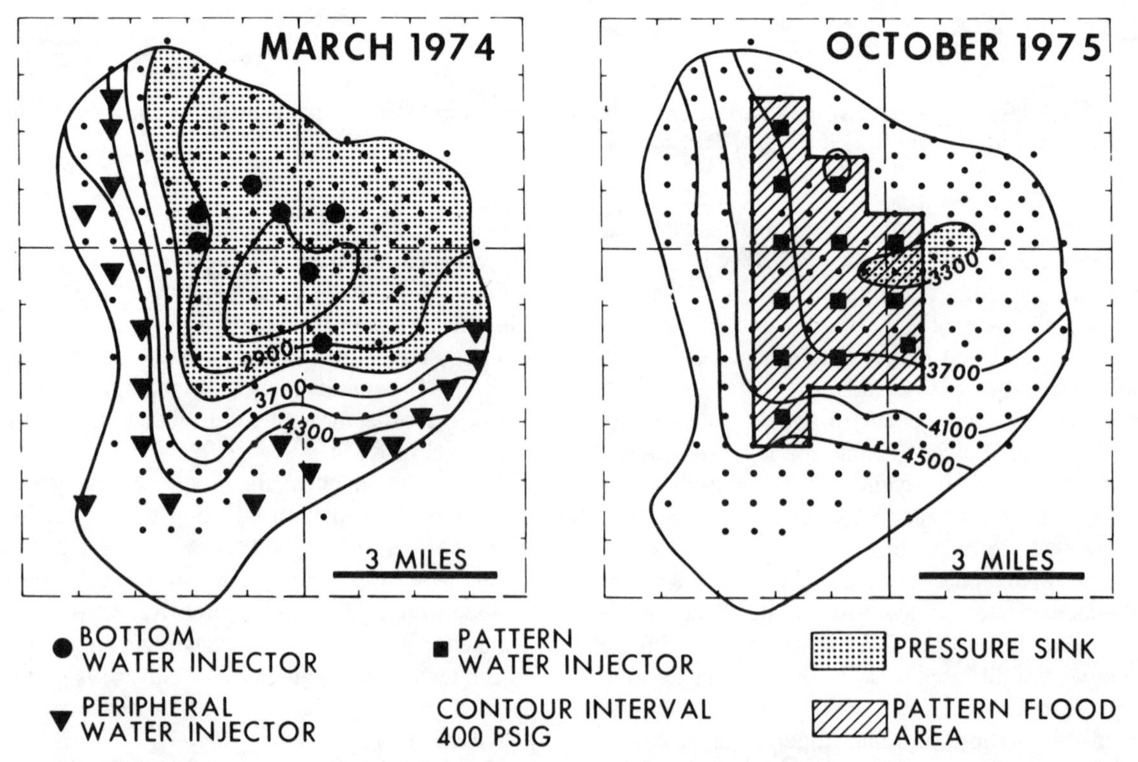

FIG. *1-44—Isobaric maps showing the result of conversion from peripheral to pattern flood, Judy Creek.*[20] *Permission to publish by The Society of Petroleum Engineers.*

than 100 b/d oil. The flood was maintained under good control with the pattern flood operational along with some water being injected into peripheral wells. Pressure was maintained at near the original reservoir pressure of 3500 psi. Additional recovery as a result of this program was projected at 10 million bbl of oil.

The Judy Creek task force study provides these guidelines for future studies in any area of the world.

1. Detailed reservoir description is the key to reservoir development and operation.

2. Because of the usual reservoir heterogeneity in both a lateral and vertical direction in carbonate reservoirs, the facies must be mapped to define detailed geometry and the most likely patterns of fluid movement under various depletion programs.

3. Reservoir description should be updated along with reservoir performance to provide a continuously upgraded reservoir model as reservoir depletion continues.

4. A study of this type should be a task force job employing engineers, geologists, log analysts, and other specialists as required. The team effort provides for greater profitability from oil and gas reservoirs, and more efficient use of valuable professional personnel. The task force or team approach usually provides a bonus in improved morale and enthusiasm of all personnel involved. In the mid-1980's a study was initiated with the objective of developing a reservoir tertiary recovery project designed to obtain additional oil recovery. In 1987-88 a CO_2 flood was planned to obtain additional oil recovery.

Detailed Reservoir Description–Key to Planning Gas-Driven Miscible Flood in the Golden Spike Reservoir[25]

A gas-driven LPG bank miscible flood failed to achieve its objectives because of incomplete information on the extent of barriers to vertical sweep in a relatively homogeneous pinnacle reef.

The Golden Spike, Leduc D 3 "A" reef reservoir, located in central Alberta, Canada was discovered in 1949. This pinnacle reef covers 1,385 acres with an average thickness of 480 ft, and originally contained 319 MMstb of oil with no gas cap or water leg. Initial production by dissolved gas drive resulted in rapid pressure decline. Pressure maintenance with gas injection was started in 1953. Analysis of available geologic data and reservoir performance to 1963, indicated a relatively homogeneous carbonate reef.

Based on this analysis of the reservoir, a miscible flood using hydrocarbon solvent was initiated in 1964. The solvent was injected as a thin bank across the top of the reef. Computer calculations, based on available reservoir description, indicated the gas-driven solvent bank (7% of hydrocarbon volume) would recover 95% of the original oil in place. This prediction was based on a model study of the apparent homogeneity of the reservoir rock and vertical sweep efficiency of the solvent.

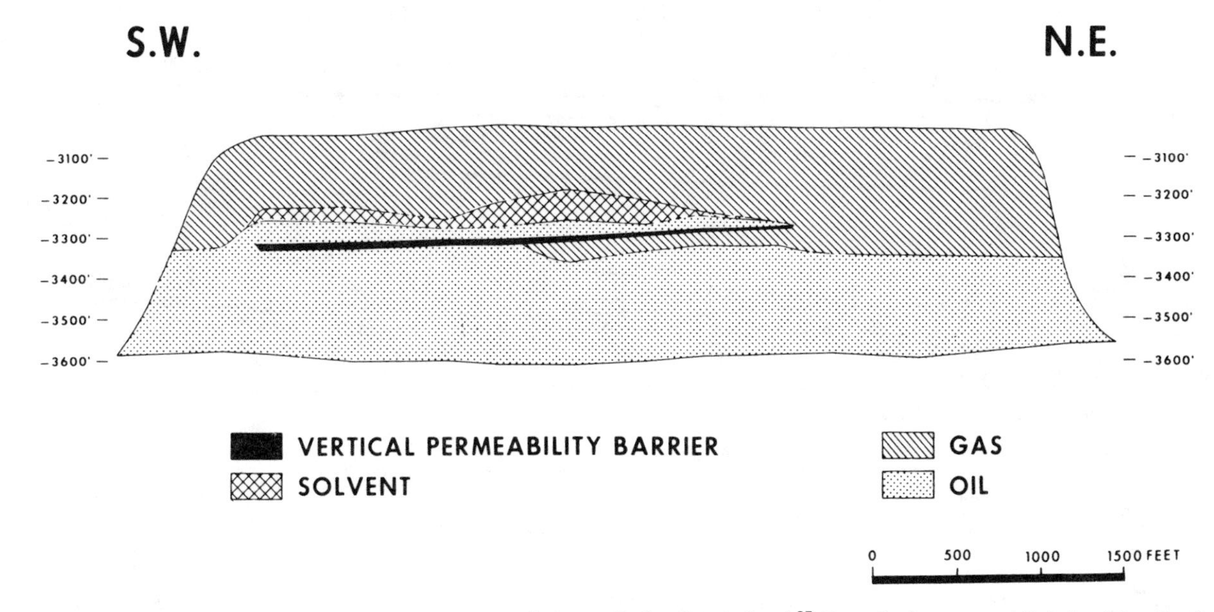

FIG. *1-45—Observed 1973 fluid distribution in Golden Spike D3 A Pool.*[25] *Permission to publish by The Society of Petroleum Engineers.*

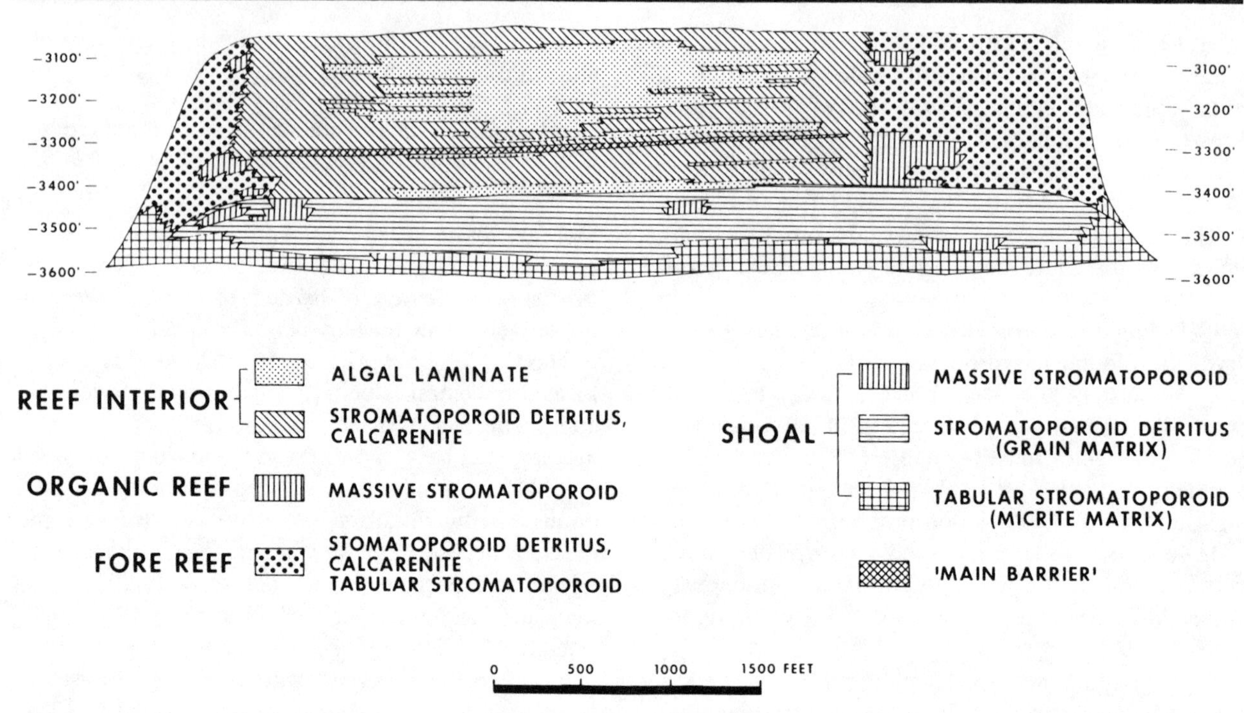

FIG. *1-46—Environmental facies of the Golden Spike reservoir.*[25] *Permission to publish by The Society of Petroleum Engineers.*

Near the end of the placement of the planned solvent bank volume in 1972, oil production of the seven producing oil wells had declined appreciably. Infill drilling was then initiated to increase oil production and to obtain additional geologic information. Observed fluid distribution (see Fig. 1-45) suggested injected gas was by-passing the solvent bank and underrunning a lateral barrier to vertical permeability.

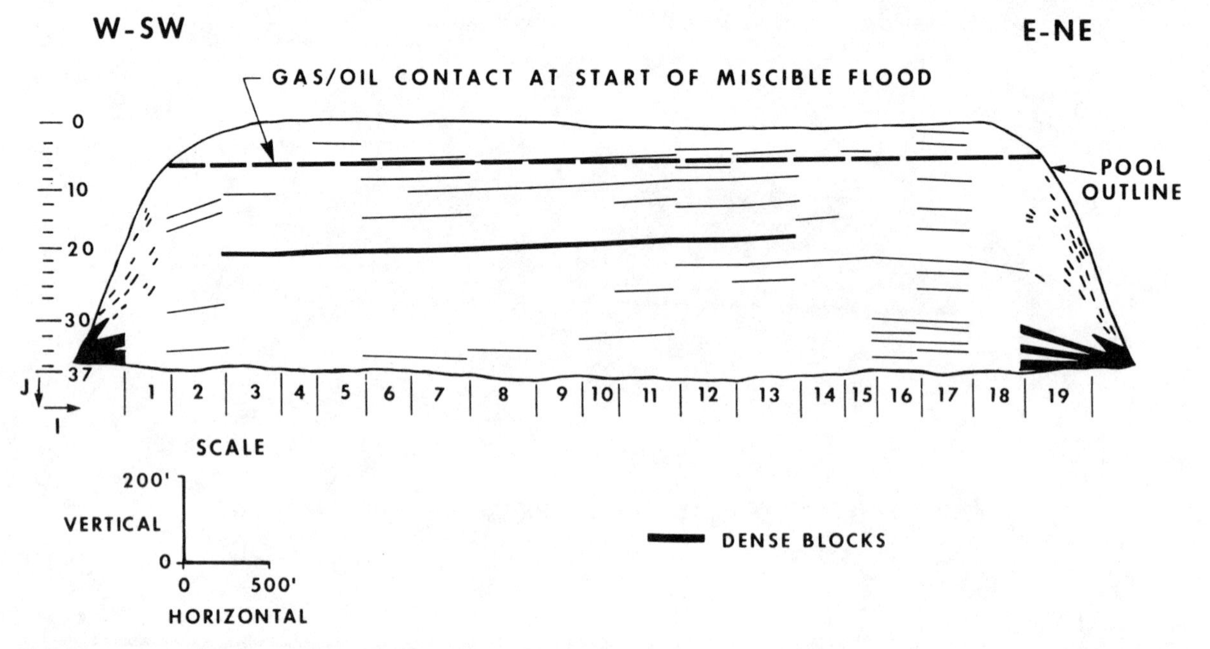

FIG. *1-47—The 2-D cross-sectional model, inluding vertical permeability barriers, Golden Spike reservoir.*[25] *Permission to publish by The Society of Petroleum Engineers.*

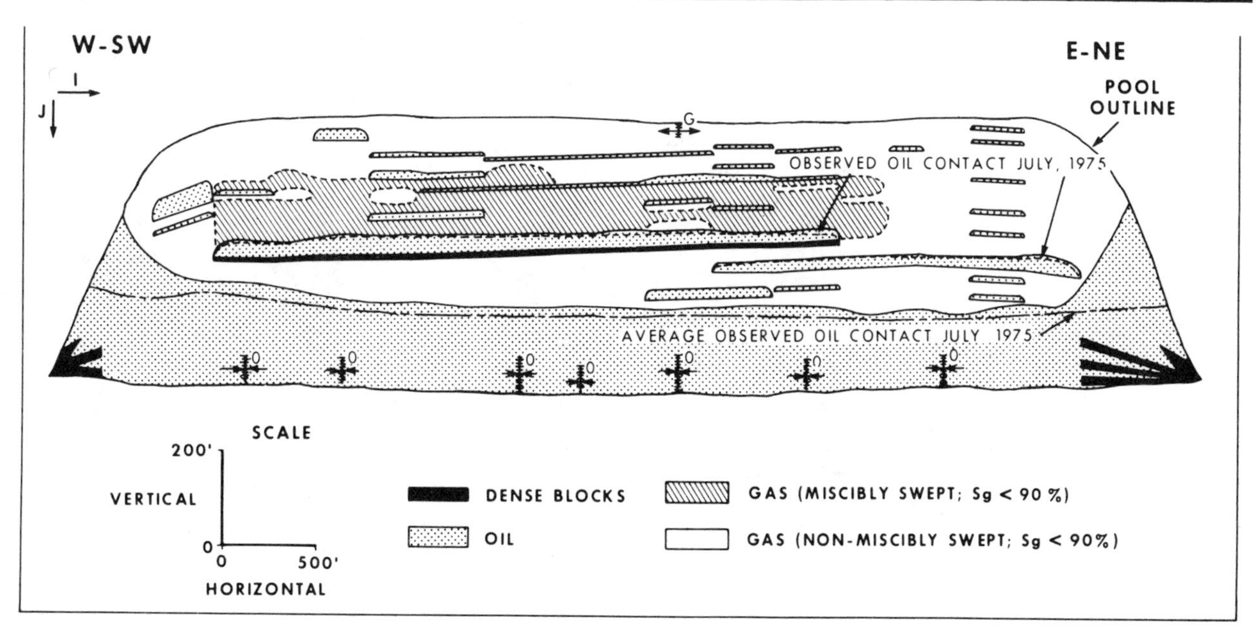

FIG. *1-48—History matched model fluid distribution in 1975, Golden Spike reservoir.*[25] *Permission to publish by The Society of Petroleum Engineers.*

Because of the continued poor performance of the miscible flood, a major reservoir description and reservoir performance study was carried out with much more core and log data being available from an extensive infill drilling program.

Figure 1-46 provides detailed environmental facies of the Golden Spike reservoir, and Figure 1-47 shows a reinterpreted model of the reservoir showing all barriers to vertical permeability. The average horizontal permeability from the 1974 model study is 2,000 md, and the average vertical permeability is 40 md.

Figure 1-48 shows a model of the production history-matched fluid distribution in 1975, based on this new detailed reservoir description, bottom hole fluid sample data, and reservoir performance data.

Because of barriers to vertical sweep, the solvent bank dispersed into the gas cap. Since the miscible bank was no longer present and effective, the solvent flood was converted to gas drive.

Predicted additional recovery by the miscible bank was reduced to 10 MMstb, compared with the original estimate of 69 MMstb. The original recovery estimate was based on the incorrect assumption that the reef was sufficiently free of barriers to permit a top to bottom sweep of the reef. The estimate of ultimate recovery, based on the 1974 study, is 67%, compared to the estimate of 95% with an efficient miscible displacement, based on the reservoir study made prior to 1964.

SUMMARY OF GEOLOGIC CONSIDERATIONS

The benefits of detailed reservoir description coupled with continued monitoring and prediction of reservoir performance is essential to optimize ultimate recovery and profits. Reservoir operation, including well completions and workovers, cannot be efficiently carried out unless an adequate reservoir description is available along with a continuously updated plan of reservoir operations and control. To aid reservoir and production engineers to reach their objectives of increasing profits, decreasing costs, and increasing oil and gas recovery, the following recommendations are made:

1. Increase your technical background in petroleum geology.
2. Develop a continuing interest in geology.
3. For each reservoir and field being considered, request from the geology department the following detailed information:

   a. Depositional environment and postdepositional history.
   b. Geologic factors controlling porosity and permeability.
   c. Accurate porosity, horizontal permeability, and vertical permeability, based on whole core analysis for vertical permeability.
   d. Other required fluid and reservoir data.

4. After obtaining an analysis of all geologic data for each reservoir, integrate geologic data with production and engineering data to develop a reservoir operation plan along with optimum well completions and workovers to implement the reservoir plan.

5. Encourage company management and geologists to obtain additional geologic data early during the development of an oil and gas reservoir. Full bore cores and full bore core analysis are badly needed—especially to determine barriers to vertical movement of fluids for reservoir simulation and well operations.

REFERENCES

1. Alpay, O. A.: "A Practical Approach to Definition of Reservoir Heterogeneity," *JPT* (July 1972) p. 841.

2. Black, H. N.; Carlile, W. C.; Coulter, G. R.; and Blalock, S.: "Lithology As a Guide to San Andres Stimulation," Presented at Southwestern Petroleum Short Course, Dept. of Pet. Eng., Texas Tech Univ. (April 17–18, 1975).

3. Black, H. N., and Stubbs, B. A.: "A Case History Study—Evaluation of San Andres Stimulation Results," SPE 5649 (Oct. 1, 1975).

4. Busch, D. A.: "Stratigraphic Traps in Sandstones-Exploration Techniques," *Memoir 21,* AAPG (July 1974).

5. Busch, D. A., and Link, D. A.: *Exploration Methods for Sandstone Reservoirs,* OGCI Publications, Tulsa, OK (1985).

6. Choquette, Phillip W., and Pray, Lloyd C.: "Geologic Nomenclature and Classification of Porosity in Sedimentary Carbonates," Reprint Series No. 5; Carbonate Rocks II: *Porosity and Classification of Reservoir Rocks,* AAPG.

7. Link, Peter K.: *Basic Petroleum Geology,* 2nd Ed., OGCI Publications, Tulsa, OK (1988).

8. Craddock, D. L., and Garza, B. T.: "A Case History—Fracturing the Morrow in Southern Blain and Western Canadian Counties of Oklahoma," SPE 10131 (Oct. 1981).

9. Craig, F. F., Jr.; Willcox, P. J.; Ballard, J. R.; and Naton, W. R.: "Optimized Recovery Through Cooperative Geology and Reservoir Engineering," SPE 6108 (1976).

10. Depositional Environments in Carbonate Rocks, S.E.P.M. No. 14 (March 1969).

11. Dickey, Parke A.: *Petroleum Development Geology,* 3rd Ed., PennWell Books, Tulsa, OK (1986).

12. Dowling, Paul L., Jr.: "Application of Carbonate Environmental Concepts to Secondary Recovery Projects," SPE 2987 (1970).

13. Elkins, L. F., and Skov, A. M.: "Some Field Observations and Heterogeneity of Reservoir Rocks and Its Effect on Oil Displacement Efficiency," SPE 282 (April 1962).

14. Flewitt, W. E.: "Refined Reservoir Description Maximizes Petroleum Recovery," S.P.W.L.A., Annual Logging Symposium (June 1975).

15. Reeckmann, A., and Friedman, G. M.: *Exploration for Carbonate Reservoirs,* John Wiley & Sons, New York (1982). Available from OGCI Publications.

16. Groult, J., and Reiss, L. H.: "Reservoir Inhomogeneities Deduced from Outcrop Observations and Production Logging," *JPT* (July 1966) p. 883.

17. Halbouty, Michel T.: "Needed; More Coordination Between Earth Scientists and Petroleum Engineers," SPE 6107 (Oct. 1976).

18. Harris, D. G.: "The Role of Geology in Reservoir Simulation Studies," *JPT* (May 1975) pp. 625–632.

19. Hewitt, H. C.: "How Geology Can Help Engineer Your Reservoir," *Oil and Gas J.* (Nov. 14, 1966) pp. 171–178.

20. Jardine, O.; Andrews, D. P.; Wishart, J. W.; and Young, J. W.: "Distribution and Continuity of Carbonate Reservoirs," *JPT* (July 1977) pp. 873–885.

21. LeBlanc, R. J., Sr.: "Distribution and Continuity of Sandstone Reservoirs," SPE 6137 (Oct. 1976).

22. LeBlanc, R. J., Sr.: "Geometry of Sandstone Reservoir Bodies," *Memoir 18,* AAPG.

23. Morgan, J. T.; Cordiner, F. S.; Livingston, A. R.: "Tensleep Reservoir Study Oregon Basin Field, Wyoming-Reservoir Characteristics," *JPT* (July 1977) pp. 886–896.

24. Pittman, G. M.: "Improved Well Completion Through Applied Core Data," API Paper (Oct. 1975).

25. Reitzel, G. A., and Callow, G. O.: "Pool Description and Performance Analysis Leads to Understanding Golden Spike's Miscible Flood," *JPT* (July 1977) pp. 867–872.

26. Robinson, Robert B.: "Classification of Reservoir Rocks by Surface Texture," Reprint Series No. 5; Carbonate Rocks II: *Porosity and Classification of Reservoir Rocks,* AAPG.

27. Sangree, J. B.: "What You Should Know to Analyze Core Fractures," *World Oil* (April 1969) pp. 69–72.

28. Sneider, R. M.; Richardson, F. H.; Paynter, D. D.; Eddy, R. E.; and Wyant, I. A.: "Predicting Reservoir-Rock Geometry and Continuity in Pennsylvanian Reservoirs, Elk City Field, Oklahoma," *JPT* (July 1977) pp. 851–866.

29. Visher, G. S.: *Exploration Stratigraphy,* PennWell Books, Tulsa, OK (1984). Available from OGCI Publications.

30. Wayhan, D. A., and McCaleb: "Elk Basin Madison Heterogeneity—Its Influence on Performance," *JPT* (Feb. 1969) p. 153.

31. Weber, K. J.: "Sedimentological Aspects of Oil Fields in the Niger Delta," Geologie En Mijnbouw (1971) Vol. 50, No. 3, pp. 559–576.

32. Weber, K. J., and Daukoru, E.: "Petroleum Geology of the Niger Delta," World Petroleum Congress, Tokyo (1975).

33. Zeito, George A.: "Interbedding of Shale Breaks and Reservoir Heterogeneities," *JPT* (Oct. 1965) p. 1223.

34. Van Everdingen, A. F., and Kriss, H. S.: "Recovery Efficiency," SPE 7427 (Oct. 1978).

35. Wadman, D. H.; Lamprecht, D. E.; and Mrosovsky, I.: "Reservoir Description Through Joint Geologic-Engineering Analysis," SPE 7531 (Oct. 1978).

36. Hartman, J. A., and Paynter, D. D.: "Drainage Anomalies in Gulf Coast Tertiary Sandstones," SPE 7532 (Oct. 1978).

37. Sneider, R. M.; Tinker, C. N.; and Meckel, L. D.: "Deltaic Environmental Reservoir Types and Their Characteristics," *JPT* (Nov. 1978) p. 1538.

38. Ghauri, W. K.: "Production Technology Experiences in a Large Carbonate Waterflood; Denver Unit, Wasson San Andres Field, West Texas," SPE 8406 (Sept. 1979).

39. Sarmiento, Roberto, and Pickel, Jack S.: *Reservoir Geology Training Manual,* Oil & Gas Consultants International, Inc., Tulsa, OK (1988).

40. Thompson, Alan: "Preservation of Porosity in the Deep Woodbine/Tuscaloosa Trend, Louisiana," SPE 10137 (Oct. 1981).

41. Robinson, Alec E.: "Faces Types and Reservoir Quality of the Rotliegendes," SPE 10303 (Oct. 1981).

42. Poston, S.; Aruna, M.; and Thakur, G. C.: "G-2 and G-

3 Reservoirs, Delta South Field, Nigeria—An Engineering Study," SPE 9513 (Sept. 1981).

43. Crocker, M. E.; Donaldson, E. C.; and Marchin, L. M.: "Comparison and Analysis of Reservoir Rocks and Related Clays," SPE 11973 (Oct. 1983).

44. Hyland, C. R.: "Pressure Coring—An Oilfield Tool," SPE 12093 (Oct. 1983).

45. Berg, Robert R.: *Reservoir Sandstones,* Prentice-Hall, Inc., Englewood Cliff, NJ (1985).

46. Miall, Andrew D.: "Reservoir Heterogenities in Fluivatile Sandstone: Lessons from Outcrop Studies," Bulletin, AAPG (1988) Vol. 72, No. 6, pp. 682–697.

47. Wagoner, Fred J.: *Geological Computer Applications Training Manual,* Oil & Gas Consultants, Inc., Tulsa, OK (1988).

48. Friedman, G. M.: *Principles of Sedimentology,* John Wiley & Sons, New York (1978). Available from OGCI Publications.

49. Moore, Clyde H.: *Carbonate Porosity Training Manual,* Oil & Gas Consultants International, Inc., Tulsa, OK (1988).

50. Lowell, James D.: *Structural Styles in Petroleum Exploration,* OGCI Publications, Tulsa, OK (1987).

51. White, David A.: *Prospect Play Assessment Training Manual,* Oil & Gas Consultants International, Inc., Tulsa, OK (1988).

52. Helander, D. P.: *Fundamentals of Formation Evaluation,* OGCI Publications, Tulsa, OK (1983).

GLOSSARY OF SELECTED GEOLOGIC TERMS USED IN THIS CHAPTER[1]

Abyssal Pertaining to the ocean environment or depth zone of 3,000 ft or deeper.

Anticline Term applies to a fold, generally convex upward, with strata which dip in opposite directions from a common ridge or axis, like a roof of a house.

Calcarenite A clastic limestone or dolomite comprising over 50 percent sand-size calcium carbonate particles cemented as a calcareous sandstone.

Calcareous Containing up to 50 percent calcium carbonate.

Calcite Calcium Carbonate mineral, CaCO3.

Clastics Rocks composed of fragmental material from pre-existing rocks. Most common clastics are sandstones and shales.

Conglomerates Water-worn fragments of rock or pebbles cemented together.

Cross-stratification Strata arranged at an angle to the original dip of the formation.

Detritus Fragmental material, such as sand, silt and clay, derived from older rocks by disintegration or abrasion.

Diagenesis Chemical, physical and biological changes in sedimentary rock before consolidation takes place.

Dip The angle at which a stratum is inclined from the horizontal.

Eolian (Aeolian) Deposits transported and deposited by the wind, such as loses and dune sand.

Evaporite Nonclastic sedimentary rock composed primarily from an aqueous saline solution due to evaporation of the solvent. Salt or anhydrite are examples of evaporites.

Fluviatile Sedimentary deposits laid down by a river or stream.

Graben Depression produced by subsidence of a strip between normal faults.

Isopach Line drawn on map through points of true equal thickness of a designated stratigraphic zone or unit.

Lithology Description of rocks.

Mineralogy The science of the study of minerals.

Neritic or **Shelf Area** Marine environment extending from low tide to a depth of about 600 feet.

Oolite A sedimentary rock, usually limestone, made up of ooliths cemented together.

Oolith A spherical to ellipsoidal body, 0.25 to 2 mm in diameter, usually calcareous, in successive concentric layers, commonly around a nucleus such as a shell fragment.

Outcrop That part of a geologic formation or structure that appears at the surface of the Earth.

Paleoslope The direction of initial dip of a former land surface, probably the slope at the time of deposition, such as a flood plain or continental slope.

Paralic Pertaining to environments of the marine borders, such as lagoonal, or shallow neritic (shelf).

Pinnacle A tall, slender, pointed mass of rock.

Progradation A seaward advance of the shoreline resulting from the nearshore deposition of river-borne deposits.

Sedimentary Basin A geologically depressed area, with thick sediments in the interior and thinner sediments at the edges.

Silt An inorganic granular material between 1/16 mm and 1/256 mm in diameter or between the size of course clay and very fine sand.

Stratum A single sedimentary bed or layer, visibly separated from other layers, regardless of thickness.

Subaerial Formed or existing on land above water.

Subaqueous Formed or existing below water.

Strike The horizontal line of intersection between a dipping surface and a horizontal plane.

Structure Contour A contour that portrays a structural surface such as a formation boundary or a fault.

Structural Trap An entrapment for oil or gas resulting from folding, faulting or other deformation.

Texture General physical appearance or character of a rock, including the geometric aspects of and the mutual relationship among its component particles or crystals. The size, shape and arrangement of the constituent elements of a rock.

Trench An elongated but proportionally narrow depression, with steeply sloping sides.

---

[1]Definitions of terms are from: "Glossary of Geology and Related Sciences with Supplement," Second Edition 1960 by The American Geological Institute.

Chapter 2 Reservoir Considerations In Well Completions

Hydrocarbon properties
Components, phases, and molecular behavior
Characteristics of reservoir rocks
Porosity, permeability, and wettability
Fluid distribution
Fluid flow in the reservoir
Pressure distribution near well bore
Radial and linear flow near well bore
Near-well-bore flow restrictions
Reservoir characteristics affecting well completion

INTRODUCTION

Oil and gas wells are expensive faucets that enable production of petroleum reserves or allow injection of fluids into an oil or gas reservoir. A prudently planned initial well completion program is the first and most important step in obtaining satisfactory producing well life to attain maximum recovery with minimum well workover. An optimum initial well completion program must consider not only geologic and fluid conditions occurring in the reservoir at time of discovery, but also changes in fluid saturations adjacent to the well as fluids are produced.

Many times the question of *where* to complete a well is allowed to overshadow the equally important problem of *how* to complete the well. This question of how to complete the well involves effective communication with all desired zones within the completion interval, effective shutoff of undesired zones within or near the completion interval, and solution of mechanical problems such as sand control. Formation damage must be a paramount consideration in any well work.

While it is desirable to minimize future workovers, conditions often occur where workovers are required to correct conditions which, at the time of initial completion, could not be, or were not foretold. In many cases workovers may be forecast as a future requirement in an optimum well completion program.

The purpose of this chapter is to briefly consider the characteristics of reservoir fluids and the flow of those fluids in the area around the wellbore, in order to tie these parameters into well completion, workover and stimulation operations. In this discussion we have borrowed frequently from the work of Norman J. Clark.[1]

HYDROCARBON PROPERTIES OF OIL AND GAS

Crude oil and gas occurring in the earth consist of a large number of petroleum compounds mixed together. These compounds are composed of hydrogen and carbon in various ways and proportions. Petroleum compounds are, therefore, called hydrocarbons, and each compound is made up of different portions of the two elements. Seldom are two crude oils found that are identical and certainly never are two crude oils made up of the same proportions of the various compounds.

Components

Hydrocarbon compounds making up petroleum can be grouped chemically into series. Each series consists of those compounds similar in their molecular makeup and characteristics. Within a given series there is a range of compounds from extremely light to extremely heavy or complex.

The most common hydrocarbon compounds are

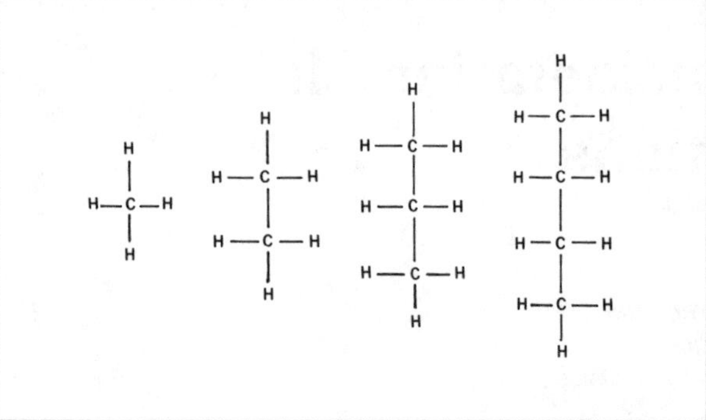

FIG. 2-1—*Structural formulas of four lightest paraffin compounds.*

those of the paraffin series which include methane, ethane, propane, butane, etc. See Figure 2-1.

Petroleum deposits include some quantity of nearly all components throughout the entire range of weights and complexities. Gas is not composed entirely of light molecules; the majority of its component molecules are light and simple; whereas, liquid crude oil is made up of a majority of heavier more complex component molecules.

Phases

Generally, all substances can exist as a solid, liquid, or gas. These three forms of existence are termed phases of matter. Whether a substance exists as a solid, liquid, or gas phase is determined by temperature and pressure conditions.

Hydrocarbon compounds, either individually or in a mixture, also change state or phase in response to changing temperature and pressure conditions. This is called "phase behavior" and many times is an important consideration in reservoir and well operations.

Molecular Behavior

The phase behavior of hydrocarbons can be explained by the behavior of the molecules making up the mixtures. Four physical factors are important. As shown in Figure 2-2 these are: (1) pressure, (2) molecular attraction, (3) kinetic energy, and (4) molecular repulsion.

Increased pressure tends to force molecules closer together so that gas will be compressed or possibly changed to a liquid. If pressure is decreased, gas expands and liquid tends to vaporize to gas. These phase changes caused by changes in pressure are termed normal or regular phase behavior.

Pressure and volume are related in that pressure results from molecular bombardment of the walls of the container or a liquid surface. Increased volume tends to reduce pressure by increasing the distance molecules must move to strike the container.

Molecular attraction acts on molecules the same as external pressure. The attraction force between molecules increases as the distance between the

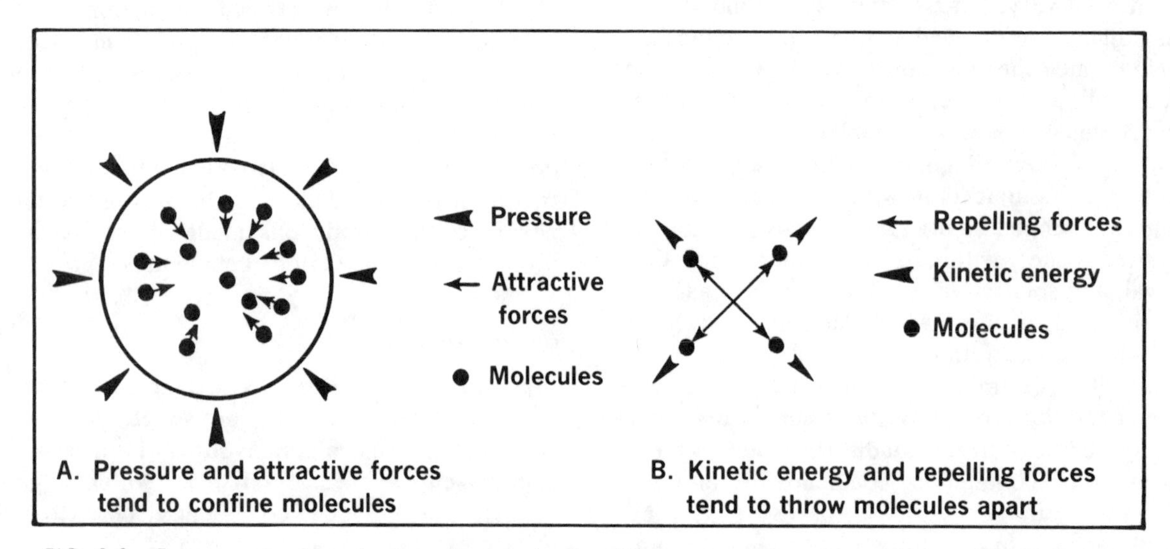

FIG. 2-2—*Forces governing hydrocarbon behavior.* Elements of Petroleum Reservoirs. *Permission to publish by The Society of Petroleum Engineers.*

molecules decreases; it also increases as the mass of the molecules increases. With smaller molecules, methane or ethane, there is less attraction between molecules and greater tendency for them to be thrown apart by their kinetic energy into gas; whereas, larger molecules, hexane and heptane, tend to be attracted together into a liquid.

Kinetic energy, or molecular motion, increases with temperature. So, the greater the temperature of a material, the greater the tendency for the material to be thrown apart and thus decrease its density (change from a liquid to a gas or a gas to expand). As temperature decreases, kinetic energy decreases, and all molecules (even the lighter molecules) tend to be attracted together into a liquid state and even frozen into a solid state. This behavior is also called normal phase behavior.

When the molecules get so close together that their electronic fields overlap, a repelling force tends to increase the resistance to further compression.

When hydrocarbon materials appear to be at rest (not expanding, contracting in volume, or changing state), the forces tending to confine the molecules balance the forces tending to throw them apart, and the material is considered to be in "equilibrium."

In petroleum reservoirs, temperature usually remains constant; therefore, only pressure and volume are altered to an appreciable degree in the reservoir during production operations. However, in the well and in surface facilities, temperature, pressure, and volume relations all become important factors.

Pure Hydrocarbons

For a single or pure hydrocarbon such as propane, butane, or pentane, there is a given pressure for every temperature at which the hydrocarbon can exist both as a liquid and a gas, Figure 2-3.

If pressure is increased without a temperature change, the hydrocarbon is condensed to a liquid state. If pressure is decreased without a temperature change, the molecules disperse into a gas. As temperature increases, kinetic energy increases, and

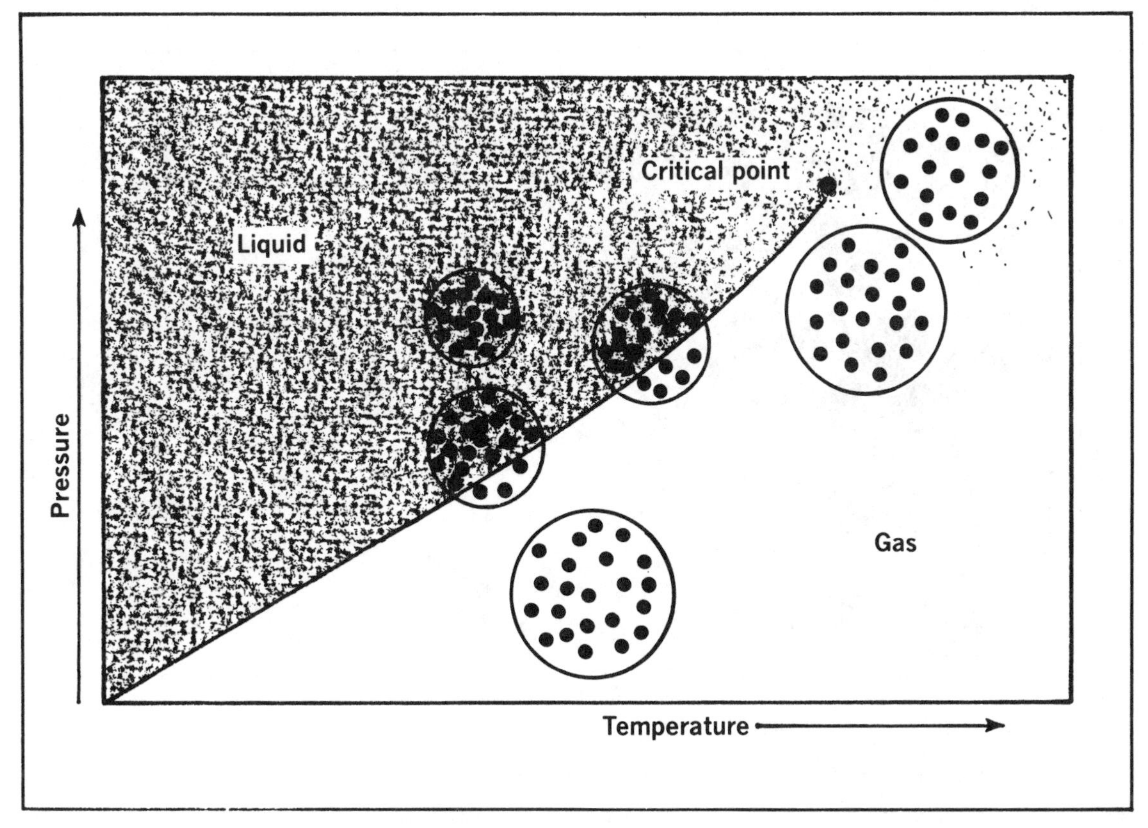

FIG. *2-3—Vapor pressure vs. temperature for a pure hydrocarbon component.* Elements of Petroleum Reservoirs. *Permission to publish by The Society of Petroleum Engineers.*

higher pressure is required for the balanced conditions at which the two phases can exist simultaneously.

The curve Figure 2-3, plotted through the pressure-temperature points where the two phases exist, is called the "vapor-pressure curve." There is a temperature above which the material will not exist in two phases regardless of the pressure. This is called the "critical point," and temperature and pressure at this point are called "critical temperature" and "critical pressure."

Material is commonly considered to be a gas when it exists at temperature and pressure below the vapor-pressure curve and as a liquid above the vapor-pressure curve. However, ranges of temperature and pressure exist in which a material can be classified as either liquid or gas.

In these ranges (shown in the upper right-hand portion of Figure 2-3) the temperature is so great that attractive forces between the molecules are not sufficiently large to permit them to coalesce to a liquid phase. Increased pressure merely causes the molecules to move together uniformly.

Hydrocarbon Mixtures

In a mixture of two components, the system behavior is not so simple. There is a broad region in which two phases (liquid and gas) co-exist.

Figure 2-4 is a diagram of the phase behavior of a 50:50 mixture of two hydrocarbons such as propane and heptane. Superimposed on the correlation are vapor-pressure curves of two components in their pure state.

The two-phase region of the phase diagram is bounded by a "bubble-point" line and a "dew-point" line, with the lines joining at the critical point. At the bubble point, gas begins to leave solution in oil with decreasing pressure. At the dew point, liquid generally begins to condense from gas with increasing pressure; however, above the critical temperature, condensation may occur at some points along the dew point curve with increasing pressure, and at other points with decreasing pressure.

At the critical point, properties of both gas and liquid mixtures are identical. Our previous definition

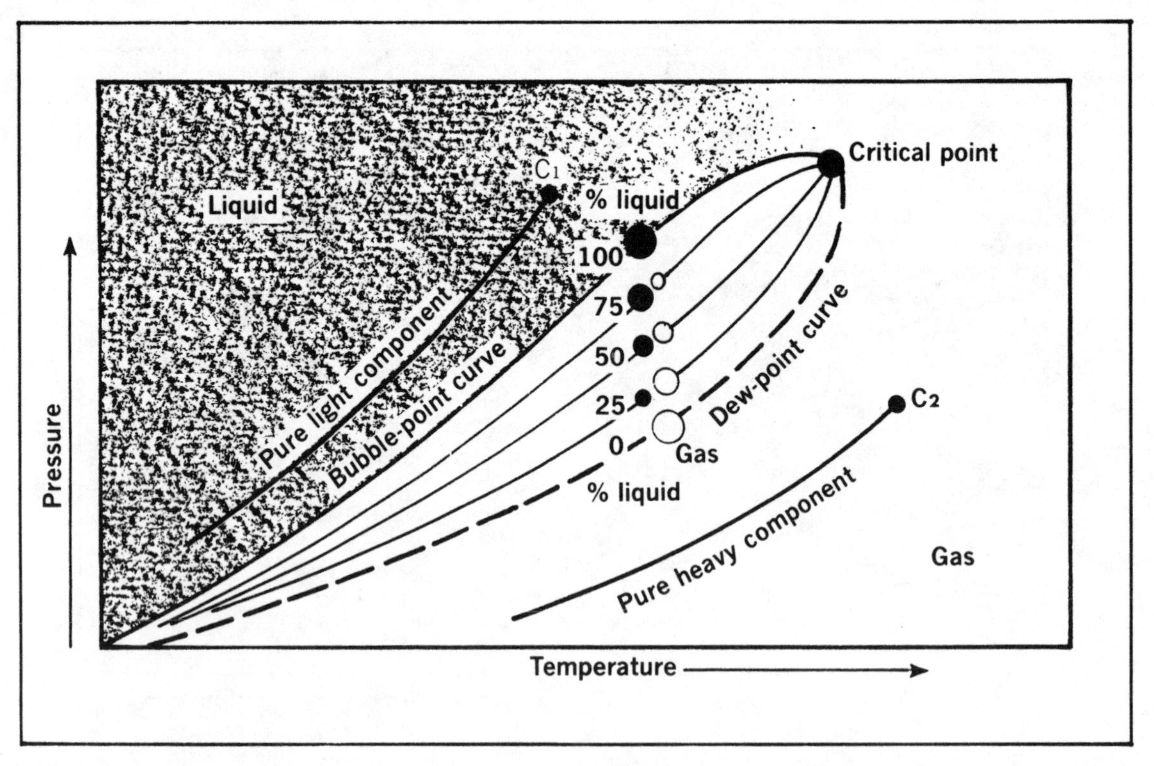

FIG. *2-4—Vapor pressure curves for two pure components and phase diagram for a 50:50 mixture of the same components.* Elements of Petroleum Reservoirs. *Permission to publish by The Society of Petroleum Engineers.*

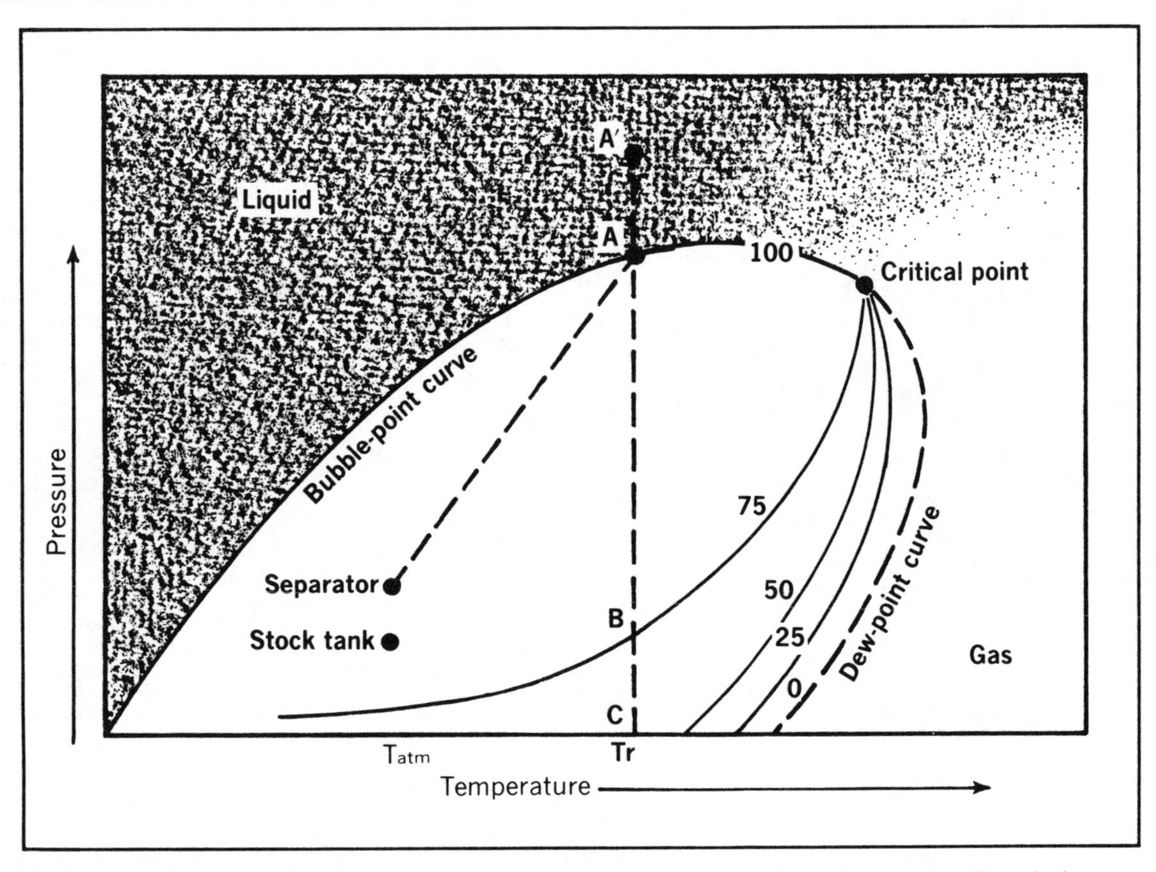

FIG. *2-5—Phase diagram of low shrinkage oil.* Elements of Petroleum Reservoirs. *Permission to publish by The Society of Petroleum Engineers.*

of "critical point" no longer applies because in a multicomponent system both liquid and gas phases exist at temperatures and pressures above the critical point. The variation may be slight in a two-component system, however, with a larger number of components, the pressure and temperature ranges in which two phases exist increase greatly.

An idealized pressure-temperature phase diagram of a crude oil in a reservoir with a temperature Tr is shown in Figure 2-5. Crude oil at its bubble point or saturation pressure is represented by Point A.

The same oil would be "undersaturated" if reservoir pressure were represented by Point A′. Separator and stock-tank temperatures and pressures are also shown in Figure 2-5.

The vertical line, A-B, represents the relative quantities of liquid and gas existing at equilibrium at a particular pressure as reservoir pressure is dropped at constant reservoir temperature. In the production process this is physically represented by gas coming out of solution in the reservoir, the amount of which is governed by the reduction in reservoir pressure.

Quantities of liquid and gas represented by location of the stock tank point in Figure 2-5, however, do not indicate what would occur in the stock tank because the composition of the original mixture changes at the separator in the production process.

Retrograde Condensate Gas

Some hydrocarbon mixtures exist in the reservoir above their critical temperature as condensate gases. When pressure is decreased on these mixtures, instead of expanding (if a gas) or vaporizing (if a liquid) as might be expected, the intermediate and heavier components tend to condense.

Conversely, when pressure is increased, they vaporize instead of condensing. The process termed "retrograde" is illustrated by temperature condition Tr in Figure 2-6. Condensation of liquids in the

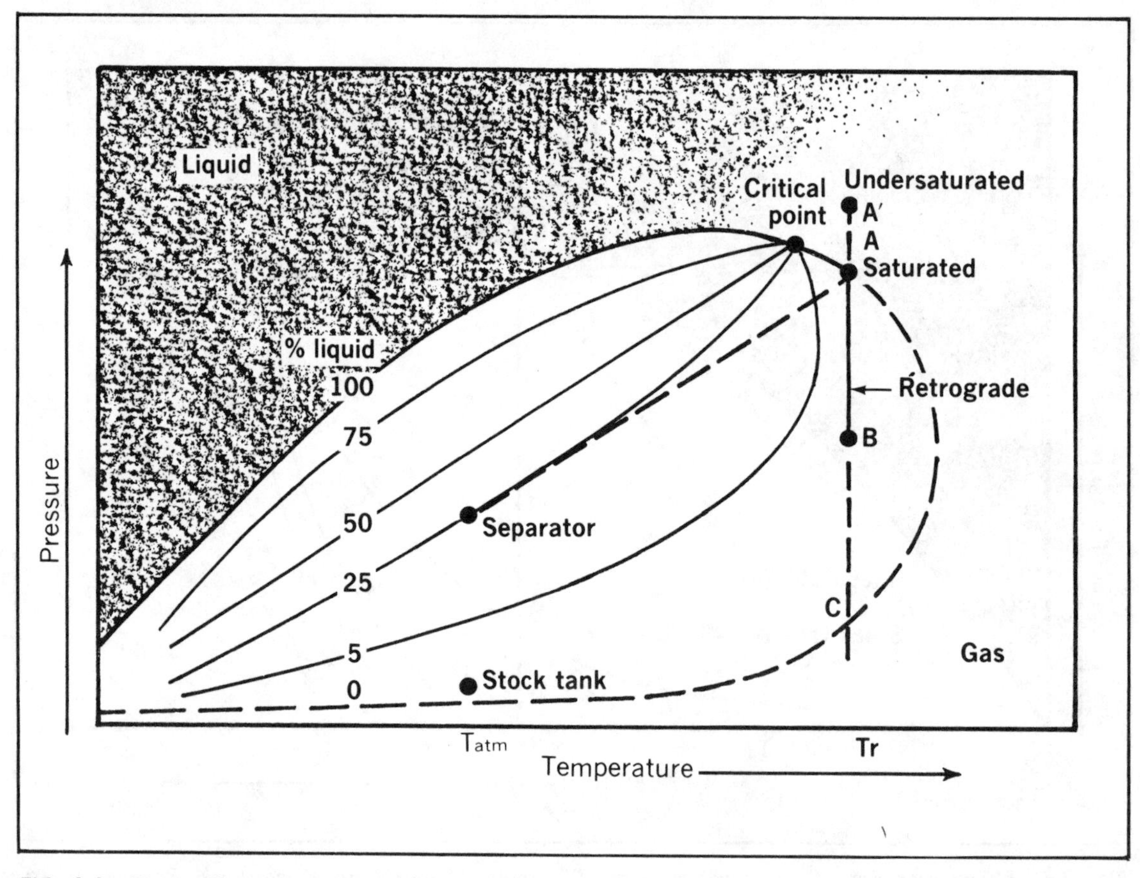

FIG. *2-6—Phase diagram of retrograde condensate gas.* Elements of Petroleum Reservoirs. *Permission to publish by The Society of Petroleum Engineers.*

reservoir alters relative permeability relationships, and usually results in loss of well productivity and also of hydrocarbon recovery.

Gas

Behavior of gases is shown in Figures 2-7 and 2-8. Hydrocarbon in a gas reservoir may be termed as "wet" gas or "dry" gas, depending upon its behavior. Both exist at temperatures above their critical temperature. When the temperature of a wet gas is reduced to stock tank temperature condensation of heavier components results. With a dry gas no condensation of liquids results at stock tank temperature.

Practical Uses of Hydrocarbon Data

The practical approach to the study of reservoir fluid behavior is (1) to anticipate pressure and temperature changes in the reservoir and at the surface during production operations, and (2) to measure by laboratory tests the changes occurring in the reservoir fluid samples. The results of these tests then provide the basic fluid data for estimates of fluid recovery by various methods of reservoir operation, and also for estimates of reservoir parameters through transient pressure testing.

Two general methods are used to obtain samples of reservoir oil for laboratory examination purposes: (1) by means of a subsurface sampler, and (2) by obtaining surface samples of separator liquid and gas. These samples are then recombined in the laboratory in proportions equivalent to gas-oil ratio measured at the separator.

Information concerning the characteristics and behavior of gas needed for work with gas reservoirs depends upon the type of gas and the nature of the problem. If retrograde condensation is involved, needed information may require numerous tests and measurements. If wet gas is involved (with no retrograde condensation) or if dry gas is involved, information is less complex.

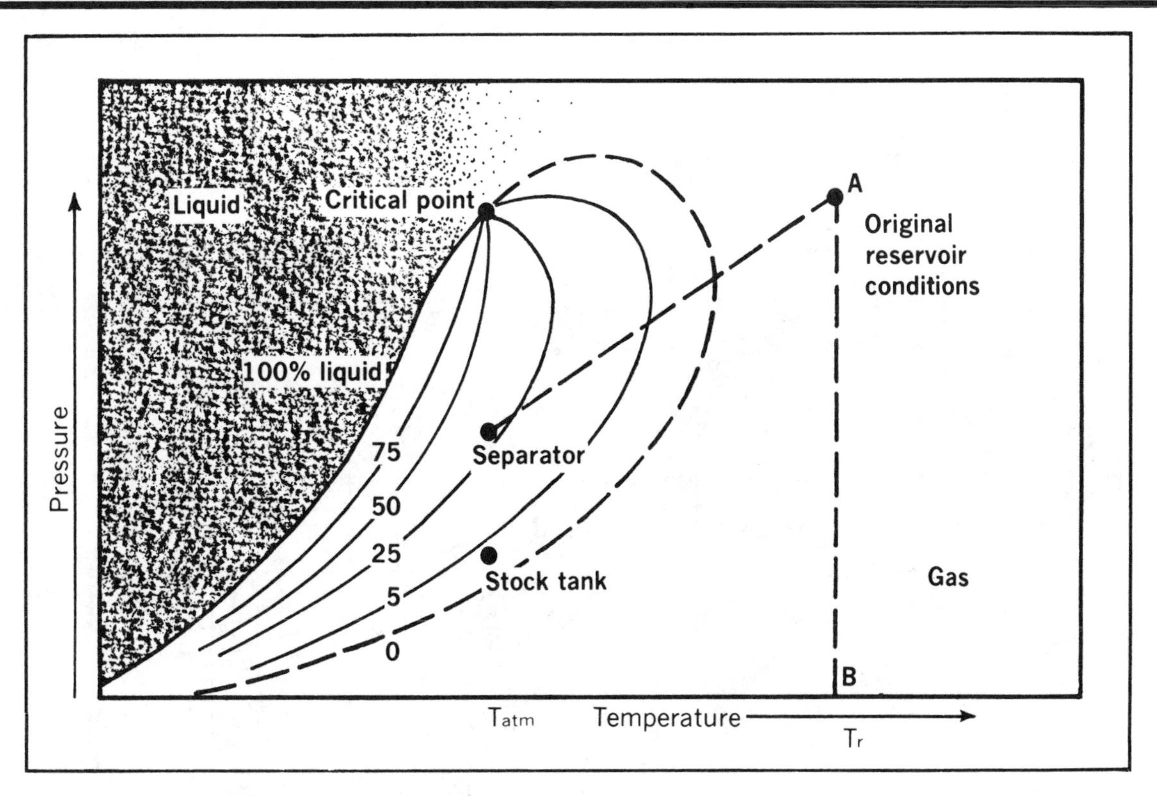

FIG. *2-7—Phase diagram of wet gas.* Elements of Petroleum Reservoirs. *Permission to publish by The Society of Petroleum Engineers.*

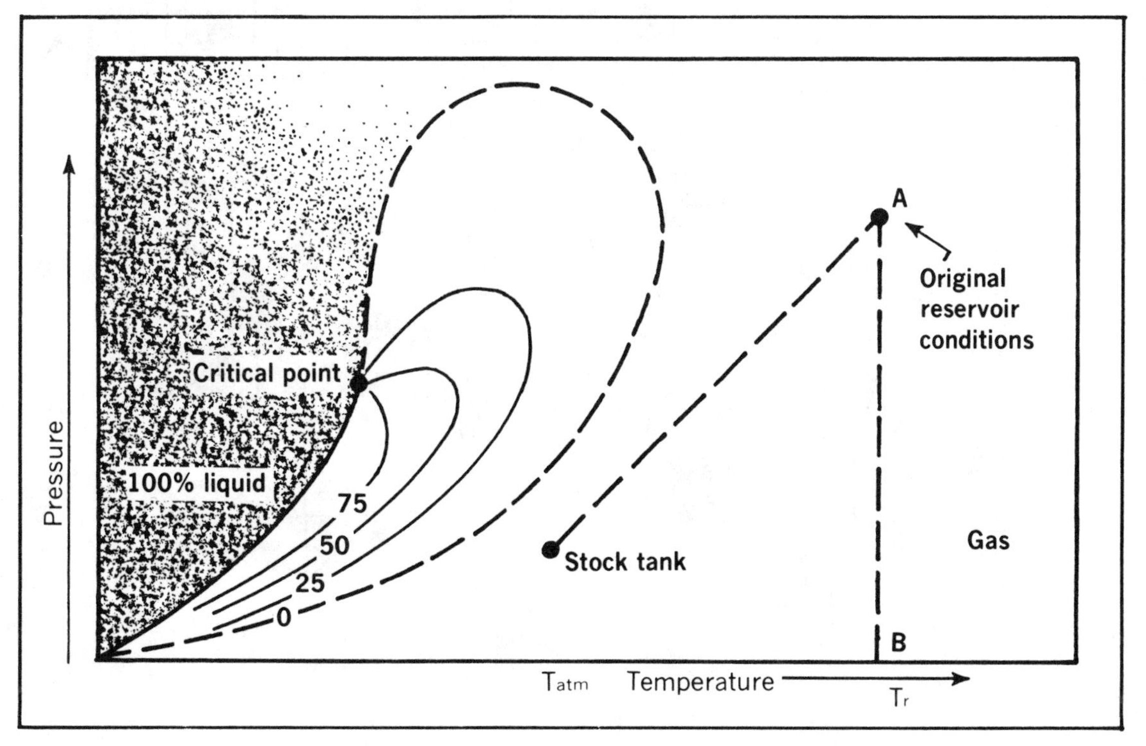

FIG. *2-8—Phase diagram of dry gas.* Elements of Petroleum Reservoirs. *Permission to publish by The Society of Petroleum Engineers.*

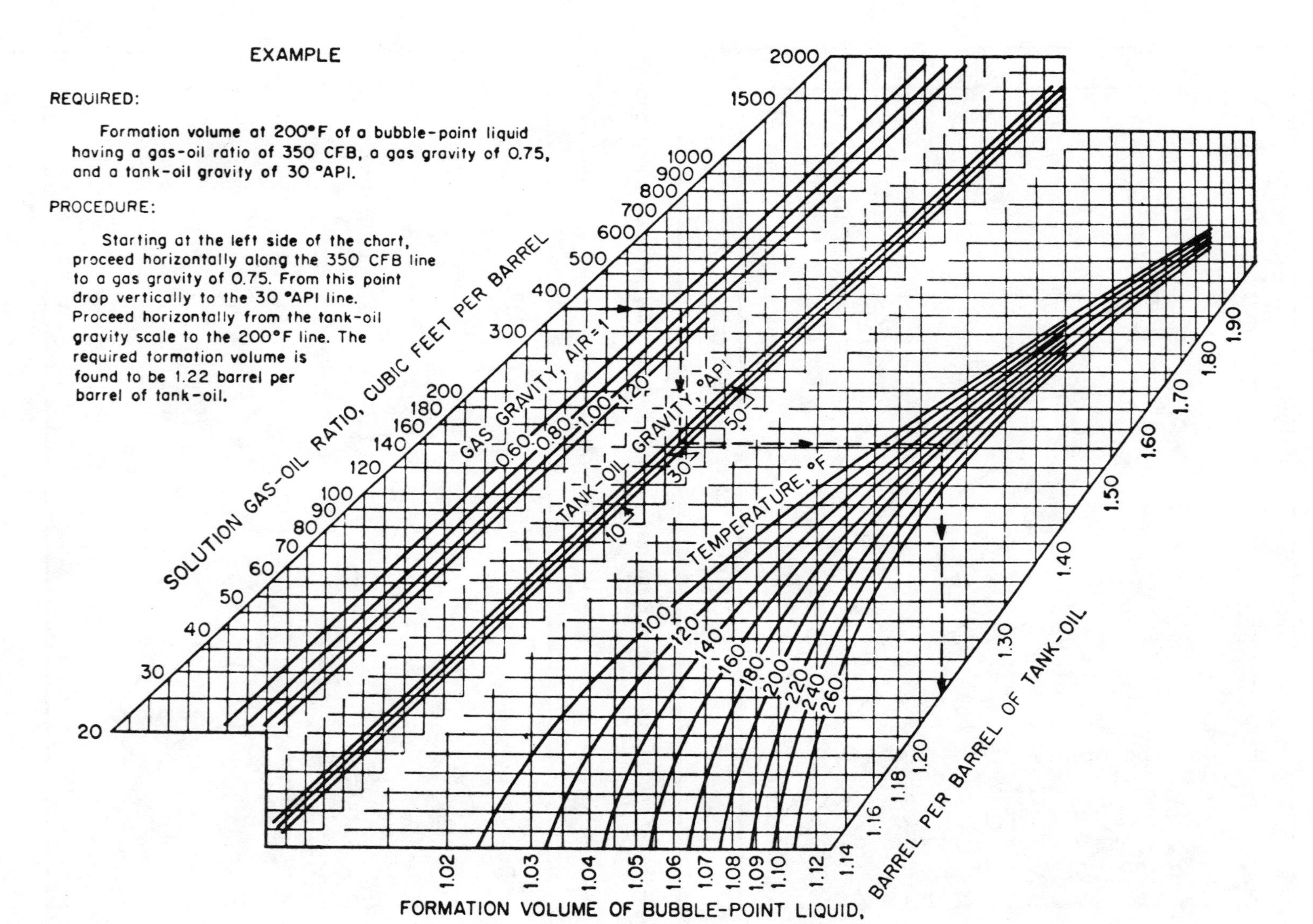

FIG. 2-9—*Calculation of oil-formation volume factor by Standing's correlation. Permission to publish by The Society of Petroleum Engineers.*

Correlation of Properties of Oils

Several generalizations of oil sample data are available, permitting correlations to be made to minimize the need for oil-reservoir sampling, testing and analysis.

These correlations are valuable for many practical, day-to-day reservoir engineering calculations. Typical of these correlations are those of Standing[2] (GOR vs. formation volume factor, bubble-point pressure and two-phase formation volume factor) and by Beal[3] and Carr et al[4] (viscosities of air, water, natural gas, crude oil, and associated gases). Common correlations are shown in Figures 2-9 and 2-10.

CHARACTERISTICS OF RESERVOIR ROCKS

Porosity

Porosity or pore space in reservoir rock provides the container for the accumulation of oil and gas and gives the rock characteristic ability to absorb and hold fluids. Most commercial oil and gas reservoirs occur in sandstone, limestone, or dolomite rocks; however, some reservoirs even occur in fractured shale. Figures 2-11 and 2-12 show some reservoir rock characteristics.

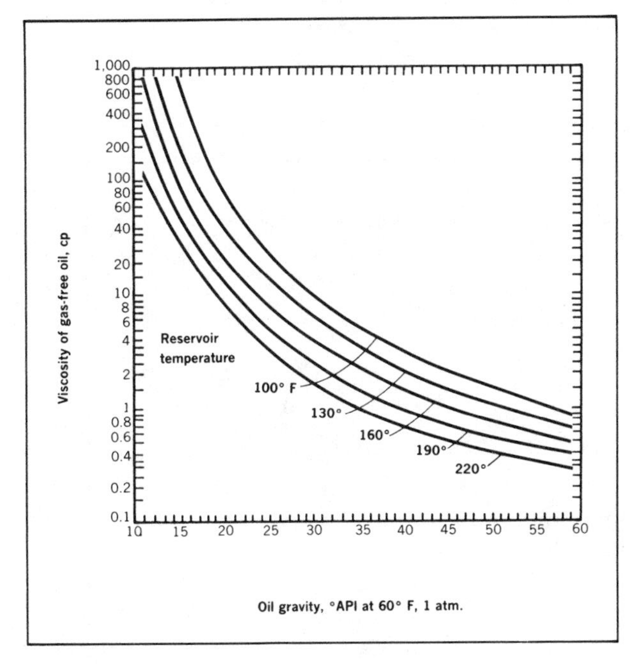

FIG. *2-10A—Viscosity of gas-free crude oil at reservoir temperature.[3] Permission to publish by The Society of Petroleum Engineers.*

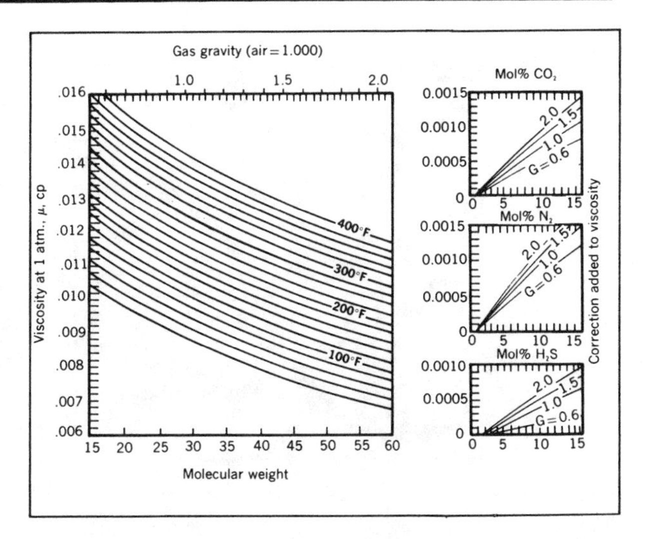

FIG. *2-10B—Viscosity of hydrocarbon gases at one atmosphere and reservoir temperatures, with corrections for nitrogen, carbon dioxide, and hydrogen sulfide.[4] Permission to publish by The Society of Petroleum Engineers.*

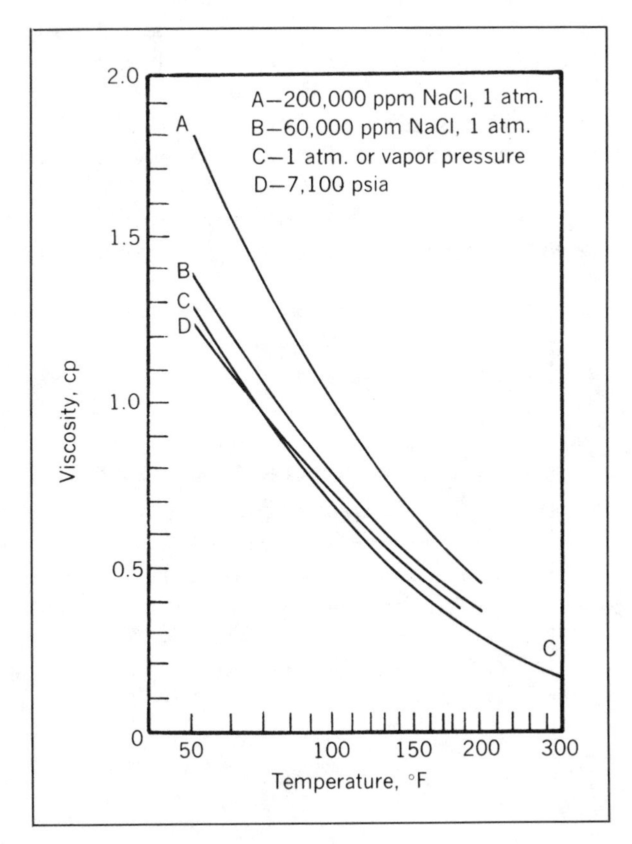

FIG. *2-10C—Viscosity of water as a function of temperature, pressure, and salinity. After Chesnut.*

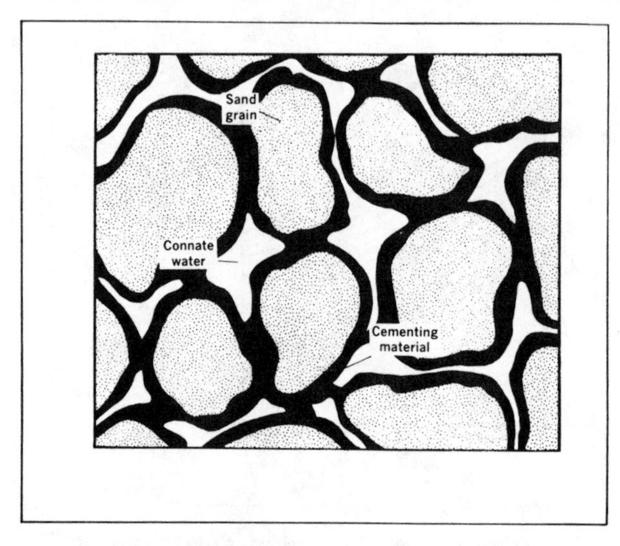

FIG. *2-11—Consolidated sandstones.* Elements of Petroleum Reservoirs. *Permission to publish by The Society of Petroleum Engineers.*

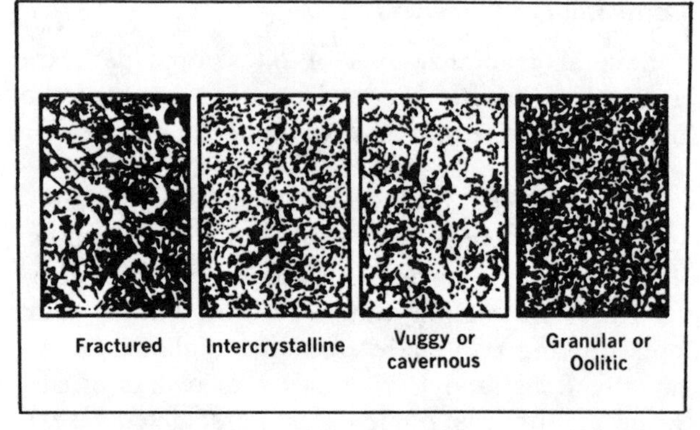

FIG. *2-12—Consolidated limestones.* Elements of Petroleum Reservoirs. *Permission to publish by The Society of Petroleum Engineers.*

Permeability

Permeability is a measure of the ease with which fluid can move through the inter-connected pore spaces of the rock. Many rocks, such as clays, shales, chalk, anhydrite, and some highly cemented sandstones, are impervious to movement of water, oil, or gas, even though they may actually be quite porous.

In 1856 the French engineer, Henry Darcy, working with water filters, developed a relation which describes fluid flow through porous rock. Darcy's Law states that rate of flow through a given rock varies directly with (1) permeability (measure of the continuity of inter-connected pore spaces), and (2) the pressure applied; and varies inversely with the viscosity of the fluid flowing.

In rock having a permeability of 1 darcy, Figure 2-13, 1 cc of a 1-cp viscosity fluid will flow each second through a portion of rock 1 cm in length and having a cross section of 1 cm^2, if the pressure drop across the rock is 1 atmosphere.

$$K = \frac{q \mu L}{A \Delta p} \quad (1)$$

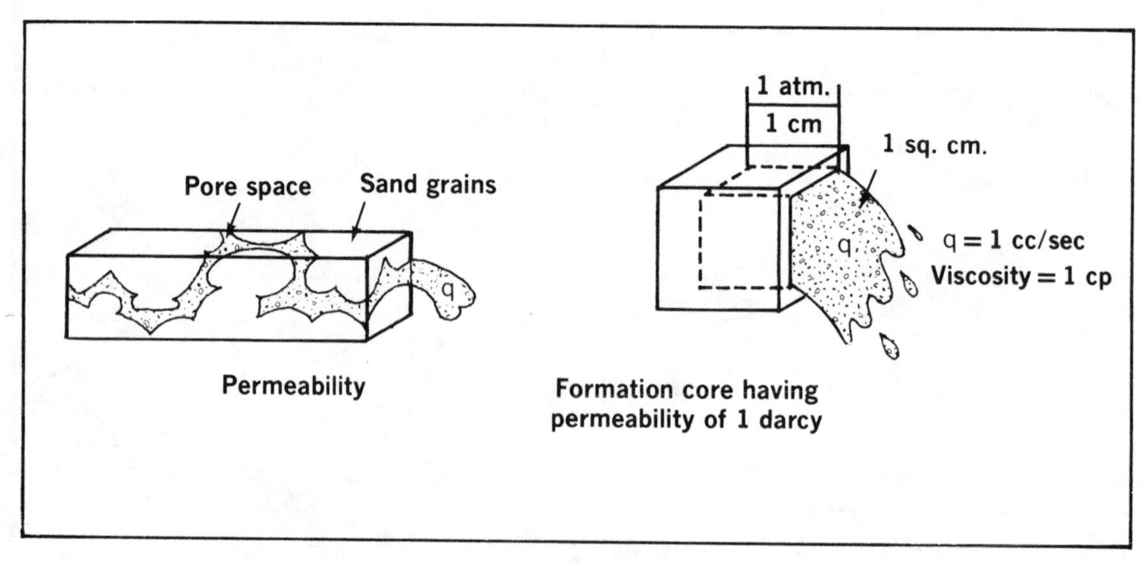

FIG. *2-13—Fluid flow in permeable sand.* Elements of Petroleum Reservoirs. *Permission to publish by The Society of Petroleum Engineers.*

In oil field units the linear form of Darcy's Law for flow of incompressible fluid through a rock filled with only one fluid is as follows:

$$q = 1.127 \times 10^{-3} \frac{kA\,(p_1 - p_2)}{B\mu L} \quad (2)$$

where:

q = flow rate, stb/day
k = permeability, md
A = flow area, ft^2
μ = viscosity, cp
L = flow length, ft
p_1, p_2 = inlet and outlet pressures, psi
B = formation volume factor, res. bbl/stb

Relative Permeability

Because two or three fluids—gas, oil, and water—can, and often do, exist in the same pore spaces in a petroleum reservoir, relative permeability relationships must be considered. Relative permeability represents the ease with which one fluid flows through connecting pore spaces in the presence of other fluids, compared to the ease with which one fluid flows when it alone is present.

Consider a rock filled only with oil at high pressure (Figure 2-14A). Gas has not been allowed to come out of solution; therefore, all available space is filled with oil, and only oil is flowing.

If reservoir pressure is allowed to decline (Figure 2-14B), some lighter components of the oil will evolve as gas in the pore spaces. Flow of oil is reduced but gas saturation is too small for gas to flow through the pores.

If pressure continues to decline, gas saturation continues to increase, and at some point (equilibrium gas saturation) gas begins to flow; oil flow rate is further reduced, (Figure 2-14C).

With further increases in gas saturation more and more gas and less and less oil flows through the pores until finally nothing but gas is flowing (Figure 2-14D). Significant amounts of oil may remain in the pore spaces, but cannot be recovered by primary means because relative permeability to oil is now zero.

This same principle governs the flow of oil in the presence of water. The saturation of each fluid present affects the ease of fluid movement or relative permeability.

Figure 2-15 shows typical oil water relative per-

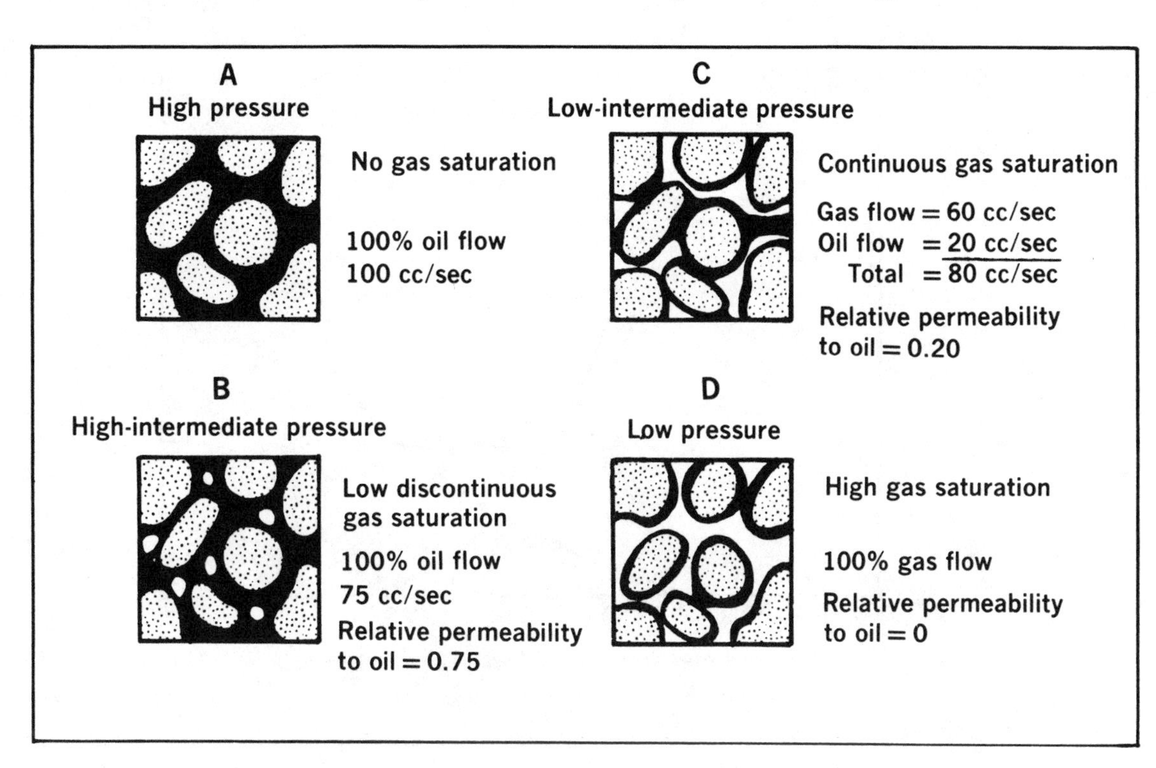

FIG. *2-14—Gas-oil relative permeability concept.* Elements of Petroleum Reservoirs. *Permission to publish by The Society of Petroleum Engineers.*

meability relations for a water wet sandstone. A reservoir represented by Figure 2-15 would have an initial or connate water saturation of about 27%. With an active water drive to maintain pressure, increases in water saturation occurring as water moved in to expel the oil would reduce the relative permeability to oil.

This would result in decreasing oil production and increasing water production. When water saturation reached 75%, relative permeability to oil would be reduced to zero and further oil flow would stop. To reach this point in a practical situation might not be feasible, since very large percentages of water would have to be produced.

The gas-oil or oil-water relative permeability relationships of a particular reservoir rock depend on the configuration of the rock pore spaces, and the wetting characteristics of the fluids and rock surfaces. In an oil-water system the relative permeability to oil is significantly greater when the rock surface is "water wet."

Where two or more fluids are present the "permeability" of equation (2) must represent the permeability of the rock to the desired fluid. This can be done by multiplying the absolute permeability of the rock (permeability to one fluid when completely filled with that fluid) by the relative permeability of the rock to the desired fluid.

$$q_o = 1.127 \times 10^{-3} \frac{k_{abs} k_{ro} A (p_1 - p_2)}{B_o \mu L} \quad (3)$$

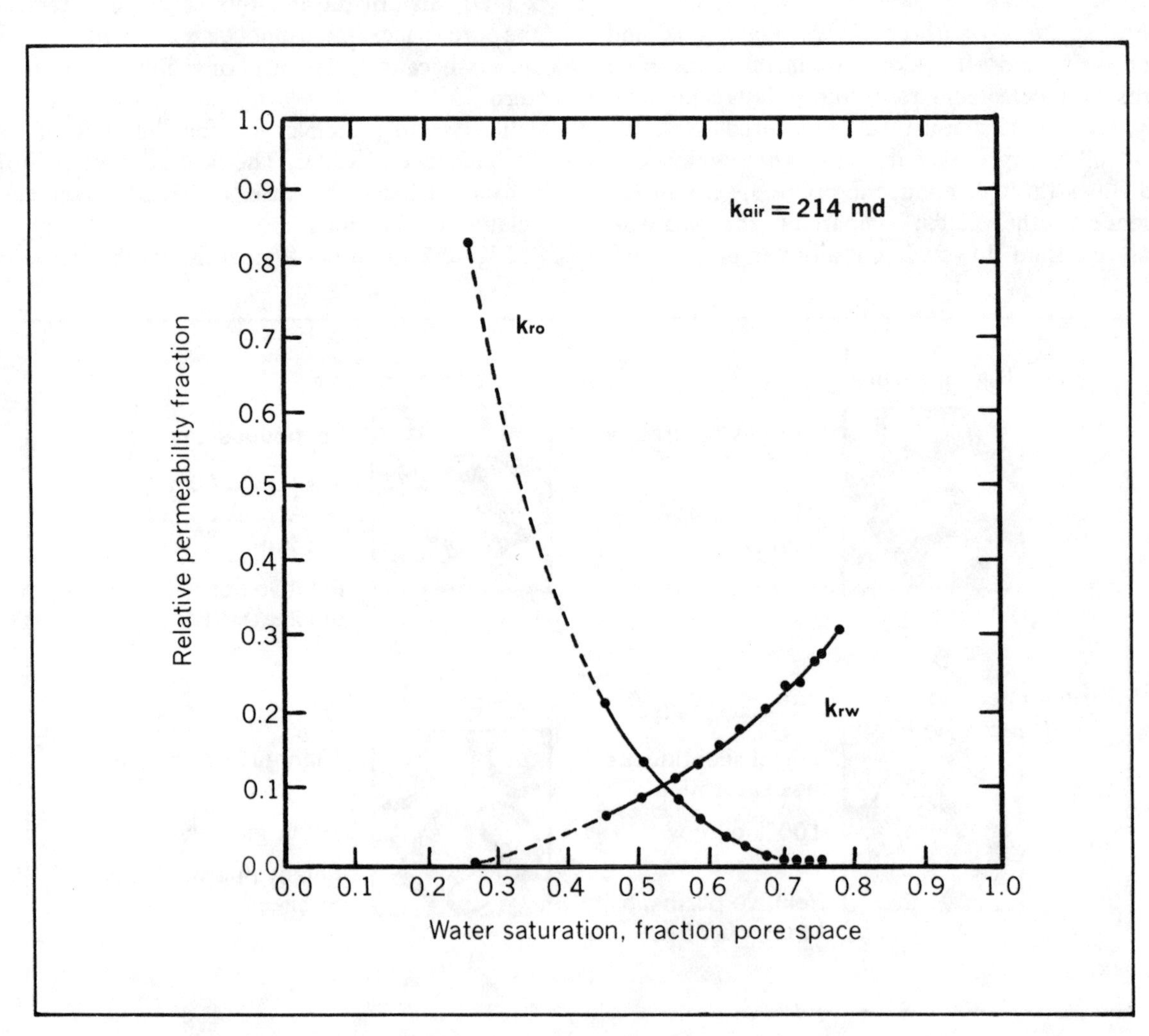

FIG. 2-15—Oil-water relative permeability (water-wet core).[7] Permission to publish by The Society of Petroleum Engineers.

q_o = oil flow rate, stb oil/day
k_{abs} = absolute permeability, md
k_{ro} = relative permeability to oil

For a well producing both water and oil, the "water cut" or the fraction of water in the total flow stream at standard conditions of temperature and pressure can be calculated by this relation:

$$f_w = \frac{1}{1 + \frac{k_o}{k_w} \times \frac{\mu_w}{\mu_o} \times \frac{B_w}{B_o}} \quad (4)$$

k_o, k_w = relative permeability
μ_o, μ_w = viscosity
B_o, B_w = formation volume factor

Wettability

Most reservoir rocks were formed or laid down in water, with oil moving in later from adjacent zones to replace a portion of the water. For this reason, most reservoir rocks are considered to be "water wet." The grains of the rock matrix are coated with a film of water, permitting hydrocarbons to fill the center of the pore spaces. Productivity of oil is maximized with this condition.

Actual wettability of a particular reservoir rock is difficult to determine because the process of cutting cores and preparing them for lab tests can, in fact, alter wettability characteristics. Further compounding the problem, most investigators currently believe that there are varying degrees of wettability between strongly water wet and strongly oil wet conditions.

From the standpoint of well completions, stimulation, and workover operations, it is important to realize that the wettability characteristics of the rock near the wellbore can be unfavorably altered by fluids placed in contact with the rock. This is discussed more completely in subsequent chapters on formation damage and surfactants.

Briefly, it is very important to tailor the characteristics of these completion, workover, and stimulation fluids, such that a strongly water wet condition is maintained to maximize relative permeability to oil in an oil-water system, and also, to prevent formation of water-in-oil emulsions in the pore system near the wellbore.

Fluid Distribution

Fluid distribution vertically in the reservoir is important. The relative amounts of oil, water, and gas present at a particular level in the reservoir determine the fluids that will be produced by a well completed at that level, and also influence the relative rates of fluid production.

If oil, water, and gas were placed in a tank, there would be sharp boundaries between the water and oil below, and between the oil and gas above. If the tank were then filled with sand, the contacts between the oil and water and the oil and gas would be quite different, because now the gas, oil, and water exist in capillary spaces. Capillary forces related to wettability and surface tension work against density differences between the fluids to significantly change the previous sharp interfaces between the fluids.

As shown in Figure 2-16, the water saturation in a water wet pore system varies from 100% below the oil zone to progressively lower percentages at points higher in the oil zone. This is because moving higher in the oil zone, the radius of the film between oil and water decreases, due to greater capillary forces. The water fits further back into the crevices between sand grains, and the quantity of water diminishes.

The zone from a point of 100% water (free water level) upward in the sand to some point above which water saturation is fairly constant is called the "transition zone." Relative permeability relations permit both water and oil to flow within the transition zone. Water saturation above the transition zone is termed "irreducible water saturation" or more commonly the "connate" water saturation. Above the transition zone only oil may flow in an oil-water system.

Connate water saturation is related to permeability. Pore channels in lower permeability rocks are generally smaller. For a given height above the free water level, capillary pressure will be the same in two pores of different sizes. Therefore, the film between the water and oil will have the same curvature, and the amount of water occurring in the crevice will be about the same. As shown in Figure 2-17, more oil is contained in the large pore space, however, and the percent of water in the small pore will be greater.

The nature and thickness of the transition zones between water and oil, oil and gas, and water and

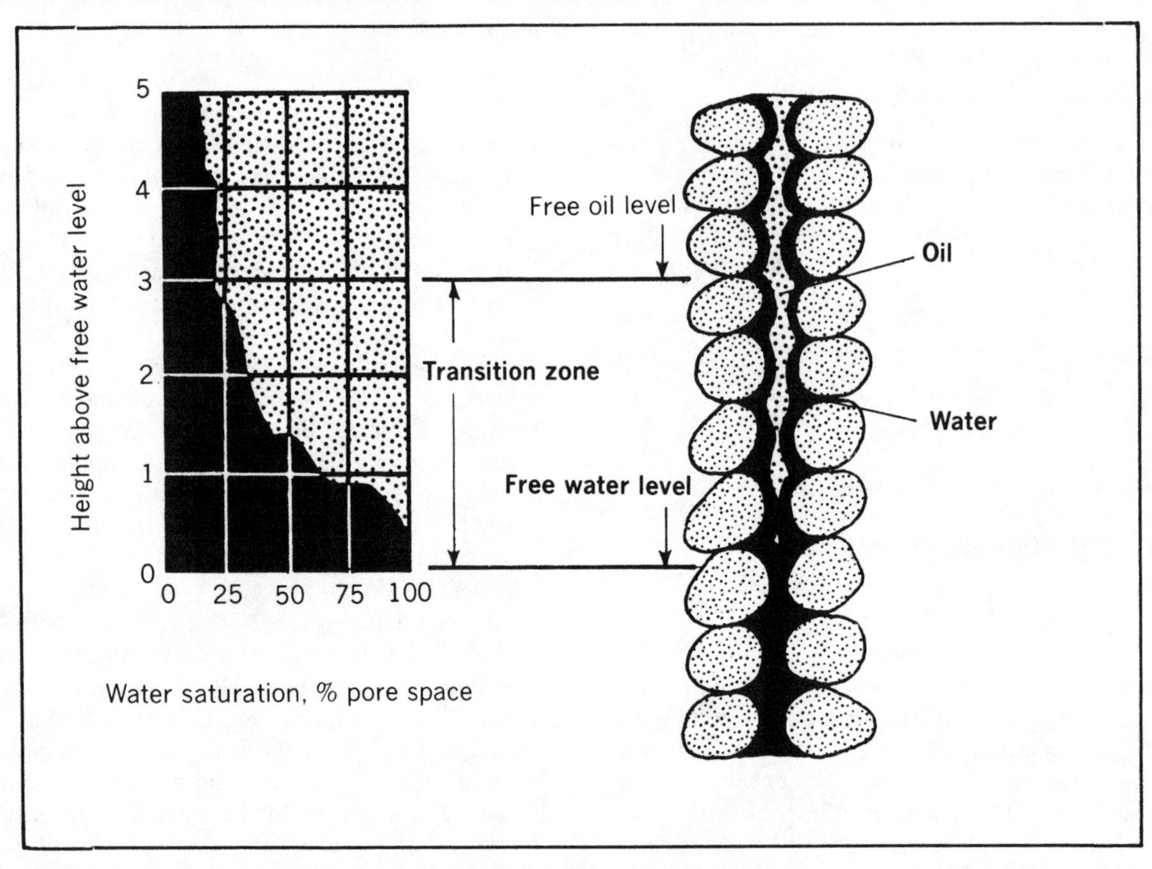

FIG. *2-16—Effects of height above free water level on connate-water content in oil sand.* Elements of Petroleum Reservoirs. *Permission to publish by The Society of Petroleum Engineers.*

gas are influenced by several factors, among which are uniformity, permeability, and wettability of the rock, and the surface tension and density differences between the fluids involved. Generally these statements can be made concerning fluid distribution:

—The lower the permeability of a given sand, the higher will be the connate water saturation.

—In lower permeability sands, the transition zones will be thicker than in higher permeability sands.

—Due to the greater density difference between gas and oil as compared to oil and water, the transition zone between oil and gas is not as thick as the transition zone between oil and water.

A well completed in the oil-water transition zone will be expected to produce both oil and water, depending on the saturations of each fluid present at the completion level. Figure 2-18 summarizes oil, water, and gas saturation in a typical homogeneous rock situation.

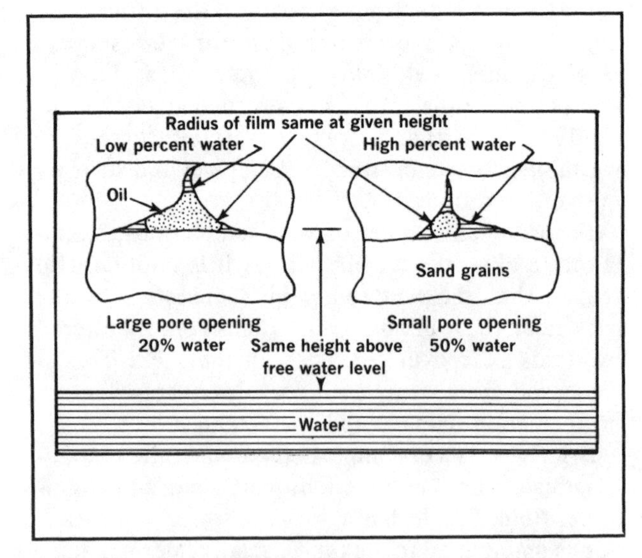

FIG. *2-17—Effect of pore size and shape on connate-water content.* Elements of Petroleum Reservoirs. *Permission to publish by The Society of Petroleum Engineers.*

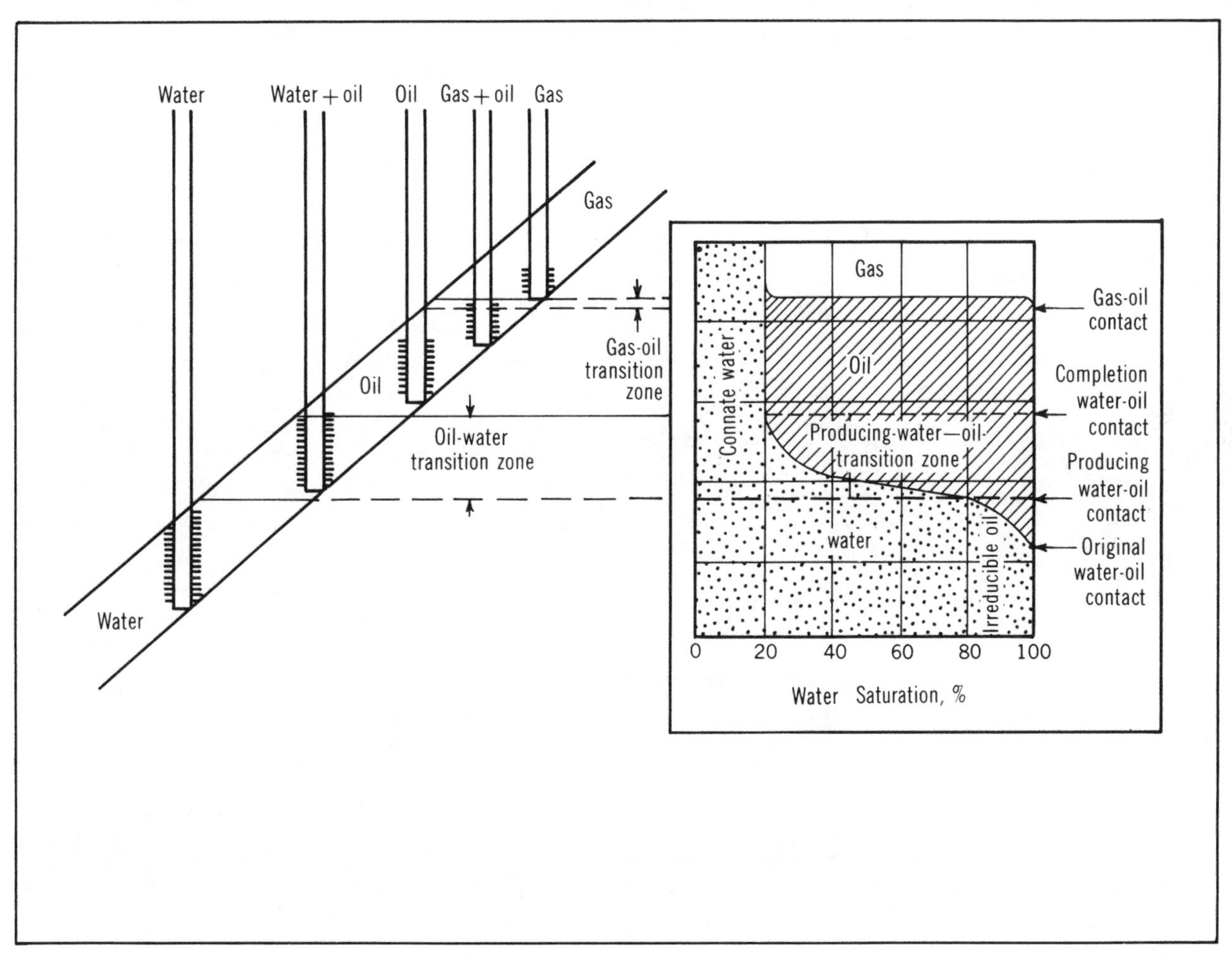

FIG. *2-18—Fluid distribution in a uniform-sand reservoir containing connate water, oil, and a gas cap.*

FLUID FLOW IN THE RESERVOIR

Oil has little natural ability to produce itself into a well bore. It is produced principally by pressure inherent in gas dissolved in oil, in associated free gas caps, or in associated aquifers.

Pressure Distribution Around the Well Bore

Pressure distribution in the reservoir and factors which influence it are of great significance in interpreting well production trends caused by presure characteristics.

Figure 2-19 shows pressure distribution around a producing oil well completed in a homogeneous zone. Some distance away from the well, pressure is assumed to be 3,000 psi. Moving nearer the well, pressure gradually declines to about 2,700 psi. From nere to a point within the well bore opposite the completion interval, pressure sharply declines to 2,000 psi. At the wellhead, influenced by hydrostatic and also frictional effects in the tubing, pressure is down to about 600 psi.

In a radial flow situation where fluids move toward the well from all directions, most of the pressure drop in the reservoir occurs fairly close to the wellbore. As shown in Figure 2-20, in a uniform sand, the pressure drop across the last 15 ft of the formation surrounding the wellbore is about one-half of the total pressure drop from the well to a point 500 ft away in the reservoir.

Obviously flow velocities increase tremendously as fluid approaches the wellbore. This area around the wellbore is the "critical area." To maximize well productivity everything possible must be done to prevent flow restriction in this critical area.

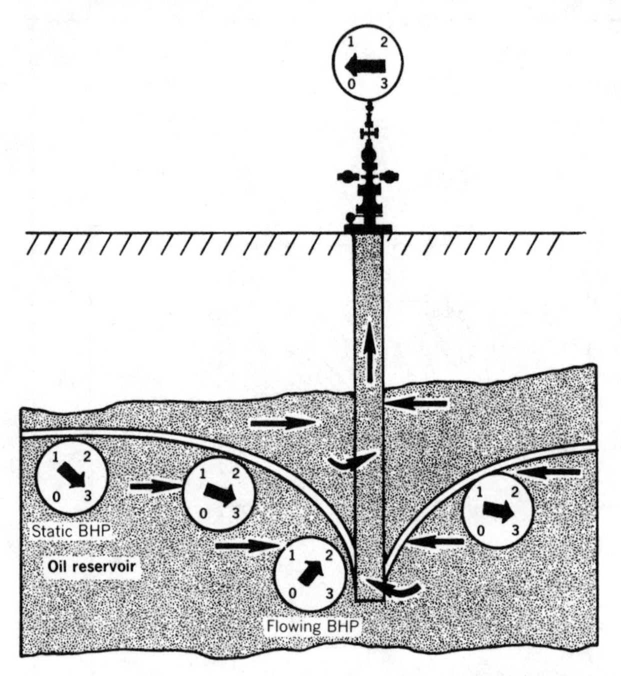

FIG. *2-19—Pressure conditions around a flowing well.* Elements of Petroleum Reservoirs. *Permission to publish by The Society of Petroleum Engineers.*

Radial Flow Around the Wellbore

Steady state radial flow of incompressible fluid is described by Darcy's Law in the oil field units of Figure 2-21:

$$q = \frac{.00708\, kh\,(p_e - p_w)}{B\mu \ln (r_e/r_w)} \tag{5}$$

Corrections are required to account for flow of compressible fluids, and for turbulent flow velocities.

For non-homogeneous zones (the usual case) permeabilities must be averaged for flow through parallel layers of differing permeabilities (Figure 2-22):

$$\bar{k} = \frac{k_1 h_1 + k_2 h_2 + k_3 h_3}{h_1 + h_2 + h_3} \tag{6}$$

Varying permeabilities in series as shown in Figure 2-23 can be averaged as follows:

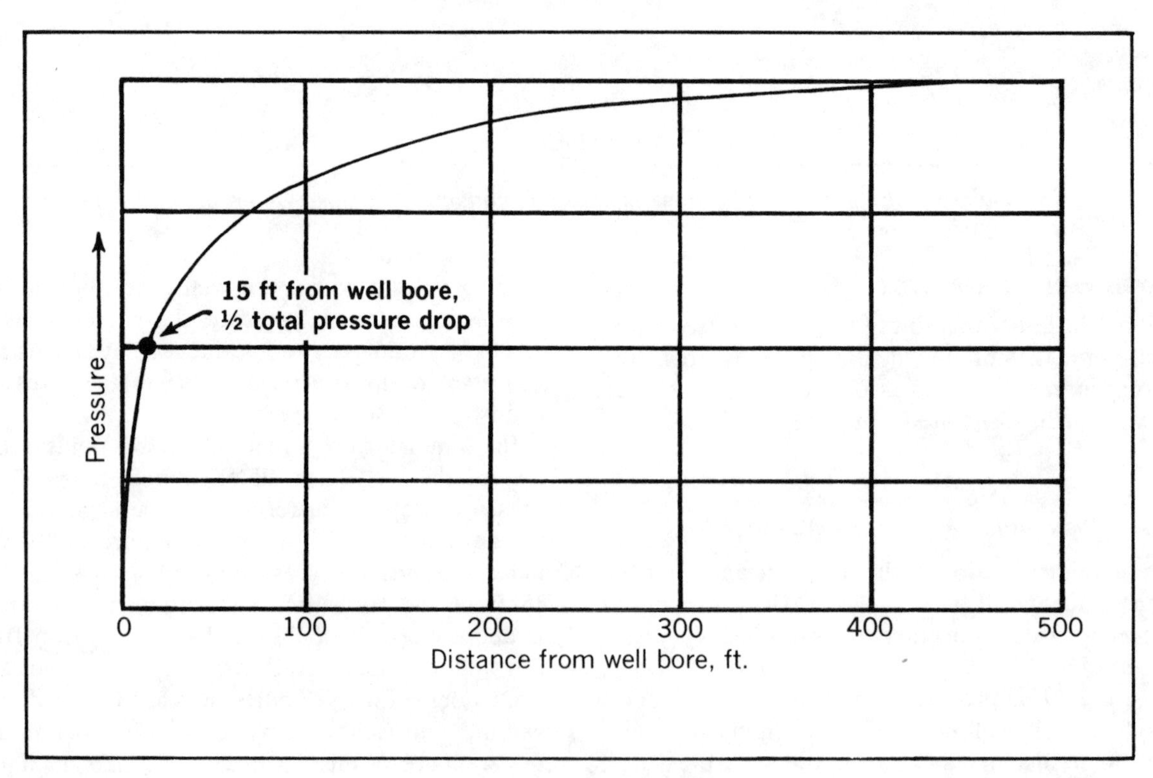

FIG. *2-20—Pressure distribution near the well in radial flow.* Elements of Petroleum Reservoirs. *Permission to publish by The Society of Petroleum Engineers.*

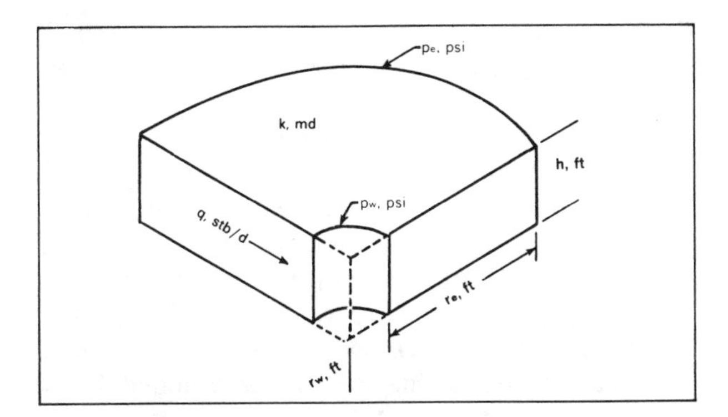

FIG. 2-21—Units for Darcy's Law equation.

$$\bar{k} = \frac{\ln (r_e/r_w)}{\frac{\ln (r_1/r_w)}{k_1} + \frac{\ln (r_2/r_1)}{k_2} + \frac{\ln (r_3/r_2)}{k_3}} \quad (7)$$

Linear Flow Through Perforations

Ideally (but usually not true) the "perforation tunnel" through the casing and cement sheath is thought to be a void space completely open to flow of fluid. In the ideal case, the perforation tunnel does not offer much restriction to flow.

In sand problem wells, highly permeable gravel used to hold the formation sand in place must fill the perforation tunnel to prevent movement of sand into the tunnel. In this case, the flow restriction of the sand or gravel-filled perforation becomes important.

Flow through the "perforation tunnel" takes on a linear, rather than radial configuration. The linear form of Darcy's Law must be corrected for the fact that turbulent flow usually exists.

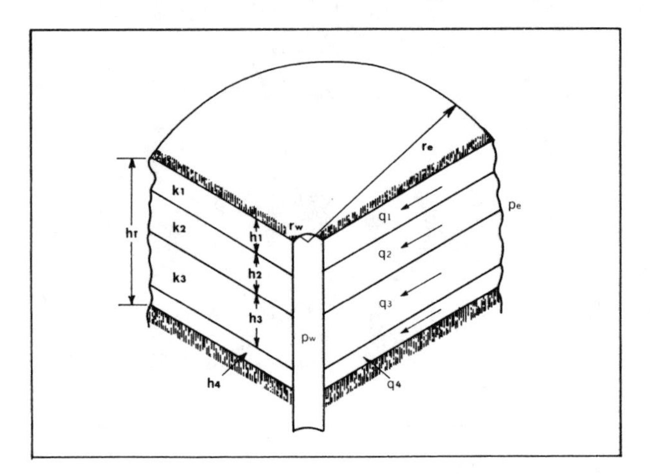

FIG. 2-22—Radial flow, parallel combination of beds.

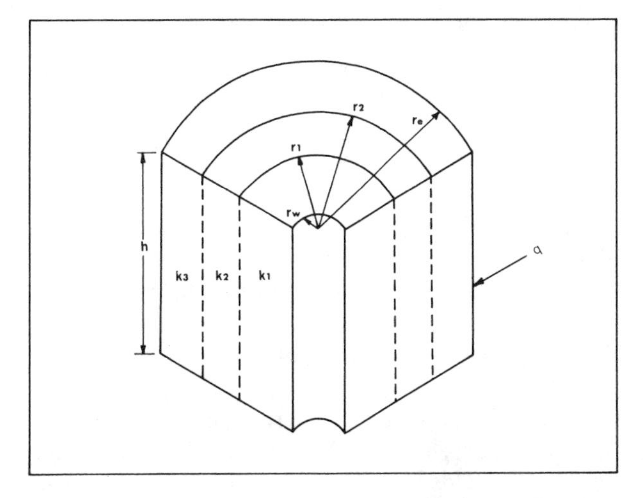

FIG. 2-23—Radial flow, series combination of beds.

Experimental measurements of pressure drop through gravel-filled perforations, compared with uncorrected linear flow Darcy Law calculations, are shown in Figure 2-24.

Curve A indicates that plugging by even high-permeability (one darcy) sand gives large pressure drop. Actual test data with very high-permeability sand, Curve B, proves turbulent flow results in higher pressure drop than Darcy's Law calculations, Curve C, predict.

Saucier[8], as well as other investigators, have provided turbulence correction factors, which can be applied to the Darcy equation, to permit calculation of pressure drop through the perforation tunnel.

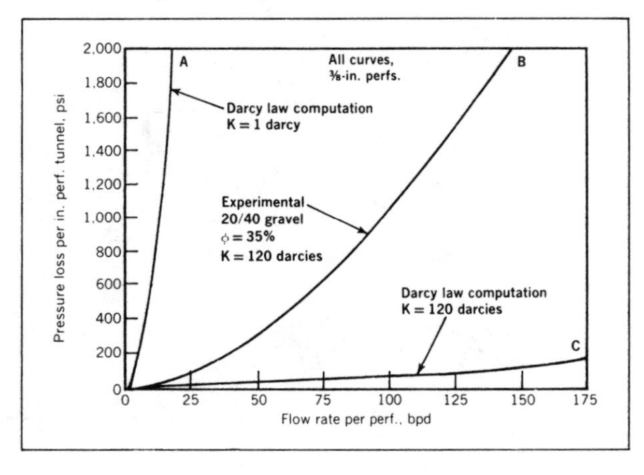

FIG. 2-24—Pressure drop vs. flow rate through perforation (R. H. Torrest, reported by Bruist).[9] Permission to publish by The Society of Petroleum Engineers.

Causes of Low Flowing Bottom-Hole Pressure

In a reservoir with uniform sand and fluid conditions and no artificial restrictions at the well bore, two factors may cause low flowing bottom-hole pressure in a well. These are permeability and producing rate as shown in Figure 2-25.

With low permeability or excessive rate of production, pressure drawdown will be appreciably higher, thus reducing flowing bottom-hole pressures and possibly requiring that a well be put on artificial lift if high rates of production are required.

Low flowing bottom-hole pressure many times occurs through damage to permeability adjacent to the well bore, caused by drilling or completion operations. This is particularly unfortunate because at this point in the reservoir, restriction is greatly magnified in effect.

Figure 2-26 shows a normal pressure sink compared to a pressure sink in a well where the formation has been damaged. Formation damage may result from any of a number of causes as discussed in detail in other sections of this manual.

The existence of a zone of reduced permeability near the wellbore can be determined through well testing and calculation techniques. Generally, average permeability of the drainage area away from the wellbore (determined by pressure buildup analysis), is compared with permeability, which is a combination of the permeability near the wellbore and that in the drainage area (determined by a productivity test).

Hurst and van Everdingen[5,6] introduced the term Skin or Skin Effect to describe the abnormal pressure drop through the damage zone. This abnormal pressure drop is in addition to the normal radial flow pressure drop, and can be calculated by:

$$\Delta p_s = \frac{141.2\, qB\mu}{kh} \times s \tag{8}$$

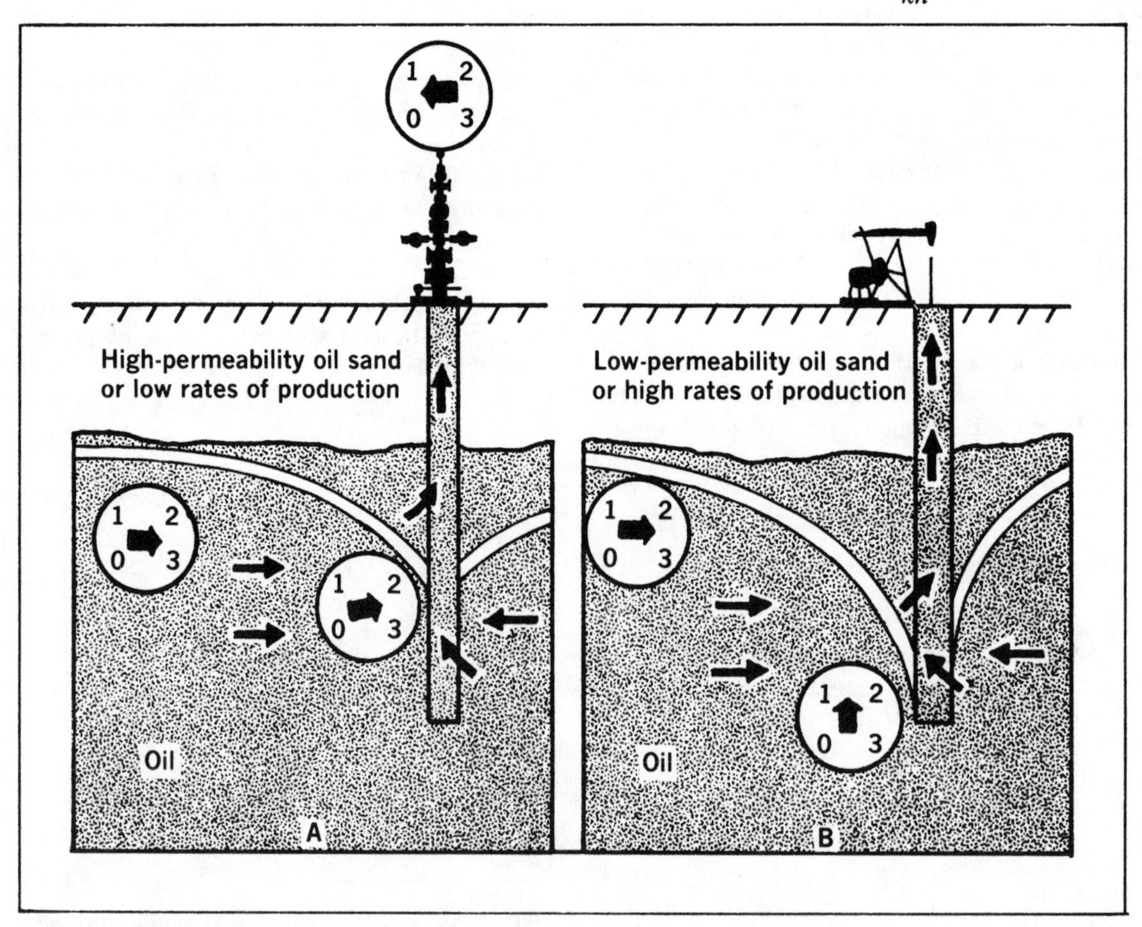

FIG. 2-25—*Effects of permeability and production rates on bottom-hole and well-head pressures.* Elements of Petroleum Reservoirs. *Permission to publish by The Society of Petroleum Engineers.*

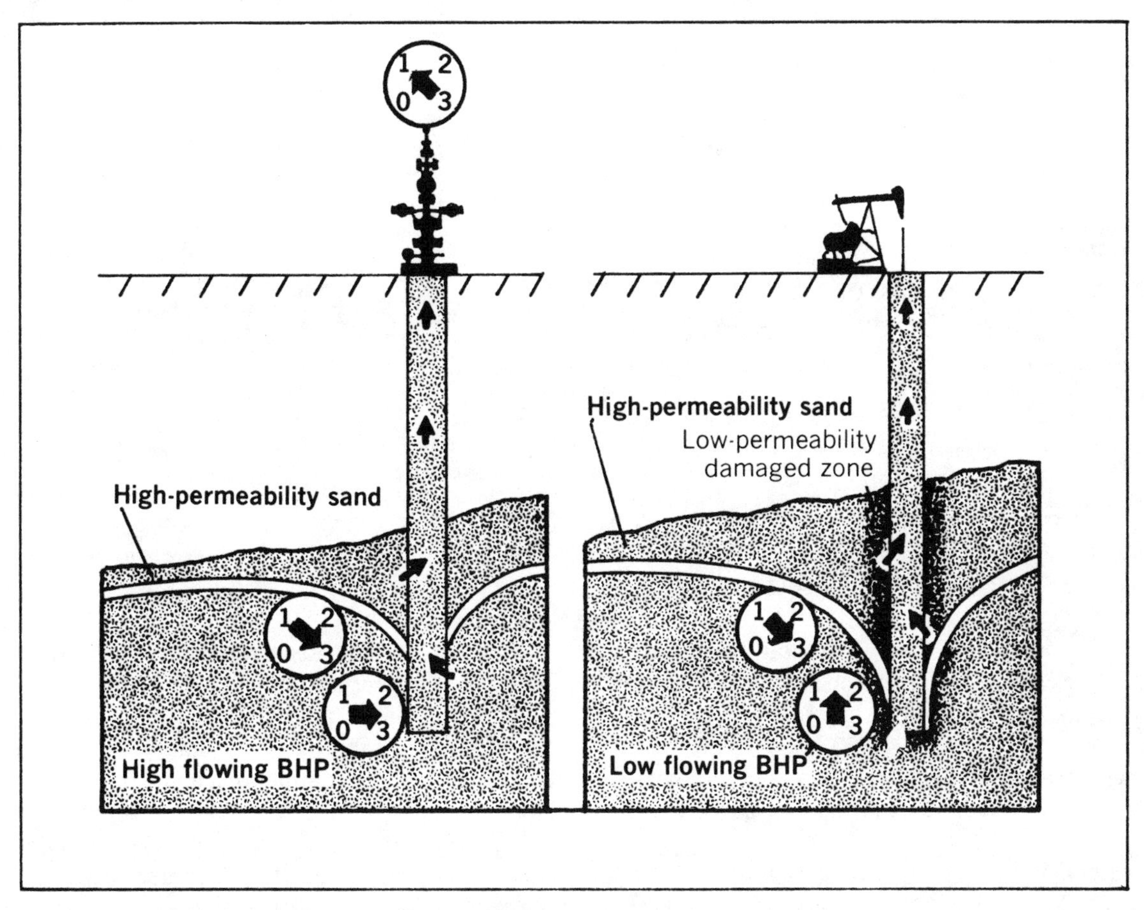

FIG. *2-26—Influence of skin effect on pressure around a well bore.* Elements of Petroleum Reservoirs. *Permission to publish by The Society of Petroleum Engineers.*

Other terms used to quantify formation damage are Damage Ratio and Flow Efficiency.

Damage Ratio:

$$\mathrm{DR} = \frac{q_t}{q_a} = \frac{\text{Theoretical flow rate without damage}}{\text{Actual flow rate observed}} \qquad (9)$$

Also:

$$\mathrm{DR} = \frac{J_{\text{ideal}}}{J_{\text{actual}}} = \frac{\bar{p} - p_{wf}}{\bar{p} - p_{wf} - \Delta p_s} \qquad (10)$$

Flow Efficiency:

$$\mathrm{FE} = \frac{J_{\text{actual}}}{J_{\text{ideal}}} = \frac{\bar{p} - p_{wf} - \Delta p_s}{\bar{p} - p_{wf}} \qquad (11)$$

In multizone completion intervals, where transient pressure-testing techniques may give questionable results concerning formation damage, production logging techniques may be helpful. Flow profiling may point out zones in an otherwise productive interval, which are not contributing to the total flowstream. The noncontributor zones are likely damaged.

EFFECTS OF RESERVOIR CHARACTERISTICS ON WELL COMPLETIONS

Reservoir Drive Mechanisms

In an oil reservoir, primary production results from the utilization of existing pressure. Basically, there are three drive mechanisms: dissolved gas, gas cap, and water drive; however, as a practical matter most reservoirs produce through some combination of each mechanism.

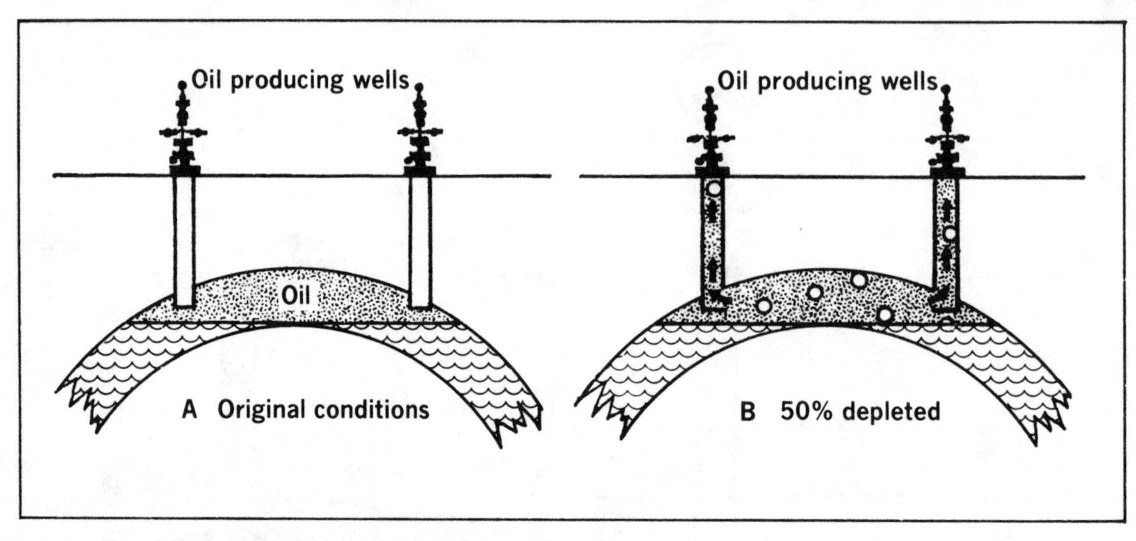

FIG. *2-27—Dissolved-gas-drive reservoir.* Elements of Petroleum Reservoirs. *Permission to publish by The Society of Petroleum Engineers.*

In a dissolved gas drive reservoir, the source of pressure is principally the liberation and expansion of gas from the oil as pressure is reduced, Figure 2-27.

A gas-drive reservoir uses principally the expansion of a cap of free gas over the oil zone, Figure 2-28.

A water drive uses principally expansion or influx of water from outside and below the reservoir, Figure 2-29.

The effect of the reservoir drive mechanism on producing well characteristics must be taken into account in making well completions initially, and later in recompleting wells to systematically recover reservoir hydrocarbons. Figures 2-30 and 2-31 show typical reservoir pressure vs. production and gas-oil ratio vs. production for the three basic drive mechanisms.

In a dissolved gas drive reservoir, (with no attempt to maintain pressure by fluid injection) pressure declines rapidly; gas-oil ratio peaks rapidly, and then declines rapidly; and primary oil recovery is relatively low. Recompletions could not be expected to reduce gas-oil ratio.

In a gas-cap-drive reservoir, pressure declines less rapidly. Gas-oil ratios increase as the gas cap expands into the up-structure well completion intervals. But recompletion or shutting in of up-structure wells provide possibilities for overall gas-oil ratio control. In a water drive reservoir, pressure remains relatively high. Gas-oil ratios are low; but down-structure wells soon begin to produce water. This must be controlled by recompletion or shutting in of these wells. Eventually even up-structure wells must produce significant amounts of water in order

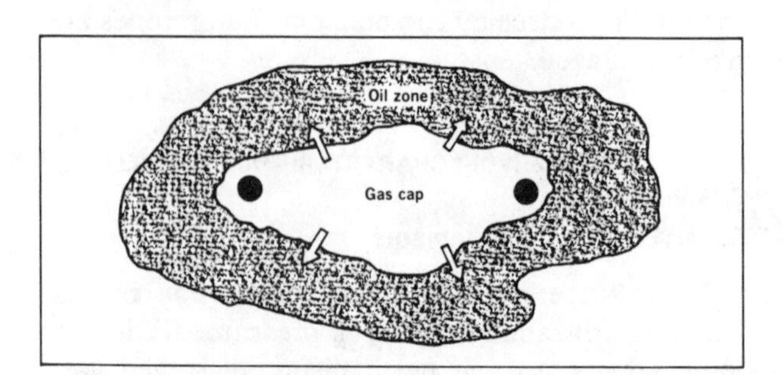

FIG. *2-28—Gas-cap-drive reservoir.* Elements of Petroleum Reservoirs. *Permission to publish by The Society of Petroleum Engineers.*

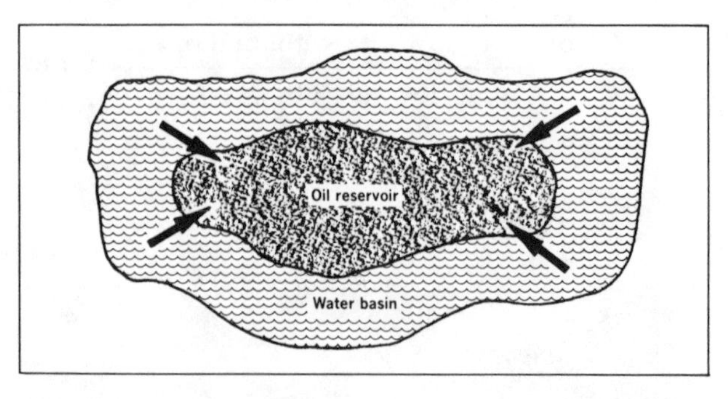

FIG. *2-29—Water-drive reservoir.* Elements of Petroleum Reservoirs. *Permission to publish by The Society of Petroleum Engineers.*

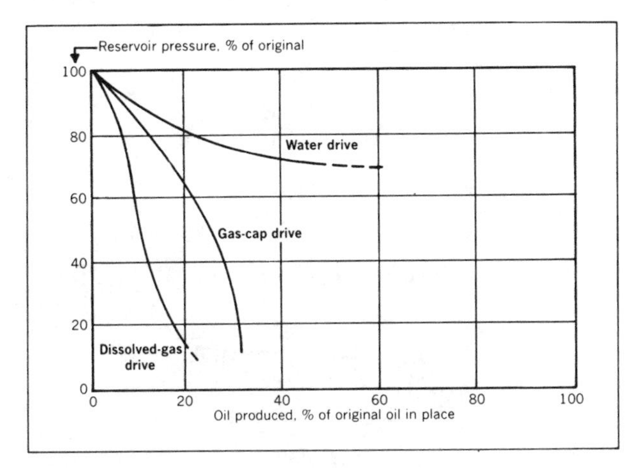

FIG. *2-30—Reservoir-pressure trends for various drive mechanisms.* Elements of Petroleum Reservoirs. *Permission to publish by The Society of Petroleum Engineers.*

to maximize oil recovery.

Obviously many factors must be considered in developing a reservoir, however, the main factors deal with the reservoir itself and the procedures used in exploitation. Well spacing, or better, well location, is one important factor. Money, time, labor, and materials consumed in drilling wells are largely non-recoverable. Therefore, if development drilling proceeds on close spacing before the drive mechanism is correctly identified the investment will already have been made when the recovery mechanism is finally determined.

This does not present an impossible problem,

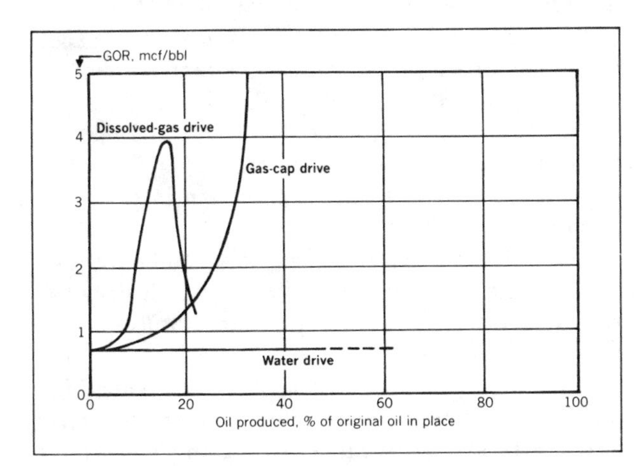

FIG. *2-31—Gas-oil ratio trends for various drive mechanisms.* Elements of Petroleum Reservoirs. *Permission to publish by The Society of Petroleum Engineers.*

even when the predominant drive cannot be determined early in the development. A certain number of wells must be drilled in any event if the field is of appreciable size. Enough wells are needed to define the reservoir—that is, to establish the detailed geologic picture regarding zone continuity and to locate oil-water and gas-oil contacts. Beyond this minimum, the number of infill wells and the well spacing can be varied in many instances.

The development program should be based on reservoir considerations and conditions, rather than on surface conditions or on some arbitrary grid pattern. The development program can be outlined schematically with subsurface stratigraphic cross-sections and a surface plan for well locations on the structure map. Detailed knowledge of the geology of the reservoir and its depositional environment is the key to an effective development plan, as is pointed out in Chapter 1, Geologic Considerations.

Many case histories are available to show the problems resulting from reservoir development without sufficient consideration of the stratigraphy of the reservoir.

With regard strictly to the effect of reservoir drive mechanism, general statements can be made as to well development patterns as outlined in the next topics.

Dissolved-Gas-Drive Reservoirs—Well completions in a dissolved-gas drive reservoir with low structural relief can be made in a regularly spaced pattern throughout the reservoir, and provided the rock is not stratified, can be made low in the reservoir bed, Figure 2-32.

A regular spacing pattern could also be used for a dissolved-gas drive reservoir with a high angle of dip, Figure 2-33.

Again the completion intervals should be structurally low because of the angle of structural dip, and exact subsurface location would vary with well location on the structure. Here it is expected that the oil will drain down-structure in time so that higher-than-usual oil recovery will be realized with minimum investment in wells. The operator must recognize the reservoir situation soon enough to eliminate drilling the structurally high wells.

Due to low recovery by the primary mechanism, some means of secondary recovery will almost certainly be required at some point in the life of the reservoir. Initial well completions need to be designed with this in mind.

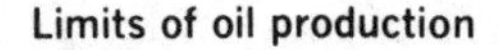

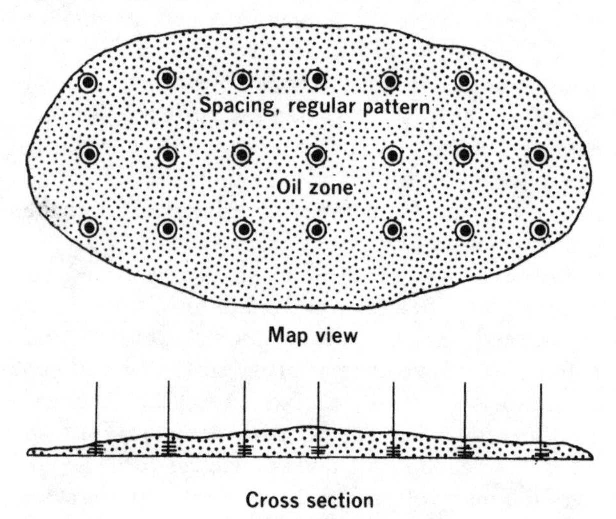

FIG. *2-32—Dissolved-gas-drive reservoir, low angle of dip.* Elements of Petroleum Reservoirs. *Permission to publish by The Society of Petroleum Engineers.*

Gas-Cap-Drive Reservoirs—Wells may be spaced on a regular pattern in a gas-cap drive reservoir where sand is thick, dip angle is low, and the gas-cap is completely underlain by oil, Figure 2-34.

Again, completions should be made low in the section to permit the gas cap to expand and drive oil down to the completion intervals for maximum recovery with minimum gas production.

A gas-cap drive reservoir in a thin sand with a high angle of dip is likely to be more efficiently controlled by having completion spaced irregularly but low on the structure to conform to the shape of the reservoir, Figure 2-35.

Because of the high angle of dip, a regular spacing pattern may cause many completions to be located

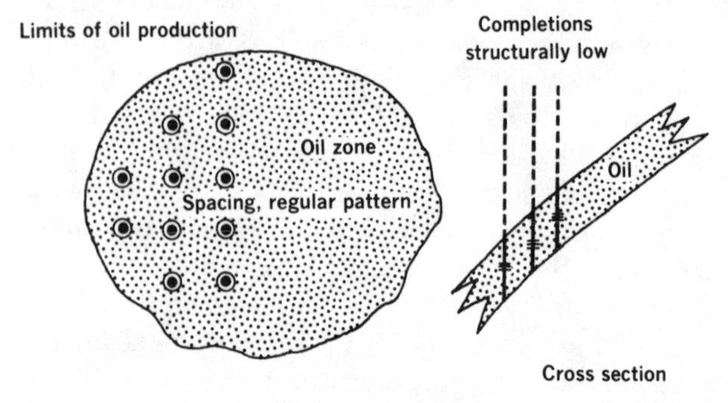

FIG. *2-33—Dissolved-gas-drive reservoir, high angle of dip.* Elements of Petroleum Reservoirs. *Permission to publish by The Society of Petroleum Engineers.*

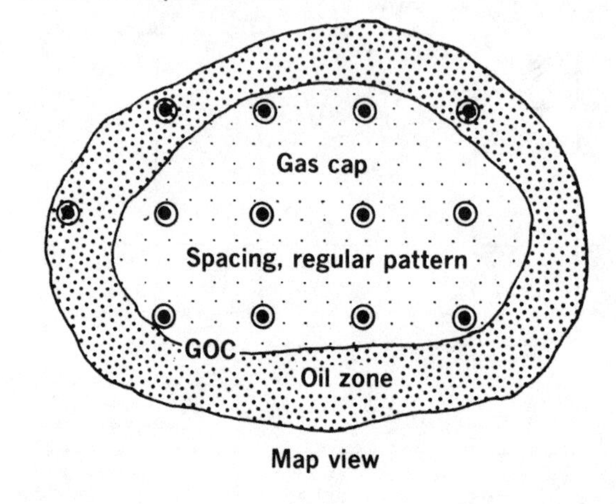

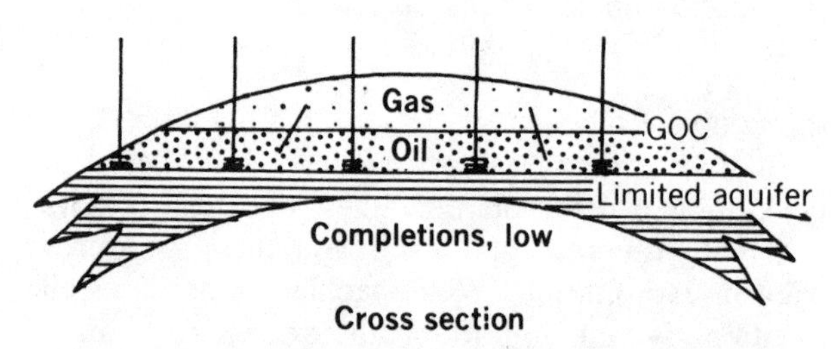

FIG. *2-34—Gas-cap-drive reservoir, low angle of dip.* Elements of Petroleum Reservoirs. *Permission to publish by The Society of Petroleum Engineers.*

too near the gas-oil contact. Such an oil reservoir is common where multiple thin sands are found on a single structure and the oil column is only a fraction of the total productive relief.

Water-Drive Reservoirs—Wells may be spaced on a regular pattern in a water drive reservoir having a thick sand and low angle of dip, Figure 2-36.

Completion intervals should be selected high on the structure to permit long producing life while oil is displaced up to the completion intervals by invading water from below.

A water-drive reservoir in a thin sand with high angle of dip may best be developed with irregular well-spacing because of the structural characteristics, Figure 2-37.

The completions, however, should be made high on the structure to delay encroachment of water into the producing wells. Spotting the wells on a regular spacing pattern not only may cause a number of wells to produce water early in the life of the reservoir and result in their early abandonment, but also may reduce the effectiveness of the water

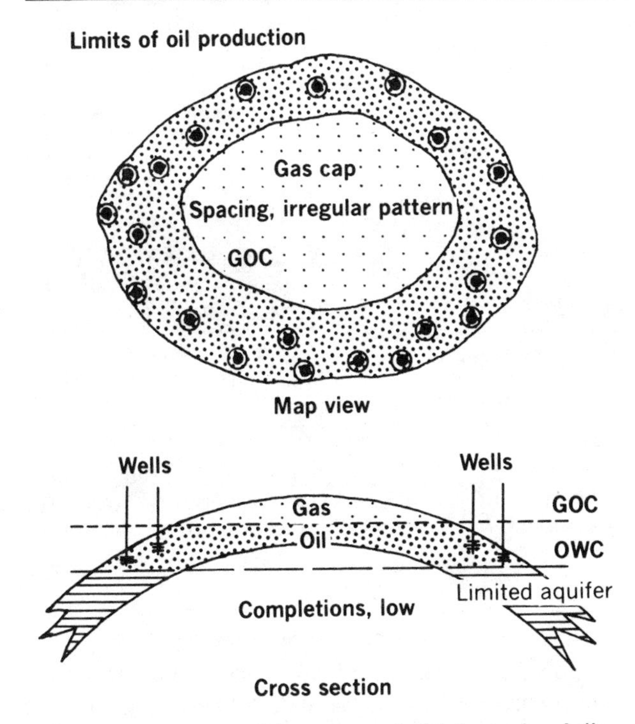

FIG. *2-35—Gas-cap-drive reservoir, high angle of dip.* Elements of Petroleum Reservoirs. *Permission to publish by The Society of Petroleum Engineers.*

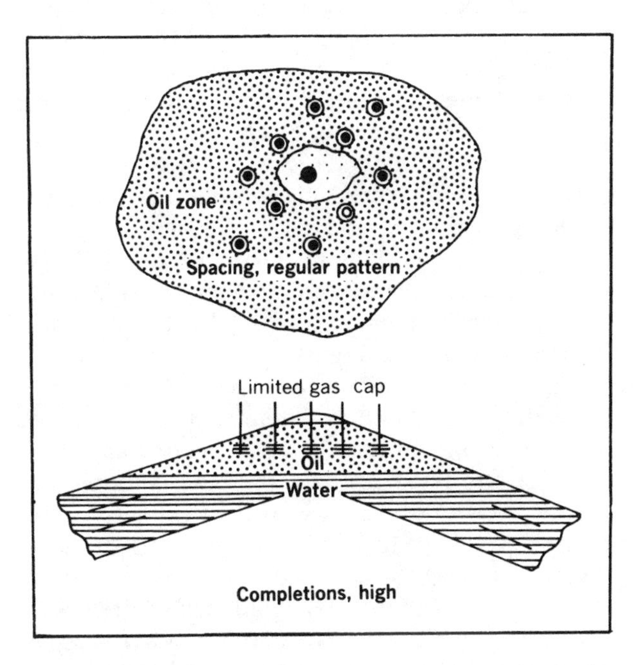

FIG. *2-36—Water-drive reservoir, low angle of dip, thick sand.* Elements of Petroleum Reservoirs. *Permission to publish by The Society of Petroleum Engineers.*

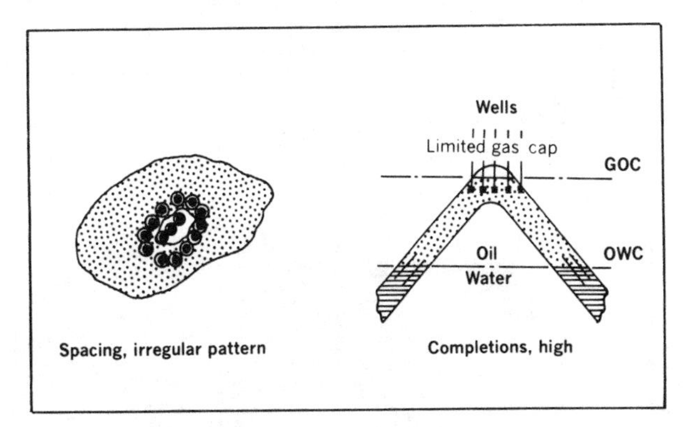

FIG. *2-37—Water-drive reservoir, high angle of dip, thin sand.* Elements of Petroleum Reservoirs. *Permission to publish by The Society of Petroleum Engineers.*

drive through excessive early water production. Fewer wells would then remain to produce the remainder of the oil, thus lengthening unnecessarily the length of time required to deplete the reservoir.

Significant amounts of water must be produced in the later life of the field in order to maximize recovery.

Reservoir Homogeneity

The general procedure, as previously described and shown by the illustrations, is to complete high for water drive and low for dissolved gas and gas-cap drive reservoirs to have an adequate number but

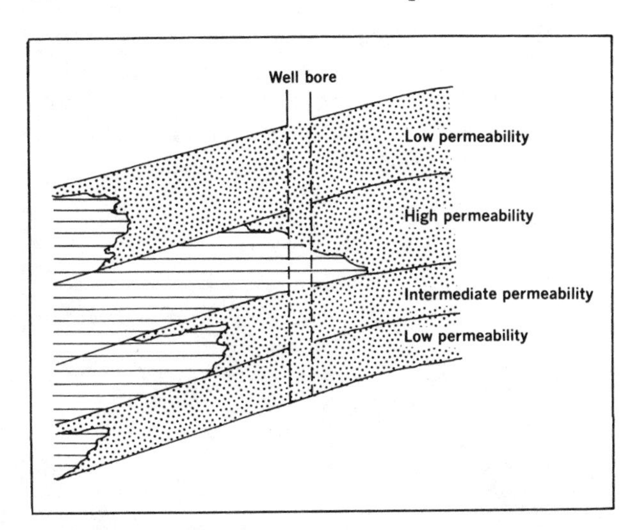

FIG. *2-38—Irregular water encroachment and premature water breakthrough in high-permeability layers of reservoir rock.* Elements of Petroleum Reservoirs. *Permission to publish by The Society of Petroleum Engineers.*

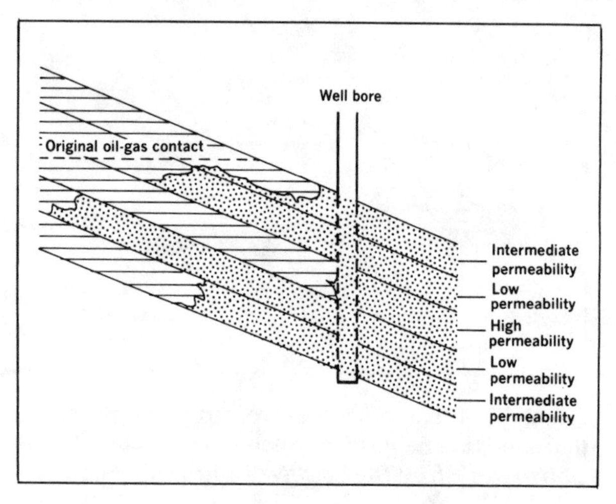

FIG. *2-39—High gas-oil ratio production caused by early encroachment of gas through high-permeability zones of stratified reservoir.* Elements of Petroleum Reservoirs. *Permission to publish by The Society of Petroleum Engineers.*

not too many wells. It would be practical, however, to make such completions only if the reservoir were quite uniform.

Most sandstone formations were originally laid down as stratified layers of varying porosity and permeability. Similar statements can be made regarding carbonate, and even reef-type reservoirs. Thus, the normal sedimentary process results in reservoirs of a highly stratified nature. Fluids flow through alternate layers with different degrees of ease, and many times impermeable zones separate the permeable beds so that no fluid can move from

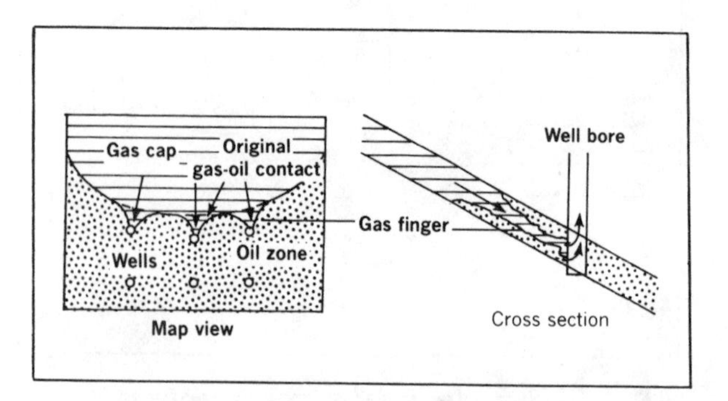

FIG. *2-40—Fingering of free gas into well along bedding planes.* Elements of Petroleum Reservoirs. *Permission to publish by The Society of Petroleum Engineers.*

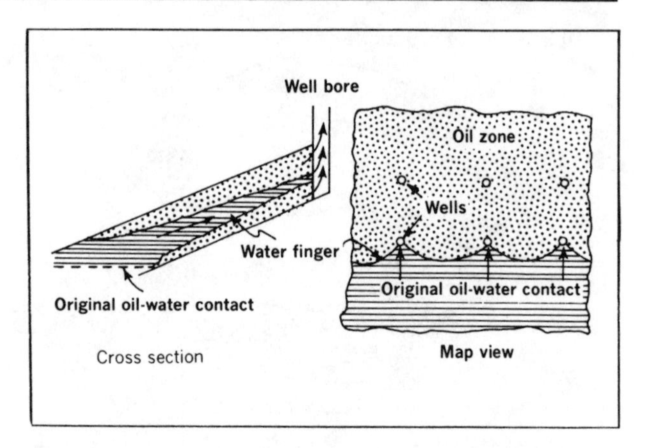

FIG. *2-41—Fingering of water into well along bedding planes.* Elements of Petroleum Reservoirs. *Permission to publish by The Society of Petroleum Engineers.*

bed to bed. This is shown in Figures 2-38 and 2-39.

In thin beds or highly stratified beds "fingering" (movement of fluid parallel to bedding planes) of free gas down from a gas cap, or water up from a water basin is always a possibility when short completion intervals combined with high rates of production are encountered.

Figures 2-40 and 2-41 illustrate these two problems. If the reservoir is stratified either by shale breaks or by variations in permeability, it probably will be necessary to stagger the completion intervals in various members of the reservoir to be sure that each member is drained. Some vertical staggering of completion intervals can be effected during development to secure proportionate withdrawals from the various strata.

Additional distribution of completions between the various members of this pay may then be made during later workovers, on the basis of experience and operational conditions.

For maximum recovery from reservoir completions, intervals should be limited to one identifiable zone wherever practical. Single-zone completions are preferred to facilitate thorough flushing for higher recoveries and to obtain flexibility in recompletion work for control of reservoir performance.

Completions comprising more than one reservoir are termed multi-zone completions. Such completions may be needed for low permeability reservoirs which require long completion intervals for obtaining economic quantities of production.

REFERENCES

1. Clark, Norman J.: "Elements of Petroleum Reservoirs," Henry L. Doherty Series, AIME, Dallas (1960) Revised 1969.

2. Standing, M. B.: *Volumetric and Phase Behavior of Oil Field Hydrocarbon Systems,* Reinhold Publishing Corp., New York (1952).

3. Beal, Carlton: "The Viscosity of Air, Water, Natural Gas, Crude Oil and its Associated Gases at Oil Field Temperatures and Pressures," *Trans.,* AIME (1946).

4. Carr, Norman L.; Kobayashi, Riki; and Burrows, David B.: "Viscosity of Hydrocarbon Gases Under Pressure," *Trans.,* AIME (1954).

5. van Everdingen, A. F.: "The Skin Effect and its Influence on the Productive Capacity of a Well," *Trans.,* AIME (1953).

6. Hurst, William: "Establishment of Skin Effect and its Impediment to Fluid Flow Into a Well Bore," *Pet. Eng.* (Oct. 1953).

7. Morgan, J. T., and Gordon, D. T.: "Influence of Pore Geometry on Water-Oil Relative Permeability," *JPT* (Oct. 1970) p. 1199.

8. Saucier, R. J.: "Gravel Pack Design Considerations," *JPT* (Feb. 1974) p. 205.

9. Bruist, E. H.: "Better Performance of Gulf Coast Wells," SPE 4777, New Orleans (1974).

10. Dowling, Paul L., Jr.: "Application of Carbonate Environmental Concepts to Secondary Recovery Projects," SPE 2987 (1970).

11. Weber, K. J.: "Sedimentological Aspects of Oil Fields in the Niger Delta," *Geologie En Mijnbouw,* Vol. 50, No. 3 (1971).

12. Jardine, D.; Andrews, D. P.; Wishart, J. W.; and Young, J. W.: "Distribution and Continuity of Carbonate Reservoirs," SPE 6139, New Orleans (Oct. 1976).

13. Reitzel, Gordon A., and Callow, George O.: "Pool Description and Performance Analysis Leads to Understanding Golden Spike's Miscible Flood," SPE 6140, New Orleans (Oct. 1976).

14. Earlougher, Robert C., Jr.: "Advances in Well Test Analysis," Monograph Volume 5, Henry L. Doherty Series, AIME (1977).

Chapter 3 Well Testing

Well production testing
Periodic production tests
Productivity or deliverability tests
Transient pressure tests
Basics of buildup, drawdown, injection, and multiple well tests
Drill stem testing
Basics of DST tools, and procedures
Pressure buildup analysis
Procedures for good test data
Eyeball interpretation of charts

Well Production Testing

INTRODUCTION

The objectives of Well Production Testing vary from a simple determination of the amount and type of fluids produced to sophisticated transient pressure determinations of reservoir parameters and heterogeneities. Briefly, Well Testing procedures are a set of tools which properly used can provide valuable clues as to the condition of production or injection wells. The Well Completion or Production Engineer needs to be able to design and conduct the simplier testing procedures, and to be familiar with the possibilities and limitations of the more sophisticated procedures.

The purpose of this section is to present a brief discussion of the basics of production well testing. A somewhat more complete discussion of Drill Stem Testing follows. References at the end of this section should be consulted for detailed explanation and example calculations using the many well testing methods.

Generally Oil or Gas Well Production Tests may be classified as:

—Periodic Production Tests
—Productivity or Deliverability Tests
—Transient Pressure Tests

Periodic production tests have as their purpose, determination of the relative quantities of oil, gas and water produced under normal producing conditions. They serve as an aid in well and reservoir operation and also in meeting legal and regulatory requirements.

Productivity or deliverability tests are usually performed on initial completion or recompletion to determine the capability of the well under various degrees of pressure drawdown. Results may set production allowables, aid in selections of well completion methods, and design of artificial lift systems and production facilities.

Transient pressure tests require a higher degree of sophistification and are used to determine formation damage or stimulation related to an individual well, or reservoir parameters such as permeability, pressure, volume and heterogeneities.

PERIODIC PRODUCTION TESTS

Production Tests are run routinely to physically measure oil, gas and water produced by a particular well under normal producing conditions. Test results may then be used to allocate total field or lease production between wells where individual well production is not monitored continuously. They provide the basis for periodic reports to legal or regulatory groups.

From the standpoint of well and reservoir operation, they provide periodic physical evidence of well conditions. Unexpected changes, such as extraneous water or gas production may signal well or reservoir problems. Abnormal production declines may mean artificial lift problems, sand fillup in the casing, scale buildup in the perforations, etc.

For oil wells results are usually reported as oil production rate, barrels of oil per day; gas-oil ratio (GOR), cubic feet per barrel; and water-oil ratio

(WOR), percentage of water in the total liquid stream. Test equipment can consist of nothing more than a gas-oil separator and a stock tank, with appropriate measuring devices such as an orifice meter for gas and a hand tape for oil and water. The current trend is toward more sophisticated systems, the ultimate providing recording meter measurement of water, and oil, as well as gas, along with automatic switching and unattended operation.

Accuracy of measurements, and careful recording of the conditions under which the test was run are of obvious importance. Choke size, tubing pressures, casing pressures, details of the artificial lift system operation, in short everything affecting the ability of the well to produce should be recorded. Problems such as measurement of emulsion, power oil fluid with hydraulic pumping systems, input gas with gas lift systems, must be recognized and properly handled. Stabilized producing conditions are of obvious importance where short test periods are used. Tests should be run at the "normal" production rate, since changes of rate often influence the relative quantities of oil, gas and water.

For gas wells routine production tests per se are less common, since gas production is usually metered continuously from individual wells. Gas production is reported in Mcfd (thousands of standard cubic feet per day) or MMcfd (millions of standard cubic feet per day). Hydrocarbon liquids or water are reported in barrels condensate per million cubic feet (BCPMM) or barrels water per million cubic feet. Stabilized flow condition, careful metering and reporting of volumes and pressures are again of obvious importance.

PRODUCTIVITY OR DELIVERABILITY TESTS

Productivity or Deliverability Tests represent the second degree of sophistication in oil or gas well production testing. They involve a physical or empirical determination of produced fluid flow versus bottom hole pressure drawdown. With a limited number of measurements they permit prediction of what the well should produce at other pressure drawdowns. They do not rely on a mathematical description of the flow process. They are successfully applied to non-darcy, below-the-bubble point flow conditions, even though fluid properties and relative permeabilities are not constant around the wellbore.

They do not permit calculation of formation permeability or the degree of abnormal flow restriction (formation damage) near the wellbore. They do, however, include the effects of formation damage. Thus, they can be used as an indicator of well flow conditions, or as a basis for a simple comparison of completion effectiveness among wells in a particular reservoir.

Deliverability tests represent stabilized producing conditions. They involve the measurement of bottom hole static and flowing pressure, as well as fluid rates produced to the surface.

Commonly used deliverability tests for oil wells may be classified as:

—Productivity Index
—Inflow Performance
—Flow after Flow
—Isochronal

Gas-well-deliverability tests are designed to establish the "absolute open flow potential," or the production rate if flowing bottom hole pressure could be reduced to zero. Termed multipoint back-pressure tests, or simply back pressure tests, they can be classified according to test procedure as:

—Flow after Flow
—Isochronal

Oil Wells

Productivity Index—The Productivity Index test is the simpliest form of Deliverability Test. It involves the measurement of shut-in bottom hole pressure; and, at one stabilized producing condition, measurement of the flowing bottom pressure and the corresponding rate of liquids produced to the surface. Productivity Index is then defined as:

$$\text{PI} = J = \frac{q}{p_i - p_{wf}} \tag{1}$$

q = Total liquids stb/d
p_i = Shut-in bottom hole pressure, psi
p_{wf} = Flowing bottom hole pressure, psi
$p_i - p_{wf}$ = Pressure drawdown, psi

Specific PI accounts for the length of the producing section:

$$\text{Specific PI} = \frac{\text{Productivity Index}}{\text{Length of producing zone}} \tag{2}$$

With a well producing above the bubble point, the PI may be constant over a wide range of pressure drawdowns. However, with flow below the bubble point, and gas occupying a portion of the pore system, PI falls off with increased drawdown.

Productivity Index also declines during the life of a well due to many factors, among which are changes in reservoir pressure, composition and properties of produced fluids, and flow restriction or formation damage near the wellbore. Productivity Index does however give the Well Completion Engineer a useful index of well and wellbore conditions, and with recognition of the limitation involved, a yardstick for comparison between wells.

Inflow Performance Test—The simple concept of Productivity Index attempts to represent the inflow performance relation of a well as a straight line function, (Figure 3-1, Well A). The true inflow performance relation or IPR usually declines at greater drawdowns as shown in Figure 3-1, Wells B and C.

An inflow-performance test should, in effect, consist of PI tests at several production rates in order to provide a better representation of the true inflow performance relation of the well. Vogel[11], based on a computer simulation of dissolved gas drive reservoirs, wherein he calculated IPR's using a wide range of reservoir and fluid parameters, proposed the general IPR curve of Figure 3-2. Often

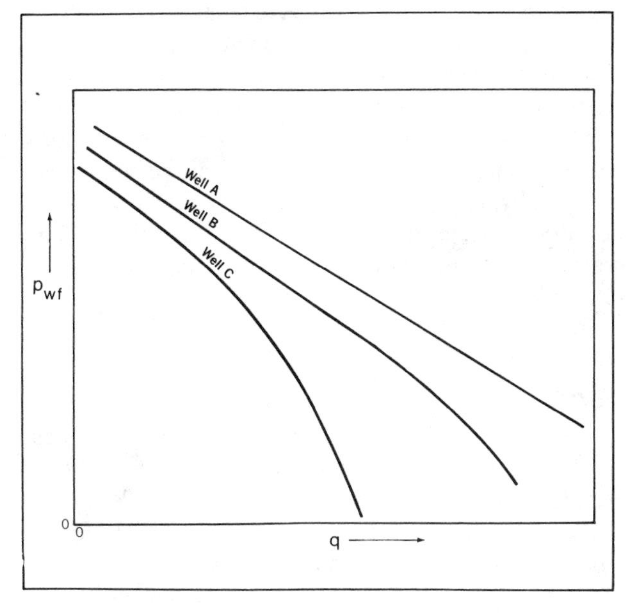

FIG. *3-1—Typical inflow performance curves.*

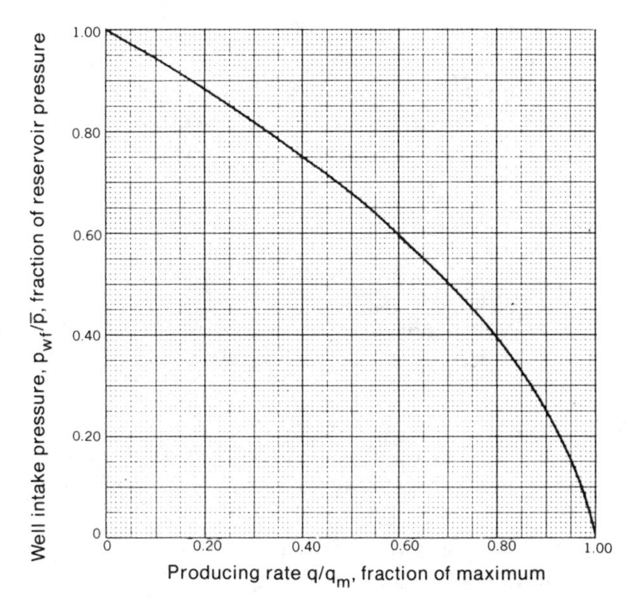

FIG. *3-2—Inflow performance relationship. (Vogel). Permission to publish by The Society of Petroleum Engineers.*

this same Vogel relation is successfully applied to other types of reservoir drive systems.

A primary advantage of the Vogel relation is that, with a value for static reservoir pressure, one well test, recording production rate and corresponding bottom hole pressure, is all that is needed to provide a reasonable value for production rate at any other flowing bottom hole pressure. In conducting the test a key point is to insure that stabilized flow conditions are established. Standing[14] extended the Vogel relation such that, if the "Flow Efficiency" could be estimated, then the effect of flow restriction or damage near the wellbore could be included in a simple graphical analysis to predict the IPR with the restriction, and also the IPR with the restriction removed. (See chapter on well-completion design.

Again, it should be pointed out that the inflow performance relation is an empirical representation, and that it changes during the life of the well. Methods for predicting the IPR at future times were suggested by Standing involving current versus future mobility ratio and formation volume factors.[14]

Flow After Flow—Back-pressure tests have been used for empirical determination of Gas Well capability for many years.[7] Essentially a plot of flow rate, versus "squared drawdown pressure," on a log-log paper provides a straight line which can be extended to predict flow rate for any drawdown.

Oil Well deliverability can be represented in the same manner.[15] The method again is particularly useful for reservoirs producing below the bubble point where mathematical description is impractical. Oil rate is related to pressure drawdown empirically as follows:

$$q_o = J' (p_i^2 - p_{wf}^2)^n \quad (3)$$

J' = productivity coefficient
n = empirically determined exponent: $0.5 < n < 1.0$

Figure 3-3 shows ideal flow rate and bottom-hole flowing pressure versus time for a properly run flow after flow test.

A log-log plot of flow rate, q, vs. $(p_i^2 - p_{wf}^2)$ should define a straight line with slope "$1/n$." This plot can then be used to predict flow rate for any possible drawdown pressure, Figure 3-4. At least four rates should be run, and each rate should continue until the well reaches a stabilized flowing pressure condition. This requirement for many wells means that long testing periods are required—and often limits the usefulness of the flow after flow method.

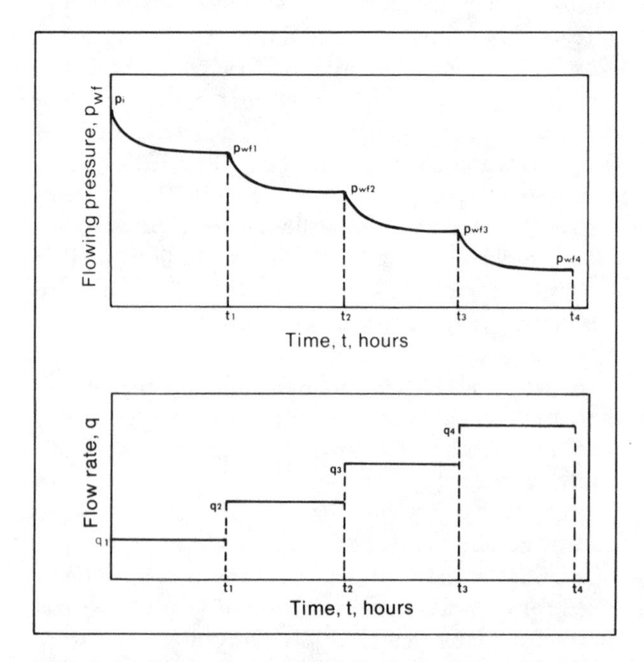

FIG. *3-3—Pressure-rate history for a flow after flow test.*[18] *Permission to publish by The Society of Petroleum Engineers.*

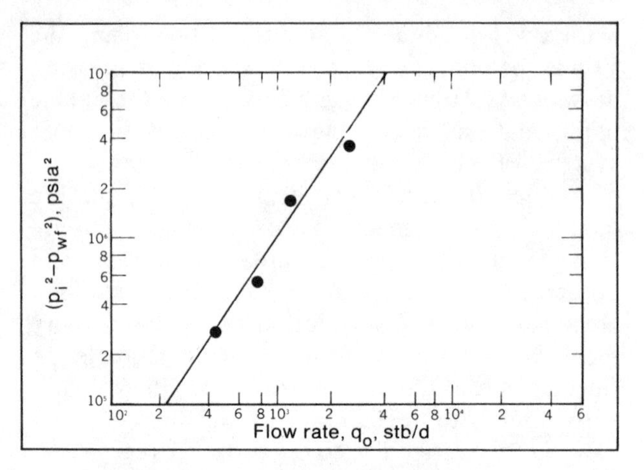

FIG. *3-4—Flow after flow deliverability curve.*[18] *Permission to publish by The Society of Petroleum Engineers.*

Modified Isochronal—To reduce testing time the Gas Well Isochronal Test procedure has also been adapted to Oil Wells.[15] The test is run as shown ideally in Figure 3-5, with a series of relatively short equal-time flow and shut-in periods, and one final flow period where flowing pressure is permitted to stabilize. Four to six hours is sufficient duration for the equal flow or shut-in period.

In making the flow rate versus squared drawdown pressure plot, each flowing period is divided into time increments; i.e. 1/4 hr., 1 hr., and 4 hrs. after the start of the flow period. A plot of q vs. $(\bar{p}^2 - p_{wf}^2)$ is made for each time increment as shown in Figure 3-6, where $\bar{p}$ = average reservoir pressure. Shut-in pressure for each calculation must be the shut-in pressure just prior to the start of that particular flow period.

The slope of the resulting plots determines the proper slope for the "stabilized" curve which is drawn through the one data point where flow conditions were actually allowed to stabilize. The stabilized curve is then used for predictive purposes.

Gas Wells

Gas-well deliverability testing was formalized through work done by the U.S. Bureau of Mines and the Railroad Commission of Texas.[6,7] The original empirical procedure was called the Multipoint Back Pressure Test. The technique requires careful measurement of flow rates and surface pressures at four stabilized flow conditions, and a surface shut-in pressure. Surface pressures

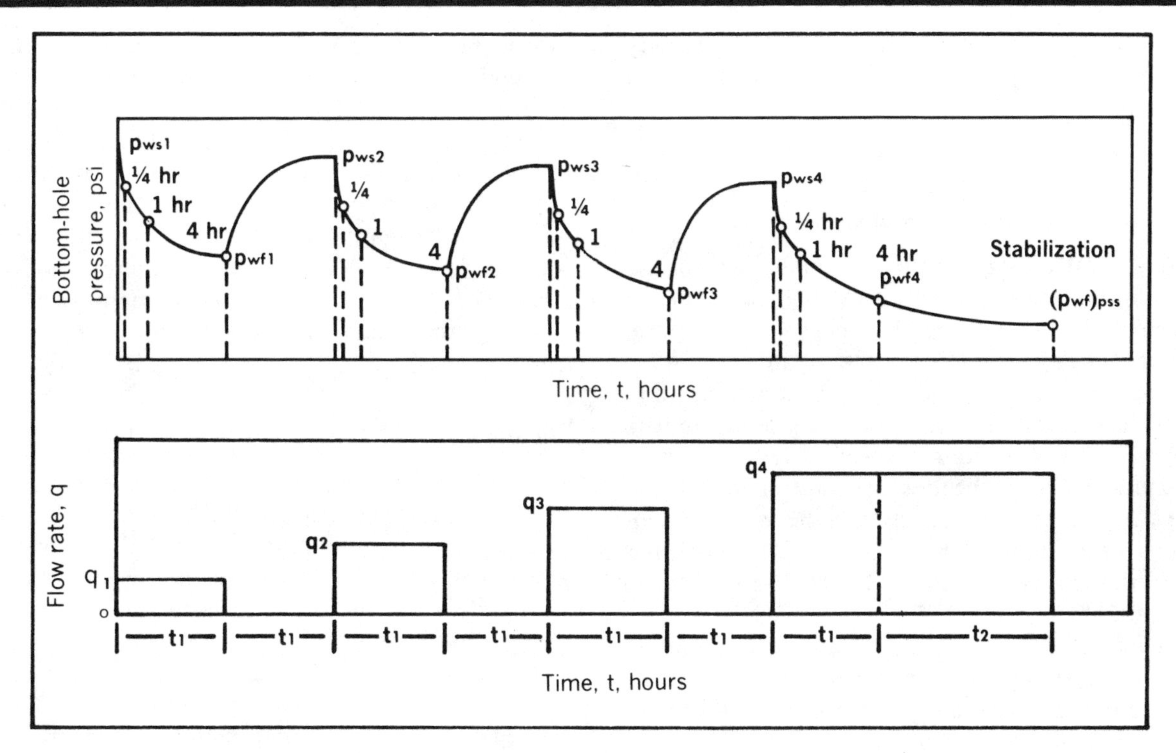

FIG. *3-5—Pressure-rate history for a modified-isochronal-flow test.*[18] *Permission to publish by The Society of Petroleum Engineers.*

are then converted to bottomhole pressures by calculation procedures.

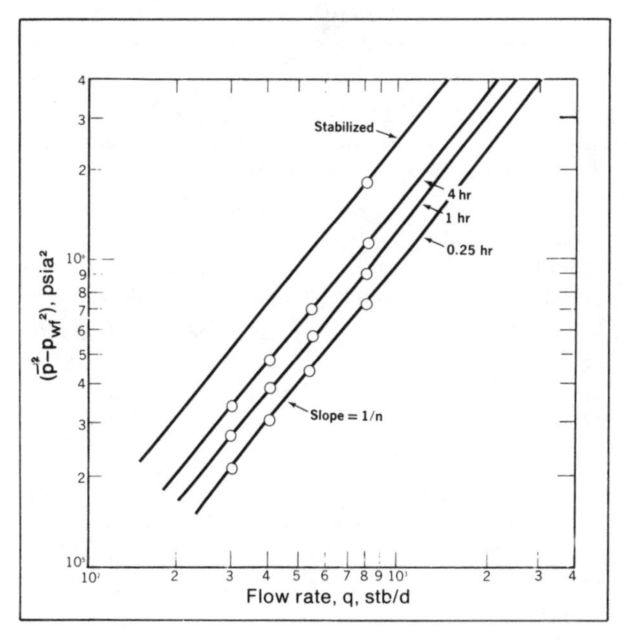

FIG. *3-6—Modified-isochronal-test data plot.*[18] *Permission to publish by The Society of Petroleum Engineers.*

A log-log plot of flow rate vs. $(p_i^2 - p_{wf}^2)$ should define a straight line which can be extended to determine the flow rate at zero bottom hole pressure (or maximum drawdown) which is termed the absolute open-flow potential of the well. The slope of the line usually varies between 0.5 and 1.0. In some cases, (for example DST's) one flowing pressure is used to estimate the AOF range, assuming that the slope is 1.0 for maximum AOF, or 0.5 for minimum AOF.

Properly carried out to stabilized conditions, the original multipoint back-pressure test procedure is a flow-after-flow test. As a practical matter, stabilization for low permeability formation may require many hours, thus, becomes impractical.

The isochronal test for gas wells was first suggested in 1955,[8] subsequently modified, is currently used as standard test procedure for many conditions. This procedure described previously for oil wells reduces the test time requirements to a practical range and apparently provides results comparable to the original flow-after-flow procedure.

Some areas where permeabilities are very low and stabilization impractical have standardized short term tests for gas wells. One such test requires that the well be shut in then flowed for twenty minutes at which time the measured flow rate indicates well capability.

TRANSIENT PRESSURE TEST

Basis for Transient Pressure Analysis

Assume that the only well completed in a reservoir is shut in until a completely stable situation is reached. If this well is then put on production, and pressure is reduced in the wellbore, a wave of reduced pressure gradually moves outward into the reservoir establishing a pressure gradient or sink toward the well. With continued fluid withdrawals from the wellbore, the pressure wave moves further outward. Each point passed by the wave experiences a continuing pressure decline. At a particular time, the maximum distance the wave has traveled is called the drainage radius of the well.

When the wave front reaches a closed boundary, pressure at each point within the boundary continues to decline, but at a more rapid rate. If the wave front encounters a boundary, which supplies fluid at a rate sufficient to maintain a constant pressure at the boundary, pressure at any point within the drainage radius will continue to decline, but at a slower rate. With either the closed boundary, or the constant pressure boundary, the pressure gradient toward the well tends to stabilize after a sufficient time. Pressure level at a particular point may continue to decline.

For the constant pressure boundary a "steady state" will be approached where both pressure gradient and pressure level do not change with time. For the closed boundary a "pseudo steady state" condition is reached where pressure gradient is constant, but pressure declines linearly with time at each point within the drainage radius.

Changes in production rate, or production from other wells will cause additional pressure wave movements, which affect pressure decline and pressure gradients at every point within the drainage radius of the first well.

The basis for Transient Pressure analysis is the observation of these pressure changes, and the fluid withdrawal or injection rates which caused them; along with mathematical descriptions of the flow process, involving properties of the rock through which the movement occurred, and the characteristics of the fluids moving within.

The diffusivity equation, describing fluid flow through reservoir rock, assumes horizontal flow, negligible gravity effects, a homogeneous and isotropic rock, and a single slightly compressible fluid.[10] Also Darcy's Law must apply, and porosity, permeability, viscosity and compressibility must be independent of pressure. With these limitations the diffusivity equation can be easily solved. Where minor variances occur, approximation methods result in reasonable solutions. Where major variances occur, numerical reservoir simulation techniques must be used in an attempt to model these variations.

Limitations and Application

While assumptions and complications of analysis techniques appear to greatly limit application of transient pressure testing for the Well Completion Engineer, under favorable circumstances—and with experience—it can be a useful tool in solving well problems. It should not be oversold—but it does provide a chance to recognize near-wellbore problems; i.e.—formation damage—which might not be detected in the composite picture shown by Deliverability Tests.

For the reservoir analyst the chance to determine average reservoir permeability and pressure, reservoir volume, and perhaps something about reservoir heterogeneities is intriguing.

Pressure transients are subject to all of the normal well completion problems of communication behind casing, partial or ineffective perforations, or partial penetration of the producing zone. Proper analysis must consider all available clues. Production tests, details of perforating and well completion or workover operations, flow profiles or other production logging data may be helpful.

Wellbore Damage or Stimulation—From the standpoint of the Well Completion Engineer wellbore damage or stimulation indicators are of practical importance. There are several ways to quantify damage or improvement. One method uses the idea of "skin" or "*skin effect*". Shown graphically in Figure 3-7.

Pressure drop across the infinitesimally thin skin, Δp_s, is added to the transient pressure drop in the reservoir to represent the wellbore pressure. The pressure drop across the skin can be calculated as follows:

$$\Delta p_s = \frac{141.2\ qB\mu}{kh} s \qquad (4)$$

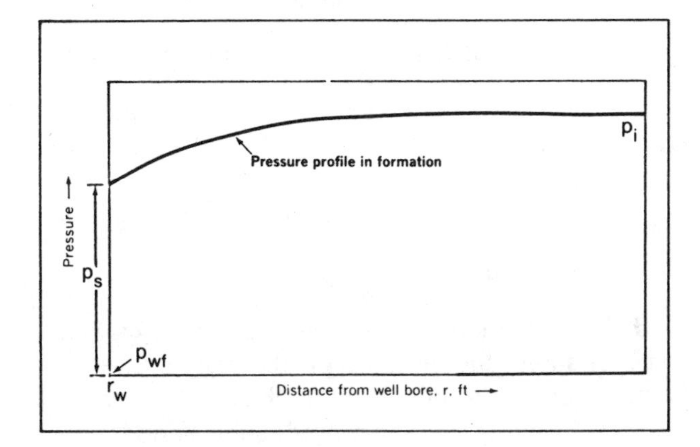

FIG. 3-7—Pressure distribution around a well with a positive skin factor.[18] Permission to publish by The Society of Petroleum Engineers.

Δp_s = pressure drop across skin, psi
B = formation volume factor, reservoir bbl/stb
μ = viscosity, cp
s = skin factor, dimensionless
k = permeability, md
h = height, ft

The value of the skin factors can vary from about −5 for a hydraulically fractured well, to ∞ for a completely plugged well. One problem with the concept of skin effect is that the numerical value of the skin "s" does not directly show the degree of damage.

Flow Efficiency (or Condition Ratio) describes the well's actual flow capacity as a fraction of its capacity with no damage.

$$FE = \frac{J_{actual}}{J_{ideal}} = \frac{\bar{p} - p_{wf} - \Delta p_s}{\bar{p} - p_{wf}} \qquad (5)$$

Damage Ratio is the inverse of flow efficiency. Skin factor, flow efficiency, or damage ratio can be determined from most of the transient pressure techniques that follow.

Wells completed with only a part of the producing zone open—through ineffective perforating or the fact that the well was not drilled completely through the zone will appear to be damaged even if there is no physical flow restriction.

Deviated holes penetrating the reservoir at an angle with no other problems will appear stimulated.

Reservoir Heterogeneities—Reservoir heterogeneities affect transient pressure behavior. Variation in rock properties due to depositional or post depositional changes—vugs—fractures—directional or layered permeability; changes in fluid properties—gas-oil, oil-water contacts—or changes due to pressure variation; changes due to stimulation effects or injection fluids—all affect pressure transients.

Multiple well tests are more severly affected by heterogeneities, and as a result can be used to define heterogeneities in simple cases. A primary difficulty is that many different conditions can give the same transient pressure response. Without other evidence, geologic, seismic, fluid flow or well performance, transient pressure tests should not be used to infer heterogeneous reservoir properties. If individual layers of a multiple layered reservoir are not isolated, meaningful individual layer information on permeability, skin factor or average reservoir pressure cannot be estimated with current technology.

Description of Transient Pressure Tests

Almost every conceivable method of creating and observing changing wellbore pressures has been employed for reservoir analysis. A reasonable classification of transient pressure tests by type is:

—Pressure buildup
—Pressure drawdown
—Multiple rate
—Injection buildup or fall-off
—Multiple well interference
—Drill-stem tests

Each type presents certain advantages and limitations, and certain factors which are particularly important for reasonable results. The following description is intended to give the Completion Engineer an acquaintance with each type. More details are provided concerning pressure buildup tests. Drill-stem testing, being more in the province of completion engineering, is handled as a separate section.

For an excellent and practical treatment of transient pressure analysis and calculation procedures for oil wells, S.P.E. Monograph No. 5, "Advances in Well Test Analysis", by Robert C. Earlougher, Jr.[18] is recommended.

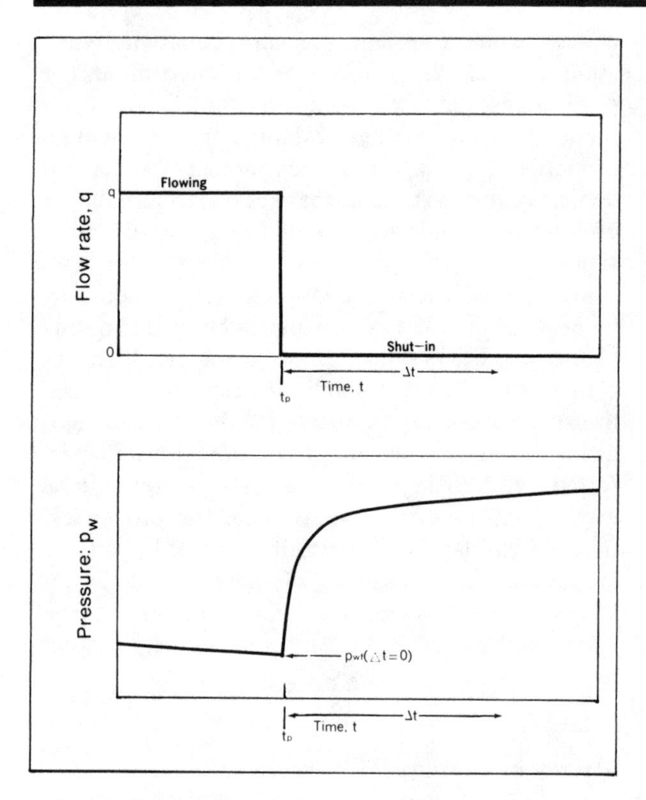

FIG. *3-8—Idealized rate and pressure history for a pressure buildup test.*[18] *Permission to publish by The Society of Petroleum Engineers.*

Pressure-Buildup Testing—Pressure-buildup testing is probably the most familiar transient well testing technique. Essentially the well is produced at constant rate long enough to establish a stabilized pressure distribution, then is shut in as shown in Figure 3-8. Stabilization is important, otherwise erroneous data will be calculated. If stabilization is impossible, other techniques, such as variable rate testing must be used.

Referring to Figure 3-8, t_p is production time, and Δt is shut in time. Pressure is measured immediately before shut in, and is recorded as a function of time during the shut-in period. The resulting pressure buildup curve is then analyzed for reservoir properties and wellbore condition.

Horner Method—After wellbore storage effects have diminished, wellbore pressure during shut in is:

$$p_{ws} = p_i - \frac{162.6\, qB\mu}{kh} \log\left(\frac{t_p + \Delta t}{\Delta t}\right) \qquad (6)$$

p_{ws} plotted versus $\log \frac{t_p + \Delta t}{\Delta t}$ is a straight line with slope $(-m)$, and intercept (p_i). The value of t_p (hrs) should be estimated as follows: cumulative production from last pressure equalization, divided by the stabilized production rate just before shut in, times 24—or $t_p = \frac{24\, V_p}{q}$.

This plot, Figure 3-9 is called the Horner Plot. The straight line portion of the curve may be extrapolated to infinite time to obtain p_i. The value of p_i is accurate for short production periods, but is somewhat too high for long production periods.

The measured slope value (m) can be used to determine reservoir permeability:

$$k = \frac{162.6\, qB\mu}{mh} \qquad (7)$$

Skin factor (s) does not appear in the Horner Equation—but does affect the early-time shape of the curve—as do wellbore storage effects. Skin factor "s" can be calculated from the following equation:

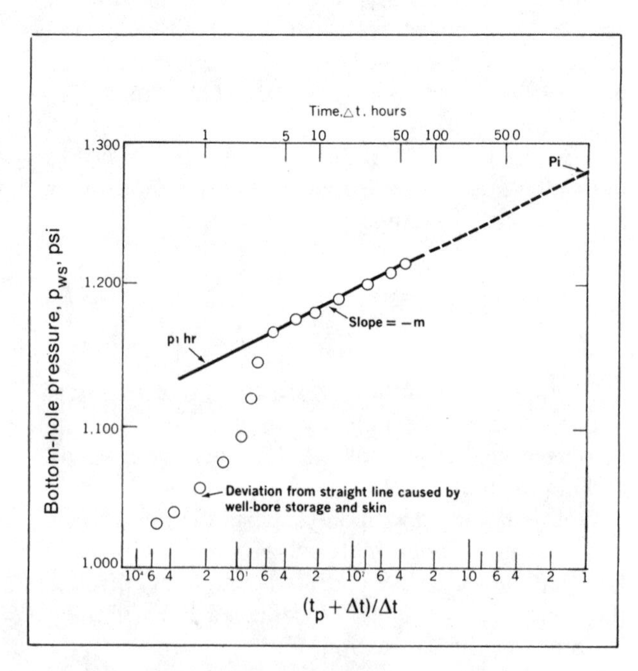

FIG. *3-9—Horner plot of pressure buildup data.*[18] *Permission to publish by The Society of Petroleum Engineers.*

$$s = 1.151 \left[\frac{p_{1\,hr} - p_{wf}(\Delta t = 0)}{m} - \log \left(\frac{k}{\phi \mu c_t r_w^2} \right) + 3.23 \right] \quad (8)$$

The value for $p_{1\,hr}$ must be taken from the straight line extrapolation.

If wellbore storage effects last so long that a semi-log straight line does not develop, "type curve" techniques presented by McKinley, Ramey, and Earlougher and Kersch, may be helpful in salvaging estimates from otherwise unusable data. Type curve methods should be considered a last resort, however.

Miller-Dyes-Hutchinson Plot—The Horner build-up plot may be simplified if the producing time is much greater than the shut-in time. The Miller-Dyes-Hutchinson (MDH) buildup equation is:

$$p_{ws} = p_{1\,hr} + m \log \Delta t \quad (9)$$

$$m = \frac{162.6\, q B \mu}{kh} \quad (10)$$

A plot of p_{ws} vs. log Δt should be a straight line—with a typical example shown in Figure 3-10. Extrapolated shut-in pressure p_i is again somewhat too high.

Since the *MDH* method is somewhat easier to use it is usually employed—except for the case of Drill Stem Tests where flow time is about the same magnitude as shut-in time. Here the Horner method must be used.

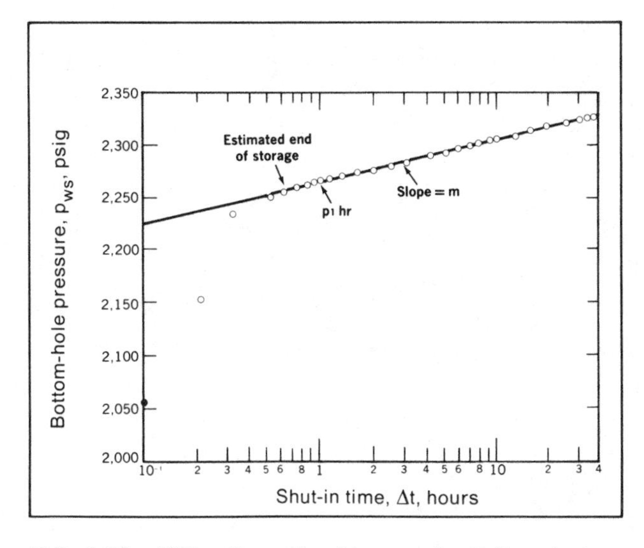

FIG. *3-10—Miller-Dyes-Hutchinson plot.*[18] *Permission to publish by The Society of Petroleum Engineers.*

Frequently pressure buildup tests are not as simple as discussed. Many factors can affect the shape of the pressure buildup plot. These include wellbore storage effects, formation damage, partial penetration, hydraulic fractures, unstabilized flow conditions, fluid and rock interfaces, water-oil or gas-oil contacts, and rock heterogeneities. Practical problems such as a leaking pump or lubricator or a bottom-hole pressure gauge or reading device in poor condition, may render otherwise good data unusable. Or if used, it may provide erroneous results.

A major cost consideration of the Pressure Build-up Test is the fact that income is deferred while the test is being run.

Pressure Drawdown Testing—Pressure draw-down tests have two advantages over pressure buildup tests. First, production continues during the test period. Second, in addition to formation permeability and formation damage information, an estimate can be made of the reservoir volume in communication with the wellbore. Thus the "Res-ervoir Limits" test can be used to estimate if there is sufficient oil—or gas (if a dry gas reservoir)—in place to justify additional wells in a new reservoir.

Multiple Rate Testing—Pressure buildup and drawdown tests require a constant flow rate, which is sometimes impossible or impractical to maintain for a sufficiently long period. Multiple rate analysis can be applied to several well flow situations: for example, uncontrolled variable rates; a series of constant rates; or constant bottom hole pressure with continually changing flow rate.

Multiple rate tests have the advantage of provid-ing transient test data without the requirement of well shut in. They minimize wellbore storage effects and phase segregation effects; thus, sometimes provide good results where buildup or drawdown tests would not. This is perhaps their greatest advantage.

Accurate flow rate and pressure measurements are essential. Rate measurements are much more critical than in constant rate tests. Rate changes must be sufficient to significantly effect the tran-sient pressure behavior. The analysis procedure is direct and simple but computations needed to make the plot are more bothersome, and are sometimes left to the computer.

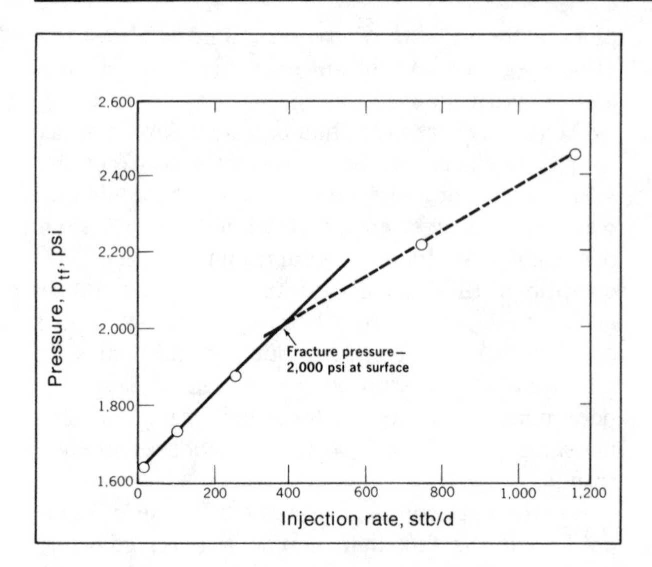

FIG. *3-11—Step-rate injectivity plot.*[18] *Permission to publish by The Society of Petroleum Engineers.*

Injection-Well Testing—Injection well transient pressure testing is basically simple as long as the mobility ratio between the injected and in-situ fluids is about unity. Injection is analogous to production—thus an Injectivity test parallels a Drawdown test, and a Pressure Falloff test parallels a Pressure Buildup test. Calculation of reservoir characteristics is similar.

A Step Rate Injectivity test, Figure 3-11, is normally used to estimate fracture pressure in an injection well. Fracturing may not be important in a water injection well, but could be critical in a tertiary flood to avoid injecting an expensive fluid through uncontrolled fractures. Fluid is injected at a series of increasing rates. Each time step should be about the same length—usually one hour with formation permeability less than 5 md—or 30 minutes with permeability more than 10 md. Six to eight rates are desirable to bracket the estimated fracture pressure. Bottom hole pressures should be plotted, but if friction loss in the flow string is small, surface pressure can be used.

Interference Tests (multiple well testing)—In an Interference test, a long-duration rate change in one well creates a pressure change in an observation well that can be related to reservoir characteristics. A Pulse test is an interference test that provides similar data by changing production rate in a cyclic manner to produce short term pressure pulses. The pressure response to the pulses is measured at one or more observation wells. Pressure responses are very small—perhaps 0.01 *psi*—thus special measuring equipment is required.

If enough observation wells are used, interference-test data can sometimes be analyzed by computer methods to give a description of the variation of reservoir properties with location. Vertical pulse testing in one well (under ideal well conditions) may indicate vertical formation continuity. Orientation and length of vertical fractures may be estimated through pulse testing and reservoir simulation techniques.

REFERENCES

1. Miller, C. C.; Dyes, A. B.; and Hutchinson, C. A., Jr.: "The Estimation of Permeability and Reservoir Pressure from Bottomhole Pressure Buildup Characteristics," *Trans.*, AIME (1950).

2. Horner, D. R.: "Pressure Buildup in Wells," Proceedings Third World Petroleum Congress, The Hague (1951).

3. van Everdingen, A. F.: "The Skin Effect and Its Influence on the Productive Capacity of a Well," *Trans.*, AIME (1953).

4. Hurst, William: "Establishment of the Skin Effect and Its Impediment to Fluid Flow Into a Wellbore," *Petr. Engr.* (Oct. 1953).

5. Gladfelter, R. E.; Tracy, G. W.; and Wilsey, L. E.: "Selecting Wells which will Respond to Production Stimulation Treatment," *Drill. and Prod. Prac.* API (1955).

6. Rawlins, E. L., and Schellhardt, M. A.: "Back Pressure Data on Natural Gas Wells and Their Application to Production Practices," Monograph 7 USBM (1956).

7. Railroad Commission of Texas, "Back Pressure Testing of Natural Gas Wells."

8. Cullender, M. H.: "The Isochronal Performance Method of Determining the Flow Characteristics of Gas Wells," *Trans.*, AIME (1955).

9. Brons, F., and Martin V. E.: "The Effect of Restricted Fluid Entry on Well Productivity," *JPT* (Aug. 1961) p. 172.

10. Matthews. C. S., and Russel, D. G.: "Pressure Buildup and Flow Tests in Wells," Monograph Series, SPE of AIME, Dallas (1967).

11. Vogel J. V.: "Inflow Performance Relationships for Solution Gas Drive Wells," *JPT* (Jan. 1968) p. 83.

12. Ramey, H. J., Jr.: "Short-Time Well Test Data Interpretation in the Presence of Skin Effect and Wellbore Storage," *JPT* (Jan. 1970) p. 97.

13. McKinley, R. M.: "Wellbore Transmissibility From After Flow-Dominated Pressure Buildup Data," *JPT* (July 1971) p. 863.

14. Standing, M. B.: "Concerning the Calculation of Inflow Performance of Wells Producing Solution Gas Drive Reservoirs," *JPT* (Sept. 1971) 1, p. 141.

15. Fetkovich, M. J.: "The Isochronal Testing of Oil Wells," SPE 4529, Las Vegas, (Sept. 1973).

16. Earlougher, Robert C., Jr., and Kersch, Keith M.: "Analysis of Short-Time Transient Test Data by Type-Curve Matching," *JPT* (July 1974) p. 793.

17. Ramey, H. J.: "Practical Use of Modern Well Test Analysis," SPE 5858 (Oct. 1976).

18 Earlougher, Robert C., Jr.; "Advances in Well Test Analysis," SPE Monograph Series No. 5, Dallas. (1977).

19. Erdle, J. C.; Archer, D. A.; Stiff, T. J.; Callihan, M. C.: "Well Test Engineering Software with Built-in Expert Advice," SPE 15309 (June 1986).

Drill Stem Testing

BACKGROUND

Objective

Drill stem testing provides a method of temporarily completing a well to determine the productive characteristics of a specific zone. As originally conceived, a Drill Stem Test provided primarily an indication of formation content. The pressure chart was available, but served mainly to evaluate tool operation.

Currently, analysis of pressure data in a properly planned and executed DST can provide, at reasonable cost, good data to help evaluate the productivity of the zone, the completion practices, the extent of formation damage, and perhaps the need for stimulation.

Many times actual well producing rates can be accurately predicted from DST data. The DST shows what the well will produce against gradually increasing back-pressure. From this a Productivity Index (P.I.), or Inflow Performance Relationship (IPR) can be established, and if the flowing pressure gradient in the tubing can be estimated, then actual producing rate can be determined.

Reservoir Characteristics

Reservoir characteristics that may be estimated from DST analysis include:

—*Average Effective Permeability*—This may be better than core permeability since much greater volume is averaged. Also, effective permeability rather than absolute permeability is obtained.

—*Reservoir Pressure*—Measured, if shut-in time is sufficient, or calculated, if not.

—*Wellbore Damage*—Damage ratio method permits estimation of what the well should make without damage.

—*Barriers—Permeability Changes—Fluid Contacts* —These reservoir anomalies affect the slope of the pressure buildup plot. They usually require substantiating data to differentiate one from the other.

—*Radius of Investigation*—An estimate of how far away from the wellbore the DST can "see."

—*Depletion*—Can be detected if reservoir is small and test is properly run.

In summary, the DST, properly applied, has become a very useful tool for the Well Completion Engineer.

Basics of DST Operations

Simply a Drill Stem Test is made by running in the hole on drill pipe a bottom assembly consisting of a packer and a surface operated valve. The DST valve is closed while the drill string is run, thus pressure inside the drill pipe is very low compared to hydrostatic mud column pressure. Once on bottom, the packer is set to isolate the desired formation zone from the mud column, and the control valve is opened to allow formation fluids to enter the drill pipe.

After a suitable period, the valve is closed, and a pressure buildup occurs below the valve as formation fluids repressure the area around the wellbore. After a suitable buildup time, the control valve usually is opened again, and the flowing and shut-in periods repeated, (several times if desired), to obtain additional data and verification. Figure 3-12 shows schematically a simple single-flow-period operational sequence.

Pressure vs Time Plot

The entire sequence of events is recorded on a pressure vs time plot shown schematically in Figure 3-13. As the tools are run in the hole, the pressure bomb records the increase in hydrostatic mud column pressure. The pressure at Point A is termed the initial hydrostatic mud pressure.

Initial Flowing and Shut-in Periods—One objective of the DST is to determine the static or shut-in reservoir pressure of the zone. To measure true

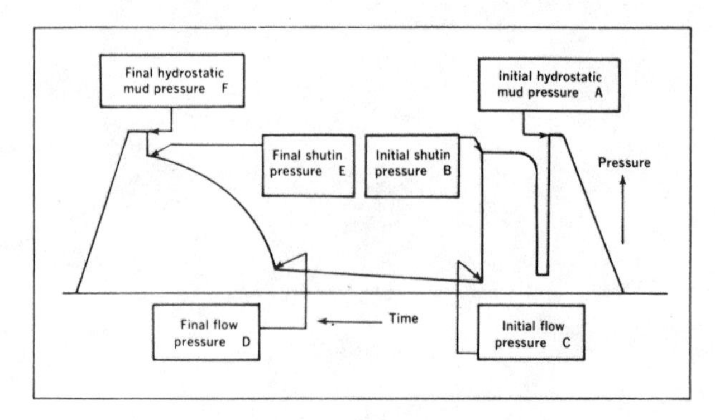

FIG. 3-13—*Sequence of events in a DST.*

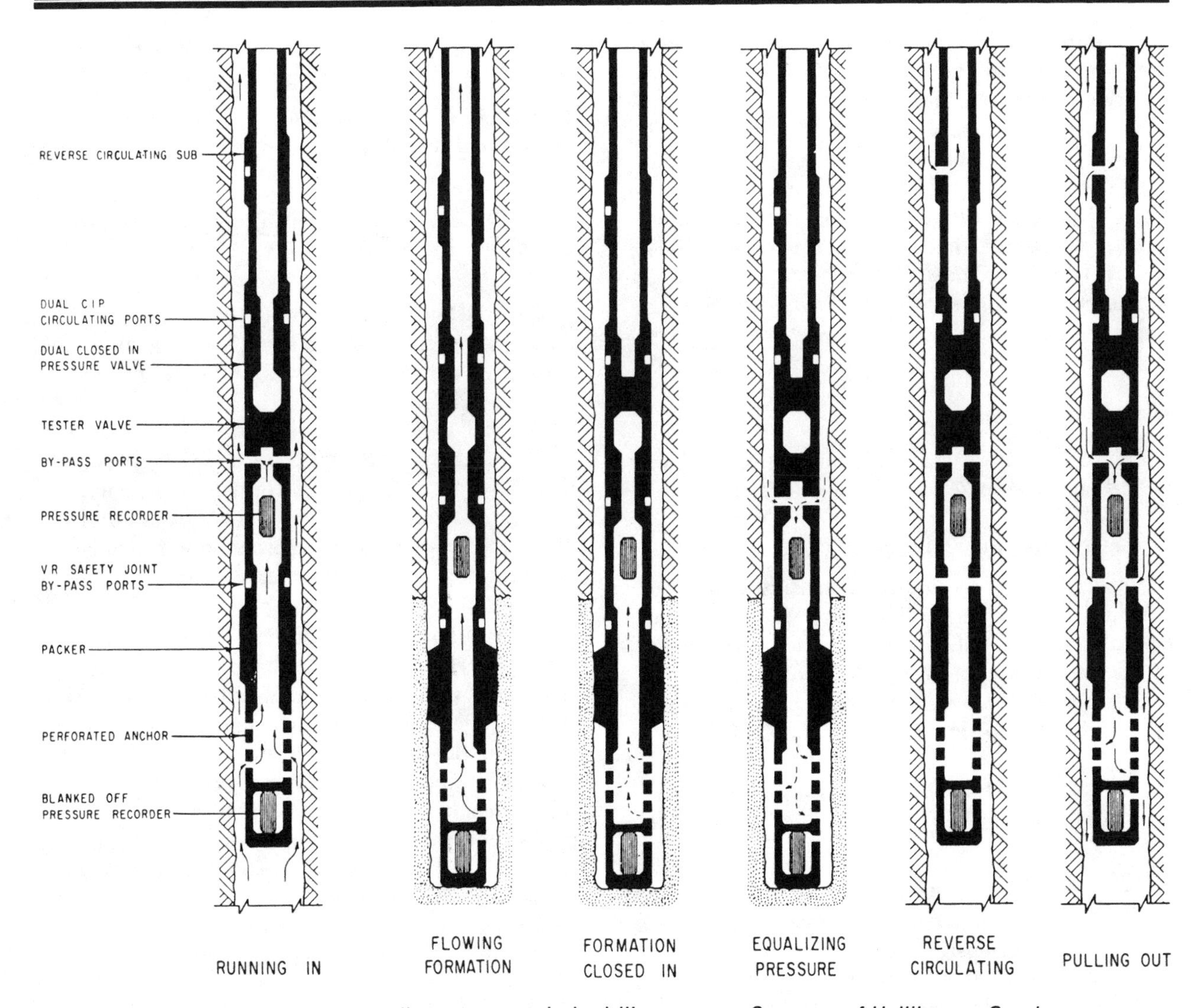

FIG. 3-12—*Fluid-passage diagram, open-hole drill stem test. Courtesy of Halliburton Services.*

static reservoir pressure, any over-pressured condition near the wellbore, due to drilling fluid filtration or fluid compression in setting the DST packer, must first be relieved by a short flowing period.

The control valve is then closed and pressure builds up toward static reservoir pressure. Depending on the length of the flowing period, the length of the shut-in period, and certain reservoir parameters, the pressure may or may not level out at the true static reservoir pressure. Thus Point B, initial shut-in pressure, may be somewhat less than true static reservoir pressure.

Second Flowing and Shut-in Periods—The objective of the second flowing and shut-in periods is to permit the calculation of reservoir parameters, as well as to determine fluid production. As the control valve is opened, pressure falls to Point C (initial flowing pressure). As fluids move into the drill pipe above the recorder, pressure increases, reflecting the increase in hydrostatic head of liquid and in back pressure, due to flow restrictions in the tools.

After a suitable period of time, depending on test objectives, hole conditions and the cost of rig time, the tool is shut in. The pressure at the moment of shut in, Point D, is termed the final flowing pressure. Recorded pressure then rises toward Point E, the final shut-in pressure. Usually the final shut-in pressure is significantly less than the static reservoir pressure, again due to limitations of time relative to reservoir parameters.

This second pressure buildup portion of the DST, following a reasonably long flowing period, usually provides data for transient pressure analysis. Sometimes DST's are designed having two reasonably long flowing periods and subsequent shut-in periods, Figure 3-14, to provide verification of calculations, and to permit more sophisticated well testing techniques.

Field-Recorded DST Pressure Chart

Figure 3-14 is typical of actual multiple-flow-period charts with key points described therein.

THEORY OF PRESSURE BUILDUP ANALYSIS

Horner Equation—Transient pressure analysis of a DST is based on the Horner pressure buildup equation. This equation describes the repressuring of the wellbore area during the shut-in period, as formation fluid moves into the "pressure sink" created by the flowing portion of the DST:

$$p_{ws} = p_i - \frac{162.6\, q \mu B}{kh} \log_{10}\left(\frac{t'_p - \Delta t'}{\Delta t'}\right) \quad (1)$$

Where

p_{ws} = measured pressure in the wellbore during buildup, psig
t'_p = flowing time, minutes
$\Delta t'$ = shut-in time, minutes
p_i = shut-in reservoir pressure, psig
q = rate of flow, stb/day
μ = fluid viscosity, cp
B = formation volume factor, reservoir bbl/stock tank bbl
k = formation permeability, md
h = formation thickness, ft

Conditions which must be assumed during the buildup period for Equation 1 to be strictly correct are: radial flow, homogenous formation; steady-state conditions; infinite reservoir; single-phase flow. Most of these conditions are met on a typical DST. Steady state flow is perhaps the condition causing the primary concern, particularly at early shut-in time.

Horner Buildup Plot—Assuming these conditions are met, then a plot of p_{ws} vs $\log_{10}\left(\frac{t'_p + \Delta t'}{\Delta t'}\right)$ should yield a straight line, and the slope (m) of the straight line should be:

$$m = \frac{162.6\, q \mu B}{kh} \quad (2)$$

The constant m is representative of a given fluid having physical properties μB flowing at a rate q through a formation having physical properties kh.

Figure 3-15 is an idealized Horner Plot. The DST pressure chart of Figure 3-15 shows very simply how flowing time t'_p, and formation pressure p_{ws} at various shut-in times $\Delta t'$, are picked from the chart and related to the Horner Plot. Usually p_{ws} is determined at 5-minute intervals along the shut-in pressure curve.

In a multiple flow period DST, selecting a value for t'_p creates some problem mathematically. However, very little error is caused by assuming that t'_p is the time of the flowing period immediately preceding the particular shut-in period. With equal flowing periods on a multiple flow period DST, this is usually done. With a very short initial flowing period, compared to a longer second flowing period, t'_p can be assumed to be the total of the flowing times with very little error.

In Figure 3-15 the slope m of the "straight line" is numerically the difference between the p'_f pressure value at $\log_{10}\frac{t'_p + \Delta t'}{\Delta t'} = 0$, and at $\log_{10}\frac{t'_p + \Delta t'}{\Delta t'} = 1.0$. If the points are plotted on semilog paper, m is the change in pressure over one log cycle.

The ideal situation of Figure 3-15, where all points line up as a straight line, is seldom seen in actual DST's, since "after flow" or wellbore storage effects cause a deviation from the straight line during the early times. As a rule of thumb, four points are needed to determine the straight line.

The length of shut-in time required to approach a steady state, or straight line condition fluctuates with reservoir and fluid characteristics and flow conditions. Experience has formulated certain rules of thumb to help determine what is sufficient shut-in time. Generally the shut-in pressure must reach at

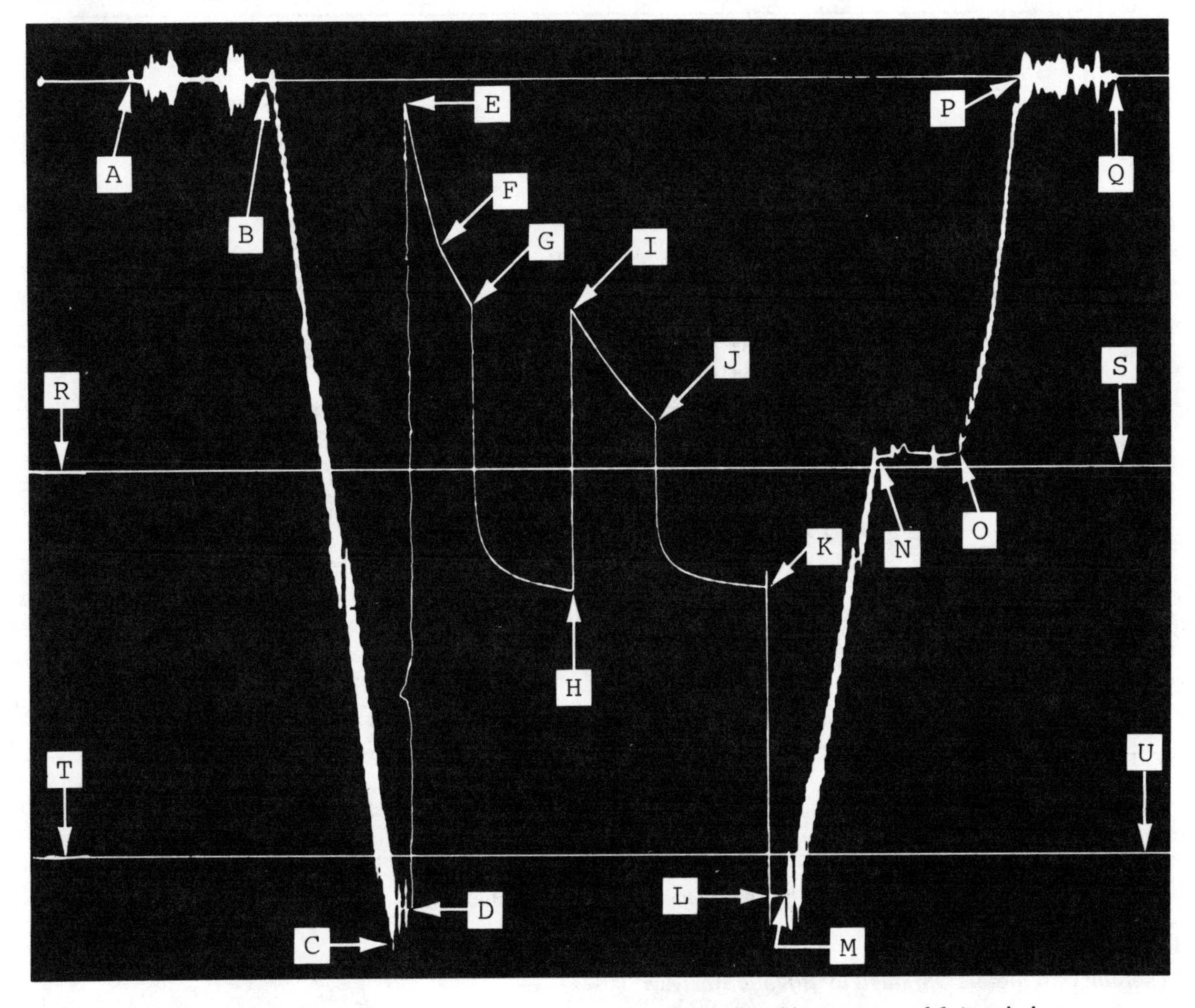

A-Q Baseline drawn by recorder.
A Recorder clock started.
A-B Tools being made up.
B-C Running in hole.
C-D On bottom—completing surface hookup—initial hydrostatic mud pressure.
D Tester valve opens.
E Beginning 1st flow period (initial flow pressure).
E-F Fluid filling small ID drill collars.
F-G Fluid filling larger ID drill pipe.
G End 1st flow period (final flow pressure).
G-H First closed-in pressure period.
H Final closed-in pressure of 1st period.
I Begin 2nd flow period (initial flow pressure).
J End 2nd flow period (final flow pressure).
K End 2nd shut-in period (final closed-in)
K-L Equalizing hydrostatic pressure across packer.
M-N Pulling out of hole.
N-O Reached top of fluid fillup in drill pipe—reversing out fluid to surface.
O-P Continued trip out of hole.
P-Q Breaking down tools.
R-S, T-U 1,000-psi lines drawn by chart interpreter.

FIG. *3-14—Typical chart for multiple-flow-period DST. Courtesy of Halliburton Services.*

least 65% of static pressure in order to produce a straight line on the Horner Plot.

Prior to work by McKinley, Ramey and others, unless the straight line portion of the buildup plot could be identified, no further DST analysis was possible. By use of type-curve methods, however, it is sometimes possible to make reasonable estimates of formation parameters even though the DST tool was not shut in long enough to indicate the Horner straight line.

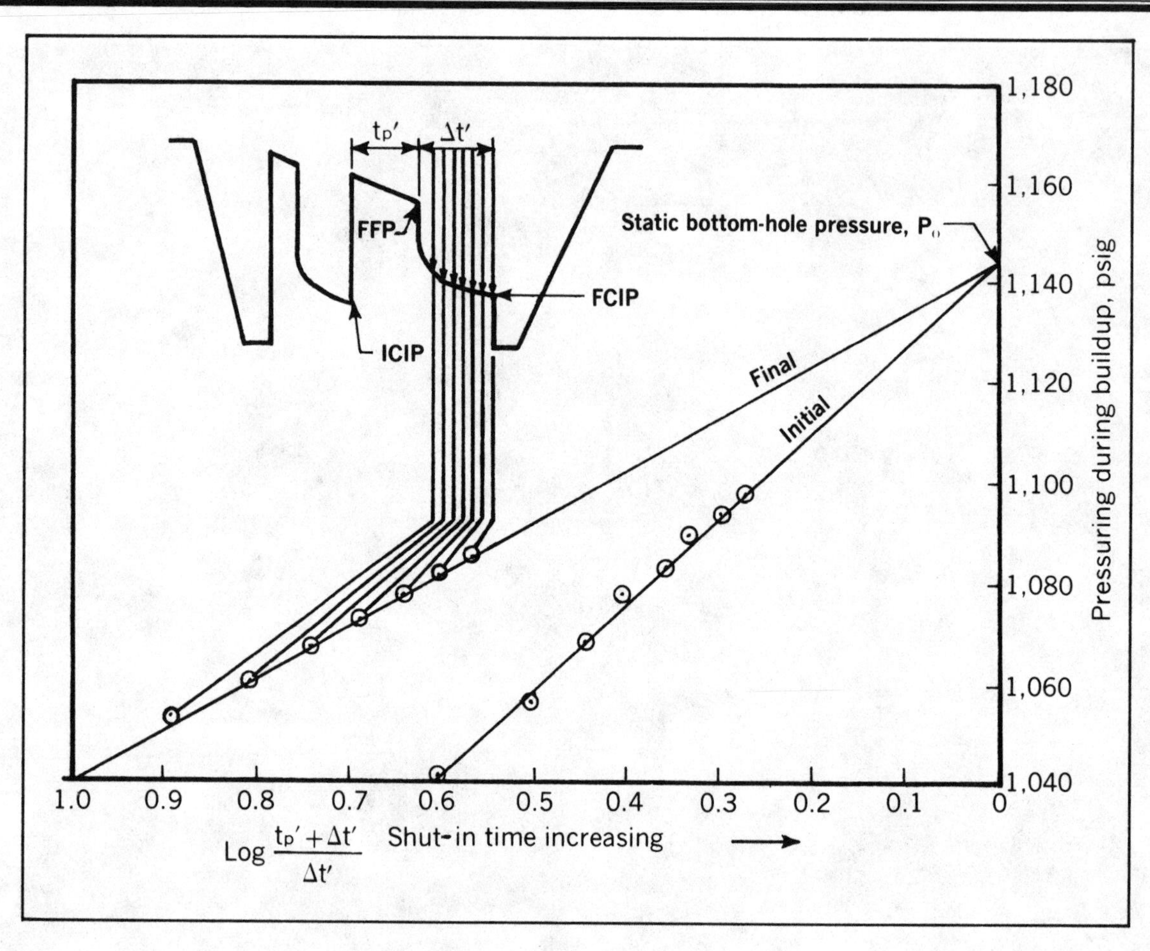

FIG. 3-15—*Idealized Horner buildup plot.*

Reservoir Parameters Obtained by Buildup Analysis

Permeability (k)—Assuming the Horner buildup plot produces a straight line such that m can be determined; then the average effective permeability k can be calculated as follows:

$$k = \frac{162.6\, q\, \mu\, B}{mh} \tag{3}$$

Parameters μ (viscosity) and B (formation volume factor) can be estimated from available correlations, if the API gravity of the crude oil and the gas-oil ratio can be determined through wellsite measurements after the test.

Formation thickness h must be the net thickness of productive zone, which should be determined from log analysis. If net thickness h is unavailable, kh or formation capacity is determined:

$$kh = \frac{162.6\, q\, \mu\, B}{m} \tag{4}$$

If all the reservoir parameters are unknown, transmissibility $\frac{kh}{\mu B}$ is determined:

$$\frac{kh}{\mu B} = \frac{162.6\, q}{m} \tag{5}$$

Static Reservoir Pressure (p_i)—Static, or shut-in reservoir pressure, is obtained by extrapolating the Horner straight line to an "infinite" shut-in time.

At infinite shut-in time, $\frac{t'_p + \Delta t'}{\Delta t'} = 1.0$; or as shown in Figure 3-15, $\log_{10} \frac{t'_p + \Delta t'}{\Delta t'} = 0$.

In Figure 3-15, both the 1st buildup plot and the 2nd buildup plot extrapolate to the same static pressure. This lends confidence to the analysis. If the 2nd buildup static pressure is lower than the 1st, then depletion of the reservoir is a possibility.

Depending on the length of the initial flowing period and reservoir parameters, the initial shut-in buildup may provide a stabilized value of static reservoir pressure that can be read directly from the pressure gauge. Pressure may be assumed to be stabilized if it holds the same value for 15 minutes or longer.

Wellbore Damage—Many times DST results are affected by formation damage. Thus, to be meaningful, the effect of flow restriction caused by the damaged zone must be accounted for in analyzing a specific DST.

Hurst and van Everdinger presented the following empirical equation for a dimensionless value "s" denoting "skin factor:"

$$s = 1.151\left[\frac{p_i - p_{ff}}{m} - \log\frac{kt'_p}{\phi\mu c r_w^2} + 2.85\right] \quad (6)$$

The skin factor is useful in comparing damage between wells, however, cannot be readily applied to a specific formation to show what that zone should make if damage were removed.

Griffin and Zak carried Equation 6 one step further introducing the concept of Damage Ratio (DR), which compares flow rate observed on a DST (q_a) to the theoretical flow rate without damage (q_t):

$$\text{DR} = \frac{q_t}{q_a} \quad (7)$$

An equation for calculation of DR based on the skin factor relation of Hurst and van Everdingen is:

$$\text{DR} = \frac{(p_i - p_{ff})}{m\left(\log\dfrac{kt'_p}{\phi\mu c r_w^2} - 2.85\right)} \quad (8)$$

Where:

p_i = shut-in reservoir pressure, psi
p_{ff} = formation pressure at flow time T, psi (final flowing pressure)
c = fluid compressibility, vol/vol/psi
ϕ = formation porosity, fraction
μ = viscosity of reservoir fluid, cp
r_w = well bore radius, inches
k = effective permeability, md
t'_p = flowing time, minutes

DR substantially greater than 1.0 indicates damage. Griffin and Zak simplified this equation by assigning average values to the formation parameters k, ϕ, c, μ, and r_w. This produced an equation for Estimated Damage Ratio:

$$\text{EDR} = \frac{(p_i - p_{ff})}{m(\log t'_p + 2.65)} \quad (9)$$

Reservoir and Fluid Anomaly Indications

Many times the assumptions of the Horner buildup equation, homogeneous formation, single phase flow and infinite reservoir, do not hold in an actual case. If changes occur within the radius of investigation of the DST, they can be detected by a change in slope of the Horner buildup plot.

Permeability or Viscosity—Examining the Horner slope equation, it is seen that if rate of flow q remains constant, then permeability k, or fluid viscosity μ, are likely suspects for change as the wave of increasing pressure travels toward the wellbore.

$$m = \frac{162.6\, q\mu B}{kh} \quad (10)$$

Permeability may change due to natural lensing or due to formation damage, Figure 3-16A. It is doubtful, however, that formation damage would affect sufficient volume of formation to be detected as a change of slope on the buildup plot. Fluid viscosity could change due to a change in fluid phase or type (i.e., gas to oil). "Seeing" the gas-liquid contact from the upstructure well of Figure 3-16B would be difficult, due to the normally short radius of investigation through a gas column. Seeing the gas-liquid contact from the downstructure well is a much more likely possibility.

Barrier—A sealing barrier such as a fault or permeability pinchout can cause a change of slope m. If the barrier is a straight line as A-A' in Figure 3-17A, then the buildup slope will change by a factor of 2, Figure 3-17B.

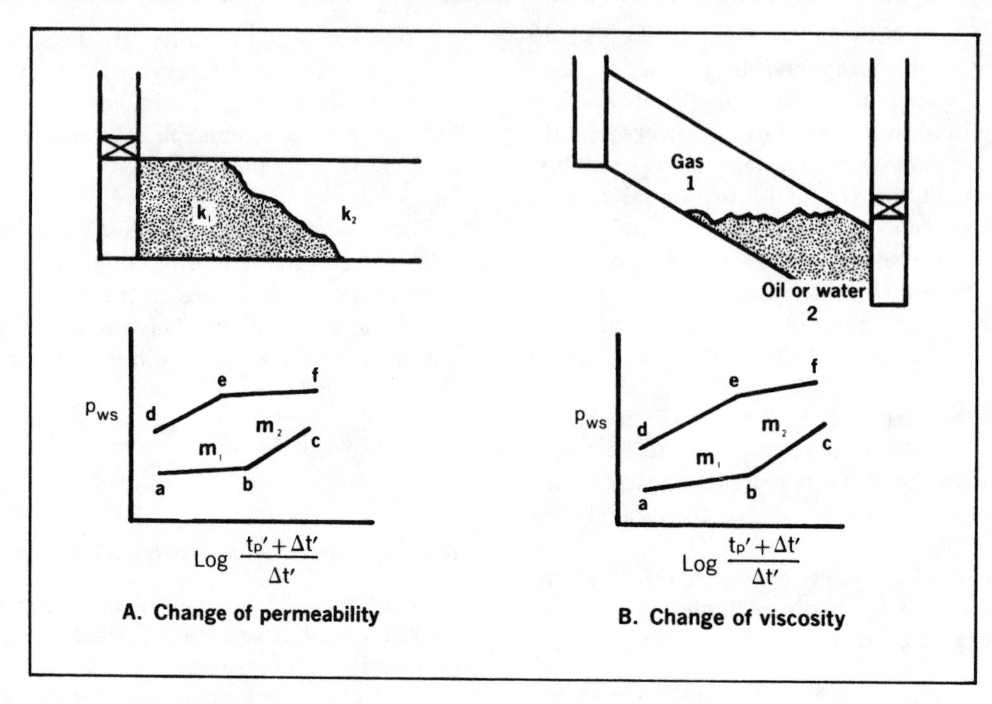

FIG. 3-16—*Effect of changing permeability and viscosity.*

In summary, a change in permeability, or viscosity, or existence of a barrier, all can cause a change in the slope of the Horner plot. Thus, the fact that a change in slope appears on the buildup plot leaves open the question of what caused the anomaly. This must be resolved through other geologic or reservoir information.

Distance to Anomaly—The distance to the anomaly, (r_a) whether it be a barrier, or change of permeability, or a fluid contact, can be calculated using Horner's equation:

$$-E_i\left(\frac{-3793 r_a^2 \phi \mu c}{k\, t_p}\right) = 2.303\, ln\left(\frac{t_p + \Delta t_a}{\Delta t_a}\right) \quad (11)$$

r_a = distance to anomaly in feet
t_p = flow time in hours
Δt_a = shut-in time at the point of slope change, hours
$-E_i$ = exponential integral value—See Figure 3–18.

Radius of Investigation

The following equation from Van Poollen may be used to estimate the radius of investigation of a particular DST in an infinite radial flow system:

$$r_i = \sqrt{\frac{k\, t_p'}{5.76 \times 10^4 \phi \mu c}} \quad (12)$$

r_i = radius of investigation, ft
t_p' = flow time in minutes
Other units as previously noted

Obviously the longer the flowing time, the further back away from the wellbore we are looking with our DST.

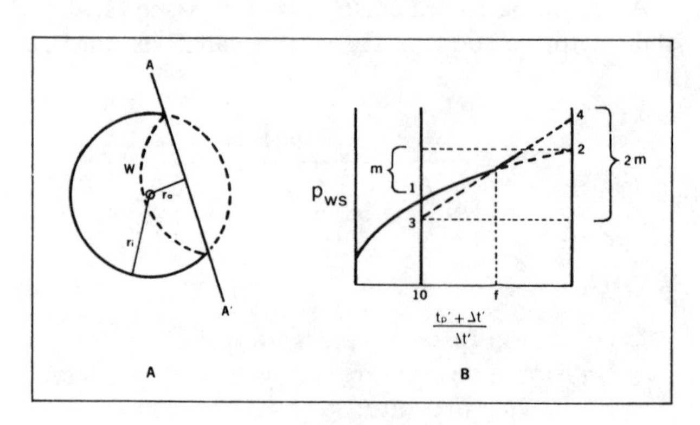

FIG. 3-17—*Effect of a fault.*

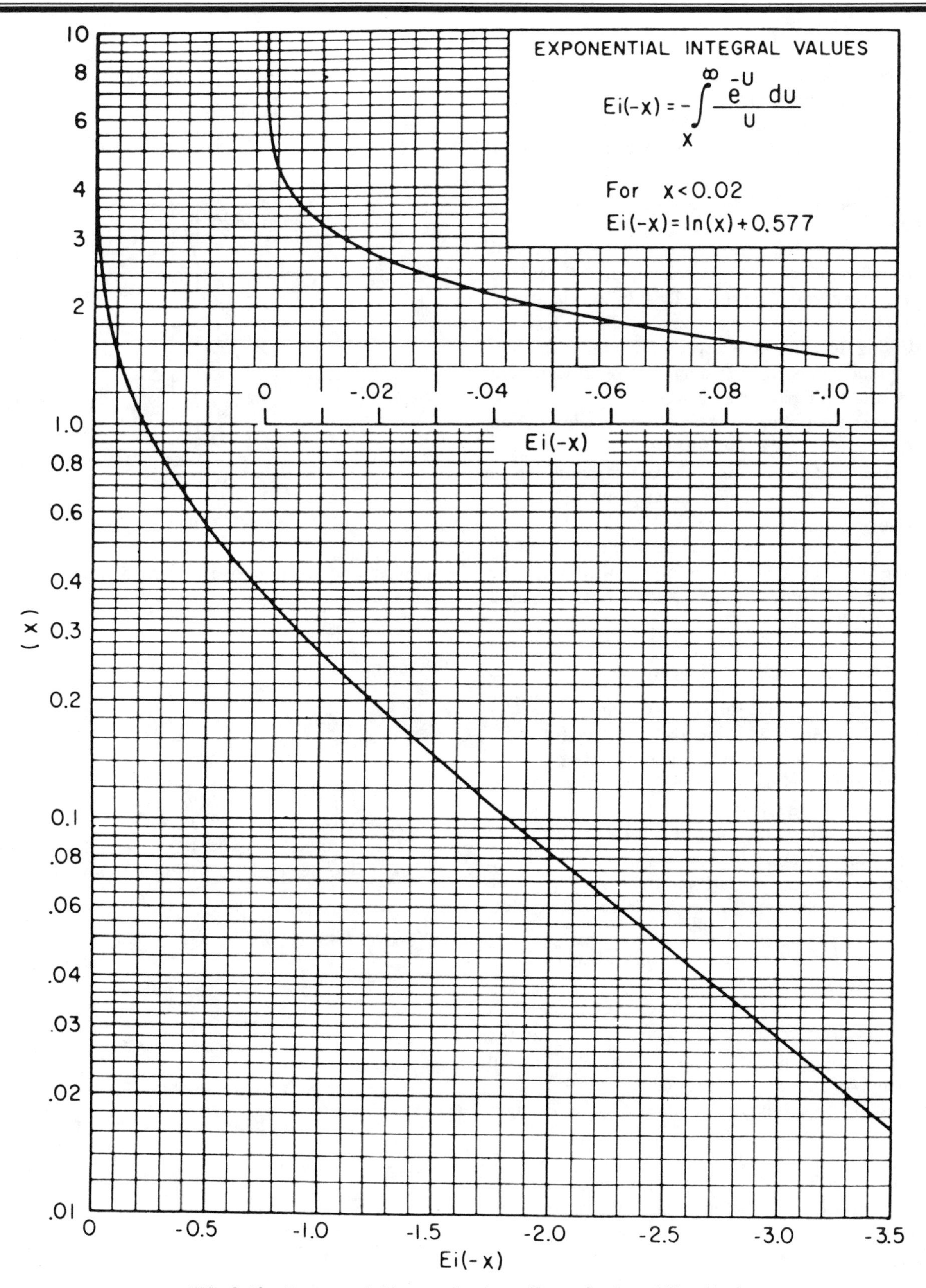

FIG. 3-18—*Exponential integral values (From Craft and Hawkins).*

Depletion

If the extrapolated or stabilized shut-in pressure from the second buildup is lower than the initial shut-in pressure, a depleting reservoir may be the cause. Obviously a reservoir must be extremely small for depletion to occur on a DST, but many field examples are available to prove that it can occur.

Another possibility, however, is that the recorded initial shut-in pressure may be higher than true shut-in reservoir pressure. This effect is called supercharge. Supercharge could be due to leak off of drilling fluid filtrate overpressuring the formation around the wellbore, or to compression of well fluid below the DST packer as it was set.

In some formations a short initial flowing period (1 to 3 minutes) is not sufficient to relieve the overpressured condition. Where this condition is suspected to exist, longer initial flowing periods (perhaps 20 minutes) should be used.

The important point is that the question of supercharge must be resolved before depletion can be diagnosed. A second DST is sometimes required to define depletion.

Reservoir Parameters (Gaseous System)

In DST's of gas zones, flow rate is usually reported in terms of cubic feet per day (or more conveniently, Mcfd) rather than barrels per day. This requires accounting for deviation of the reservoir gas from the Perfect Gas Law using the gas deviation factor (Z) and the absolute formation temperature (°R).

For the Horner buildup plot, the square of the formation pressure (p_{ws}) during the buildup is plotted versus $\left(\frac{t'_p + \Delta t'}{\Delta t'}\right)$ as shown in Figure 3-19.

Values of Z and μ for gas may be obtained from the literature knowing the specific gravity of the gas. Equations assume that compressibility and viscosity of gas remain reasonably constant over the range of temperature and pressure changes occurring during the flow period. Large pressure drawdown between the wellbore and the external boundary of the reservoir such as may occur in a low permeability zone may render this assumption invalid. If this is true, a bottom-hole choke should be used to reduce drawdown.

Equations for permeability, estimated wellbore damage and absolute open flow potential for a gas zone are:

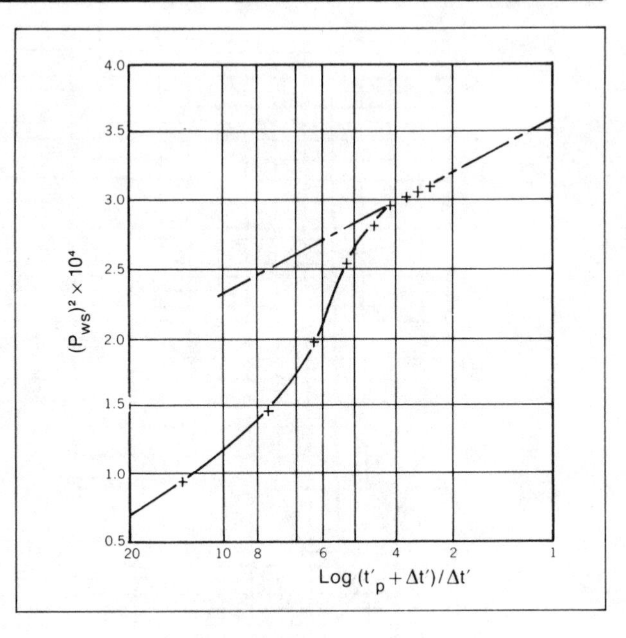

FIG. 3-19—Horner plot for gas well.

Permeability

$$k = \frac{1637\, q_g T_f \mu Z}{m_g h} \qquad (13)$$

Z = gas deviation factor
q_g = rate of flow mcf/d
T_f = formation temperature, °R = (°F + 460)
m_g = Horner buildup slope for gas well

Wellbore Damage (EDR)

$$\text{EDR} = \frac{p_i^2 - p_{ff}^2}{m_g (\log t'_p + 2.65)} \qquad (14)$$

Absolute Open Flow Potential (MCF/Day)

Using the single-point back-pressure test method:

$$\text{AOF} = \frac{q_g (p_i^2)^n}{(p_i^2 - p_{ff}^2)^n} \qquad (15)$$

Where:

n is an exponent varying between 0.5 and 1.0.

$$\text{If } n = 1.0: \quad \text{Max. AOF} = \frac{q_g p_i^2}{p_i^2 - p_{ff}^2} \qquad (16)$$

$$\text{If } n = 0.5: \quad \text{Min. AOF} = \frac{q_g p_i}{\sqrt{p_i^2 - p_{ff}^2}} \qquad (17)$$

Analysis of DST Data Using Type-Curve Methods

Several "type-curve methods" are available for analyzing early time data from pressure transient tests. Although these methods have perhaps greatest application in longer term tests, they can sometimes be used in DST analysis to salvage at least some information from a test where sufficient data is not available to obtain a "straight line" on the Horner plot. Ramey[11], McKinley[12], and Earlougher-Kersch[14] methods have application, with McKinley reportedly easier to use, but Earlougher-Kersch, and Ramey perhaps more accurate.[15]

The important point is that where the Horner plot can be made it should be used. In this case, type-curve methods may be helpful in picking the correct straight line on the Horner plot by indicating when wellbore storage or afterflow effects have ended.

RECOMMENDATIONS FOR OBTAINING GOOD TEST DATA

The key to DST evaluation is obtaining and recording good data. The DST must be planned to fit the specific situation.

Recording surface events, both character and time, is important to chart analysis. For example: What types and amounts of fluids were recovered? What were the characteristics of each fluid (salinity and perhaps resistivity of water, was it water cushion, mud filtrate or formation water, how much of each; API gravity and gas-oil ratio of recovered oil, etc.)? What size chokes were used? When were they changed? When did fluid come to the surface?

Time intervals allotted to each of the basic DST operations should ideally be adjusted during the test based on surface observations. The "closed chamber" method of analyzing the initial flowing period (discussed later) provides very early indications of formation fluid types and rates for use by the on-site supervisor in running the DST. Experience is a good teacher—past experience in the area should be studied in planning subsequent tests.

The Initial Flowing Period—This must be sufficient to relieve the effect of supercharge or overpressure in the formation immediately surrounding the wellbore. Normally 5 to 20 minutes is sufficient; however, longer times may be desirable in low productivity reservoirs in order to positively differentiate supercharge from depletion.

Some DST analysts prefer that the initial flowing period be extended to a time equal to the final flowing period, in order that the initial build-up curve can be analyzed the same as the final build-up curve. Thus, two sets of calculations are available for comparison. In this case, the static reservoir pressure probably cannot be actually measured by the pressure gauge, but can be obtained by extrapolation of the Horner plot.

Initial Shut-in Period—With no previous experience, the length of Initial Shut-In Period may be based on statistical studies as follows:

—With 30 minute shut-in, only 50% of tests reached static reservoir pressure.

—With 45 minute shut-in, 75% of tests reached static reservoir pressure.

—With 60 minute shut-in, 92% of tests reached static reservoir pressure.

Final Flow Period—This should be at least one hour of good to strong blow. The longer the flow period, the deeper the radius of investigation. If fluid reaches surface, additional time is desirable to obtain accurate volume gauges and gas-oil ratios. If blow quits, nothing is gained by continuing flow test.

Final Shut-In Period—This is the most important portion of the DST as far as formation evaluation is concerned. The length of shut-in should be based on events during the flow period.

—If formation fluid surfaces, FSI period should be one-half the flowing time (but never less than 30 minutes).

—With good to strong blow, FSI period should equal the flowing time (but never less than 45 minutes).

—With poor blow FSI period should be twice the flowing time (120 minutes if possible).

Closed Chamber DST Technique

This technique, in effect, permits a more scientific look at the bubbles in the "bubble bucket." It gives a good indication, in the first minutes of the initial flowing period, what fluid (gas, water, or perhaps oil) is entering the drill pipe, and an estimate of the fluid production rate.

To use the technique, a conventional downhole DST assembly is run. Conventional surface equipment is also used; however, we must be able to shut in the drill pipe at the surface, and we must be able to measure surface pressure with a reasonably accurate gauge. Figure 3-20 shows a suitable surface hook up.

In conducting the closed chamber test, the DST tool is opened conventionally, but as soon as bubbles appear in the bubble bucket, giving positive indications that the tool is open, the bubble hose is shut in. As formation fluids move into the drill pipe, surface pressure rises. This rate of pressure buildup is recorded for 5 to 10 minutes and provides the clues needed to estimate fluid type and entry rate. With gas entry, even at low rates, surface pressure buildup is quite rapid.

Figure 3-21 shows gas flow rate versus rate of surface pressure increase for various drill pipe internal capacities. With water entry, even at high rates, surface pressure buildup is quite slow.

Figure 3-22 compares surface pressure rise with water and gas entry at various rates. Entry of crude oil, due to gas breakout, usually shows a relatively slow initial pressure rise which increases more rapidly as gas breakout continues. Experience permits differentiation between oil and water.

With a properly run closed chamber technique,

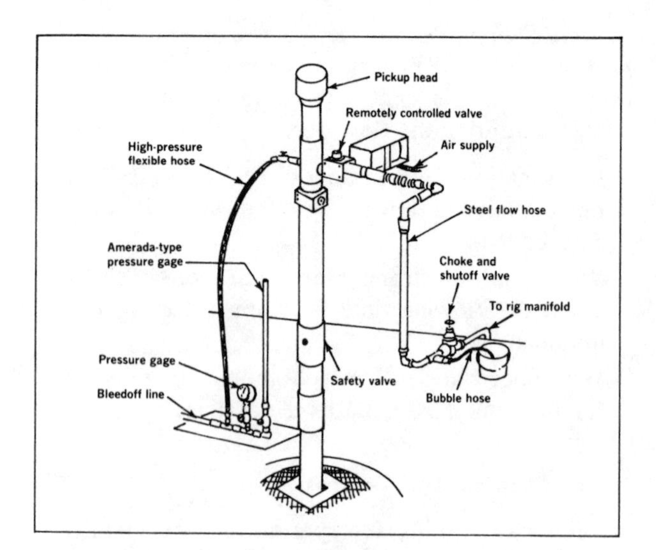

FIG. *3-20—Surface equipment for closed-chamber testing.*[16] *Permission to publish by The Society of Petroleum Engineers.*

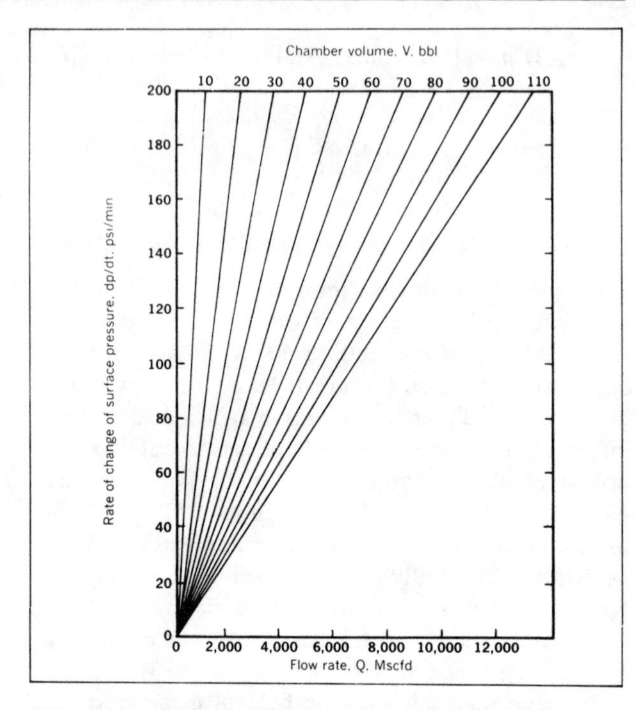

FIG. *3-21—Rate of change of surface pressure vs flow rate for various chamber volumes.*[16] *Permission to publish by The Society of Petroleum Engineers.*

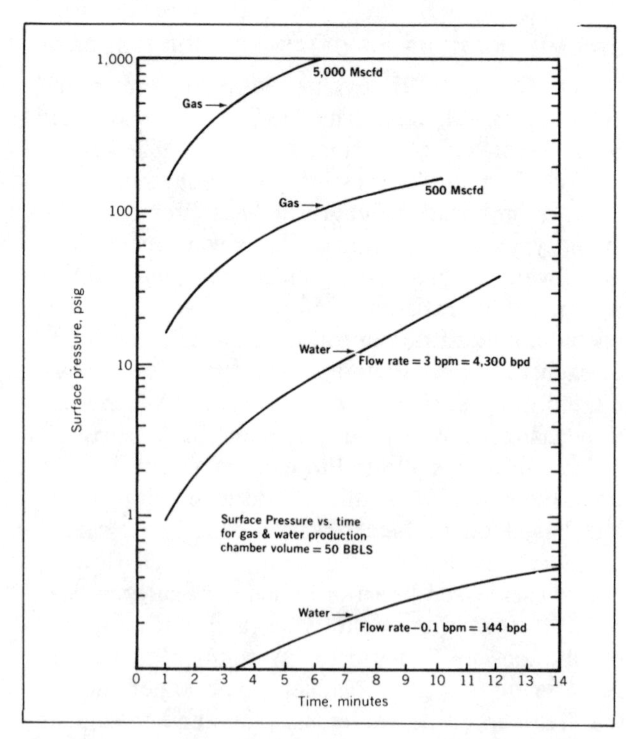

FIG. *3-22—Surface pressure rise with water and gas entry at various rates.*[16] *Permission to publish by The Society of Petroleum Engineers.*

the well site supervisor should reasonably know fluid types and rates in the first 5 to 10 minutes of the DST, such that he can perhaps terminate the DST at that point—or adjust subsequent flow and shut-in periods to maximize the usefulness of the data obtained.

DST's in High Capacity Wells

Meaningful interpretation of formation characteristics in high capacity zones is difficult with conventional DST tools, since flow restrictions within the tools prevent flowing the zones at a high enough rate to provide sufficient drawdown.

Where this is a problem, DST assemblies are available, which provide straight through flow patterns with a minimum flow diameter of $2^1/_4$ inches. Where desirable, perforating guns can be run through the DST assembly. Stimulation can also be performed below these DST tools.

DST's from Floating Vessels

The incentive for effective formation evaluation of offshore exploratory wells is obvious and DST's from floating vessels are becoming more and more common.

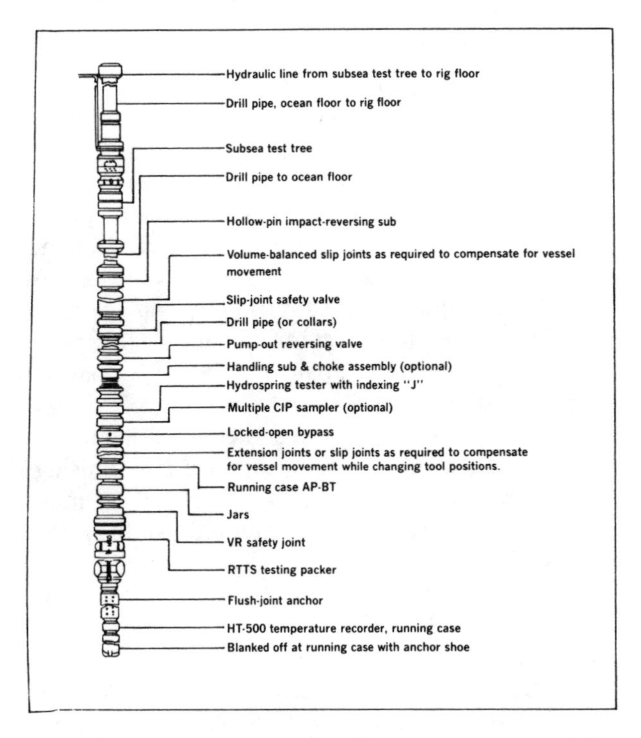

FIG. *3-24—Floating vessel test string operated by drill-pipe reciprocation.*

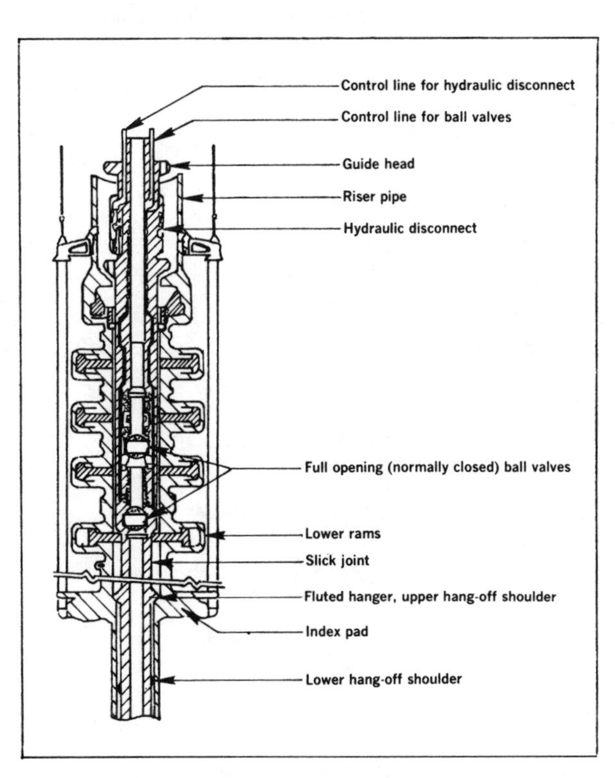

FIG. *3-23—Subsea test tree.*

Safety considerations are paramount in conducting DST's from a floating vessel. The subsurface test tree (Figure 3-23) was the first step in this regard. Typically, it provides a fail-safe means of shutting in the well at the ocean floor, and a hydraulic disconnect to allow removal of the section of drill pipe above the ocean floor BOP, if necessary to move the vessel with the DST tool on bottom.

Two systems of DST tool operation have been developed. In the first system, operation of the tools is strictly by reciprocation of the drill pipe, Figure 3-24. Tools are similar to those of a land operation, except that slip joints (also serving as a shut-in valve when closed) are required to compensate for vessel motion. Drill collars usually provide the string weight below the slip joints to keep the DST packer set.

In the second, opening and closing of the tester valve is controlled by annulus pressure (annular pressure responsive). Thus, all pipe movement is eliminated during the test, from the time the packer is set and the subsea test tree is spaced out, until the packer is pulled loose after the test is finished.

Wireline Formation Tester

The wireline formation tester is essentially a logging type device, which permits confirmation of formation fluid, indications of productivity and formation pressure. A recent tool design essentially consists of a packer, which can be forced against the wall of the borehole to isolate the mud column, a hydraulic piston used to create pressure drawdown, and two sample chambers (2¾ gal capacity is typically used) to collect formation fluid samples.

On one trip in the hole, any number of formation pressure measurements can be made in different zones, using the piston device to create drawdown, and two fluid samples can be obtained in promising zones. Thus, a rapid method of "looking" at multiple zones is available to aid and confirm log analysis.

The Schlumberger RFT tool is shown schematically in Figure 3-25. The pretest chambers provide an indication of whether or not a packer seal has been obtained and if so, an estimate of flow rate from the zone. Pressures are recorded during flow and buildup.

In a high permeability zone, buildup occurs very rapidly as shown in Figure 3-26.

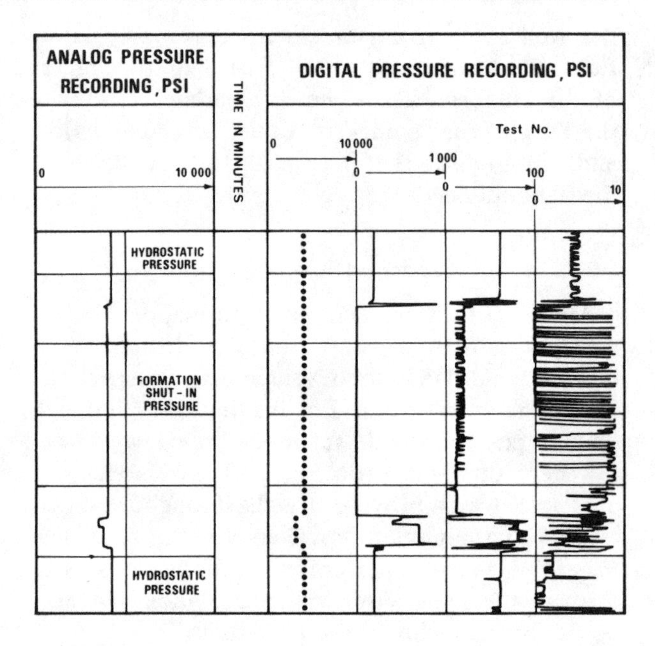

FIG. *3-26—High-permeability zone. Courtesy of Schlumberger.*

Figure 3-27 shows analog and digital pressure recording of a test in a low permeability carbonate zone. When the pretest chamber was opened, pressure dropped from hydrostatic of 4426 psi to a flow

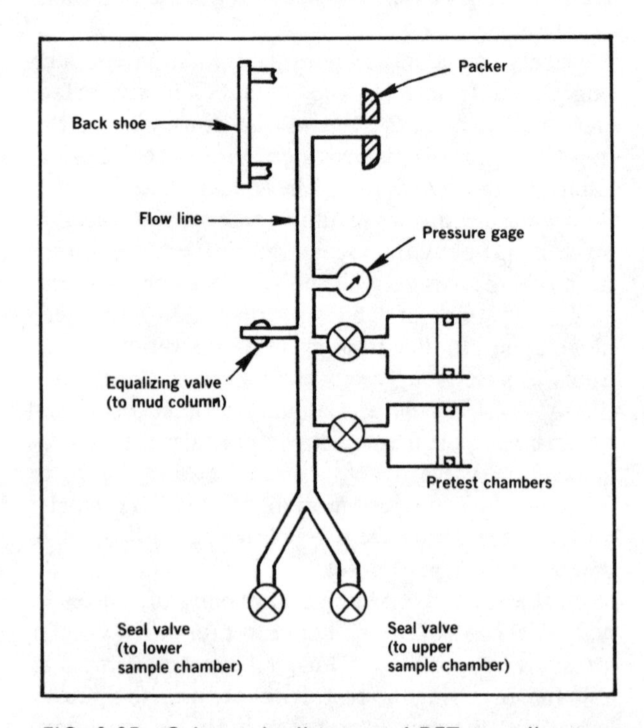

FIG. *3-25—Schematic diagram of RFT sampling system. Courtesy of Schlumberger.*

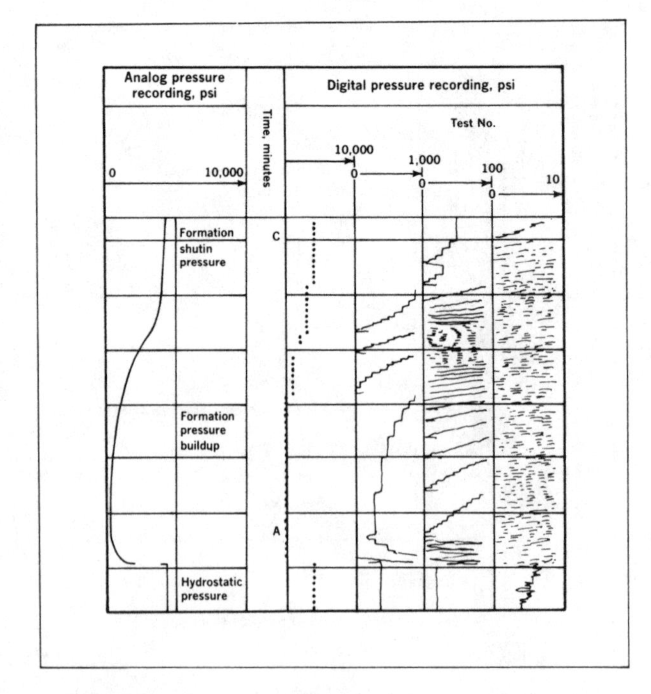

FIG. *3-27—Low-permeability zone. Courtesy of Schlumberger.*

pressure of 290 psi (point A). Buildup began immediately and reached a nearly stabilized maximum of 4060 psi (point C) 5½ minutes later. Total time required to test this zone was 8 minutes.

"EYEBALL" INTERPRETATION OF PRESSURE CHARTS

Present day DST tools provide very accurate pressure vs time data for use in evaluating a formation with the Horner calculation procedures. The first and the most important step before calculations, however, is a careful examination of the pressure charts. Preliminary checks should be made to compare pressures and chart indications with known data:

—Flowing pressures should be indicative of drill pipe fillup. Final flow pressure should equal the hydrostatic head of recovery in the drill pipe.

—Initial flow pressure should reflect the use or the absence of a water cushion.

—Possibility of mechanical malfunctions should be investigated by careful study of chart indications. Tie down the cause of all the important "wiggles."

—Compare indications of top and bottom charts to show clock and tool plugging problems.

Even where calculation procedures are not possible, there may be significant information which can be interpreted by "Eyeball" methods.

The actual DST charts on the following pages show examples of DST problems[8] which restrict calculation possibilities. Included also are example situations which can be reasonably interpreted by "eyeball" methods.

Base Line Drawn Incorrectly—The base line is the basis for all pressure measurements on a formation test chart. Figure 3-28 shows the base line inconsistent with the initial pressures of the test.

Leaking Drill Pipe—At points of delay, while going into the hole or at total depth, a decrease in recorded pressure is indicative of a loss in the hydrostatic head. There are two possibilities:

—The hole may be taking fluid.

—The drill pipe may be leaking, Figure 3-29.

Stair-Stepping Gauge—A "stair-stepping" appearance in a buildup curve Figure 3-30, may be caused by:

—The chart drum lugs and/or inner case runners may be dirty.

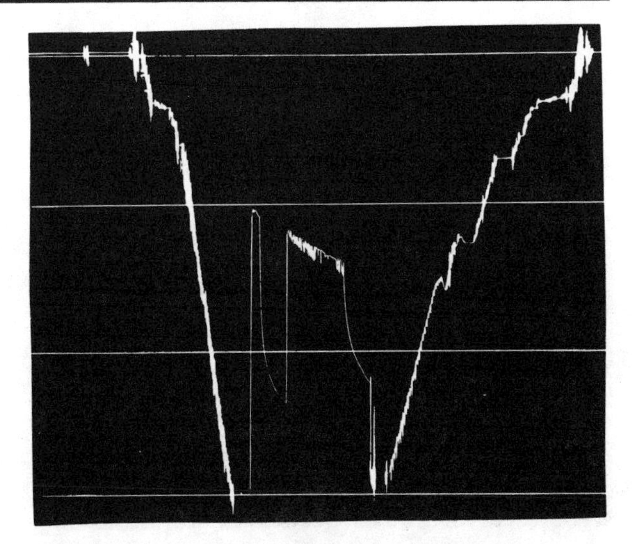

FIG. 3-28—*Base line drawn incorrectly.*[8]

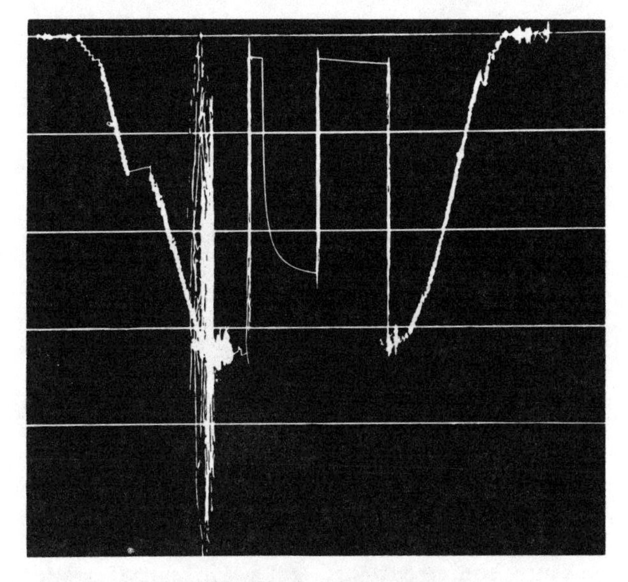

FIG. 3-29—*Leak in drill pipe.*[8]

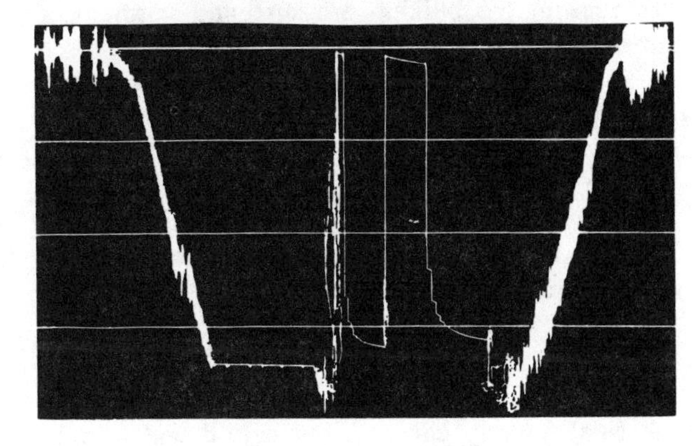

FIG. 3-30—*Stair-stepping.*[8]

—The lead screw may not be straight.
—The inner case or cover may be crooked or rough.

Clock Failure—This problem is characterized by time discontinuance, as shown in Figure 3-31.

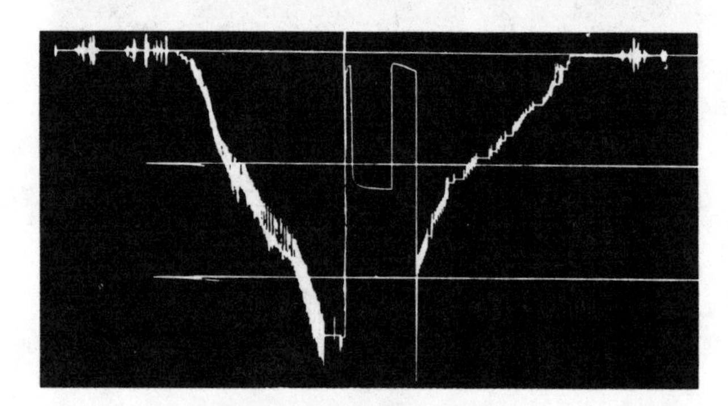

FIG. *3-31—Clock failure.*[8]

Clock Running Away—A clutch spring malfunction may cause the clock to "run away;" this occurrence is usually caused by excessive rough treatment and does not damage the clock, Figure 3-32.

FIG. *3-32—Clock running away.*[8]

Leaking Closed-In Pressure Valve—Closed-in pressure equipment that leaks will cause erratic buildup curves. Leakage is characterized by a decrease in the buildup pressure and normally a subsequent rise, Figure 3-33.

False Buildup—A common point of great concern in a dual closed-in operation of formation testing is the frequent difference in the initial and final buildup pressures. This may be attributed to a short first flow period resulting in a higher buildup. Supercharge is the most common condition resulting in an abnormally high initial buildup of pressure. This is shown in Figures 3-34 and 3-35.

Limited Reservoir—A depleting reservoir will also exhibit a high initial buildup curve compared to the final buildup curve. Flow period will usually indicate a decreasing rate of flow. The difference

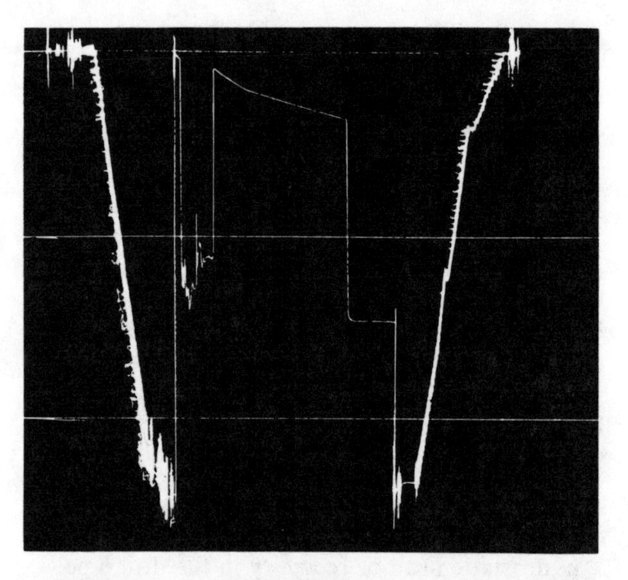

FIG. *3-33—Leaking closed-in pressure valve.*[8]

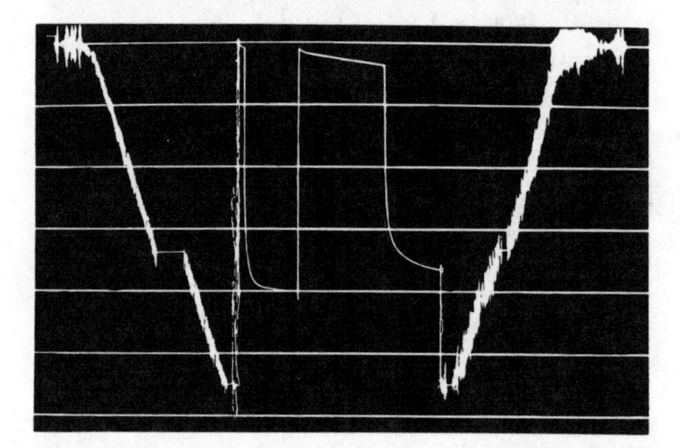

FIG. *3-34—Chart indicating supercharge.*[8]

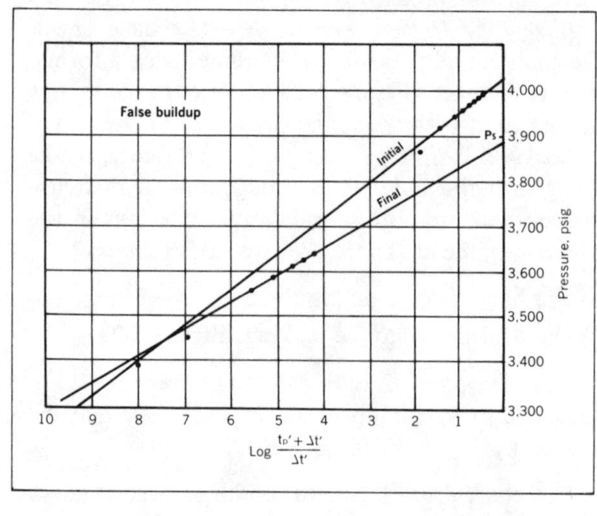

FIG. *3-35—False buildup, due to supercharging.*[8]

in the extrapolated static reservoir pressures of the initial and final buildup curves indicates the loss in reservoir pressure during the second flow period. These conditions are indicated in Figures 3-36 and 3-37 for a depleting liquid and gas reservoir.

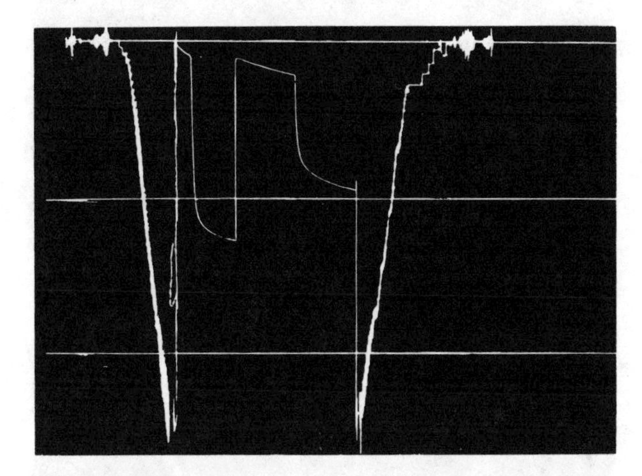

FIG. 3-36—*Depleting liquid reservoir.*[8]

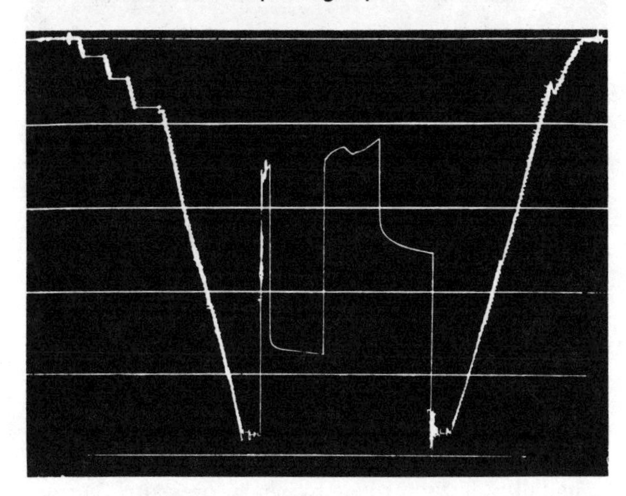

FIG. 3-37—*Depleting gas reservoir.*[8]

Barrier—A barrier may be detected in the plot of the buildup curve of a formation test. The final buildup curve may have a peculiar appearance as compared to the initial as shown in Figures 3-38 and 3-39.

Formation Damage (Liquid Zone)—Formation damage in an oil zone is usually indicated by:

—The very sharp rise after shut-in.

—A short radius curve.

—A reasonably flat slope.

—A high differential pressure between closed-in and final flow pressure.

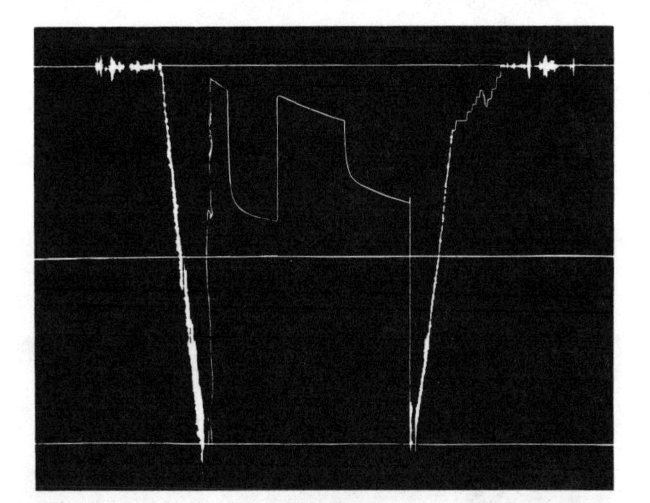

FIG. 3-38—*Barrier within radius of investigation.*[8]

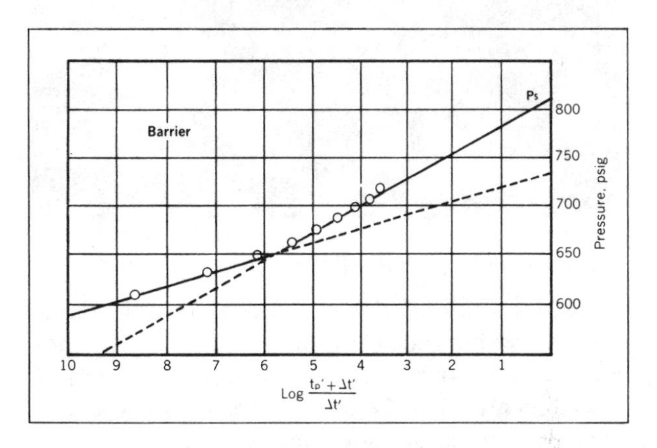

FIG. 3-39—*Extrapolation of a buildup curve, indicating a barrier.*[8]

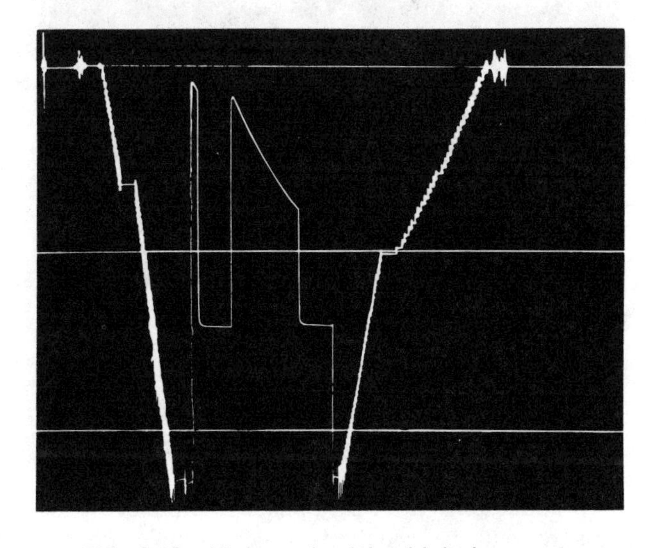

FIG. 3-40—*High productivity, high damage.*[8]

Figure 3-40 shows high productivity, high damage.

Formation Damage (Gas Zone)—High-capacity gas wells exhibit damage characteristics due to the back-pressure through the bottomhole choke, therefore, damage may be present in a high or low production formation, as shown in Figure 3-41.

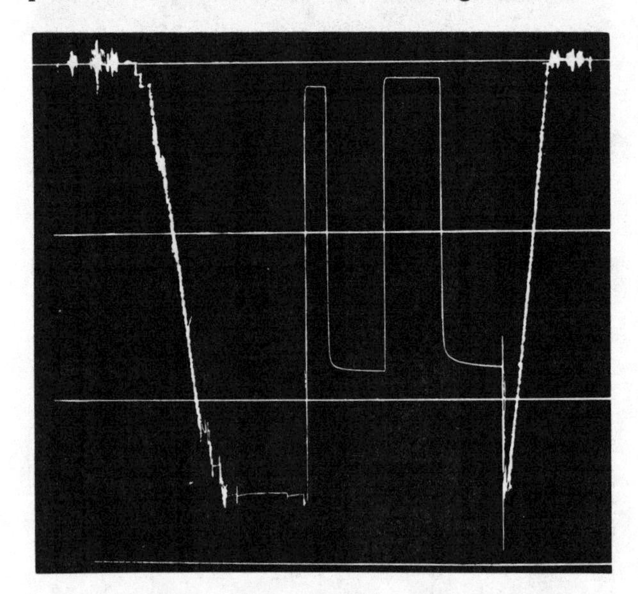

FIG. 3-41—Low productivity, high damage.[8]

Cleanup of Formation Damage—The initial buildup may exhibit damage; however, damage may no longer be present after the final flow period. This indicates that the damage was removed during the flow period, as shown in Figure 3-42.

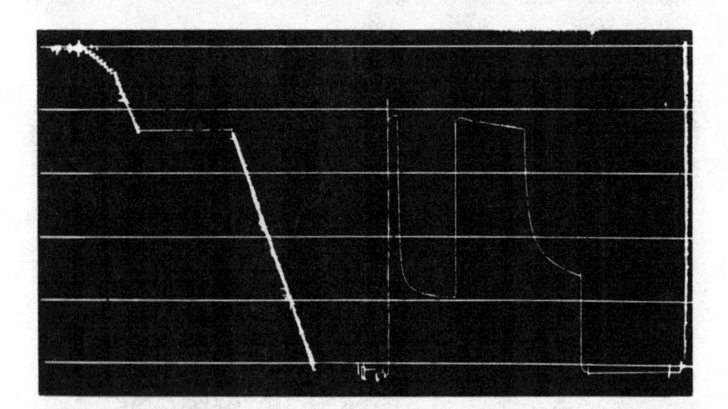

FIG. 3-42—Formation damage indicated on initial buildup; no damage indicated on final buildup.[8]

Buildup Closure Time—Approximately 75% closure must be attained for an accurate extrapolation plot of a buildup curve. This figure will vary with well conditions. If a single phase is produced, less closure will be required while multiple phases usually require more closure. Figures 3-43 and 3-44

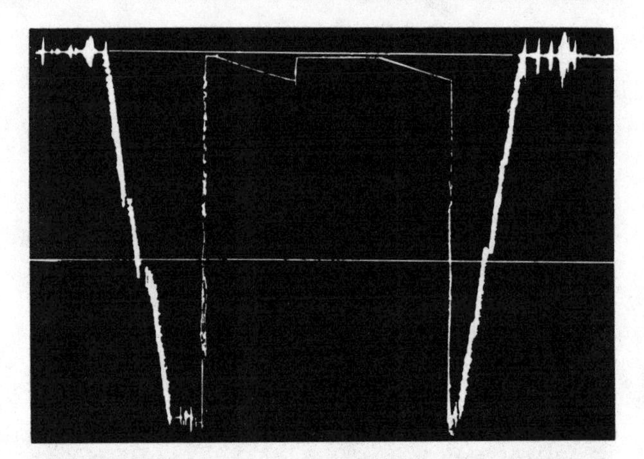

FIG. 3-43—Low-permeability formation with a low reservoir pressure.[8]

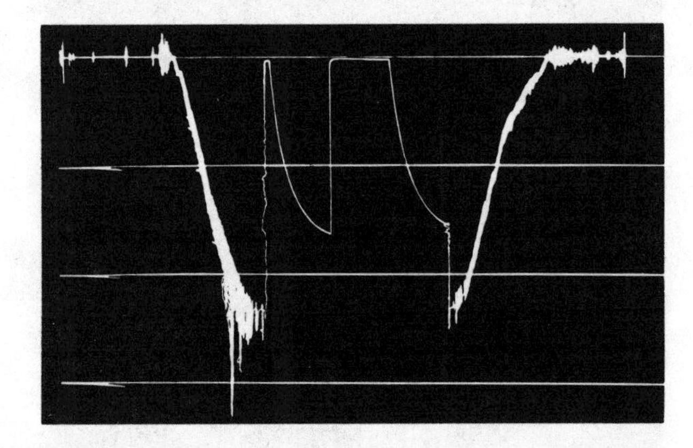

FIG. 3-44—Low-permeability formation with a high reservoir pressure.[8]

show low permeability and varying bottom hole pressure.

Plugging in Test Tool—Plugging, one of the most common mechanical problems in formation testing, is usually characterized by sharp pressure fluctuations, if the plugging alternately plugs and frees itself. However, when the plugging is sustained momentarily the action will appear as small buildup curves, as shown in Figure 3-45.

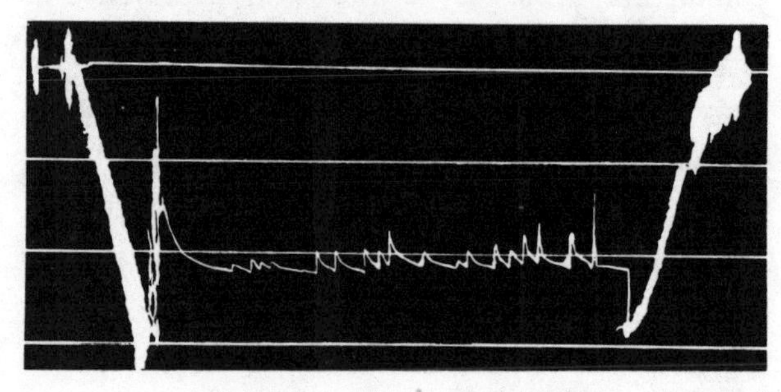

FIG. 3-45—Plugging-in tool.[8]

Plugging in the Flow Perforation—Plugging flow perforations are evident when Figure 3-46 indicates little or no change in flowing pressures, while the blanked-off chart, Figure 3-47, indicates an increased pressure.

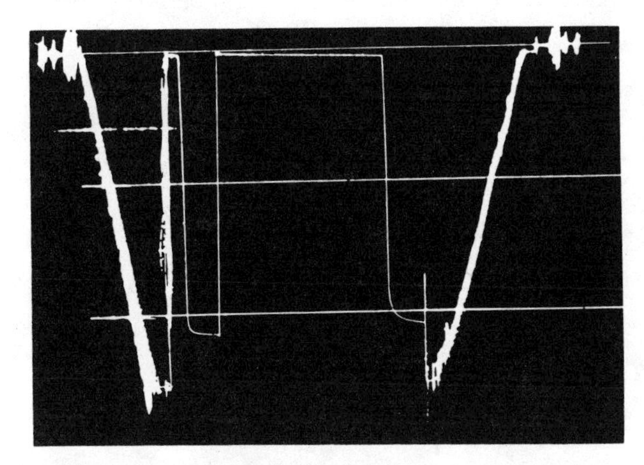

FIG. *3-46—Top gauge—plugging in the flow perforations.*[8]

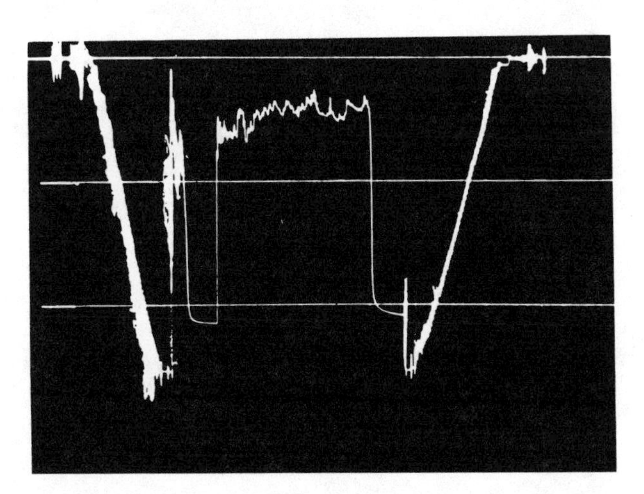

FIG. *3-47—Bottom gauge—plugging in the flow perforations.*[8]

Changes in Flow Rate—An abrupt change of rate in the flow period can usually be attributed to a change in the ID of the pipe. Ordinarily this change to a slower rate will indicate when the fill-up has cleared the drill collars, Figure 3-48.

Swabbing—Some wells are swabbed during a formation test. This action will cause a drawdown in the flow period outlined by a decreasing sequence of small "inverted fishhooks" as shown in Figure 3-49. A subsequent flow and swabbing action may very often be noted.

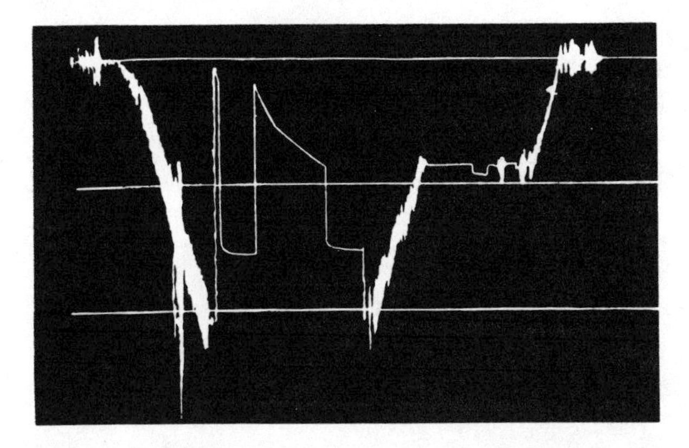

FIG. *3-48—Fillup transition from drill collar to drill pipe.*[8]

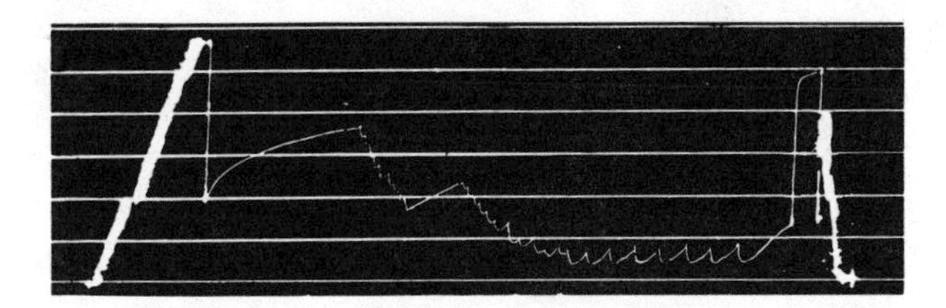

FIG. *3-49—Swabbing during a formation test.*[8]

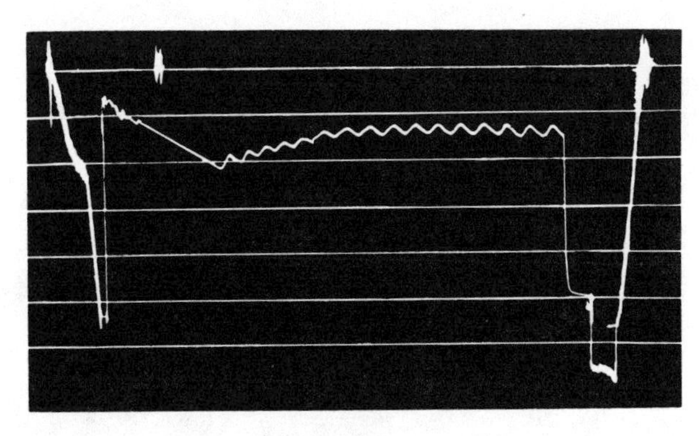

FIG. *3-50—Flowing in heads.*[8]

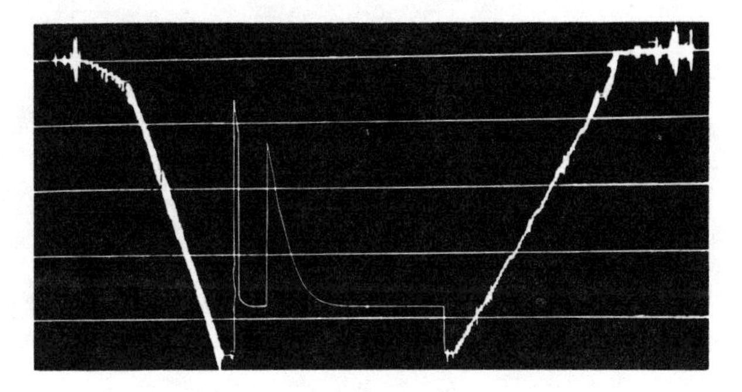

FIG. *3-51—Equalized flow.*[8]

Flowing In Heads—The flow pressures of a single phase flow are usually very uniform; however, a combination of phases will result in an irregular pattern. If gas is breaking through the fluid, this pattern may be very "lumpy" in appearance. A uniform "rippling" appearance occurs when a well is flowing in heads, as shown in Figure 3-50.

Equalized Flow—The second flow period should be of such duration as to establish a uniform rate of flow. A longer flow period can spoil the buildup curve by allowing the well to equalize before attempting a closed-in pressure, Figure 3-51.

REFERENCES

1. Horner, D. R.: "Pressure Buildup in Wells," Proc. Third World Pet. Congr., Sect. II, E. J. Brill, Leiden, Holland (1951).

2. van Everdingen, A. F.: "The Skin Effect and Its Influence on the Productive Capacity of a Well," *Trans.*, AIME (1953) Vol. 198, p. 171.

3. Miller, Dyes & Hutchinson: "The Estimation of Permeability and Reservoir Pressure From Bottom Hole Pressure Buildup Characteristics," *Trans.*, AIME, (1950) Vol. 189, pp. 91–104.

4. Zak, A. J., Jr., and Griffin, P., III: "Here's a Method for Evaluation DST Data," *Oil and Gas J.* (April 1957).

5. van Poollen, H. K.: "Radius-of-Drainage and Stabilization-time Equations," *Oil and Gas J.* (Sept. 14, 1964) Vol. 62, No. 37.

6. Maier, L. F.: "Recent Developments in the Interpretation and Application of DST Data," *JPT* (Nov. 1962) p. 1213.

7. McAlister, J. A.; Nutter, B. P.; and Lebourg, M.: "Multi-Flow Evaluator Better Control and Interpretation of Drill Stem Testing," *JPT* (Feb. 1965) p. 207.

8. Murphy, W. C.: "The Interpretation and Calculation of Formation Characteristics from Formation Test Data," Hallburton Services.

9. Johnston Testers, "Review of Basic Formation Evaluation."

10. Matthews, C. S., and Russell, D. G.: "Pressure Buildup and Flow Tests in Wells, Monograph No. 1, H. L. Doherty Series, AIME (1967).

11. Ramey, H. J., Jr.: "Short-Time Well Test Data Interpretation in the Presence of Skin Effect and Wellbore Storage," *JPT* (Jan. 1970) p. 97.

12. Milner, E. E., and Warren, D. A., Jr.: "Drill Stem Test Analysis Utilizing McKinley System of After-Flow Dominated Pressure Buildup," SPE 4123, Annual Meeting, San Antonio, TX (Oct. 1972).

13. Edwards, A. G., and Shryock, S. H.: "A Summary of Modern Tools and Techniques Used in Drill Stem Testing," Halliburton Services (Sept. 1973).

14. Earlougher, R. C., Jr., and Kersch, K. M.: "Analysis of Short-Time Transient Test Data by Type-Curve Matching," *JPT* (July, 1974) p. 793.

15. Sinha, B. K.; Sigmon, J. E.; and Montgomery, J. M.: "Comprehensive Analysis of Drill-Stem Test Data With the Aid of Type Curves," SPE 6054, Fall Meeting, New Orleans (Oct. 1976).

16. Alexander, L. G.: "Theory and Practice of the Closed-Chamber Drill-Stem Test Method," SPE 6024, Fall Meeting, New Orleans (Oct. 1976).

17. Erdle, James C.: "Current Drillstem Testing Practices: Design, Conduct and Interpretation," SPE 13182, (Sept. 1984).

Chapter 4 Primary Cementing

INTRODUCTION

Probably the single most important factor in the well completion operation is obtaining a satisfactory primary cementing job. An effective primary cement job is the necessary starting point for all subsequent operations. With a defective primary cement job all remaining operations are adversely affected. While primary cementing is often the responsibility of the drilling group, it is the completion, production, and workover groups who are most affected by, and perhaps should be most interested in, primary cementing.

Figure 4-1 shows the usual procedure for placing cement and details much of the equipment needed to facilitate a typical cementing operation. This section is primarily concerned with cements and cementing practices, and less concerned with casing and cementing equipment.

CEMENTING MATERIALS

Function of Cement in Wells

In well completion operations cements are almost universally used to fill the annular space between casing and open hole. Two principal functions of primary cement are:

—To restrict fluid movement between formations.
—To support the casing.

Cement materials properly placed around the casing, having permeabilities less than 0.1 md, and compressive strengths greater than 100 to 300 psi should be satisfactory for these functions.

The key to success is proper placement of the cement completely around the casing effectively displacing all of the annular drilling mud.

Cementing compositions can be broadly classified as neat or tailored. Properties of neat cement are relatively inflexible and tailored cements are almost always used both to reduce the cost of the cementing operation, and to improve the capability of cementing materials as well requirements become more rigorous.

Manufacture, Composition, and Characteristics of Cement

Portland cements are finely-ground mixtures of calcium compounds. They are made from limestone (or other high calcium carbonate materials) and clay or shale. Some iron and aluminum oxides may be added if necessary. These materials are finely ground and mixed, then heated to 2600°–2800°F. in a rotary kiln. The resulting clinker is ground with a controlled amount of gypsum to form cement.

Typical cement particle size distribution is such that 85% passes a 325-mesh screen (44 microns), 90% passes a 200-mesh screen (74 microns), and 100% passes a 150-mesh screen (100 microns).

Principal compounds formed in the burning process and their functions are:

Tricalcium silicate (C_3S) is the major compound in most cement and is the principal strength-producing material. It is responsible for early strength (1 to 28 days).

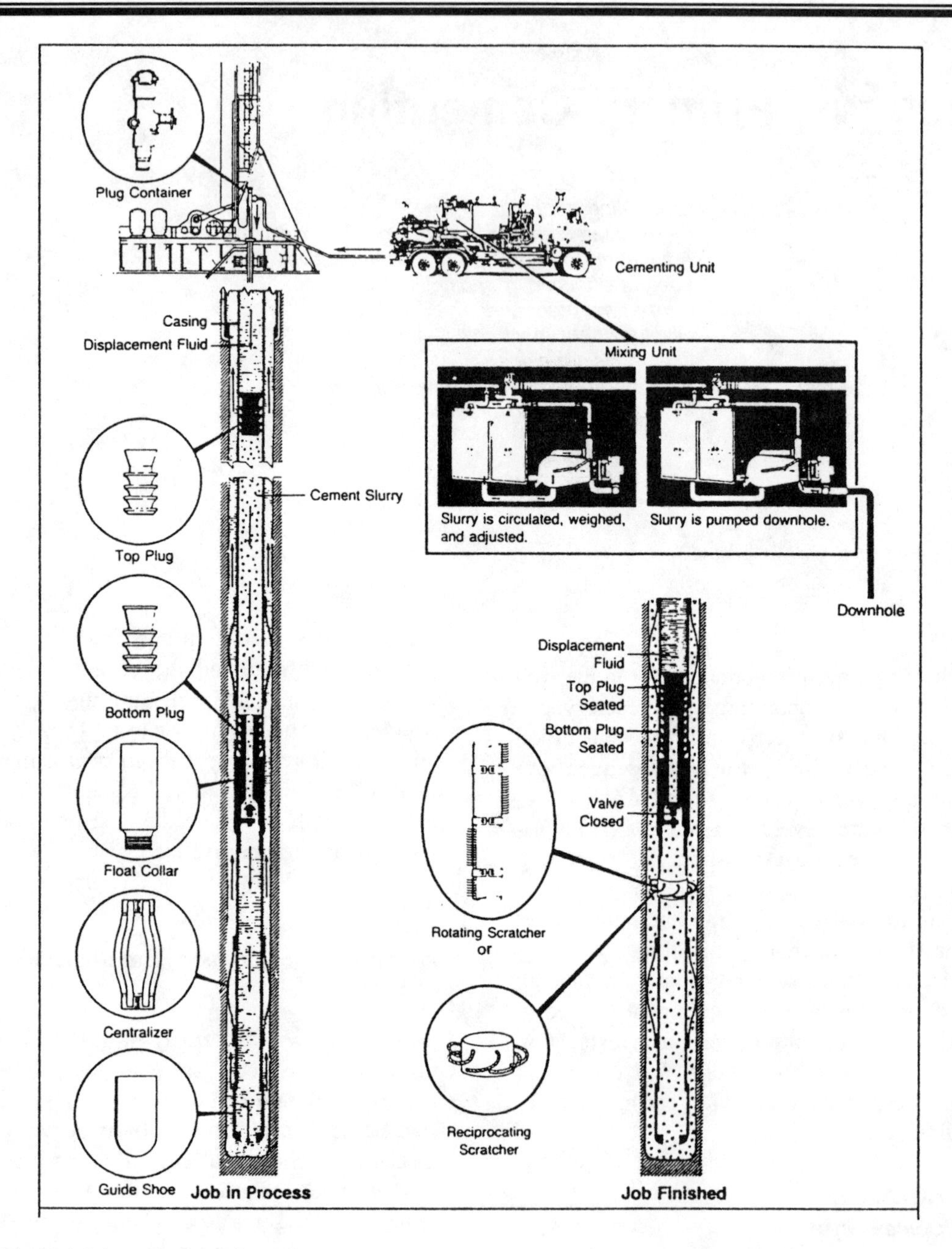

FIG. 4-1—*Typical primary cementing job.*

Dicalcium silicate (C_2S) is the slow hydrating compound and accounts for the gradual gain in strength which occurs over an extended period.

Tricalcium aluminate (C_3A) promotes rapid hydration and controls the initial set and thickening time. It affects the susceptibility of cement to sulfate attack; high sulfate resistant cement must have 3% or less C_3A.

Tetracalcium aluminoferrite (C_4AF) is the low-heat-of hydration compound in cement. It gives color to the cement. An excess of iron oxide will increase the amount of C_4AF and decrease the amount of C_3A in the cement.

All API classes of Portland cements are manufactured in essentially the same way from the same ingredients, but proportions and particle size are ad-

TABLE 4-1
Typical Composition of Portland Cement

API Class	Compounds					Fineness, Sq cm/gram
	C_3S	C_2S	C_3A	C_4AF	$CaSO_4$	
A	53	24	8	8	3.5	1600-1900
B	47	32	3	12	2.9	1500-1900
C	58	16	8	8	4.1	2000-2400
G	52	32	3	12	3.2	1400-1600
H	52	32	3	12	3.3	1200-1400

Several other cementacious compositions are used to meet special requirements.

justed to give the desired properties. The water requirement of each type of cement varies with the fineness of grind or surface area. High early strength cements have high surface area (fine grind). Table 4-1 compares typical compositions and grinds of API classes of Portland cement.

Cement sets through a crystal growth process. Figure 4-2 shows a typical arrangement of C_2S, C_3S, C_3A and C_4AF crystals in set cement. During the setting process disturbances such as gas bubbles percolating through a cement column may leave "worm holes" for subsequent communication of low-viscosity fluids.

Once the cement is in place around the casing, it is important that this crystal growth process proceed as quickly as possible to reduce exposure time to "disturbance mechanisms."

Pozmix cement combines Portland cement with pozzolan. Pozmix cement consists of Portland cement, a pozzolanic material and about 2% bentonite. By definition a pozzolan is a siliceous material which reacts with lime and water to form calcium silicates having cementitious properties. Since Portland cement releases about 15% free lime when it reacts with water, the addition of pozzolan reacts with this free lime to form a more durable mass of calcium silicates. The pozmix composition is less expensive than other basic cementing materials because more mix water is used per weight of material.

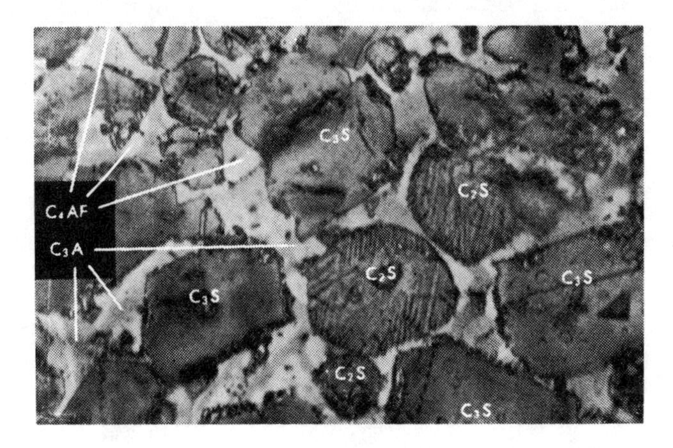

FIG. 4-2—*Typical arrangement of crystals in set cement.*

Calcium aluminate cements or Refractory cements (trade names: Cement Fondu, Luminite Cement) are manufactured by heating bauxite and limestone until liquefied, then cooling and grinding. High-alumina cements are used to cement casing through the hot zone in in-situ combustion wells, where temperatures may range from 750° to 2000°F. as the fire front passes. High-alumina cements resist attack by sulfates—and their quick setting characteristics sometimes recommend their use where formation temperatures are low. Limitations include high cost, and questionable long term strength.

Gypsum cements, usually a hemihydrate form of gypsum ($CaSO_4 \cdot 1/2H_2O$), set very rapidly, expand significantly (0.3%) on setting, but tend to deteriorate in contact with water. They are not used very often, except in connection with Portland cement.

Permafrost cements, a blend of gypsum cement with Portland cement, have low heat of hydration, and will set at 15°F before freezing. They are used in cementing through frozen formations in Arctic areas.

Selection of Cement For Specific Well Application

The problem of selecting a cementing material for a specific well application is one of designing an economical slurry that:

1. Can be placed effectively with the equipment available.

2. Will achieve satisfactory compressive strength soon after placement.

3. Will thereafter retain the properties necessary to isolate zones and to support and protect the casing.

Most API cements have been replaced by Class G or H (Basic) cement. API classification of cements for various depth and temperature conditions, with recommended mixing water quantity and resulting slurry weight, are shown in Table 4-2.

API Class A and B cements are more economical than premium types. Class B is resistant to sulfate attack. Class B per se is in limited use, being replaced by high sulfate resistant Class C, G, or H.

API Class C (High Early) cement has higher compressive strength than Class A in the first 30 hours. However, Class A with calcium chloride will give better early strengths than Class C without accelerators. Class C is available only in limited areas in the U.S.A.

API Class G & H (Basic) cements are similar to Class B, but are manufactured to more rigorous chemical and physical specifications resulting in a more uniform product. They contain no accelerators, retarders, or viscosity control agents. Class H is similar to Class G except it has slightly coarser grind providing slightly greater retarding effect for deeper, hotter conditions. Class G or H cements are compatible with accelerators or retarders for use over the complete range of API conditions; thus, with proper additives, Class G or H can be used in any cementing situation.

Worldwide, Class G with additives (if needed) is replacing most other API cements.

Mix water quantity affects slurry. Mix water quantities shown in Table 4-2 are selected to provide pumpable viscosity and minimum free water (less than 1%). The effects of reducing mix water quantities are:

—Increased slurry density and compressive strength.
—Increased slurry viscosity.
—Decreased pumpability time.
—Decreased slurry volume per sack of cement.

Increasing mix water quantities has just the opposite effect. Well conditions sometimes dictate variation in mix water quantities to obtain desired properties. If mix water proportions are increased above recommendations, bentonite (2 to 4%) should be used to "tie up" resulting free water.

Early compressive strength is affected by curing conditions. Compressive strength of neat cement increases over a period of several weeks. The compressive strength at any particular time during the curing process depends heavily on mixing and curing conditions. Typical values after 24 hours are shown in Table 4-3.

The pumpability time, or thickening time of a cement slurry, is an important factor in slurry design. Generally these statements can be made regarding thickening time:

—Higher the temperature—faster the set. This is a primary factor affecting thickening time.
—Higher the pressure—faster the set. This effect is more pronounced below 5000 psi, but continues up through limits of lab test equipment.
—Loss of water from slurry accelerates set.
—Shutdown during placement results in cement gelation which will shorten pumpability and accelerate set and strength development.

The effect of temperature on thickening time of various cement types is shown typically in Table 4-4.

The effect of extreme pressure on the thickening time of a particular cement slurry is shown in Table 4-5.

Lab tests must simulate placement conditions. In the laboratory, thickening time (elapsed time between

TABLE 4-2
API Cement Classification

API Class	*Mixing water gal/sk*	*Slurry weight, lb/gal*	*Well depth, ft*	*Static temperature, °F.*
A	5.2	15.6	0–6,000	80–170
B	5.2	15.6	0–6,000	80–170
C	6.3	14.8	0–6,000	80–170
G	5.0	15.8	0–8,000	80–200
H	4.3	16.3	0–8,000	80–200

TABLE 4-3
Typical Compressive Strength—psi @ 24 Hours

Curing Conditions temp.	press.	Class A	Class C	Class G	Class H
60°F.	0 psi	615	780	440	325
80°F.	0 psi	1,470	1,870	1,185	1,065
95°F.	800 psi	2,085	2,015	2,540	2,110
110°F.	1,600 psi	2,925	2,705	2,915	2,525
140°F.	3,000 psi	5,050	3,560	4,200	3,160
170°F.	3,000 psi	5,920	3,710	4,830	4,485

(a)—Not recommended at this temperature

TABLE 4-4
High-Pressure Thickening Time (Hours:Minutes)

Circulating temperature	Class A	Class C	Class G	Class H
91°F	4:00+	4:00+	3:00+	3:57
103°F	3:36	3:10	2:30	3:20
113°F	2:25	2:06	2:10	1:57
125°F	1:40	1:37	1:44	1:40

application of pressure and temperature and the attainment of a slurry viscosity of 100Bc) is measured using the Thickening Time Tester according to recommended practices of API Spec 10.

With no shut-down or dehydration, thickening time of a particular slurry depends on formation temperature gradient, depth, displacement rate, and hydrostatic and circulating pressures: i.e., the temperature-pressure history of the cement slurry as it is placed. API Spec 10 Well Simulation Schedules present guidelines for temperature-pressure history vs. depth and type of job (casing, squeeze, liner).

Schedule 7g, 12,000 feet Casing-Cementing Well-Simulation Test is shown in Table 4-6.

Thickening time values reported by Service Companies are often based on modified test schedules to fit individual well conditions rather than on API Spec 10 schedules.

TABLE 4-5
Effect of Pressure on Thickening Time
(Class H Cement with Retarder)

Depth, ft	Temperature °F. static	Temperature °F. circulating	Pressure, psi	Thickening time, hours: minutes
10,000	230	144	5,000	2:10
			10,000	1:34
			15,000	1:18
14,000	290	200	10,000	5:50
			15,000	4:30
			20,000	3:20
16,000	320	248	10,000	4:11
			15,000	3:39
			20,000	2:30
			25,000	2:08

TABLE 4-6
Casing-Cementing Well-Simulation Tests

Schedule 7g

Depth 12,000 ft (3660 m) **Mud density: 14 lb/gal (1.7 kg/L)**

1	2		3		4		5		6		7		8	
			Temperature Gradient, F/100 ft depth (°C/100 m depth) Temperature F (°C)											
	Pressure		*0.9 (1.6)*		*1.1 (2.0)*		*1.3 (2.4)*		*1.5 (2.7)*		*1.7 (3.1)*		*1.9 (3.5)*	
Time, min	*psi*	*(kPa)*	*F*	*(°C)*	*F*	*(°C)*	*F*	*(°C)*	*F*	*(°C)*	*F*	*(°C)*	*F*	*(°C)*
0	1,500	(10300)	80	(27)	80	(27)	80	(27)	80	(27)	80	(27)	80	(27)
2	1,900	(13100)	84	(29)	84	(29)	85	(29)	85	(29)	86	(30)	88	(31)
4	2,300	(15900)	87	(31)	88	(31)	89	(32)	91	(33)	93	(34)	96	(36)
6	2,700	(18600)	91	(33)	92	(33)	94	(34)	96	(36)	99	(37)	105	(41)
8	3100	(21400)	94	(34)	95	(35)	99	(37)	102	(39)	106	(41)	113	(45)
10	3,500	(24100)	98	(37)	99	(37)	103	(39)	107	(42)	112	(44)	121	(49)
12	3,900	(26900)	101	(38)	103	(39)	108	(42)	113	(45)	118	(48)	129	(54)
14	4,300	(29600)	105	(41)	107	(42)	113	(45)	118	(48)	125	(52)	137	(58)
16	4,700	(32400)	108	(42)	111	(44)	117	(47)	124	(51)	131	(55)	145	(63)
18	5,100	(35200)	112	(44)	115	(46)	122	(50)	129	(54)	138	(59)	154	(68)
20	5,500	(37900)	115	(46)	119	(48)	127	(53)	135	(57)	144	(62)	162	(72)
22	5,900	(40700)	119	(48)	122	(50)	131	(55)	140	(60)	150	(66)	169	(76)
24	6,300	(43400)	122	(50)	126	(52)	136	(58)	145	(63)	157	(69)	176	(80)
26	6,700	(46200)	126	(52)	130	(54)	141	(61)	151	(66)	163	(73)	183	(84)
28	7,100	(49000)	129	(54)	134	(57)	146	(63)	156	(69)	170	(77)	190	(88)
30	7,500	(51700)	133	(56)	138	(59)	151	(66)	162	(72)	176	(80)	197	(92)
32	7,800	(53800)	136	(58)	142	(61)	156	(69)	167	(75)	183	(84)	204	(96)
34	8,200	(56500)	140	(60)	146	(63)	161	(72)	172	(78)	189	(87)	211	(99)
36	8,600	(59300)	143	(62)	150	(66)	166	(74)	177	(81)	195	(91)	218	(103)
38	9,000	(62100)	146	(63)	153	(67)	171	(77)	182	(83)	202	(94)	228	(109)
40	9,400	(64800)	149	(65)	157	(69)	176	(80)	187	(86)	207	(97)	232	(111)
42	9,800	(67600)	152	(67)	161	(72)	180	(82)	192	(89)	212	(100)	239	(115)
44	10,200	(70300)	155	(68)	165	(74)	185	(85)	197	(92)	217	(103)	247	(119)
Heating Rate, F/min			1.70		1.93		2.39		2.66		3.11		3.80	
(°C/min)			(0.94)		(1.07)		(1.33)		(1.48)		(1.73)		(2.11)	

Note: For all schedules the final temperature and pressure should be held constant to completion of test within ±2°F (±1°C) and ±100 psi (±700 kPa) respectively.

For critical cementing situations (static temperature above 260–275°F.) special care should be taken to insure that lab test schedules fit actual pressure-temperature-time conditions, and that cementing materials and mix-water samples are similar to those that will actually be used on the job.

Limitations of the standard thickening time tests should be considered in applying test results. Primary limitations are:

1. Test simulates continuous pumping—no shut downs

2. Temperatures and pressures used are estimates of actual downhole conditions

3. Inherent variation of the test itself—uncertainty limits increase with increasing test temperature

Regarding **dry blending of additives** into bulk cement, most field tests show that a minimum of four pneumatic transfers from one vessel to another are needed for adequate blending.

Variation between thickening time results from laboratory tests and field tests of bulk blended samples has been studied many times. Usually, the variations are blamed on (1) improper proportioning or blending of additives or (2) improper sampling of the blended cement.

Recent studies suggest that even though significant

variation in lab test results (±15%), or field test results (±40%) must be anticipated, thickening time tests remain the best criteria of the suitability of a given field blend for a specific well application. The best way to establish criteria appears to be to set a Target Thickening Time based on the "maximum estimated placement time" multiplied by a safety factor (1.7 to 2.3). Acceptability windows are then estimated as: Target time ±15% for a lab pilot test; or Target time ±40% for a bulk blended field sampled test.

Gel Strength may develop very quickly in a cement slurry allowed to remain motionless in the casing or annulus irrespective of the thickening time. Even with a thickening time of 6 hours enough static gel strength can develop within 15 minutes to add many hundreds of psi to the bottomhole pressure needed to break circulation. New techniques of delaying gel strength development (Halliburton Services "GasStop") may offer a solution to this problem.

Required Properties and Characteristics of Oil Well Cements

Viscosity—Should be low for better flow properties and mud removal. Cement is a non-Newtonian fluid; thus viscosity is a function of shear rate. Fann Viscometer must be used to determine viscosity characteristics.

Thickening Time—Two and one-half to 3 hours generally provides necessary placement time plus safety factor. Pressure-Temperature Thickening Time Tests should be used with samples of actual cement and mix water where placement conditions are critical.

Water Separation—Set volume of cement should be equal to slurry volume. Free water should be less than 1%. In highly deviated holes water separation and collection along the high side of the hole may be a serious problem. Free water tests should be run at elevated temperature using API Spec 10 test procedures.

Strength Development—Compressive strength of 500 psi is more than sufficient to support pipe. Compressive strength is the basis for most WOC regulations, and with proper cement and accelerators WOC time can be as short as 3 to 6 hours. Laboratory specimens should be pressure-temperature cured to simulate downhole conditions.

Sulfate Resistance—Low C_3A (Class B, C, G, or H) cement is best. The basis for laboratory measurement is to cure cements in 5% sodium sulfate solution and observe cracking and disintegration.

Pozzolans help sulfate resistance of Class A cement. Dense slurry gives better resistance. High water ratio slurry has poor resistance.

Sulfate resistance of cement is primarily dependent upon the tricalcium aluminate (C_3A) content of the cement. C_3A reacts with sulfates in mix water or formation water to form calcium sulfoaluminate. Resulting enlargement of crystals causes cracks to form providing paths for further attack. Cement should have three percent or less C_3A to be highly sulfate resistant. Some cements have no C_3A (Maryneal Incor, El Toro 35, Dykerhoff Class B).

Maximum sulfate reaction takes place at low temperature (80°–100°F.). *Above* 180°F. sulfate action is nil. Sulfate concentration in formation water less than 1,500 ppm should not harm set cement.

Density—Most slurries range from 11 to 18.5 lb/gal. Pozzolans or bentonite are effective means for reducing density—any water increasing material will reduce density. For accuracy cement slurries should be weighed under pressure to negate effect of air entrainment.

Storage Stability—Dry stored cements will remain good for long periods of time. However, slight changes can occur in humid conditions which will affect thickening time on critical cementing situations.

Mix Water—Fresh water is preferable, but any drinkable water is satisfactory for cement. Inorganic compounds usually accelerate set of cement. Organic compounds, such as, mud thinners, fluid loss agents, corrosion inhibitors, and bactericides, retard set. Few waters cause "flash setting." Sea water is satisfactory but thickening time should be checked. Carbonates and bicarbonates have an unpredictable effect on thickening time; thus, waters containing high carbonates or bicarbonates (greater than 2,000 ppm) should be avoided.

Effect of Acid on Cements—HCI reacts with cement to form silica gel which resists further attack. Communication after an acid job could be due to a mud channel since acid could shrink mud particles and open channel. HF acid may affect cement to a limited extent.

Heat of Hydration—Heat developed by cement setting in a 2-inch annulus will increase temperature 30°–40°F. above formation temperature. In washouts, more cement means higher temperature. In cementing through ice lenses in permafrost regions, heat of hydration becomes an important factor since heat liberated by cement in setting tends to melt the ice lenses and prevent bonding. Gypsum-blend cement (Permafrost cement) has a low heat of hydration (18–20 btu/lb slurry) compared to high alumina cement (60–90 btu/lb slurry).

Fluid Loss Control—Desirable for squeeze cementing. Measurement is made on filter press at 1,000 psi using 325-mesh screen. Neat cement has a fluid loss greater than 1,000 cc in 30 minutes at 1,000 psi. For primary cementing, fluid loss value of 150 to 400 cc is desirable.

Where gas communication through a cemented annulus is a concern, as in gas storage wells or high pressure gas wells, use of very low fluid loss (20 cc in 30 minutes at 1,000 psi on 325 mesh screen) appears to reduce the problem. See "Cementing Through Gas Zones." For squeeze cementing, fluid loss control is a primary concern, 50 to 200 cc is a desirable range.

Permeability—Most set cement slurries have very low values—less than any producing formations. If, however, cement is allowed to freeze when setting, or if strength retrogression has occurred due to a high temperature environment, permeability will be much higher (5–10 darcies). Disturbance of cement in the setting process by gas percolation may also provide communication for low viscosity fluids.

CEMENT ADDITIVES

Functions of Cement Additives

Well cement slurries usually contain additives to modify basic properties in some way. Additives can:

—Vary cement density.
—Increase or decrease strength.
—Accelerate or retard setting time.
—Control filtration rate.
—Reduce slurry viscosity.
—Bridge for lost circulation control.
—Improve economics.

Many additives affect more than one property of the cement slurry; thus, type and concentration must be chosen with care based on laboratory tests simulating actual well conditions.

Additives are usually free flowing powders which can be blended in bulk plants to obtain uniform distribution. Most additives can be used in mix water if bulk blending facilities are lacking. Text material shows Halliburton trade names. The Appendix lists corresponding products for other Service Companies. Quantity of additives are usually specified in percent additive by weight of cement (where lb/sk is used the weight of a sack is assumed to be 94 lb).

Cement Accelerators

Cement having a compressive strength of 50 psi or greater adequately supports casing. Many operators, however, base WOC time on the period required for the cement to reach a compressive strength of 500 psi. Where cementing temperatures range from 40° to 100°F. WOC time becomes significant. For example, at 40°F., Class A cement requires 48 hours to reach a compressive strength of 500 psi.

Several accelerators added to bulk cement or mixing water are available i.e.: calcium chloride, sodium chloride, and seawater. Calcium chloride is most frequently used. Densified cement slurries (using a friction reducer to permit lower water ratios) provide faster strength buildup.

Calcium chloride is available in regular 77% grade, or as anhydrous 96% grade. Normally 2.0% regular grade or 1.6% anhydrous is used. Above 3.0% little additional acceleration advantage is gained. (Table 4-7).

Sodium chloride in low concentrations accelerates cement setting, but acts as a retarder in high concentrations. 2.0 to 2.5 wt% NaCl in cement produces optimum acceleration. Somewhat more should be used with higher water ratio bentonite cement. Table 4-8 shows thickening time and compressive strength.

TABLE 4-7
Effect of $CaCl_2$ on Thickening Time and Strength (Class A Cement)

Thickening Time, Hours:Minutes

Calcium Chloride Percent	Simulated Well Depths — Feet							
	API Casing — Cementing				API Squeeze—Cementing			
	1000′	2000′	4000′	6000′	1000′	2000′	4000′	6000′
0.0	4:40	4:12	2:30	2:25	3:30	3:29	1:52	0:58
2.0	1:55	1:43	1:26	1:10	1:30	1:20	0:54	0:30
4.0	0:50	0:52	0:50	0:58	0:48	0:53	0:37	0:23

Compressive Strength, psi

Curing Time Hours	Calcium Chloride Percent	Atmospheric Pressure			API Curing Pressure	
		40°F	60°F	80°F	95°F 800 psi	110°F 1600 psi
6	0	N. S.	20	75	235	860
12	0	N. S.	70	405	1065	1525
18	0	5	620	1430	2210	2750
24	0	30	940	1930	2710	3680
36	0	185	1500	2490	3640	4925
48	0	505	2110	3920	4820	5280
6	2	N. S.	460	850	1170	1700
12	2	65	785	1540	2360	2850
18	2	170	1810	3080	3250	4300
24	2	415	2290	3980	4450	5025
36	2	945	2900	4810	5770	6000
48	2	1460	4205	6210	6190	5680

TABLE 4-8

SODIUM CHLORIDE CEMENT ACCELERATOR
Portland Cement — API Class A
Thickening Time — Hours:Minutes
Pan American Pressure Thickening Time Tester
Water Ratio — 5.2 Gals/Sk Slurry Weight — 15.6 Lbs/Gal

Sodium Chloride Percent	API Casing — Cementing Simulated Well Depths — Feet			
	1000′	2000′	4000′	6000′
0.0	4:40	4:12	2:30	2:25
2.0	3:05	2:27	1:52	1:13
4.0	3:05	2:35	1:35	1:20

COMPRESSIVE STRENGTH — PSI

Curing Time Hours	Sodium Chloride Percent	Atmospheric Pressure		API Curing Pressure	
		60°F	80°F	95°F 800 psi	110°F 1600 psi
12	0	70	405	1065	1525
24	0	940	1930	2710	3680
48	0	2110	3920	4820	5280
12	2	290	960	1590	2600
24	2	1230	2260	3200	3420
48	2	3540	3250	3900	4350
12	4	280	1145	1530	2575
24	4	1390	2330	3150	3400
48	4	3325	3500	3825	4125

Seawater contains small amounts of sodium, magnesium, and calcium chlorides (20,000 to 40,000 ppm); thus, provides some acceleration, but not nearly as much as 4.0% $CaCl_2$.

Densified cement slurries (low water ratio) have particular application for plugs where short pumping time and fast strength buildup are desired. At higher formation temperatures a retarder may be needed to provide sufficient pumping time. Also 2% $CaCl_2$ can be used with low water ratios to obtain even faster set and strength buildup. Table 4-9 shows the effect of various water ratios.

Cement Retarders

Increasing well depths and higher formation temperatures have led to the development of cement retarders to extend pumpability time. Most retarders are compatible with all API Class cements. The current trend, however, is toward use of a Class G or H basic cement with the addition of retarder to fit individual well conditions.

The primary factor governing the use of retarders is well temperature. Additives which require higher water ratios also require additional retarder, as the increased water dilutes the retarder.

Above static bottomhole temperatures of 260–275°F., thickening time tests should be run with the actual materials to be used on the job.

Halliburton retarders include HR-4, HR-5, HR-6L, HR-7, HR-12, HR-20 and HR-13L.

HR-4 is a lignin-type (calcium lignosulfonate) retarder for use up to static temperatures of 260 to 290°F. HR-4 should not be used with high bentonite (12–25%) cement since it does not act as a dispersant (Fig. 4-3). HR-5 is a special processed lignin designed to override viscosity effects found in some cements.

HR-12 and HR-20 are blends of HR-4 and an organic acid. HR-12 can be used where static well temperatures are greater than 260°F. HR-20 is similar to HR-12 but contains more organic acid (Fig. 4-4).

HR-6L and HR-13L are liquid retarders. HR-6L is suitable for bottom hole static temperatures up to 260°F. HR-13L is an organic acid effective up to 500°F.

TABLE 4-9
API Class H Densified Cement

Slurry properties

Water ratio, gal/sk	*CFR-2 %*	*Slurry weight, lb/gal*	*Slurry volume, cu ft/sk*	*Thickening time[a] hours: minutes*
4.3	1.0	16.5	1.05	3:03
3.8	1.0	17.0	0.99	2:51
3.4	1.0	17.5	0.93	2:12

12-hour Compressive strength—psi

Slurry weight, lb/gal	*API curing conditions*			
	110°F. 1,600 psi	*140°F. 3,000 psi*	*170°F. 3,000 psi*	*200°F. 3,000 psi*
16.5	2,075	4,000	7,800	9,035
17.0	2,850	6,535	8,375	10,025
17.5	3,975	6,585	8,550	10,675

[a] 8,000 ft squeeze-cementing schedule BHST–200°F.: BHCT–150°F.

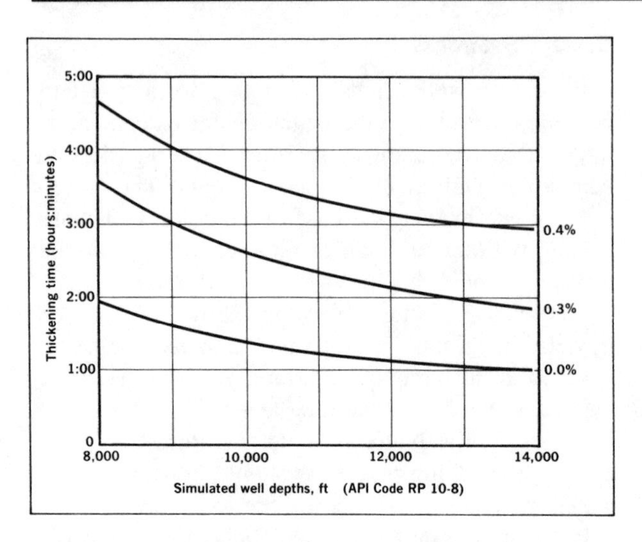

FIG. *4-3—Typical retardation with HR-4 retarder, Pozmix cement, casing schedules.*

HR-7 (Lignox or Kembreak) acts as a dispersant as well as a retarder and is the only retarder that should be used with high bentonite cement. HR-7 can be substituted for HR-4 where static formation temperature is less than 260°F (Table 4-10).

Saturated saltwater (3.1 lb sodium chloride/gal of water) when used as the mixing water will moderately retard setting time. Retardation should be sufficient for placement of API Class A cement to depths of 10,000 ft (230°F. static). Defoamer may be required to prevent foaming when saltwater and cement are mixed.

Evaporate salt is not pure NaCl but contains contaminates such as sulfates and magnesium chloride. Sufficient concentration of NaCl can probably not be obtained to have significant retarding effect.

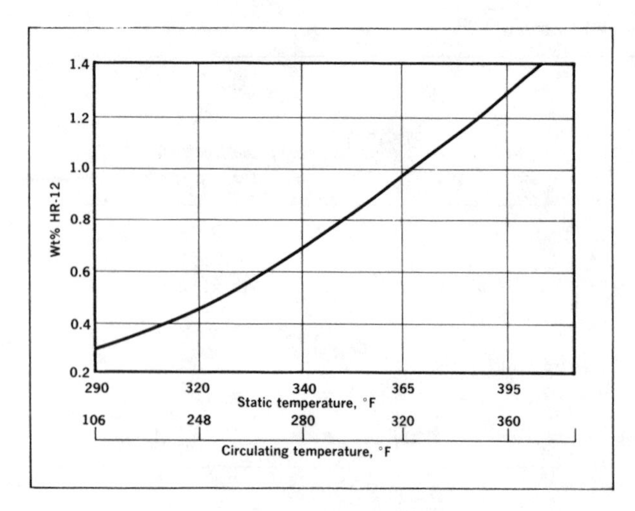

FIG. *4-4—HR-12 concentration to approximate a four-hour thickening time for casing conditions.*

TABLE 4-10
Percent HR-7 Retarder
To Approximate A 3-Hour Thickening Time

Well depth, ft	*Bottom-hole static Temperature, °F.*	*12% Bentonite*	*25% Bentonite*
Less than 4,000	Under 140	0.0	1.0
4,000–6,000	140–170	0.3–0.4	1.2
6,000–10,000	170–230	0.4–0.6	1.4
10,000–12,000	230–260	0.6–0.8	1.6

Fluid Loss Control for Primary Cementing

In cementing long sections, at relatively deep depths, high pressures and high temperatures, loss of water from the slurry as it moves past permeable formations may be detrimental from the standpoint of bridging and alteration of slurry properties.

After the cement is in place, loss of filtrate from the "setting" slurry below a bridged zone may reduce hydrostatic head, and allow entry of formation fluids into the wellbore. This is particularly true where gas zones are involved.

Where these conditions are present, use of additives to reduce fluid loss of the primary cement slurry may be justified. Additives function by forming micelles, or filmes, and improving particle size distribution.

Several materials are used to reduce fluid loss: CFR-2, CFR-3, Diacel LWL, Halad 9, Halad 14. Most have other effects also, such as reduced viscosity and retardation.

CFR-2 or 3, although primarily friction loss reducers, also provide moderate fluid loss control through a dispersing action. They have no retarding effect and can be used from 60°F. to 350°F. Their dispersing action permits use of lower water-cement ratios while maintaining reasonable slurry viscosities; thus, this densifying action provides reduced fluid loss. CFR-2 works best with fresh water, whereas CFR-3 can be used in fresh water or saturated saltwater.

Diacel LWL, CMHEC, is primarily a water loss control material. It is not effective in concentration below 0.3%. Above 0.7%, water ratio must be increased to offset slurry viscosity increase. At low temperature an accelerator (Diacel A) is needed to offset retardation. Maximum temperature limit is 170°F.

Halad 9, primarily designed for fluid loss control, has moderate retarding effect and can be used from 60° to 350°F. At low temperatures, 2% $CaCl_2$ may be needed to provide acceleration. It tends to increase slurry viscosity somewhat, thus slightly more mix water may be desirable (5.65 gal/sk rather than 5.20 gal/sk).

Halad 14 is designed for use at high temperatures, 170° to 400°F. It has the effect of reducing slurry viscosity and providing retardation as well as reducing fluid loss. Since it is normally used under critical conditions of depth and temperature, specific testing should be run to establish its properties with the particular cement composition to be used. Table 4-11 shows Halad 14 characteristics under conditions where it might be used. Note the effect of temperature on fluid loss.

Halad 22A is designed for use at very high temperatures, from 80°F. to in excess of 400°F. BHST. It can be used in salt cement slurries up to 18% NaCl or 5% KCl, where Halad 9 or 14 are less effective.

Slurry Viscosity Control

Use of cement friction-reducing additives aids removal of annular mud by the cement slurry on the basis of:

TABLE 4-11
Fluid Loss Class H Cement

Water—5.0 gal/sk

Slurry weight—15.8 lb/gal

Low temperature fluid loss

Halad-14 %	*Slurry consistency, Bc 140°F.*		*Fluid loss cc/30 min 325-mesh screen 1,000 psi Room temperature*
	Initial	*20 minute*	
0.75	1	2	64
1.00	1	2	38
1.25	1	2	30

High-temperature fluid loss

Halad-14 %	*Slurry consistency, Bc 80°F. to 190°F.*		*Fluid loss cc/30 min 325-mesh screen 1,000 psi Temperature, 190°F.*
	Initial	*20 minute*	
0.75	1	9	107
1.00	2	9	84
1.25	2	8	64

—Reducing displacement rate to provide turbulent flow in some portion of the annular cross section;

—At the same displacement rate, to increase the cross-sectional area of the annulus affected by turbulent flow;

—Friction reducers also lower placement pressures where restricted annular clearances are involved.

Friction reducers (Halliburton's CFR-2 or CFR-3) are essentially dispersing agents which reduce the apparent viscosity of the slurry. Other materials such as salt (NaCl) or Halad 9, or HR-7 which act as dispersants, also act to some extent as friction reducers.

Effect of Halliburton's CFR-2 is shown in Table 4-12. Flow rates shown are the minimum required to initiate turbulence.

TABLE 4-12
Effect of CFR-2 on Turbulent Flow Rate

COMPOSITION	Per Cent CFR-2	Hole Size (Inches) 6 3/4	6 3/4	8 3/4	9 7/8
		Casing Size OD (Inches) 2 7/8	4 1/2	5 1/2	7
		FLOW RATE—BPM			
Pozmix A Cement 14.1 lbs/gal	0.00	23.14	16.16	29.01	30.58
	0.50	8.51	6.86	11.18	12.17
	0.75	2.55	2.72	3.67	4.25
Pozmix S Cement 14.5 lbs gal	0.00	22.58	15.91	28.38	29.98
	0.50	10.40	8.09	13.49	14.58
	0.75	3.61	3.42	4.99	5.68
API Class A 15.6 lbs gal	0.00	18.18	13.58	23.29	24.93
	0.50	14.32	11.28	18.66	20.21
	0.75	6.57	5.86	8.91	9.93
API Class A 4% Gel 14.1 lbs gal	0.00	25.17	17.58	31.54	33.26
	0.50	15.50	11.21	19.65	20.88
	0.75	6.58	5.30	8.65	9.41
API Class A 12% Gel 12.8 lbs gal	0.00	23.55	16.45	29.51	31.12
	0.75	14.08	10.26	17.89	19.05
	1.00	2.93	2.88	4.10	4.67
API Class E 16.25 lbs gal	0.00	10.95	9.38	14.66	16.19
	0.50	4.14	4.46	5.97	6.94
	0.75	1.10	1.55	1.73	2.14

Light-Weight Additives

Basic cements are ground to a fineness which, when mixed with the recommended amount of water, produce slurry weights in excess of 15 lb/gal. Light-weight additives are used both to permit the use of longer columns of cement without formation break-down, and reduce the cost of the cementing material even where low weight is not a requirement.

Three methods of reducing weight are used:

1. Additives such as bentonite, attapulgite, pozzolans, or sodium silicate will permit use of increased mix water and at the same time prevent water separation.

2. Hollow ceramic spheres having low specific gravity can be used to provide reduced density slurries where total exposure pressure is less than 6,000 psi.

3. Foam cement involves the addition of Nitrogen

plus a surfactant to maintain the foam, to create slurry densities as low as 7 to 8 lbs/gal.

Wyoming Bentonite (API Spec 10 for use in Cement)—Bentonite has been used for many years to reduce slurry cost, slurry weight, and water separation. High percentages of bentonite reduce compressive strength and decrease thickening time. At temperatures above 230°F. bentonite promotes strength retrogression.

For use with cement, bentonite must meet API Spec 10 which is designed to insure that the bentonite does not contain soda ash or acrylates to increase bentonite yield.

Water requirements are about 1.3 gal for each 2% bentonite up to 8–12%. At the higher bentonite percentages, dispersants are usually used which give considerable latitude in water ratio. Average slurry properties of Wyoming bentonite in API Class H cement are shown in Table 4-13.

Prehydrating Bentonite and Attapulgite—The technique of prehydrating bentonite or attapulgite in fresh water or seawater reduces clay requirements and assures uniform blending where dry blending facilities are not available. One pound Wyoming bentonite prehydrated in fresh water provides about the same slurry properties as 3.6-lb dry blended.

Prehydrating in seawater, use of European bentonite, or dispersing with CFR-2 rather than HR-7 increases the bentonite needed to limit free water to 1%. Table 4-14 details water and clay requirements and shows slurry properties for various combinations.

Sodium Silicate—Small amounts (2–3%) of liquid sodium silicate added to Class G or H cement permit use of two to three times the normal mix water ratio, providing slurry densities less than 12.0 lb/gal and slurry yields greater than 2.5 cu ft/sk. For a given slurry density, sodium silicate slurries provide higher compressive strengths than other high water ratio slurries.

Manufactured Light-Weight Cements—Cements such as Trinity Lite-Wate or Halliburton Light-Weight contain pozzolans or shale, sometimes with the addition of bentonite, to permit use of increased mix water without exceeding the desired limit of 1% free water after set.

Table 4-15 compares properties of various light density slurries. Also shown is the longer term compressive strength buildup of a typical sodium silicate slurry.

Hollow Spheres—Thin-wall spheres of fused ceramic material have effective particle densities ranging from 0.639 gm/cc at atmospheric pressure to 1.0 gm/cc at 6,000 psi. These microspheres can be added to cement slurries to reduce slurry density to the range of 9 to 12 lb/gal. Particle size ranges from 60 to 140 mesh. Water absorbency of the microspheres is low; thus, reduced densities are obtained without the loss of compressive strength attendant with high water ratio slurries. Figure 4-5 compares slurry density and yield versus concentration of beads per sack. Figure 4-5 also shows the effect of pressure on slurry density.

Foam Cement—Ideally foam cement consists of a gas (usually nitrogen) chemically and physically stabilized as microscopic cells within an ordinary cement slurry. Critical factors to a satisfactory situation are: (1) a surfactant and stabilizer compatible with the high calcium ion and high pH environment of a cement slurry, and (2) an effective mechanical foam generating device. Figure 4-6 shows a typical equipment hookup. Responsive BOP equipment is, as usual, an important consideration, particularly if foam is to be circulated back to the surface.

TABLE 4-13
API Class H Cement With Bentonite

SLURRY PROPERTIES

Bentonite Per Cent	API Water Requirements Gal./Sk.	Cu. Ft./Sk.	Slurry Weight Lbs./Gal.	Lbs./Cu. Ft.	Slurry Volume Cu. Ft./Sk.
0	4.30	0.58	16.4	123	1.06
2	5.49	0.73	15.5	115	1.22
4	6.69	0.89	14.7	110	1.38
6	7.88	1.05	14.1	105	1.55
8	9.07	1.21	13.6	101	1.73
10	10.27	1.37	13.2	99	1.90
12	11.46	1.53	12.9	96	2.07
14	12.66	1.69	12.6	94	2.24
16	13.86	1.85	12.4	93	2.41

COMPRESSIVE STRENGTH — PSI

Bentonite Per Cent	Curing Time Hours	95°F. 800 psi	110°F. 1,600 psi	140°F. 3,000 psi	170°F. 3,000 psi	200°F. 3,000 psi
0	8	500	1200	2500	4000	5450
	24	3000	4050	5500	6700	8400
2	8	250	720	1400	2000	2500
	24	1550	2350	3250	3630	3800
4	8	130	450	830	1200	1550
	24	980	1490	2000	2250	2400
6	8	90	380	560	800	1050
	24	650	1000	1400	1650	1800
8	8	75	200	380	560	750
	24	430	700	1025	1150	1250
10	8	74	150	260	380	500
	24	325	500	700	825	900
12	8	70	120	200	280	360
	24	225	355	500	600	675
14	8	60	95	150	200	250
	24	160	270	400	490	550
16	8	50	80	110	170	220
	24	130	245	350	400	475

TABLE 4-14
Prehydrated Bentonite and Attapulgite

Dry Blend % gel equiv.	Water gal/sk	Class G Cement lb/bbl water Bentonite or attapulgite			Slurry weight, lb/gal	Slurry volume, cu ft/sk
Fresh water		*Wyo.*	*Euro.*	*Atta.*		
4	8.0	6.0	8.0	—	14.0	1.55
6	8.9	7.3	9.7	—	13.6	1.67
8	10.0	8.5	11.3	—	13.2	1.82
10	11.3	9.8	13.0	—	12.8	2.00
12	12.8	11.0	14.6	—	12.4	2.20
16	17.1	13.5	18.0	—	11.6	2.79
Sea water		*Wyo.*		*Atta.*		
4	8.0	9.0 (12.0)		2.5 (5.0)	14.1	1.55
6	8.9	11.0 (14.0)		3.5 (5.5)	13.7	1.67
8	10.0	13.5 (16.5)		4.5 (6.0)	13.3	1.82
10	11.3	15.5 (18.5)		5.5 (6.5)	12.9	2.00
12	12.8	18.8 (21.8)		6.5 (7.0)	12.5	2.20
16	17.1	23.0 (26.0)		8.5 (8.5)	11.7	2.79

[1]Add ( ) lb bentonite or attapulgite if HR-7 is used.
[2]With bentonite add 3 more lb bentonite if CFR-2 is used to replace HR-7.
[3]CFR-2 and HR-7 should not be used together when prehydrating bentonite in seawater. European bentonite should not be prehydrated in seawater.
[4]Wyoming or European bentonite can be prehydrated in 1/2 the required volume of fresh water—then seawater added to make up the remaining water requirement.

Downhole slurry density control must be properly handled. Cement slurry pump rate, surfactant injection rate and the nitrogen delivery rate must be carefully controlled. Where longer cement columns are used; and thus, nitrogen additions must be varied in stages to achieve a reasonably constant downhole density with changing hydrostatic pressures, the hole size must be calipered to permit adequate job design. Obviously channeling of the nitrafied slurry could lead to a blowout situation.

Compressive Strength—of lightweight slurries is reduced compared with a normal cement slurry. This is shown in Figure 4-7; however, except in extreme cases (density less than 8 lb/gal) sufficient compressive strength is developed to provide pipe support and to prevent annular movement of fluids. Obviously, sonic bond log appearance will be less than normal due to the fact that lower compressive strength cement does not inhibit sound transmission through the casing as well as higher compressive strength cement.

Heavy Weight Additives

A suitable weighting material for cement should have these characteristics:

—Low water requirement.
—No strength reduction of cement.
—No reduction of pumping time.

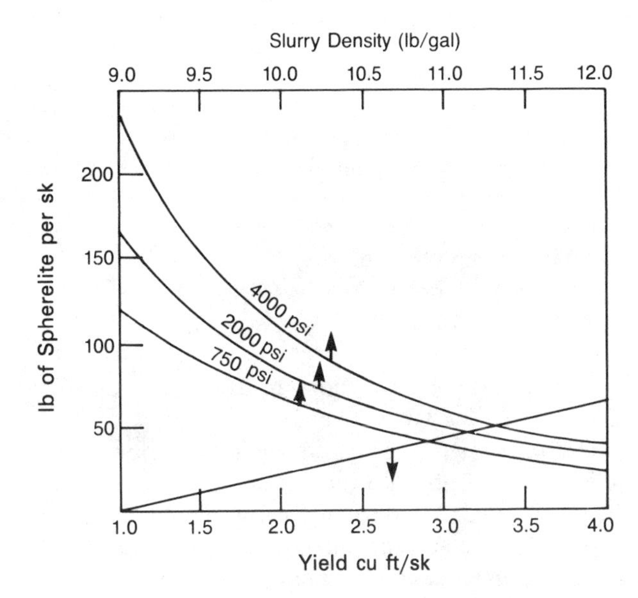

FIG. 4-5—*Effect of hollow spheres on reducing slurry density and increasing slurry yield. Courtesy Halliburton Services.*

TABLE 4-15
Properties of Low-Density Slurries

Water, gal/sk	Percent sodium silicate	Slurry Viscosity Initial	Slurry Viscosity 20 min	Free water, percent	Slurry density, lb/gal	Yield, cu ft/sk
API Class H Cement						
4.3	0	6	8	2.50	16.4	1.06
5.2	0	3	4	8.00	15.6	1.18
9.0	2	10	8	0.36	13.4	1.68
11.8	2	8	8	0.96	12.5	2.06
14.7	2	7	7	1.36	11.8	2.45
17.5	2	6	6	1.76	11.4	2.82
17.5	3	3	3	0.68	11.4	2.82
API Class A Cement						
5.2	0	4	9	2.30	15.6	1.18
9.0	2	7	8	0.00	13.4	1.68
11.8	2	6	6	0.52	12.5	2.06
14.7	2	3	2	0.68	11.8	2.45
17.5	2	2	2	1.50	11.4	2.82
17.5	3	3	3	1.00	11.4	2.82
Halliburton Light Cement						
7.7	0	7	14	0.60	13.6	1.54
8.8	0	5	9	0.60	13.1	1.69
9.9	0	4	6	1.30	12.7	1.84
10.9	0	5	7	1.30	12.4	1.97
10.9	1	11	12	0.40	12.4	1.97
12.9	1	8	8	0.56	11.8	2.21
14.9	1	5	5	0.72	11.4	2.46
16.9	1	3	4	2.10	11.1	2.74
16.9	2	7	8	0.24	11.1	2.74

Compressive stength: Class H w/2% Sodium Silicate and 11.0 gal/sk water

Curing temperature, °F	Curing time (days) 1	3	7	14	28	60
60	Set	75	530	585	755	1070
80	185	330	640	750	970	1380
95	310	700	705	950	1065	1215
140	445	615	750	1105	900	765

—Uniform particle size.
—Minimum slurry volume increase.
—Chemically inert.

The most suitable weighting materials are Hematite Ore, Barite, and Ottawa Sand.

Hematite ore (Sp. gr. 5.05) can produce slurry densities up to 22 lb/gal (Table 4-16). It is chemically inert and requires little additional water. Ilmenite ore (Sp. gr. 4.6) emits some degree of radioactivity and is less desirable than hematite.

Barite has a specific gravity of 4.2 (Table 4-17). Particle size is very fine; thus, additional water requirement (0.2 lb water per lb barite) tends to offset higher specific gravity. 18.0 lb/gal slurry is about maximum.

Sand has a specific gravity of 2.65 and a low water requirement. Maximum slurry weight is about 17.5 lb/gal (Table 4-18).

Salt saturation of cement slurry can add 0.5 to 1.0 lb/gal density.

Mix-water ratios can be reduced to increase density about 2.0 lb/gal. A friction reducer such as Halliburton CFR-2 or CFR-3 will then be required to provide a pumpable viscosity.

Where high slurry densities and low mix-water ratios are desired, a recirculating-type mixer will provide better density control than the jet-type mixer.

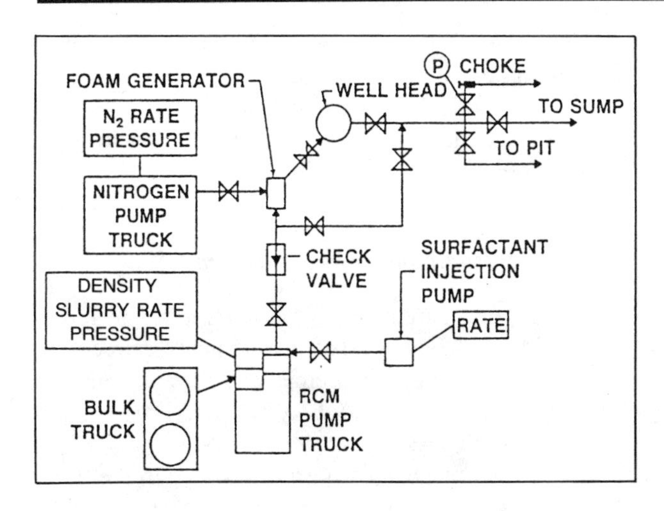

FIG. 4-6—Equipment layout—foam cement. Courtesy Halliburton Services.

Lost Circulation Additives

Lost circulation materials for cement may be classified as:

Granular—Inert materials designed to bridge at the surface of the borehole or within the formation. Effectiveness depends on proper particle size distribution to yield a low permeability bridged mass. Granulars work better in high solids fluid system, such as cement, than in lower solids drilling mud system. Typical granular materials are Gilsonite and Tuf-Plug (walnut hulls). Fibrous wood materials retard setting time and are not used.

Lamellated—Inert flake-type materials designed to mat at the face of formation. Cellophane flakes are commonly used.

Semi-Solids and Flash Setting Cements—By either chemical or physical action these materials thicken rapidly to form plug. Examples are Bentonite-Diesel Oil Slurry, Thixotropic Cement, or Silicate gels.

In laboratory tests granular materials work best in unconsolidated formations. Lamellated flake-type materials work best for slots. Semi-solids or flash-setting materials work best in large vugs or fractures.

Gilsonite is the most effective granular material. Optimum quantity is about 12 to 15 lb/sk cement.

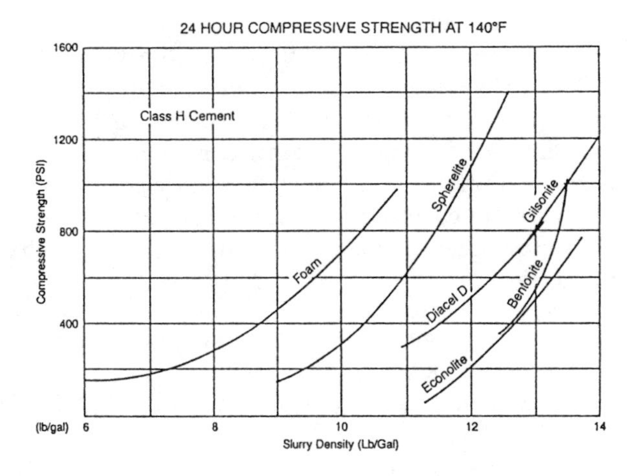

FIG. 4-7—Compressive strength of various lightweight slurries. Courtesy Halliburton Services.

TABLE 4-16
Slurry Properties API Class H with Hematite

Cement, pounds	Hematite, pounds	Water, gal/sk	Slurry volume, cu ft/sk	Slurry weight, lb/gal
94	—	4.5	1.08	16.2
94	12	4.5	1.12	17.0
94	28	4.5	1.17	18.0
94	46	4.5	1.23	19.0
94	63	4.5	1.30	20.0

TABLE 4-17
Slurry Properties API Class H with Barite

Cement, pounds	Barite, pounds	Water, gal/sk	Slurry weight, lb/gal	Slurry volume, cu ft/sk
94	—	4.5	16.2	1.08
94	22	5.1	17.0	1.24
94	55	5.8	18.0	1.46
94	108	7.1	19.0	1.83

TABLE 4-18
Slurry Properties API Class G with Sand

Cement, pounds	Sand, pounds	Water, gal/sk	Slurry weight, lb/gal	Slurry volume, cu ft/sk
94	—	5.2	15.60	1.18
94	5	5.2	15.76	1.21
94	10	5.2	15.91	1.24
94	15	5.2	16.07	1.27
94	20	5.2	16.20	1.30
94	32	5.2	16.50	1.37
94	56	5.2	17.00	1.52
94	85	5.2	17.50	1.70
94	123	5.2	18.00	1.93

Walnut Hulls are used in quantity on 2 to 8 lb/sk. Other nut hulls should be checked for effect on cement setting properties.

Cellophane Flakes (3/8-in. size) are used in quantity of 1/4 to 2 lb/sk.

Thixotropic cements develop gel strength quickly when not agitated. Thixotrophy is obtained by addition to Class G cement of about 10% calcium sulfate and 4–16% bentonite, or about 1% of a sodium silicate-like polymer.

Sodium Silicate has been used as the first stage of a two-stage process to control lost circulation in drilling or subsurface water flows in producing wells. The silicate in contact with formation or synthetic saltwater forms a stiff gel to temporarily plug the zone. Sand or other granular material can be used to promote bridging. The second stage, an accelerated cement slurry, then forms a permanent seal.

A recent approach to controlling lost circulation by reducing the hydrostatic fluid column pressure involves injecting a carefully calculated volumn of Nitrogen into a portion of the mud column just before the cement is mixed and pumped into the casing.

CEMENT BONDING

Thinking on the subject of cement bonding has undergone changes in recent years. The paramount objective in primary cementing is to eliminate fluid communication between zones. Originally it was thought that "bonding" of the cement to the casing and to the formation was closely related to this objective. Bonding was studied in the lab on the basis of steel surface characteristics and pipe expansion effects, and downhole, through sound transmission indications.

Initially the primary concern was with the bond at the casing-cement interface, because this could be "looked at" with the sonic bond log. Later it was realized that the bond between the cement and formation (largely dependent on mud filter cake properties) was more difficult to achieve, and more likely to be the source of interzonal communication problems.

Currently it appears that *effective displacement of annular mud by the cement to eliminate mud channels is the most important consideration in eliminating fluid communication between zones*. As shown in Figure 4-8, with a thin tough mud filter cake fluid communication is unlikely. But as the mud filter cake grades into a softer thicker cake, and further to what might be termed a mud channel, communication becomes more likely.

Thus, the subject of cement bonding per se, while still important, is not the primary concern it was originally thought to be. The real objective is to prevent fluid movement in the annulus—rather than to get a "good" bond log.

Bonding Measurements in the Laboratory

Bonding measurements in the laboratory are indicative of factors which affect the contact between the cement and casing, and cement and formation, *but should be considered in light of the overpowering importance of eliminating mud channels.*

Shear bond mechanically supports pipe in the hole. It is determined by measuring the force required to initiate pipe movement, divided by cement casing contact surface area (Fig. 4-9).

Hydraulic bond blocks migration of fluids. It is determined by applying liquid pressure at pipe-cement or formation-cement interface until leakage occurs.

Gas bond is measured in the same manner as hydraulic bond.

Pipe-Cement Bond

Higher cement compressive strength increases both shear and hydraulic bond in lab tests.

Pipe finish typically affects bond strength as shown by lab tests in Table 4-19. With mill varnish on casing, bond strength is time-dependent (particularly below 140°F.) Lowest bond strength occurs about two days after cementing. This effect can be seen on sonic bond logs. Oil wet pipe surfaces reduce hydraulic and shear bond strength.

The micro annulus situation as sometimes seen on the sonic bond log is primarily a function of pipe expansion or contraction due to application of heat or pressure. A true micro annulus condition does not necessarily mean fluid communication (see Production Logging chapter). The following points are significant in regard to the tightness of contact between the cement and casing, however:

—Leaving pressure on casing during WOC time is harmful to "bond."
—Pumping plug down with light fluid is helpful to "bond."
—Heat of hydration of cement is harmful to "bond" due to casing expansion.
—Circulating while WOC is helpful to "bond" due to cooling.

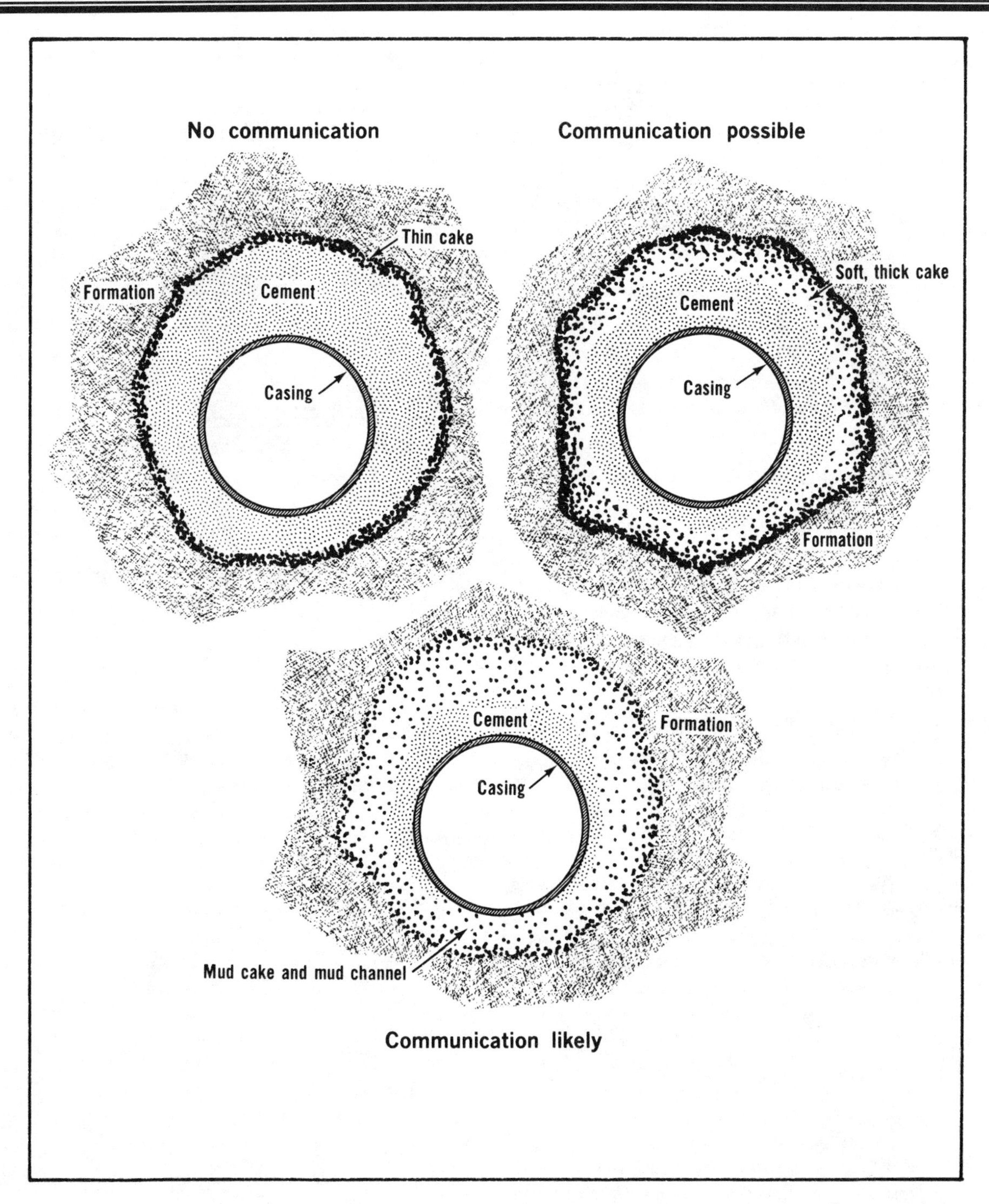

FIG. *4-8—How mud filter cake affects possibility of fluid communication between zones after cementing.*

—After cement is set, increased pressure inside casing increases "bond."

Hydraulic bond failure is time-dependent and is a function of the viscosity of the fluid causing failure. With water, field tests indicate that bond failure progresses at a rate of about 1.1 to 1.3 ft/min.

Formation-Cement Bond

Laboratory tests show that cement-formation hydraulic bond is largely influenced by the presence of mud filter cake. A tough mud cake slightly improves hydraulic bond strength compared with a soft mud cake. With clean dry formation hydraulic bond strength can

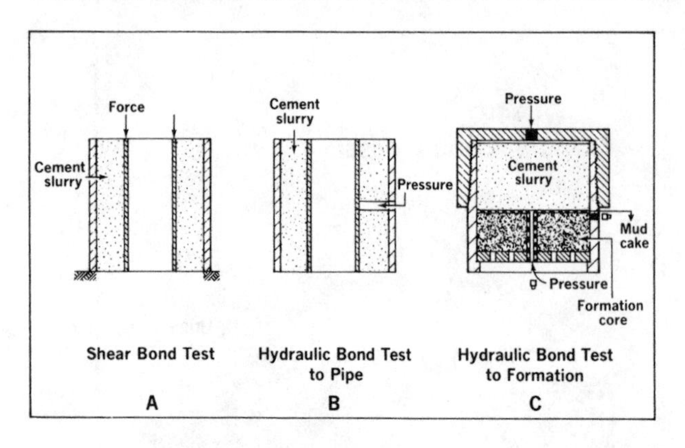

FIG. 4-9—*Laboratory bonding measurements.*[5] *Permission to publish by the API Production Department.*

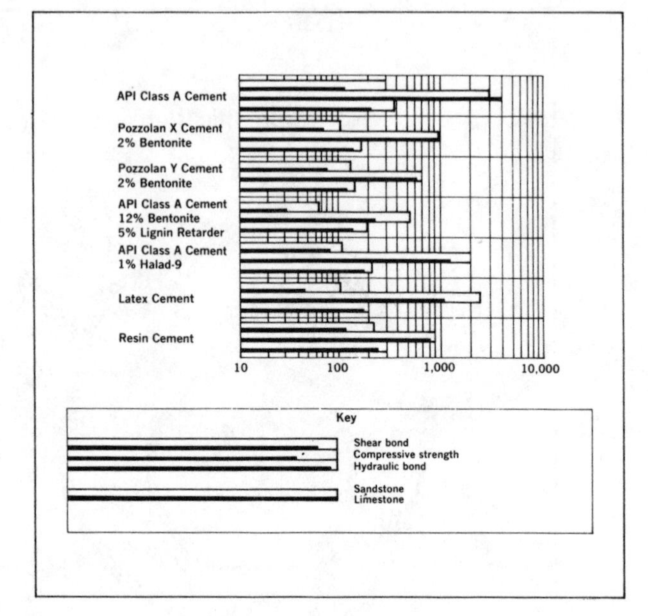

FIG. 4-10—*Bonding properties of cement to formation, cement-squeezed, walls not cleaned.*[5]

exceed formation strength. With filter cake bond strength is greatly reduced.

Higher hydraulic bond is obtained on more permeable formations since cement slurry loses water to formation and develops a higher compressive strength.

Results of lab tests of formation cement bond on Indiana Limestone (1 md) and Berea Sandstone (100 md) are shown in Figure 4-10.

FLOW PROPERTIES OF PRIMARY CEMENTS

Cement flow properties during primary cementing are important because of their effect on:

—Efficiency with which the cement displaces the annular mud column.
—Frictional pressure drop in the annulus, which adds to the hydrostatic pressure exerted on the formation.
—Hydraulic horsepower required to place the cement in a given time period.

TABLE 4-19
Bond Strength vs. Pipe Finish

Type of finish	*Bond strength*		
Steel	*Shear psi*	*Hydraulic psig*	*Gas psig*
New (Mill Varnish)	74	200–250	15
New (Varnish Chemically Removed)	104	300–400	70
New (Sandblasted)	123	500–700	150
Used (Rusty)	141	500–700	150
New (Sandblasted—Resin-Sand Coated)	2400	1100–1200	400+

Cement slurries are usually non-Newtonian fluids; thus, viscosity is a function of shear rate. In order to calculate pressure losses in a system, a mathematical description of "viscosity" or the shear stress vs. shear rate relation must be developed.

The Power Law concept more nearly describes the flow properties of cement, rather than the Bingham Plastic concept used to describe muds.

Power Law Equation: $S_s = K'(S_r)^{n'}$

S_s = shear stress (lbs/ft^2)
S_r = shear rate (sec^{-1})
n' = slope of log S_s vs. log S_r curve.
K' = intercept of log S_s vs. log S_r curve on shear stress axis.

In the Power Law concept, two slurry parameters must be determined in order to estimate frictional pressure loss and predict the flow velocity required to initiate turbulence. These are: flow behavior index (n'), and consistency index (K').

Fann dial readings taken at 600, 300, 200, and 100 rpm are converted to shear stress (pounds force/square foot), and plotted on the Y axis of log-log paper vs. Fann rpm's converted to shear rate (sec^{-1}) on the X axis. (n') is the slope of a straight line through the four points, and (K') is the intercept of the straight line at unity rate of shear (1.0 sec^{-1}).

$$\text{Shear Stress}\left(\frac{\text{lb force}}{\text{sq ft}}\right) = \frac{\text{Fann Dial Reading} \times N \times 1.066}{100}$$

$$\text{Shear Rate (sec}^{-1}) = \text{Fann rpm} \times 1.703$$

On the field model Fann Viscometer, with only 300 and 600 rpm, n' and K' can be calculated as follows:

$$n' = 3.32 \times \left(\log_{10}\frac{\text{600 rpm Dial Reading}}{\text{300 rpm Dial Reading}}\right)$$

$$K' = \frac{N \times (\text{300 rpm Reading}) \times 1.066}{100 \times (511)^{n'}}$$

N = range extension factor of the Fann torque spring (usually 1.0)

Flow Behavior Index (n') and Consistency Index (K') values have been published for most cement slurries. Power Law curves for several slurry are shown in Figure 4-11.

Where the Bingham Plastic parameters (Plastic Viscosity, PV, and Yield Point, YP) are known:

$$n' = 3.32 \times \log_{10}\left(\frac{2\text{PV} + \text{YP}}{\text{PV} + \text{YP}}\right)$$

$$K' = \frac{N \times (\text{PV} + \text{YP}) \times 1.066}{100 \times (511)^{n'}}$$

Newtonian cements do a better job of displacing mud. If $n' = 1.0$, the slurry is Newtonian, and viscosity is constant at all flow rates. For many cement slurries (particularly high gel type slurries) the n' value is considerably less than 1.0. Thus, apparent viscosity increases as flow velocity decreases. For most effective mud displacement by the cement slurry it is desirable that the cement slurry be as Newtonian as possible; i.e., n' should be close to 1.0.

Formulae For Making Flow Calculations

1. Apparent Viscosity

$$\mu_a = \frac{4.788 \times 10^4 K'}{(\text{Shear Rate})^{1-n'}}$$

μ_a = apparent viscosity, centipoise
K' = consistency index
n' = flow behavior index

$$\text{Shear Rate} = \frac{96\,V}{D},\ \text{sec}^{-1}$$

2. Displacement Velocity

$$V = \frac{17.50\,Q_b}{D^2} = \frac{3.057\,Q_{cf}}{D^2}$$

V = velocity, feet per second
Q_b = pumping rate, barrels per minute
Q_{cf} = pumping rate, cubic feet per minute
D = inside diameter of pipe, inches

For annulus,

$D = D_o - D_i$; or,

$$D = \frac{4 \times \text{Area of Flow}}{\text{Wetted Perimeter}}$$

$D^2 = D_o^2 - D_i^2$
D_o = outer pipe id or hole size, inches
D_i = inner pipe od, inches

3. Reynolds Number

$$N_{Re} = \frac{1.86\,V^{(2-n')}\rho}{K'\,(96/D)^{n'}}$$

N_{Re} = Reynolds number, dimensionless
ρ = slurry density, pounds per gallon

4. Frictional Pressure Drop

$$\Delta P_f = \frac{0.039\,L\rho\,V^2 f}{D}$$

ΔP_f = frictional pressure drop; psi
L = length of pipe, feet
f = friction factor, dimensionless (see Fig. 4-12)

For $N_{Re} < 2100$: $f = \frac{16}{N_{Re}}$

For $N_{Re} > 2100$ (Non-newtonian fluid):
$f = 0.00454 + 0.645\,(N_{Re})^{-0.7}$

5. Velocity at Which Turbulence May Begin (N_{Re} = 2100)

$$V_c^{2-n'} = \frac{1129\,K'\,(96/D)^{n'}}{\rho}$$

$$V_c = \left[\frac{1129\,K'\,(96/D)^{n'}}{\rho}\right]^{1/(2-n')}$$

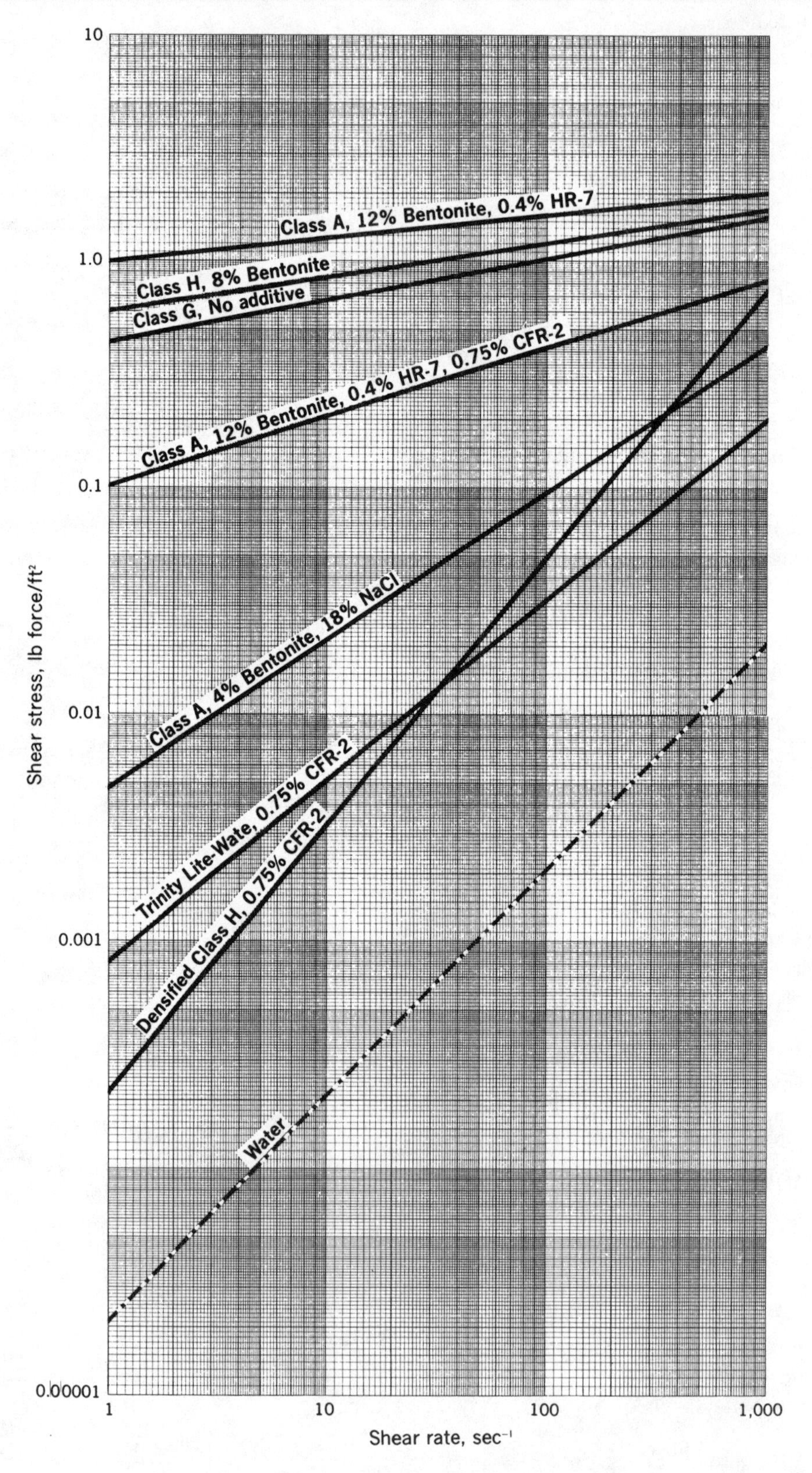

FIG. 4-11—*Flow curves for typical cement slurries.*

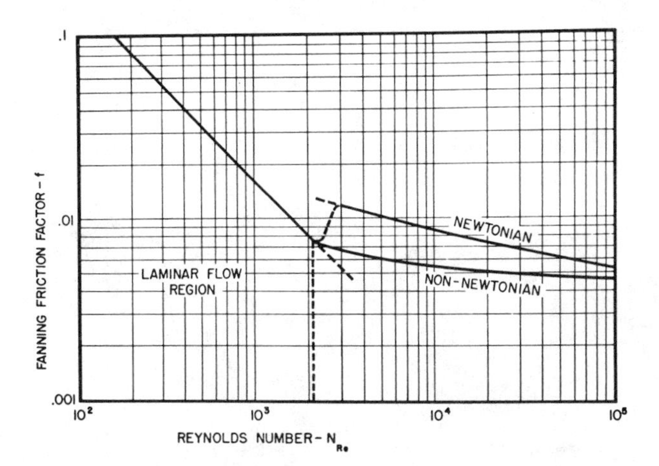

FIG. *4-12—Reynolds number-friction factor correlation.*

FIG. *4-13—Gelled mud remaining after cement was circulated in place in laboratory model study (Halliburton Services).*

V_c = critical velocity, feet per second

6. *Hydrostatic Pressure*

$P_h = 0.052\ \rho H$

P_h = hydrostatic pressure, psi

H = height of column, feet

DISPLACEMENT MECHANICS

The paramount objective in primary cementing is to place cement so as to eliminate mud channels. The most predominant cause of cementing failure appears to be channels of gelled mud remaining in the annulus after the cement is in place.

If mud channels are eliminated, almost any cement will provide an effective seal. Conversely, if mud channels remain after the cement is in place, no matter what the quality of the cement, there will not be an effective seal between formations (Fig. 4-13).

Forces helping to displace mud are:

1. Drag stress of cement upon mud due to difference in flow rates and flow properties.
2. Drag stress of pipe upon mud and cement due to pipe motion—either rotation or reciprocation.
3. Buoyant forces due to density differences between mud and cement.

Factors Affecting Annular Flow and Mud Displacement

In evaluating factors affecting displacement of mud, it is necessary to consider the flow pattern in an eccentric annulus condition with the pipe closer to one side of the hole than the other. Flow velocity in an eccentric annulus is not uniform. Highest flow rate occurs in the side of the hole with the largest clearance as shown in Figure 4-14.

With a non-Newtonian fluid it is possible to have turbulent flow in the wide side of the annular cross section, and laminar or even a stagnant zone on the narrow side.

—As flow rate increases, the annular cross-sectional area affected by turbulent flow increases.

—As the viscosity of the fluid decreases, the annular cross-sectional area affected by turbulence increases.

—As casing standoff decreases the displacement rate needed to prevent a stagnant area on the narrow side of the annulus increases.

—As the viscosity and gel strength of the mud increases, the displacement rate needed to prevent a stagnant area increases.

—The more "Newtonian" the displacing fluid, the more effective the displacement.

—With the casing close to the wall of the hole it may not be possible to displace cement at a rate high enough to develop turbulent flow throughout the entire annular cross section.

Laboratory Studies Point the Way

A number of laboratory studies have been conducted in recent years to determine the relative importance of the forces helping to displace mud. The study reported in References 22, 24, and 30 most closely simulates well conditions and temperature and wall cake effect.

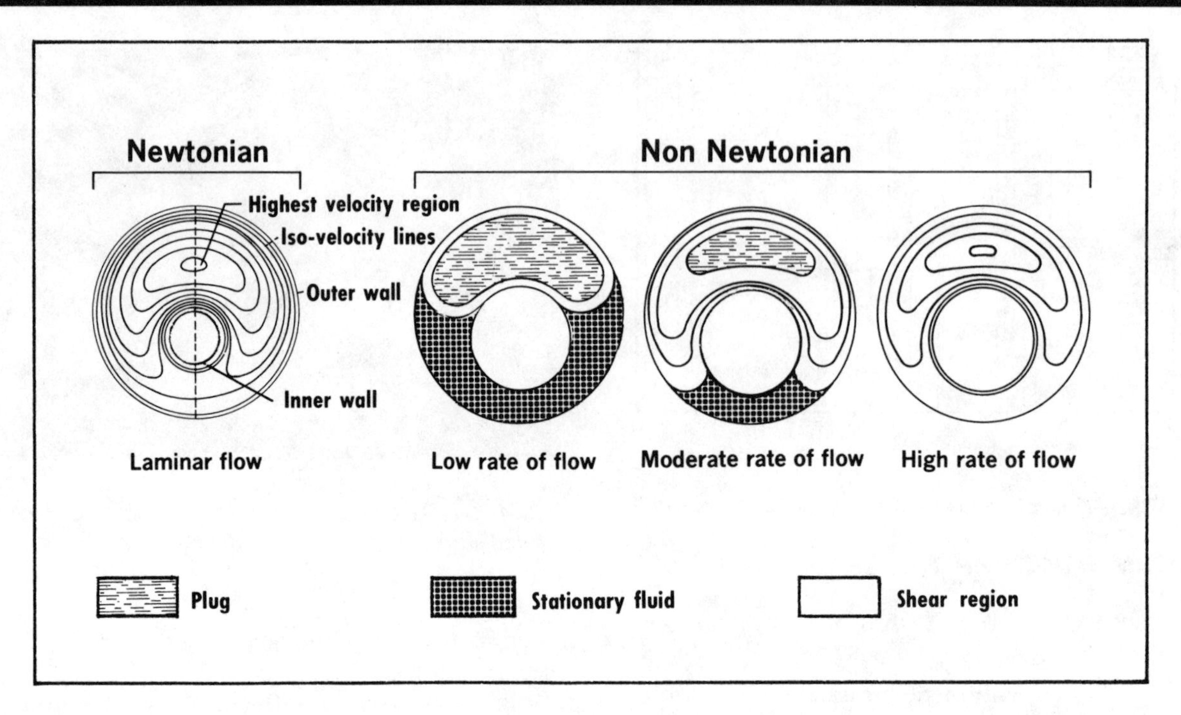

FIG. *4-14—Annulus flow patterns. After Piercy et al.*

Figure 4-15 shows the effect of pipe standoff (100% standoff = centered casing; 0% standoff = casing against the wall of the hole). Total annular mud is percent of the annular cross section filled with mud. It is channel mud plus filter cake mud. Channel mud is circulatable mud.

Figure 4-16 shows the effect of mud properties. "Thick mud" is a fresh water mud having a Plastic Viscosity of 46 cp and a Yield Point of 21 lb/100 sq ft. "Thin mud": PV = 20 cp; YP = 3 lb/100 sq ft. Laminar flow velocity = 90 ft/min; turbulent flow velocity = 255 ft/min; super turbulent flow velocity = 455 ft/min.

These data make the point that a "thin" mud can

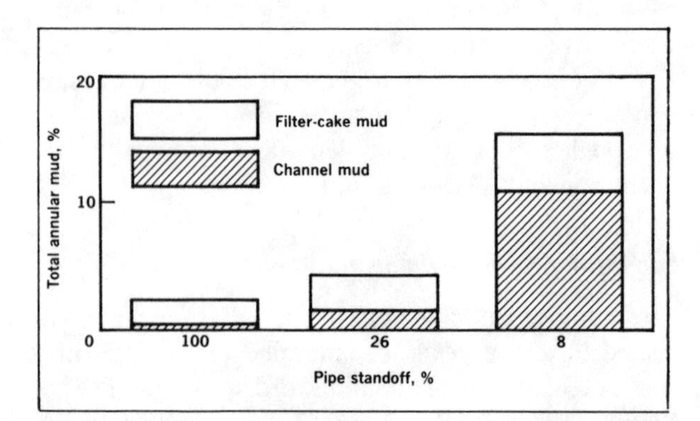

FIG. *4-15—Evaluation of pipe standoff.[24] Permission to publish by The Society of Petroleum Engineers.*

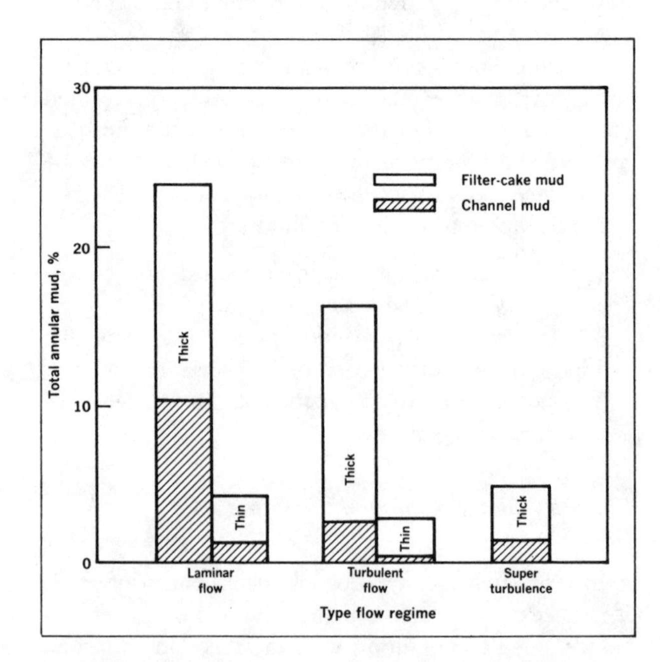

FIG. *4-16—Effect of mud properties.[24] Permission to publish by The Society of Petroleum Engineers.*

be displaced more readily, even at laminar flow rates, than a "thick" mud at super turbulent flow rates.

Figure 4-17 evaluates the effect of flow rate on displacement of a "thick mud" (PV = 43 cp; YP = 21 lb/100 sq ft) by a thin cement slurry. With other conditions equal high flow rates do a more effective job of mud removal.

Figure 4-18 shows the effect of rotation with and without scratchers. Note that pipe movement is a very important factor. Reciprocation provides essentially the same results as rotation. Further, scratchers, which basically act to break the gel of the mud, are a significant aid in mud removal, particularly in washed out sections.

Figures 4-19 and 4-20 show the importance of *mud mobility factor* and flow velocity on mud displacement in laboratory tests *where the pipe was not moved.* The mud mobility factor is inversely related to the ten minute gel strength and the API filtrate loss volume of the mud. Loss of filtrate liquid from the mud system acts to concentrate the remaining mud solids making mud removal much more difficult. Zero filtrate loss resulted in 100 percent mud removal. Zero formation permeability (thus, no loss of liquid from the mud) also produced 100 percent mud removal.

In these laboratory tests, simulating deep liner cementing situations (where it was assumed that the liner could not be moved), the mud mobility factor appeared to be the all-important variable masking out such effects as density difference between the cement and mud, laminar or turbulent flow, and cement rheology. Figure 4-20 also shows that water pre-flush

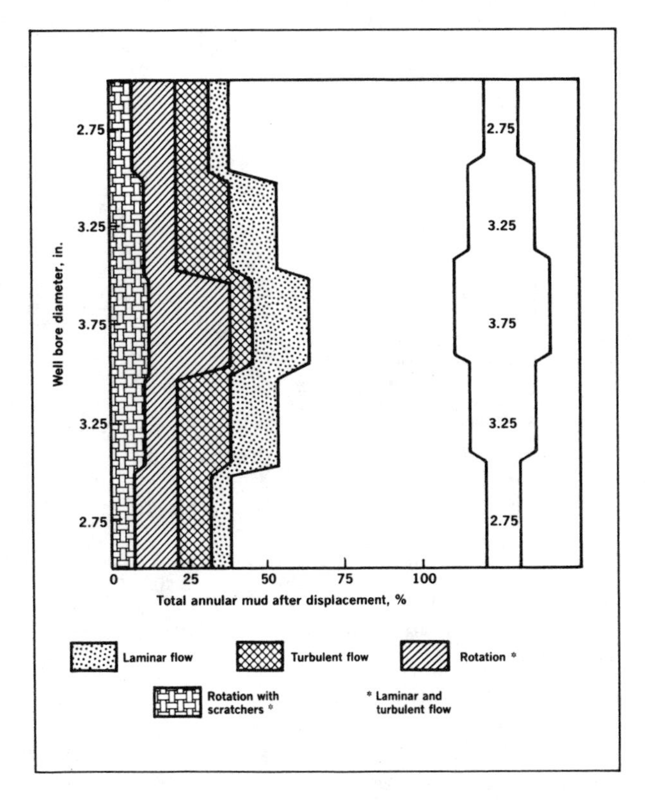

FIG. *4-18—Effect of rotation with and without scratchers.*[24] *Permission to publish by The Society of Petroleum Engineers.*

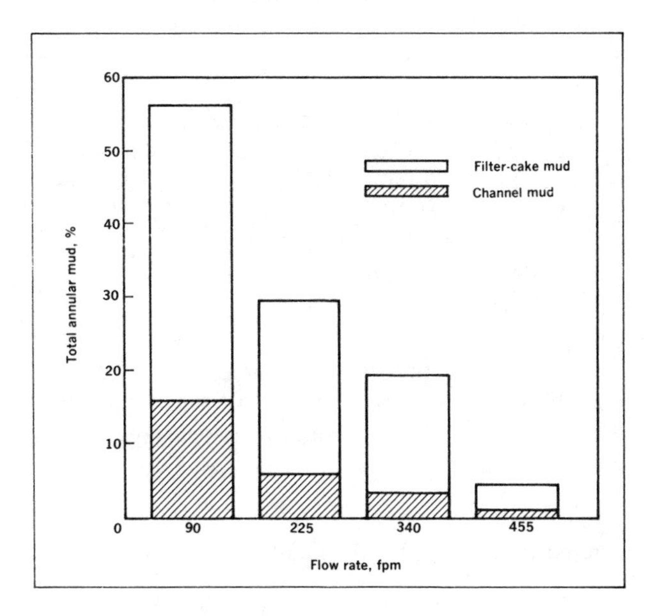

FIG. *4-17—Effect of flow rate.*[24] *Permission to publish by The Society of Petroleum Engineers.*

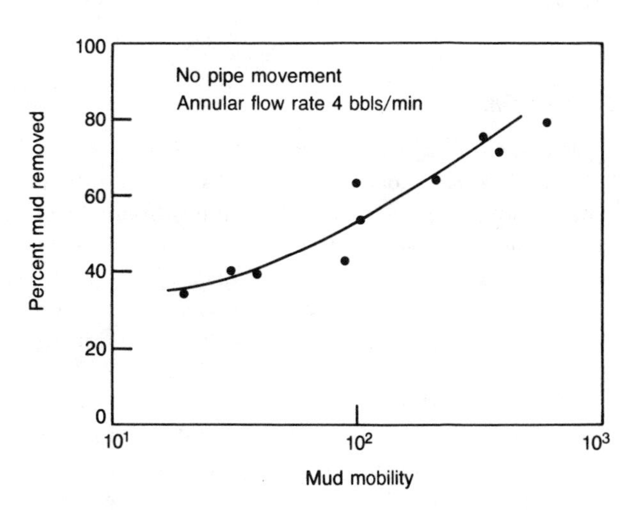

FIG. *4-19—Effect of mud mobility on mud displacement.*[33] *Permission to publish by The Society of Petroleum Engineers.*

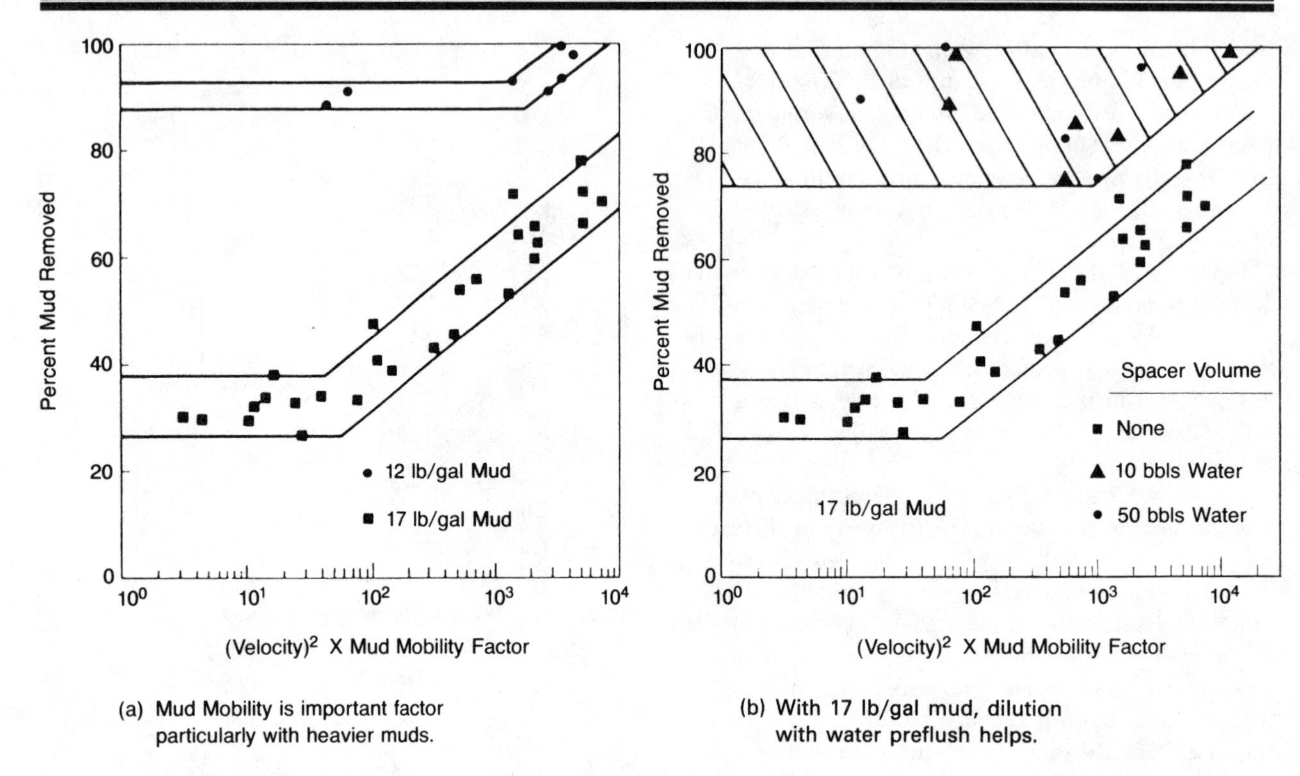

(a) Mud Mobility is important factor particularly with heavier muds.

(b) With 17 lb/gal mud, dilution with water preflush helps.

FIG. *4-20—Lab evaluation of mud displacement efficiency where pipe is not moved.*[33] *Permission to publish by The Society of Petroleum Engineers.*

ahead of cement significantly improves mud removal with the 17 lb/gal mud, apparently because of the dilution effect. In lab tests, preflush volume was not significant, probably because of the short length of the model. In field situations, volume should be a factor.

Contact Time

Field studies[15] and laboratory studies[22,24] have shown the importance of "contact time," or the length of time cement moves past a point in the annulus in *turbulent flow*. Thus, if a mud channel is put in motion, even though its velocity is much lower than cement flowing on the wide side of the annulus, given enough time the "mud channel" will move above the critical productive zone.

Lab studies[24] show that contact time is less effective with the cement in laminar flow conditions where apparently the cement does not exert sufficient drag stress on the mud to start the mud channel moving.

On a practical basis, contact time should exceed 10 minutes. At a given displacement rate contact time is directly proportional to volume of cement.

Computer Solutions

Computer solutions are helpful for planning of difficult primary cementing jobs—or for improving the economics where several similar jobs will be performed.

Slurry Flow Plan—With accurate well data on cement and mud properties, slurry flow plan provides following factors for a number of possible conditions:

1. Flow rate required for turbulence.
2. Hydraulic horsepower required.
3. Fill-up factor.
4. Contact time.
5. Circulating pressure opposite various formations, to determine the possibility of formation breakdown for various mud and cement densities, flow properties, and displacement rates.

Summary of Important Factors

All laboratory and field work performed in recent years point to the same factors as contributing to the success of primary cementing. All of the factors are aimed at removing mud from the annular cross section.

—Pipe centralization significantly aids mud displacement.
—Pipe movement, either rotation or reciprocation, is a major driving force for mud removal. Pipe motion with scratchers substantially improves mud displacement particularly in areas of hole enlargement.
—A well conditioned mud having high mobility factor (inversely related to ten minute gel strength and the API fluid loss volume) greatly increases mud displacement efficiency. Dilution of mud with a preflush helps.
—High displacement rates promote mud removal. At equal displacement rates a thin cement slurry in turbulent flow is more effective than a thick slurry in laminar flow.
—Contact time (cement volume) aids in mud removal if cement is in turbulent flow in some part of the annulus.
—Buoyant force due to density difference between cement and mud is a relatively minor factor in mud removal.

Practicalities

In a given situation it may not be possible, or even necessary, to maximize each of these factors. To some extent one factor may compensate for another. In the above list, however, centralization, pipe movement and mud conditioning are primary factors.

Careful evaluation of cementing success is the key.

If cementing failures are experienced, examination of cementing practices in light of these important factors should show where improvements in cementing practices should be initiated.

COST OF PRIMARY CEMENTING

Overall well cost should be optimized considering such factors as:

1. Cost of cement slurry—should be on a volume basis (cu ft of slurry behind pipe).
2. Cost of pumping equipment used in placing cement.
3. Cost of rig time in performing cementing operation and subsequent WOC time.
4. Cost of future remedial operations as a result of inadequate primary cementing.

Typical bulk cement and additive costs (Domestic U.S. August, 1988):

API Class G & H Cements	$ 6.08/sk
Pozmix A	3.02/sk
0–4% Gel/sk (cu ft)	0.37/cu ft
5–8% Gel/sk (cu ft)	0.74/cu ft
9–12% Gel/sk (cu ft)	1.11/cu ft
HR-4	1.10/lb
HR-7	1.05/lb
HR-12	2.65/lb
CFR-2	3.70/lb
Hi-Dense No. 3	14.20/100 lb
Gilsonite	0.33/lb
Calcium chloride	25.25/100 lb
Sodium chloride	9.40/100 lb
Gas stop additive	28.40/lb.

Comparative costs of typical Cement Slurries used in the Deep Anadarko Basin are shown in Table 4-20.

SPECIAL PROBLEM SITUATIONS—NEW DEVELOPMENTS

Cementing High Temperature Wells

Deep wells (below 15,000 feet) having static formation temperatures above 230°–250°F. are considered to be critical situations. Major problems involve displacement of mud by cement slurry, design of slurry to provide adequate rheological properties and pumpability time, attainment of desired slurry properties during the mixing process, and control of subsequent strength retrogression. The following factors should be considered:

Formation Temperature—Accurate knowledge of formation static or bottom-hole circulating temperature is the starting point for slurry design. A particular design problem exists with a long cement column due to temperature difference between the top and bottom of the slurry. Overstating temperature to provide "safety factor" is poor practice. Safety factor should be provided by adjusting pumpability time.

Slurry History—Anticipated temperature-pressure-time history of the cement slurry as it is mixed and pumped into place must be established as a basis for running Pressure Temperature Thickening Time Tests to fix additive requirements.

Laboratory Tests—Slurry design tests must be run with cementing materials, additives, and mix water which will actually be used on the job. Use of Class G or H basic cement promotes uniformity, but labo-

TABLE 4-20
Typical Cement Slurries—Deep Anadarko Basin

Slurry type	Properties		Cost	
	Wt-lb/gal	Yield cu ft/sk	sk	ft³
Class H 2% $CaCl_2$	15.6	1.18	7.10	6.01
*HLC 10% NaCl, 2% $CaCl_2$ 10 lb Gilsonite	12.7	1.93	10.34	5.36
Class H 10% NaCl, 2% $CaCl_2$	15.8	1.2	7.72	6.45
Class H (Densified) 10% NaCl, 1% CFR-2	16.4	1.10	10.38	9.42
50-50 Pozmix Cement 10% NaCl, 1% CFR-2	15.5	1.07	8.66	8.09
*HLC 5% NaCl	12.4	1.97	6.21	3.16
HLC 10% NaCl, .4% HR-4 10 lb Gilsonite, 0.75% CFR-2	12.5	2.14	12.61	5.90
HLC 4% gel, 0.5 CFR-2 6 lb Gilsonite 0.2% HR-4	11.5	2.81	10.06	3.57
Class H 30% silica flour 1% CFR-2; 1.2–1.5% HR-12 0.25 lb/sk NF-P	16.0–16.2	1.40	17.56	12.55
**80-20 Pozmix cement 2% gel, 18% NaCl 30% silica flour 2% Halad 14; 0.25 lb/sk NFP	14.5	1.84	21.71	11.79

*Halliburton Light Cement
**80% cement-20% Pozmix cement

ratory slurry tests should be checked by field tests shortly before the job with materials from the location.

Fluid Loss—Slurry design should provide for controlled fluid loss (100 to 175 cc API 1,000 psi, 325-mesh screen at 190°F). Viscosity reduction to permit turbulent flow at reasonable displacement rates should be considered.

Slurry Mixing—Batch mixing of cement slurry promotes uniformity of mixing, and permits actual tests of slurry properties and, if necessary, adjustments of properties before pumping slurry into well.

Strength Retrogression—To inhibit strength retrogression where formation temperatures are above 230°F., 30 to 40 percent by weight silica flour should be used. Water requirement for normal cement grade silica flour (less than 200-mesh particle size) is 4.8 gal/100 lb silica. For high weight cement slurries, coarse silica (60–140 mesh) can be substituted to reduce the additional water needed.

Mud Displacement—Adequate displacement of mud by the cement slurry is a major problem due to high temperature gelation of the mud, small annular clearances, and difficulty in moving a long heavy string of casing or liner. Aids in displacement of mud include:

1. Centralization of casing
2. Movement of casing
3. Reduction of mud plastic viscosity and yield point.
4. Laboratory model test simulating high temper-

ature conditions show that invert-emulsion mud (having inherent low API fluid loss and ten minute gel strength) is displaced much more effectively than water base mud where high temperature gelation is a problem.

5. Slurry displacement rates as high as possible into the turbulent flow range.

6. Use of sufficient volume of cement to insure that slower moving cement on the "narrow side" of the annulus rises to the desired height in the hole.

Steam injection wells normally have low temperature during placement of cement. Injection of steam (400° to 700°F.) later creates extreme stresses down hole on casing and cement. Best practices include:

1. Use of heavier casing—K-55, N-80, or P-105—plus special threads in some cases.
2. Injection of steam down tubing leaving empty annulus as an insulator.
3. Circulate cement to surface.
4. Cement bottom section of casing—pull tension on casing—cement remainder of casing through DV tool.
5. Use API Class A. G, or H cement (or Pozzolan cement) with 40% silica flour to prevent strength retrogression (Table 4-21).

Geothermal steam wells present very high temperature conditions during placement of cement (300° to 700°F.). It is desirable that cement have good insulating properties to reduce heat loss to formation.

A satisfactory slurry can be designed using API Class G basic cement, with 40% silica flour to prevent strength retrogression and sufficient retarder to allow placement. Slurry should be pretested in laboratory to determine retarder requirements (Table 4-22).

Expanded Perlite can be added to improve insulating properties (reduce "K" value) but it reduces compressive strength (Table 4-23).

Fire flood wells are cemented under low temperature conditions, but can be subjected to very high temperature (750° to 2,000°F.).

If temperatures higher than 750°F. are anticipated, and cracking of cement will be detrimental, a refractory-type cement (calcium-aluminate) containing 40 to 60% silica flour should be used through the hot zone. A cheaper cement can be used 150 ft above the hot zone (Table 4-24).

TABLE 4-21
API Class G Cement with Silica Flour
(Cured 3 days @ 80°F., thereafter @ 600°F.)

% Silica flour	Compressive strength—psi			
	0 days	1 day	14 days	28 days
40	2,610	3,380	3,375	3,165
50	2,385	3,212	3,015	2,925
60	2,160	2,950	2,780	2,865

TABLE 4-22
API Class G Cement
For Geothermal Steam Wells
(Cured at 440°F.: 1, 3, 7 Days)

% Silica flour	Perlite, cu ft/sk	Compressive strength—psi		
		1 day	3 days	7 days
0	0	545	545	425
40	0	7,330	11,025	10,010
40	1	3,690	3,580	3,975
40	3	1,690	1,734	1,825

Cementing in Cold Environments

Cementing in frozen formations such as the Permafrost in the northern areas of Canada, the Arctic Islands and the Alaskan North Slope presents problems depending on the type of permafrost.

Consolidated Formation—A frozen consolidated formation that is unharmed by thawing can usually be cemented with a variety of slurries that will adequately set up at the curing temperature available.

Accelerated API Class A, C, or G Cement with a slurry temperature of 50°F. to 60°F., preferably densified, could be used. Subjected to 20°F. formation temperature, unheated Class G Cement with up to 4% $CaCl_2$ freezes before it sets up.

TABLE 4-23
Thermal Insulating Values
API Class A Cement with Various Additives

Water, gal/sk	5.20	6.80	12.50	24.30	14.60
Bentonite %	—	—	2.0	2.0	—
Silica flour %	—	40.	30.	30.	30.
Perlite, cu ft/sk	—	—	1.0	3.0	—
Vermiculite %	—	—	—	—	30.
"K" Value—btu/sq ft/hour/°F/ft					
24 hours	1.068	.955	.535	.653	.324
48 hours	1.005	.828	.420	.515	.268
120 hours	.803	.475	.400	.332	.324
144 hours	—	—	.319	.333	—
329 hours	—	—	—	—	.320

TABLE 4-24
Calcium-Aluminate Cement
With Silica Flour for Fire Flood Application

% Silica flour	Compressive strength—psi 7 days curing time at oven temperature 700°F.	1,000°F.	1,500°F.
40	1,240	1,130	1,530
60	1,120	950	1,020

Subsequent thawing and hydration of the slurry results in great loss of strength and increase in permeability; thus slurry heating to maintain 40°F. is justified.

Figure 4-21 shows slurry temperature resulting from mixing cement of a particular temperature with water having a particular temperature. In situations where it is not possible to maintain slurry temperature above 40°F., gypsum-cement blend (Permafrost Cement) should be used.

High-aluminite cement (Ciment Fondu) has been used. Aluminite cement is quite expensive but will set up at low temperatures (20°F.–25°F.) to provide minimum compressive strengths of 300–500 psi in 24 hours. If allowed to freeze, however, permeability will be very high.

Unconsolidated Formations—In areas where the shallow formations consist of ice lenses, frozen muskeg, and unconsolidated sands and gravels, requirements for successful cementing are more difficult:

—Hole enlargement due to melting and erosion during drilling must be prevented.
—Cement must set at low temperatures without requiring excessive heating of mix water.
—Cement must set without excessive heat of hydration to prevent melting of permafrost.
—Cement sheath must be thick enough to provide insulation to prevent thawing of permafrost by circulation of mud during subsequent drilling.
—Displacement of mud by cement is very essential to eliminate freezable fluids in annulus.

Aluminite cement (Ciment Fondu) does not work well under these conditions because it liberates a large quantity of heat (90 btu/lb of slurry) even when diluted with 50% fly ash.

The best answer to cementing unconsolidated permafrost appears to be a gypsum-cement blend (Halliburton tradename—Permafrost Cement). Heat of hydration is less than 20 btu/lb of slurry. Salt can be added to lower freezing temperature to about 20°F. Slurry properties can be adjusted to give two hours minimum pumpability at temperatures from 30°F. to 60°F. with a 16-hour compressibility strength of 500 psi at 20°F (Fig. 4-22).

Permeability of any cement is increased significantly if it is allowed to freeze before setting. Ciment Fondu allowed to freeze has a permeability to water of 50 to 60 md. Permafrost Cement under similar conditions has a permeability of about 25 md; thus, where permeability is important, the slurry must not be allowed to freeze.

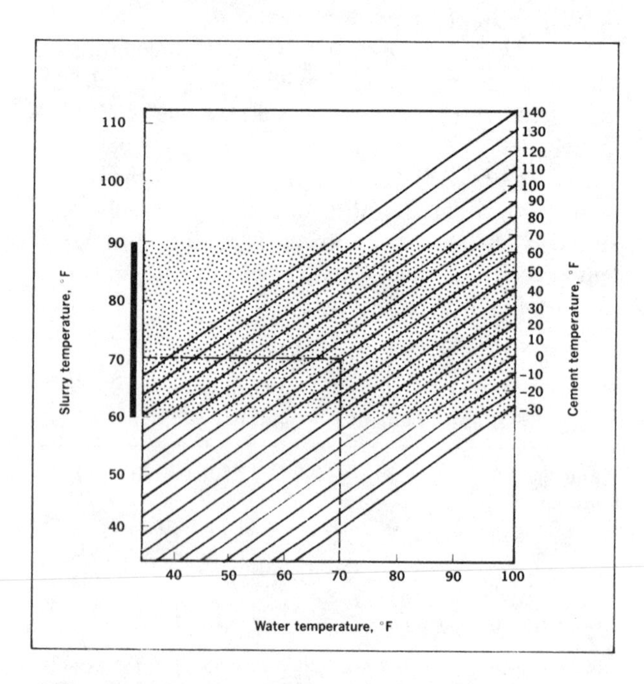

FIG. 4-21—Slurry temperature for various temperatures of water and cement.[19] Permission to publish by The Society of Petroleum Engineers.

Cementing Through Gas Zones

Gas communication through a cemented annulus was first recognized in the completion of gas storage wells. The problem became more evident in deep well completions across gas intervals, causing a pressure buildup in the annuli of the production and intermediate casings, and on occasion a blowout.

Laboratory model studies simulating downhole conditions have been helpful in understanding the occurrence of gas leakage and determining some of the factors responsible for gas migration in wells. These

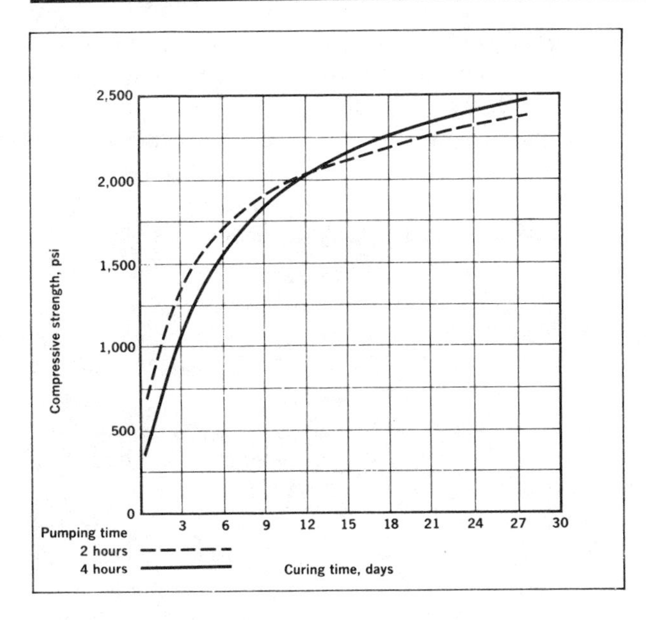

FIG. 4-22—*Compressive strength of Alaskan permafrost cement, 20° to 80°F. Courtesy of Halliburton Services.*

studies show that leakage is dependent upon hole conditions, mud-cement density, temperature, pressure differential, and cementing composition.

Leakage appears to be related to the inability of the cement column to effectively transmit full hydrostatic pressure to the gas zone.

One explanation may be that (a) local dehydration of cement slurry opposite uphole permeable zones, or (b) higher temperatures off-bottom due to the fluid circulation, may accelerate the initial setting process in a section of the cement column above the gas formations. As this upper portion of the cement column takes its initial set it begins to partially support the hydrostatic column above.

As cement filtrate is lost into lower zones the hydrostatic pressure on the gas zone is reduced allowing gas bubbles to enter the annular cement column. Due to the low density and high compressibility of gas, it migrates upward expanding and disturbing the setting cement and developing communication "worm holes" for passage of additional gas.

A companion explanation is that in the transition from a slurry to a solid, growing cement crystals begin to interfere with one another and form a mechanical structure. For a period of time they act as grains of sand in a porous rock matrix with permeability through the pore spaces between the cement crystals. This porosity is filled with the cement mix water. The mechanical interference between crystals prevents the downward movement of the cement column needed to maintain hydrostatic pressure; thus, as mix water filters into a permeable formation, or is taken up in the hydration process, the hydrostatic pressure drops. As the cement setting process continues, permeability in the pore spaces between the cement crystals is reduced, finally resulting in a very low permeability material. But for a period during the transition stage a low viscosity fluid such as gas can invade the setting cement if the effective hydrostatic pressure in the annulus is less than the gas zone pressure.

Buildup of gel strength in the cement column provides a similar scenario to that of interference between growing cement crystals—and based on current knowledge is probably the most important effect. Gel strength buildup begins shortly after the cement is in place. As it reaches a value of 100 lbs/100 sq ft, downward movement of the cement column to restore liquid lost to the formation is seriously restricted and hydrostatic pressure begins to drop. At this point the properties of the cement slurry are such that gas can move into the cement column; thus, creating the gas leakage situation. The gelation process continues and as the gel strength reaches about 500 lb/100 sq ft, the consistency of the cement is sufficient to prevent gas movement into the cement column. With this scenario, either mechanically breaking the gel strength (with a vibrator) or chemically delaying the gel strength buildup, such that hydrostatic pressure is maintained for a longer time in the transition phase, would be helpful in reducing the gas leakage problem. Toward the end of the transition period build up of mechanical strength begins and the possibility of gas invasion is reduced.

Figure 4-23 shows the **results of a field study**[34] of the gas leakage problem wherein pressure and temperature sensors were placed on the outside of the casing to record annular pressures and temperatures at various depths both during cement placement and the subsequent transition period as the cement changed from a pressure transmitting slurry to a low permeability solid material. As cement, moving upward in the annulus, passed each sensor, pressure began increasing at that sensor. At the lower sensors the slight loss of pressure before the increase probably indicates the passage of the 180–200 foot water spacer ahead of the cement. After the cement passed the top sensor, it became apparent that the cement top was rising faster than expected because of channeling, and that the pressure at the bottom sensor was nearing the formation fracturing pressure. Mixing and pumping of cement was halted at that time, the top plug was re-

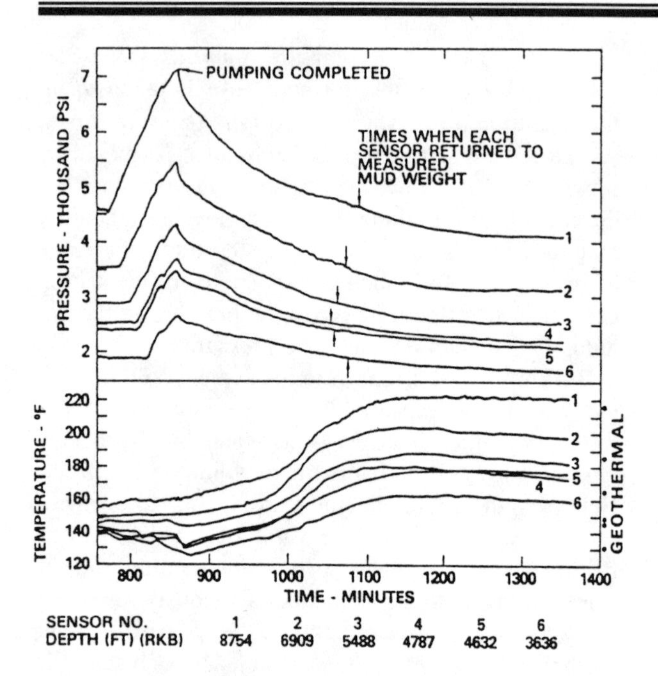

FIG. *4-23—Annular pressure and temperature—Well A. Permission to publish by Society of Petroleum Engineers.*

leased, displacement of the cement to the bottom of the casing was completed, and the pumping operation shut down. Initially the pressure dropped rapidly (probably indicating closure of an induced fracture), and then began to decline at a slower rate as the mix water moved into the formation or was absorbed in hydration. At 890 minutes on the time scale, 24 minutes after shutdown, the blowout preventors were closed and 100 psi pressure was applied to the annulus at the surface. This pressure was not detected even by the top sensor, located about 2400 feet below the top of the cement, because in this 24 minute time period the cement had developed sufficient gel strength to prevent pressure transmission. The arrows between 1000 and 1100 minutes indicate the time when the hydrostatic pressure dropped to that of the drilling fluid column prior to cement placement. Using the temperature sensors as indicators of the progress of the hydration process it is seen that hydration was essentially complete at this time, thus, formation gas probably did not enter the cement column. Note, however, that pressures continue to decline probably equalizing with the formation pressure at each sensor.

To minimize gas leakage, consideration should be given to:

1. Sufficient mud circulation prior to cementing to insure that dispersed gas is not trapped in the hole.
2. Sufficient hydrostatic overload on formation pressure, even though the hydrostatic pressure of the cement column drops during transition. This can be accomplished by closing the BOP's and maintaining annular pressure—*if pressure is applied before significant gel strength is developed* (5 to 10 minutes).
3. Mixing cement slurry as heavy as hole conditions will tolerate.
4. Using filtration control slurry to reduce loss of mix water to the formation.
5. Increasing the viscosity of the mix water such that its mobility is reduced.
6. Avoiding over-retard cement slurry. Obtain accurate well temperature information so that the slurry can be designed to set as quickly as possible after placement.
7. Addition of aluminum powder to the cement (0.5 to 1.0 lb/sk), which in contact with calcium hydroxide produced by the setting cement forms hydrogen gas. This creates a compressible phase to maintain annular pressure during the transition stage. Nitrogen in foam cement creates a similar effect.
8. A recent laboratory study shows that vibrating the casing, to continually break the gel strength of the cement, allowing it to transmit pressure for a longer time during the transition period, will minimize the gas leakage problem.
9. Probably the most promising recent development is that of delaying the buildup of gel strength on the cement column using an additive termed GasStop (Halliburton Services). Delaying gel strength development allows the cement column to transmit essentially full hydrostatic pressure, and thus prevent gas entry, for an extended period of time. The delayed gel strength phase is then followed by a rapid progression to the set state. Figure 4-24 illustrates this concept.

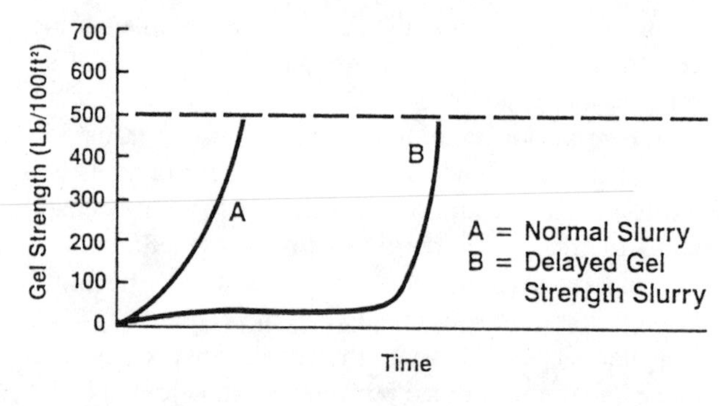

FIG. *4-24—Comparison of gel strength development.*

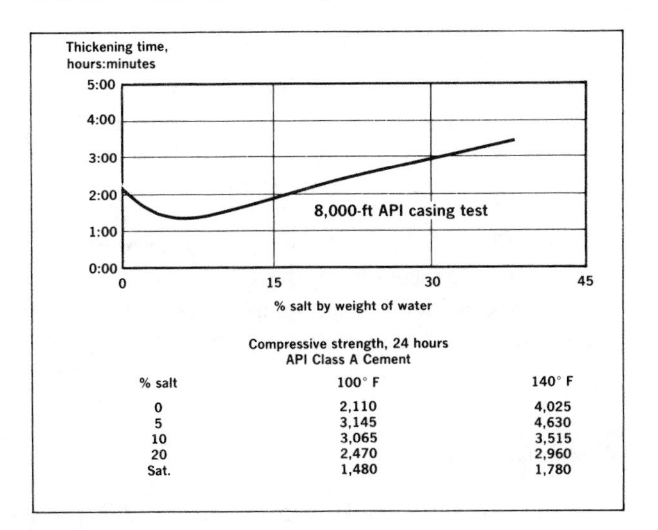

Compressive strength, 24 hours
API Class A Cement

% salt	100° F	140° F
0	2,110	4,025
5	3,145	4,630
10	3,065	3,515
20	2,470	2,960
Sat.	1,480	1,780

FIG. *4-25—Effect of salt on thickening time and 24-hour compressive strength. Courtesy of Halliburton Services.*

Salt Cement

Cementing Through Salt Sections—Salt-saturated cement (NaCl) has been used for many years to improve bonding through massive salt sections. Fresh water slurries will not bond to salt since water from the slurry dissolves away the salt at the interface.

Shale Sections—Currently there is a trend toward use of salt cement (10% to 18%) to improve bonding through shale sections where contact with fresh water from the slurry may cause disintegration of shales. Normally inhibition of clay disintegration can be obtained with less sodium chloride than that required for saturation.

Sixteen to eighteen percent sodium chloride appears to be optimum since this concentration will inhibit clay hydration, but will not significantly affect thickening time or reduce the 24-hour compressive strength of the cement (Fig. 4-25).

TABLE 4-25
Effect of Salt on Fluid Loss—Class A Cement

	Filtration Rate (cc/30 minutes) 325-mesh screen at 1000 psi	
% Salt	*1.0% Additive*	*1.2% Additive*
0	80	35
5	84	44
10	80	40
15	78	48
Sat.	197	58

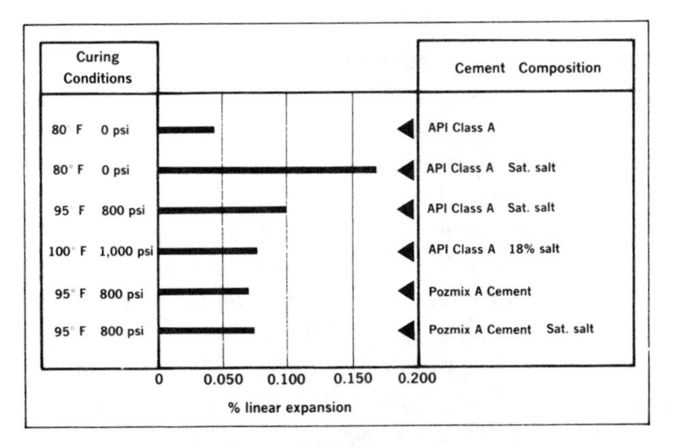

FIG. *4-26—Linear expansion of salt cement, 28-day curing. Courtesy of Halliburton Services.*

Compatibility With Other Additives—Sodium chloride is generally compatible with all light weight and heavy weight additives, accelerators and retarders, and other special additives. Some fluid-loss additives are adversely affected by the chloride ion; thus, to obtain minimum fluid loss a slight increase in fluid-loss additive is required (Table 4-25).

Expansion—Salt-saturated cement undergoes a slightly greater expansion upon setting than API Class A cement, as shown on Figure 4-26.

Rheology Improved—Sodium chloride acts as a dispersing agent to improve rheological properties and to reduce the displacement rate required for turbulent flow. Sodium chloride also increases slurry weight and volume slightly (Fig. 4-27).

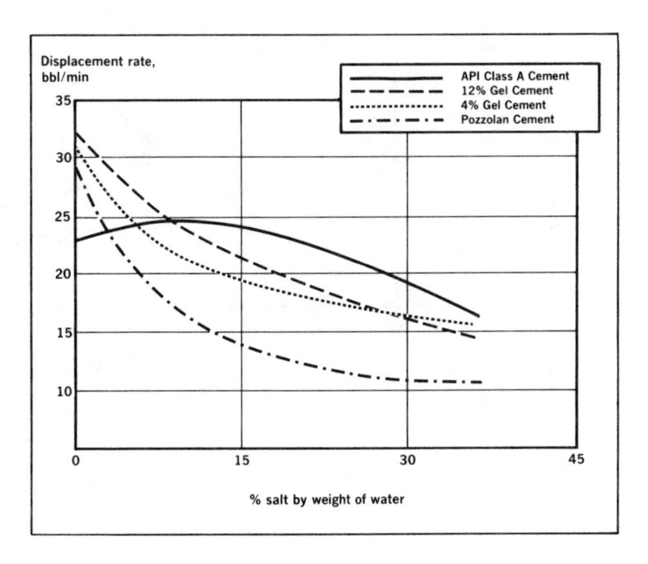

FIG. *4-27—Displacement rate for turbulent flow ($8^3/_4$-in. hole, $5^1/_2$-in. casing. Courtesy of Halliburton Services.*

Blending Problems—Bulk blending of dry granulated salt with cement greatly simplifies mixing problems. A defoamer is recommended to minimize foaming. Mixing cement with saltwater causes excessive foaming; if dry blending equipment is not available a defoamer should be used.

Fine granulated salt (20 to 100-mesh size) should be used in bulk blending. For a salt-saturated slurry 3.1 lb of NaCl/gal of fresh mix-water is required. Figure 4-28 can be used to determine the amount of salt for other degrees of saturation. Normally speaking every 1/2 lb of salt added/gal of water is equivalent to 6 wt % salt in water.

Substitution of Potassium Chloride—Potassium chloride can be substituted for sodium chloride in cement slurries to provide similar effects at lower concentrations. At high concentrations (above 10%) excessive slurry viscosities may result, however.

Delayed Setting Cement

With proper fluid loss control and retarder, cement slurry can be designed to remain fluid for periods up to 36 hours. Delayed set permits pipe to be run into unset cement. Principal applications are:

1. Multiple tubingless completions to improve primary cement job between strings.
2. Liner cementing where small clearance and small volume of cement makes satisfactory cementing difficult.

Expansive Cements

API Class A, G, or H cement cured under moist conditions and recommended water ratios expand slightly when set. Pressure decreases expansion,

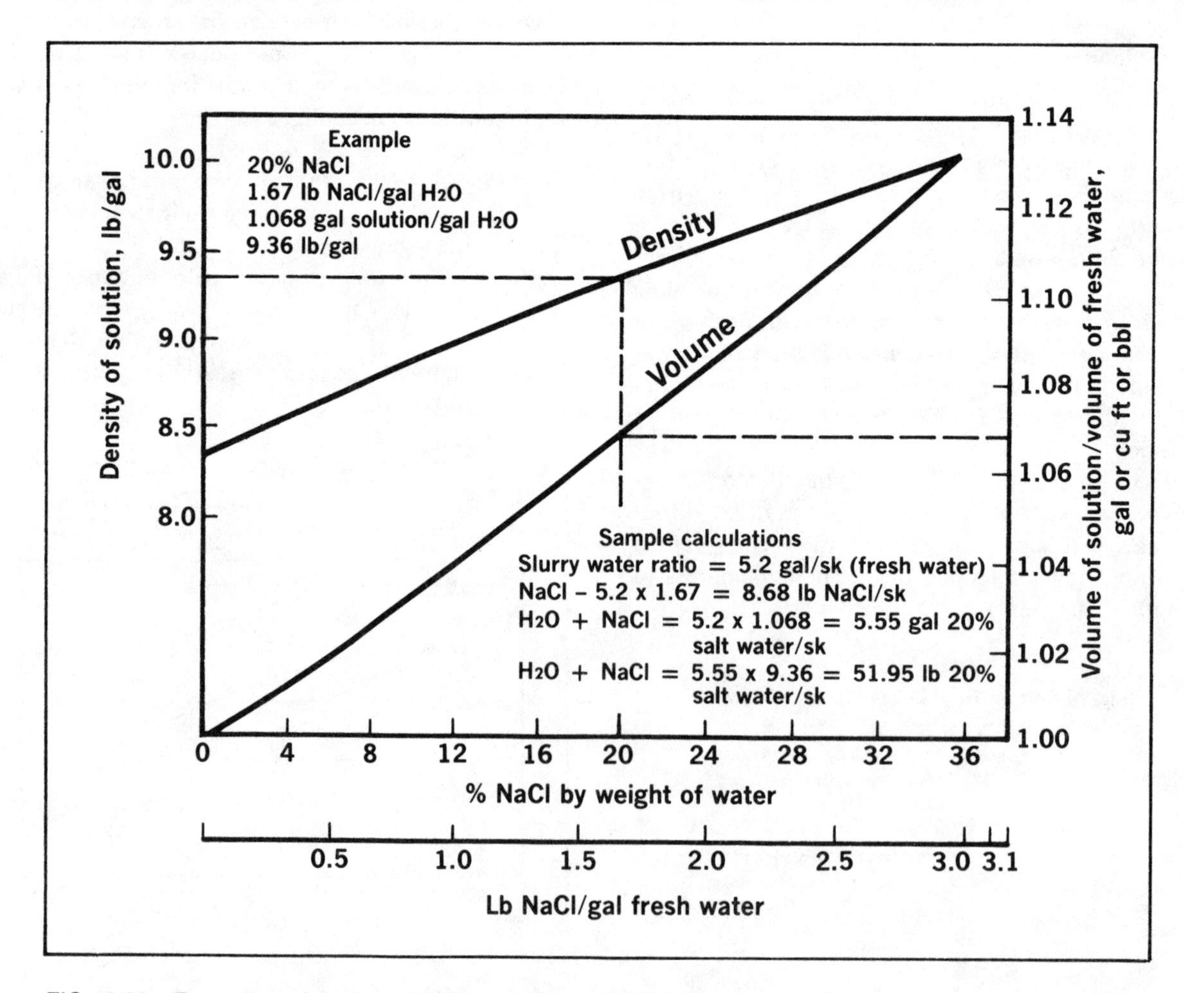

FIG. *4-28—Properties of sodium chloride solutions at 68°F, based on percent salt by weight of water for salt cement slurries. Courtesy of Halliburton Services.*

whereas temperature increases expansion, particularly from heat of reaction.

Chemical additives beneficial in increasing cement expansion include salt (NaCl), calcium or sodium sulfates, and pozzolan. Gas-forming agents (aluminum, iron, zinc, or magnesium powder) may be beneficial at low hydrostatic pressure.

For well use, salt cement (18% to saturation) appears to be a good way to realize benefits from expansion plus providing other benefits (Fig. 4-26). Cement plus gypsum and Pozmix cement also have application in wells.

Densified Cement

Densified cement—or cement mixed with less than normal water (above 3.5 gal/sk rather than 5.2 gal/sk) and with friction reducer (CFR-2 or CFR-3) added to improve pumpability—has several desirable properties including:

—High compressive strength.
—Relatively low fluid loss.
—Resistance to mud contamination.

Typical properties are shown in Table 4-26.

Ability to tolerate mud contamination is shown by the data in Table 4-27.

Thixotropic Cement

Thixotropic cement consists of Class A, G, or H cement with the addition of bentonite, gypsum and calcium chloride to provide rapid buildup of gel strength as soon as movement of the slurry stops.

Primary application is in shallow low temperature (50°–120°F.) zones where natural or induced fractures otherwise permit loss of whole cement to the formation. Typical properties are shown in Table 4-28.

TABLE 4-26
API Class A-B-G or H Cement

Weight, lb/gal	% CFR-2	Fluid loss, 1000 psi-screen	Compressive strength, 100°F.—3000 psi
15.6	0.0	1000 cc+	1465 psi
15.6	1.0	286 cc	1445 psi
16.5	1.0	192 cc	3625 psi

TABLE 4-27
Effect of Mud Contamination on Strength of Cement

% Mud contamination	12 hr-compressive @ 230°F.	
	15.6 lb/gal	17.6 lb/gal*
0	2910	7919
10	2530	5005
30	1400	2910
60	340	2315

*Densified cement w/CFR-2

PRIMARY CEMENTING PRACTICES

Practical Considerations Before Cementing

Mud Conditioning—Mud should be conditioned before running casing at pumping rate equal or greater than when drilling in order to clean the hole. Drill pipe rotation aids hole cleaning. Plastic viscosity and yield strength of mud should be as low as possible.

After casing is on bottom, mud should be circulated and pipe rotated or reciprocated to break gel strength. Circulation should continue until "mud caliper" tests indicate 90–95% of the fluid in the hole is actually moving.

Centralization—Effective centralization is a critical factor in obtaining a good primary cement job.

In straight hole use one centralizer per joint, 200 ft above and below pay zone. Use one centralizer every

TABLE 4-28
Properties of Thixotropic Cement—Class A
(Thickening Time 60 Min)

Water gal/sk	Bentonite %	Cal-Seal lb/sk	$CaCl_2$ %	Slurry wt lb/gal	Compressive strength psi @ 80°F. 6 hr	12 hr	24 hr
7.8	4	8	2.0	14.4	160	475	1120
13.0	12	10	1.0	12.7	45	145	460
15.6	16	10	1.0	12.3	30	110	205

TABLE 4-29
API Specifications for Centralizers

Casing size, in.	*Wt. of casing, lb/ft*	*Minimum restoring force, lb*[(a)]
$4^1/_2$	11.6	464
$5^1/_2$	15.5	620
7	26.0	1,040
$9^5/_8$	40.0	1,600

[(a)] at 0.67 standoff ratio.

third joint in remainder of cemented zone. See Table 4-29 for API specifications for centralizers.

Centralizers should be placed in gauge sections of the hole as determined by caliper or other logs.

In crooked hole centralizer placement depends on hole deviation (Fig. 4-29). Movement of centralizers on body of casing should be limited by clamp rings.

Starting force—Maximum force should be less than weight of joint of 40-ft casing between centralizers.

Scratchers—Wall cleaners or scratchers probably help in eroding wall cake and improving "bond", although the need for wall cleaners in this regard is difficult to prove in the field.

Recent lab tests closely simulating downhole conditions indicate that the important function of scratchers is not to "remove the wall cake," but to break the gel of the mud and to mix up the mud with the cement, thus reducing channeling.

On this basis scratchers should be given serious consideration. They should perhaps be renamed "stirrers" instead of scratchers to more properly reflect their actual downhole effect.

Floating Equipment—Float shoe and float collar (with back pressure valve) minimizes derrick strain, and prevents cement back-flow when pressure on casing is released after cement is in place.

Float collar should be located one or two joints above the float shoe to prevent mud-contaminated cement (which may collect below the top wiper plug) being placed outside the bottom casing joint.

Differential fillup equipment can often justify ad-

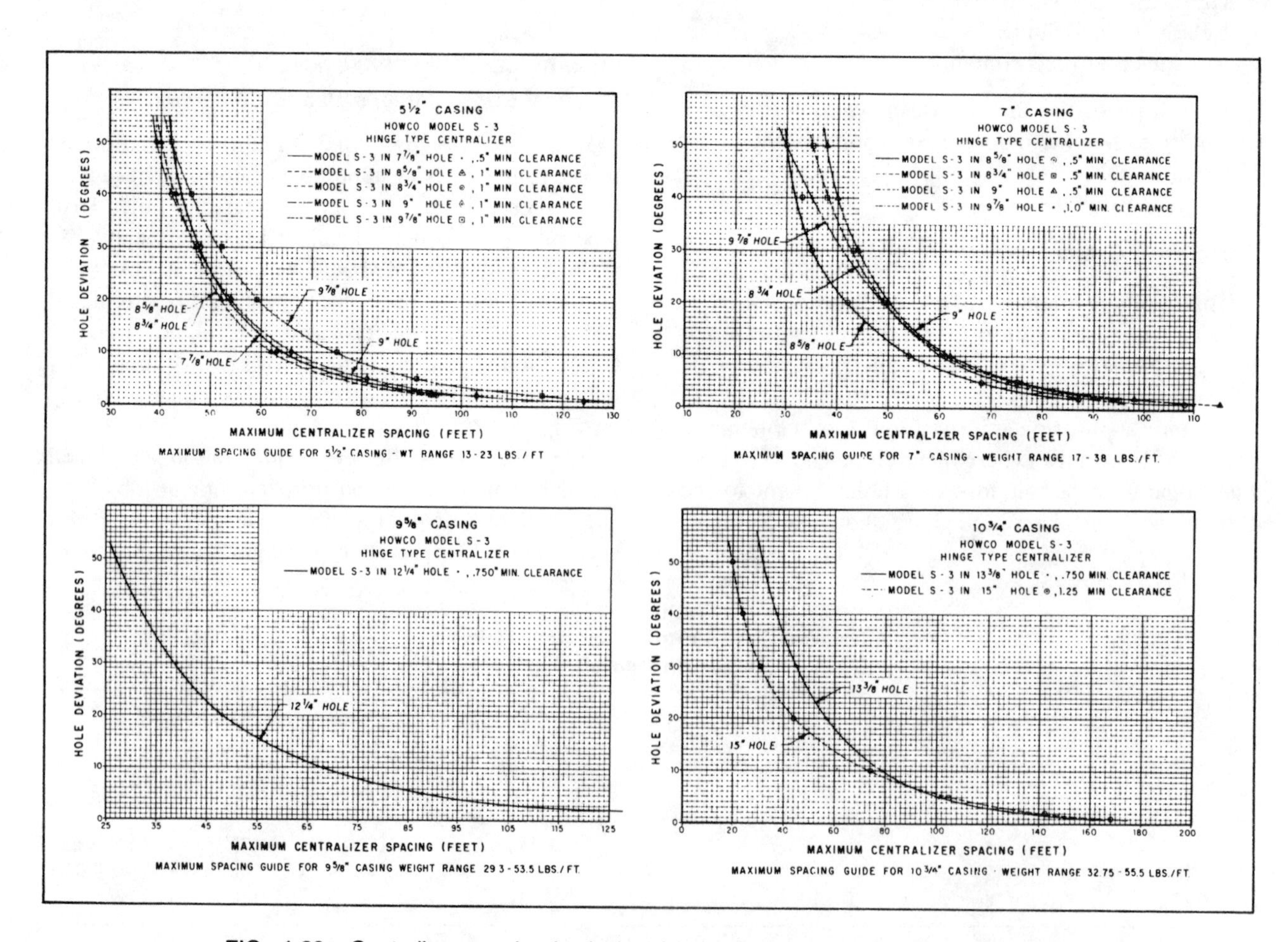

FIG. *4-29—Centralizer spacing in deviated hole. Courtesy of Halliburton Services.*

ditional cost by reduced downtime to fill casing, and reduced pressure surge on formation.

Casing-Running Practices—Where lost circulation is not a problem casing can be run at a rate of 1,000 ft/hour or faster. Magnitude of pressure surges should determine running time where lost circulation is a possibility.

Volume of Cement—Circulate cement at least 500 ft above top producing formation. Corrosion protection or lost circulation zones may dictate more cement. Where possible a larger volume of cement, placed in turbulent flow, reduces possibility of mud channels remaining in the productive zone. Use of lower density slurries permit longer cement columns.

Hole washout must be considered to determine the cement volume to provide a desired length of fillup. With a caliper survey to determine annular volume between casing and hole, add 10% to 15% for cement volume. With no caliper survey, add 25% to 100% depending on experience.

Considerations During Cementing

Cement Mixing—Quality of the cement mixing operation correlates with job success. Weight of cement slurry should be monitored (and recorded) to insure that the correct quantity of mix water is used.

Where densified or high weight slurries are used, particularly at displacement rates less than 5 bbl/min, a recirculating-type jet mixer improves uniformity of slurry compared with the standard jet mixer. Figure 4-30 compares these two types of mixers.

For critical cementing jobs, batch mixing provides much greater slurry precision, and permits measuring and adjusting rheological or other properties of the slurry before it is pumped into the well. Several large batch tanks permit continuous operation.

Washes Ahead of Cement—Water if used in sufficient quantity is an excellent wash since it is cheap, easy to put into turbulence, and doesn't affect cement setting time. Fifty bbl of water (300–500 ft of annular fill) should be used unless hydrostatic head is reduced excessively.

Some mud thinners (quebracho, lignosulfonates) added to water may retard (or inhibit entirely) cement setting.

Dilute mix of Portland or Pozzolan cement for scavenging purposes is an excellent preflush since it is easy to put in turbulent flow and solid particles promote erosion of gelled mud and filter cake.

Acetic acid (10%) with corrosion inhibitor and surfactant may be beneficial. Acetic acid should not cause pitting corrosion.

HCl (5% to 10%) with corrosion inhibitor and surfactant is sometimes used; but pitting corrosion is possible. Acid should be separated from cement by water spacer.

Non-acid washes, such as mud thinner and surfactant (Halliburton Mud Flush), may be beneficial where large volumes of water would reduce hydrostatic head.

Thickening behavior can be caused when cement and some invert oil systems come in contact. Diesel oil with a water wetting surfactant should be used as spacer, followed by a flush in sufficient quantity to water-wet the pipe.

Cementing Wiper Plugs—Top plug separates mud and cement, and provides shut-off when cement is in place.

Bottom plug should be used to wipe mud off casing ahead of cement, as well as to separate mud and cement. Without bottom plug, mud film wiped by top plug accumulates ahead of top plug as shown in Table 4-30.

Two-plug containers should be used to facilitate release of plugs without delay.

Displacing Fluid Behind Top Plug—Mud is normally used on surface or intermediate casing although fresh water may be better, depending on mud program for drilling deeper.

For the production casing freshwater, saltwater, or seawater could be desirable, depending on completion program. Selection should minimize formation damage and completion time. Diesel oil might be used to reduce swabbing time.

Sugarwater or retarding additive is sometimes placed immediately above top plug in small diameter casing to inhibit setting of cement that may have bypassed top plug.

Casing Movement During Cementing—Casing movement rates high on the list of factors affecting successful mud displacement. Ideally casing should be reciprocated (or rotated) until the top plug reaches bottom. Pipe movement may be desirable after cement is in place.

Frictional drag, weight of the pipe, and differential sticking are factors acting to prevent casing movement. Differential sticking, often the important factor, is a function of:

—Contact area between pipe and permeable borehole wall
—Pressure difference between mud column and formation
—Available pulling force
—Sticking coefficient.

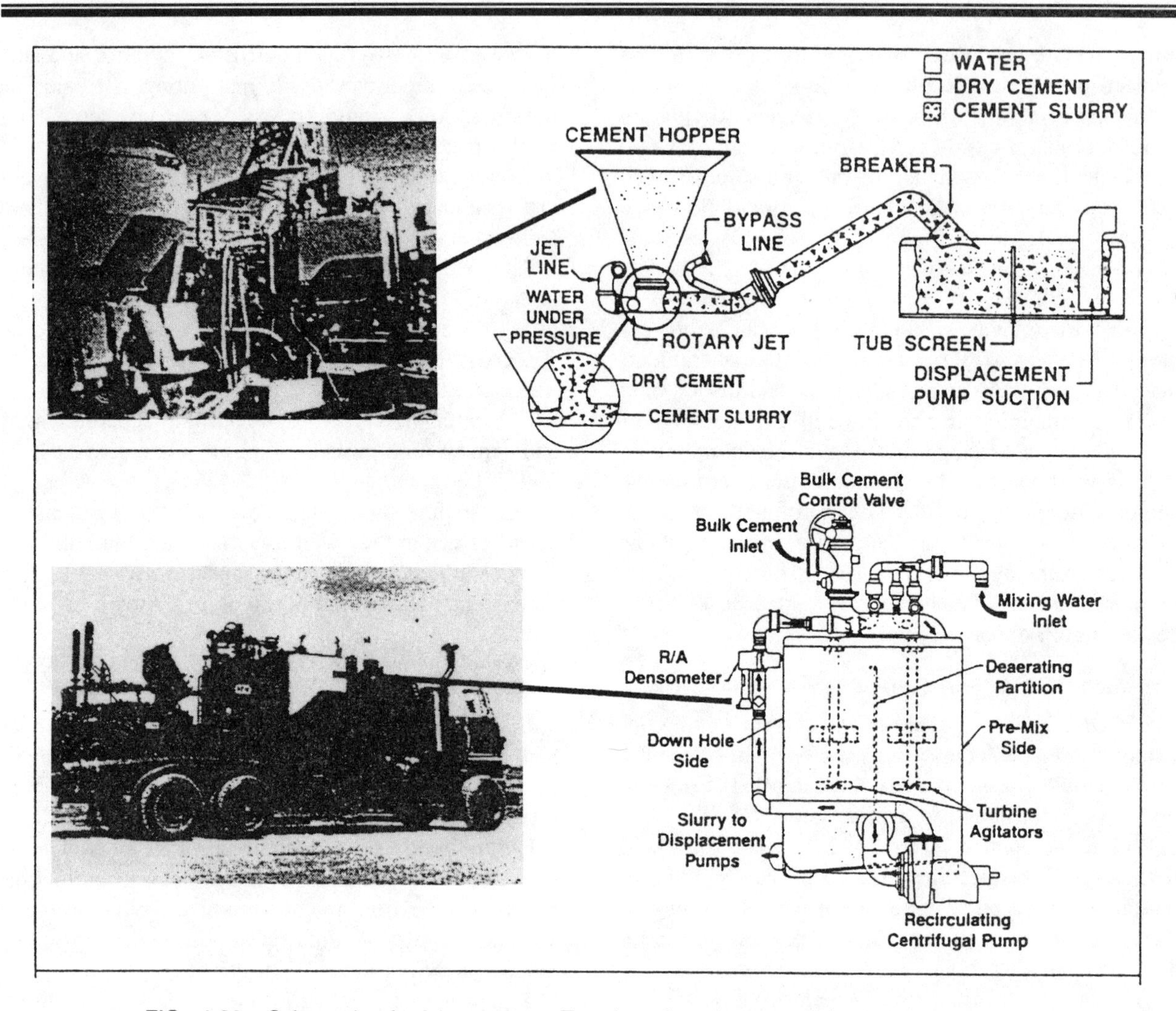

FIG. *4-30—Schematic of mixing system. Top: jet mixer; bottom: recirculating mixer.*

Sticking coefficient is a function of the mud properties (primary water loss) and the time that the casing remains stationary against a permeable formation. By reducing "stationary time" through fast cementing head hook-up procedures after casing is on bottom, differential sticking can often be eliminated.

TABLE 4-30

Mud film thickness, in.	Fillup/1,000 ft in casing ahead of top plug — 5½-in. casing	7-in. casing
1/16	50.6 ft	40.0 ft
1/32	25.5 ft	20.1 ft
1/64	12.6 ft	9.9 ft

Casing movement (reciprocation or rotation) during cement displacement should be slow as the cement first reaches bottom but should increase as the cement is displaced and the top plug reaches bottom. Rate of reciprocation should be on a two-minute cycle over 15 to 20-ft intervals. Rotation can be done with hydraulic casing tongs to limit torque to that used to make up casing.

Considerations After Cementing

Casing Pressure—Casing pressure should be released after it is determined that the back pressure valves in float collar and guide shoe are holding. This reduces casing expansion, and microannulus effect on the sonic bond log.

In small diameter casing (tubingless completions)

it may be desirable to hold pressure on casing, as a means of applying additional tension (to prevent subsequent buckling) if casing is landed before cement takes initial set. The sonic bond log may be adversely affected, but hoop expansion due to differential pressure is low in small diameter pipe.

Waiting-On-Cement Time—The reason for WOC time is to permit cement to attain sufficient strength to:

—Anchor the pipe and withstand shock of subsequent operation.
—Seal permeable zones, (and confine fracture pressures).

Tests show that tensile strength of 8 psi (compressive strength of 50 psi) is sufficient for anchoring in most situations. See Table 4-31.

Compressive strength of 130 psi with Portland cement provides sufficiently low permeability for sealing.

WOC time is often based on time required for cement to attain 500 psi compressive strength. This allows a safety factor of 2 to 5.

Lab and field correlations show that generally Portland cement achieves tensile strength of 8 psi in 1.5 times the period required to reach maximum temperature due to heat of hydration of the cement. As a practical method, temperature increase of the fluid inside the casing due to the setting cement can be observed on the rig floor by flow back of fluid from the casing.

TABLE 4-31
Length of Casing Supported by a 10-ft Column of 8-psi Tensile Strength Cement

Casing OD, in.	Casing weight, lb/ft	Length supported by 10 ft of cement, ft
4½	9.5	367
	11.6	301
5½	15.5	275
	17.0	251
	20.0	213
7	20.0	271
	26.0	209
	32.0	170
10¾	40.5	206
	51.0	163

(After Bearden and Lane)

Casing Landing Practices—Casing landing practices based on API Study Group Recommendations are outlined below:

In general, it is recommended that all casing be landed in the casing head at exactly the position in which it was hanging on the hook when the cement plug hit bottom. In other words, land the casing as cemented with the only movement of the pipe being that necessary to transfer the weight of the casing to the well head or casing hanger.

This recommendation applies to all wells where mud weights do not exceed 12.5 lb/gal and where:

—Standard design factors are used in tension and collapse.
—Wellhead equipment is available to permit hanging weight equal to the tensile strength of the casing on the hanger without damage to the casing.
—The joint strength, in compression, of the top section of the surface casing is sufficient to withstand the loads imposed by landing the casing as cemented plus the weight of the tubing, plus induced loads that may be brought about by future operations.

The ability of the surface formations surrounding the surface casing to help support the loads imposed on the top of the surface casing is an important factor to be considered in designing the top section of the surface casing to withstand these loads.

Special cases are those where casing is to be set:

—In extremely deep wells where the standard design factors are of necessity reduced.
—Where extreme top-hole operating pressures are anticipated.
—Where excessive mud weights are necessary.
—Where other unusual circumstances dictate special consideration.

In these cases, it is the opinion of the committee that the landing practice should be based on calculations as outlined by Lubinski. Several operators have used Lubinski's work to arrive at rather simple formulas that are considered adequate for their use on an area or field-wide basis.

Testing of Primary Cement Job—A temperature survey, run two to four hours after cementing, is an excellent method of locating cement top.

Bond logs should normally be used only to make special evaluation studies.

Assume a cement job is satisfactory unless there is good evidence to the contrary.

REFERENCES

1. Morgan, B. E. and Dumbauld, G. K.: "A Modified Low Strength Cement," *Trans.,* AIME (1950) Vol. 192.

2. Owsley, W. D.: "Twenty Years of Oil Well Cementing," *JPT* (Sept. 1953) p. 17.

3. Smith, D. K., and Carter, Greg: "Properties of Cementing Compositions at Elevated Temperatures and Pressures," *Trans.,* AIME (1958) Vol. 213.

4. Ostroot, G. W., and Walker, W. A.: "Improved Compositions for Cementing Wells with Extreme Temperatures," *JPT* (March, 1961) p. 277.

5. Evans, G. W. and Carter, G. L.: "Bonding Studies of Cementing Compositions to Pipe and Formations," *Drill. Prod. Prac.* (1962).

6. Walker, W. A.: "Cementing Compositions for Thermal Recovery Wells," *JPT* (Feb. 1962) p. 139.

7. Bleakley, W. B.: "A Really Engineered Cement Job," *Oil and Gas J.* (Feb. 12, 1962).

8. Sage, Knox A.: "Rheological Design of Cementing Operations," *JPT* (March, 1962) p. 323.

9. Pettiette, Roy, and Goode, John: "Primary Cementing of Multiple Tubingless Completions," Southern District API, (March, 1962).

10. Sage, K. A., and Smith, D. K.: "Salt Cement for Shale and Bentonitic Sands," *JPT* (Feb. 1963) p. 187.

11. Harris, Francis and Carter, Greg: "Effectiveness of Chemical Washes Ahead of Squeeze Cementing," Mid-Continent District API, (March 1963).

12. Carter, L. G., and Evans, G. W.: "A Study of Cement Pipe Bonding," *JPT* (Feb. 1964) p. 157.

13. Buster, John L.: "Cementing Multiple Tubingless Completions," Southwestern District API, (March 1964).

14. Carter, L. G., and Evans, G. W.: "New Technique for Improved Primary Cementing," Southwestern District API, (March 1964).

15. Brice, J. W., Jr., and Holmes, R. C.: "Engineering Casing Cementing Programs Using Turbulent Flow Techniques," *JPT* (May 1964) p. 503.

16. Parker, P. M.; Ladd, B. J.; Ross, W. M.; and Wahl, W. W.: "An Evaluation of a Primary Cementing Technique Using Low Displacement Rates," SPE, Denver (1965).

17. McLean, R. H.; Manry, C. W.; and Whitaker, W. W.: "Displacement Mechanics in Primary Cementing," *JPT* (Feb. 1967) p. 251.

18. Childers, Mark A.: "Primary Cementing of Multiple Casing," *JPT* (July 1968) p. 751.

19. Maier, L. F.; Carter, M. A.; Cunningham, W. C.; and Bosley, T. G.: "Cementing Practices in Cold Environments," *JPT* (Oct. 1971) p. 1215.

20. Garvin, Tom, and Sage, Knox A.: "Scale Displacement Studies to Predict Flow Behavior During Cementing," *JPT* (Sept. 1971) p. 1081.

21. Carter, Greg, and Sage, Knox A.: "A Study of Completion Practices to Minimize Gas Communication," *JPT* (Sept. 1972) p. 1170.

22. Clark, Charles R., and Carter, Greg L.: "Mud Displacement with Cement Slurries," *JPT* (July 1973) p. 775.

23. Graham, Harold L.: "Rheology-Balanced Cementing Improves Primary Success," *Oil and Gas J.* (Dec. 18, 1972).

24. Carter, Greg L., and Cook, Clyde: "Cementing Research in Directional Gas Well Completions," SPE 4313, London (April, 1973).

25. Clark, Charles R., and Jenkins, Robert C.: "Cementing Practices for Tubingless Completions," SPE 4609, Las Vegas (Oct. 1973).

26. Running and Cementing Liners in the Delaware Basin, Texas," API Bulletin DI 7 (Dec. 1974).

27. Holley, J. A.: "Field Proven Techniques Improve Cementing Success," *World Oil* (Aug. 1, 1976).

28. Smith, Dwight K.: "Cementing," Monograph Vol. 4, Henry L. Doherty Series, SPE of AIME (1976).

29. Levine, D. C.; Thomas, E. W.; Bezner, H. P.; and Tole, G. C.: "Annular Gas Flow After Cementing: A Look at Practical Solutions," SPE 8255 (Oct. 1979).

30. Haut, R. C., and Crook, R. J.: "Primary Cementing: The Mud Displacement Process," SPE 8253 (Oct. 1979).

31. Haut, Richard C., and Ronald James Crook. "Laboratory Investigation of Lightweight, Low-Viscosity Cementing Spacer Fluids," SPE 10305 (Oct. 1981).

32. Cowthran, J. C.: "Technology Used to Improve Drilling Performance and Primary Cementing Success in Katy Field," SPE 10956 (Sept. 1982).

33. Sabins, Fred L., and Sutton, David L.: "The Relationship of Thickening Time, Gel Strength, and Compressive Strengths of Oilwell Cements," SPE 11205 (Sept. 1982).

34. Cooke C. E., Jr.; Kluck, M. P.; Medrano, R.: "Field Measurements of Annular Pressure and Temperature During Primary Cementing," SPE 11205 (Sept. 1982).

35. Keller, S. R.; Crook, R. J.; Haut, R. C.; Kulakofsky, D. S.: "Problems Associated with Deviated Wellbore Cementing," SPE 11979 (Oct. 1983).

36. Hartog, J. J.; Davies, D. R.; Stewart, R. B.: "An Integrated Approach for Successful Cementations," *JPT* (Sept. 1983) p. 1600.

37. Gerke, R. R.; Simon, J. M.; Logan, J. C.; Sabins, F. L.: "A Study of Bulk Cement Handling and Testing Procedures," SPE 14196 (Sept. 1985).

38. Crook, R. J.; Keller, S. R.; Wilson, M. A.: "Solutions to Problems Associated with Deviated Wellbore Cementing" SPE 14198 (Sept. 1985).

39. Cooke, C. E., Jr.; Gonzales, O. J.; Broussard, D. J.: "Primary Cementing Improvement by Casing Vibration During Cement Curing Time" SPE 14199 (Sept. 1985).

40. Unger, K. W.; Howard, D. C.: "Drilling Techniques Improve Success in Drilling and Casing Deep Overthrust Belt Salt," SPE *Drlg. Engr.* (June 1986).

41. Sykes, R. L., Logan, J. L.: "New Technology in Gas Migration Control" SPE 16653 (Sept. 1987).

42. Specification for Materials and Testing for Well Cements," API Specification 10, 4th Ed. (Aug. 1, 1988).

43. Beirute, R. M.; Wilson, M. A.; and Sabins, F. L.: "Atten-

uation of Casing Cemented with Conventional and with Expanding Cement Across Heavy Oil and Sandstone Formations" SPE 18027 (Sept. 1988).

44. Wilson, M. A.; Beirute, R. M.; and Sabins, F. L.: "Factors Affecting Cementing Efficiency Across Heavy Oil and Other Unconsolidated Formations in Vertical and Inclined Holes" SPE 18026 (Sept. 1988).

45. Goodman, M. A.; Mitchell, R. F.; Wedelich, H.; Galate, J. W.; and Presson, D. M.: "Improved Circulating Temperature Correlations for Cementing," SPE 18029, Houston (Oct. 1988).

Appendix

Cement Additives Comparison Chart

Product Classification	Halliburton	Dowell	B.J.	Western	Chemical or Material Description
	Product Trade Names				
Accelerators	$CaCl_2$	S-1	A-7	$CaCl_2$	Calcium Chloride
	HA-5	D-43	A-8	WA-4	Blend of Inorganic Accelerators
	D-12	A-2	Diacel A	Diacel A	Diacel A
	Salt	Salt	A-5	Salt	Sodium Chloride-Granulated
H-TLW Blends	1:1 (etc.)	1:1 (etc.)		1:1 Talc	
Fluid Loss	Halad 9, 11, 14	D-60	Aquatrol 13, 15	CF-1	Low Temp. Fluid Loss Control
		D-59		CF-2	Low Temp. Fluid Loss Control
	Diacel LWL	D-8	R-6	Diacel LWL	Carboxymethyl Hydroxyethyl Cellulose
Liquid		D-73			
Turbulence Inducers	CFR-2	D-65, 45	Turbo-Mix D-16	TF-4	Polymer
	CFR-3		Turbo-Mix D-30	TF-5	
Weighting Material	Barite	D-31	W-1	Barite	Barite
	Hi Dense 3	D-76		WM-2	Hematite
	Hi Dense 2	D-18	W-3	Hmenite	Hmenite
Spacers & Washes	Mud-Flush	CW7	Mud-Sweep	WMW-1	Mud Thinner-Spacer
	Sam 4	Oil-Base Mud Spacer	J-22 & D-4	ASP-4	Oil Base Spacer
				ASP-4	Water Base Spacer
Latex	LA-2	D-15, D-78	D-5	CLX-1	Latex Cement
Extenders	Howco Gel	Bentonite	B.J. Gel	Bentoment	Bentonite
	Gilsonite	Kolite	D-7	Gilsonite	Gilsonite
	Econolite		Lo-Dense	Thrifty-Lite	Anhydrous Sodium Meta-Silicate
	Pozmix A	Litepoz 3	Diamix A, G, M	Pozment A	Artificial Pozzolans
		Litepoz 1	Diamix A, M, G	Pozment N	Natural Pozzolans
	Pozmix 140	Litepoz 180	Thermoset		Pozzolan—Lime Mixtures
	Howcolight HLC	D-79	Lo Dense	Thrifty-Ment	H-Poz Blends
Anti-Foam	D-Air 1	D-46	D-6	AF-4	Powdered Anti-Foam Agent
	D-Air 2	D-47	D-6	AF-L	Liquid Anti-Foam Agent
Mud Decontaminant	Mud Kil-2	K-21	Firm Set II	Shur Set II	Mud Kill Patented by Gulf Oil
Silica Sand	Silica Flour (Reg.)	J-84	D-8	SF-3	325 Mesh Silica Flour
	Silica Flour (Coarse)	D-30		SF-4	Okla. #1 Sand
Thixotropic Cement	Thixotropic Cement	Reg. Fill-Up Cmt.		Thixoment	Thixotropic Slurries
Lost Circulation	Gilsonite	Kolite D2Y	D-7 Gilsonite	Gilsonite	
	Cellophane Flakes	D-29 Jel Flakes	Cello-Flake	Cell-O-Seal	
				Kwik-Seal	
Retarders	Kembreak	Kembreak	Kembreak	WR-1	Low Temp. Retarder
	HR-4	D-22	Retroset 2	WR-2	Low Temp. Retarder (Calcium Lignosulfonate)
	HR-7	D-13	Retroset 5	WR-4	Low Temp. Retarder
	HR-12	D-28	Retroset 8	WR-6	High Temp. Organic Retarder
	Diacel LWL	D-8	Retroset 6	Diacel LWL	Diacel LWL Carboxymethyl Hydroxethyl Cellulose
	HR-20	D-99	R10, R11		High Temp. Retarder
		D-93		WR-7	Borax

Chapter 5 Well Completion Design

Factors influencing design
Conventional tubing configurations
Unconventional tubing configurations
Sizing production tubulars
Completion interval selection
Tubingless completion techniques

The individual well is much more than "just an expensive faucet." It is our only communication with the reservoir. The effectiveness of that communication is a large factor in reservoir drainage as well as overall economics. Wells represent the major expenditure in reservoir development. Oil wells, gas wells, and injection wells present unique problems depending on the specific operating conditions. The individual well completion must be designed to yield maximum overall profitability on a field basis.

FACTORS INFLUENCING WELL COMPLETION DESIGN

The ideal completion is the lowest cost completion (considering initial and operating costs) that meets or nearly meets the demands placed upon it for most of its life. To intelligently design a well completion, a reasonable estimate of the producing characteristics during the life of the well must be made. Both reservoir and mechanical considerations must be evaluated.

Reservoir Considerations

Reservoir considerations involve the location of various fluids in the formations penetrated by the wellbore, the flow of these fluids through the reservoir rock, and the characteristics of the rock itself.

Producing rate to provide maximum economic recovery is often the starting point for well completion design. Among other factors producing rate should determine the size of the producing conduit.

Multiple reservoirs penetrated by a well pose the problem of multiple completions in one drilled hole. Possibilities include multiple completions inside casing separated by packers, or several strings of smaller casing cemented in one borehole to provide in effect separate wells. Other possibilities include commingling of hydrocarbons from separate reservoirs downhole, or drilling several boreholes from one surface location.

Reservoir drive mechanism may determine whether or not the completion interval will have to be adjusted as gas-oil or water-oil contacts move. A water drive situation may indicate water production problems. Dissolved gas drive may indicate artificial lift. Dissolved gas and gas drive reservoirs usually mean declining productivity index and increasing gas-oil ratio.

Secondary recovery needs may require a completion method conducive to selective injection or production. Water flooding may increase volumes of fluid to be handled. High temperature recovery processes may require special casing and casing cementing materials.

Stimulation may require special perforating patterns to permit zone isolation, perhaps adaptability to high injection rates, and a well hookup such that after the treatment the zone can be returned to production without contact with killing fluids. High temperature stimulation again may require special cementing procedures, casing and casing landing practices.

Sand control problems alone may dictate the type of completion method and maximum production rates. On the other hand, reservoir fluid control problems may dictate that a less than desirable type of sand control be used. Sand problem zones always dictate a payoff from careful well completion practices.

Workover frequency, probably high where several reservoirs must be drained through one wellbore, often dictate a completion conducive to wireline or through-tubing type recompletion systems.

Artificial lift may mean single completions even where multiple zones exist, as well as larger than normal tubulars.

Mechanical Considerations

The mechanical configuration or "well hookup" is often the key to being able to deplete the reservoir effectively, monitor downhole performance, and modify the well situation when necessary.

The mechanical configuration of the well is the key to being able to do what ought to be done in the well from the standpoint of controlling the flow of reservoir fluids, oil, gas, and water.

Formation damage is related to the well hookup, both minimizing damage initially and relieving the effects of damage later.

Mechanically, well completion design is a complex engineering problem. Basic philosophy is to design to specific well conditions, field conditions, and area conditions.

1. *Maximize profit* considering the time value of money. Economics are sometimes best served by delaying expenditures, particularly in wells where servicing is frequent. The isolated well is the one you can afford to provide with maximum flexibility for the future.
2. *Keep the installation simple,* both from equipment and procedural standpoints—consider level of operator skill available.
3. *Overall reliability* depends on reliability of individual components and the number of components. Design out maintenance, limit moving parts, avoid debris traps. As complexity increases, provide alternatives.
4. *Anticipate all operating conditions,* and associated pressure and temperature forces.
5. *Safety* must be designed into the well. In offshore, populated, or isolated areas, automatic shut-in systems and well pressure control methods must be considered.

Basic decisions to be reached in designing the well completion are: (a) the method of completion, (b) the number of completions within the wellbore, (c) the casing-tubing configuration, (d) the diameter of the production conduit, and (e) the completion interval.

Method of Completion

Basically there are two methods of completing a well, Openhole where casing is set on top of the producing interval and Perforated Casing, where casing is cemented through the producing interval and communication is established by perforating. Each method has areas of predominate usage depending on formation characteristics. Generally openhole has greater application in carbonate zones. But each has inherent advantages and limitations.

Open-Hole Completion—The casing set on top of producing zone. See Figure 5-1.

Advantages:

1. Adaptable to special drilling techniques to minimize formation damage, or to prevent lost circulation into the producing zone.
2. With gravel pack, provides excellent sand control method where productivity is important.
3. No perforating expense.

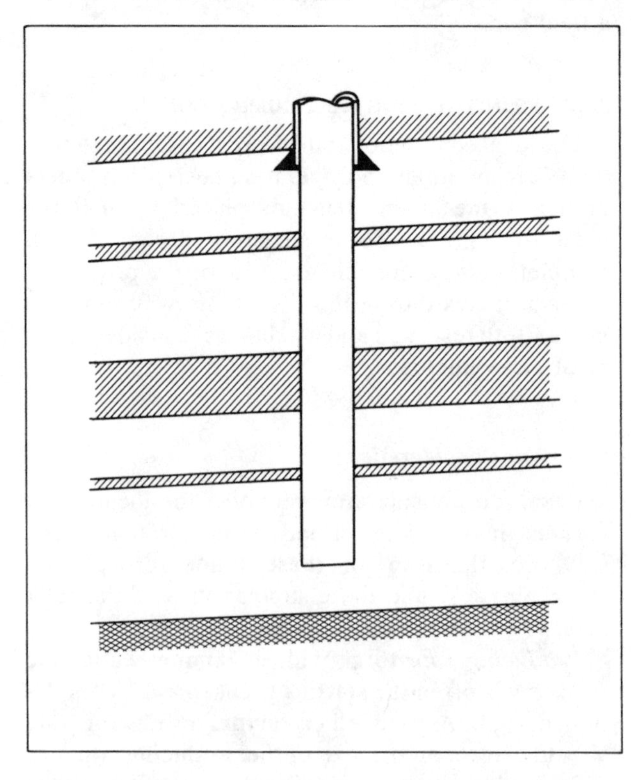

FIG. 5-1—*Open-hole completion.*

4. Log interpretation is not critical since entire interval is open.
5. Full diameter opposite pay.
6. Can be easily deepened.
7. Easily converted to liner or perforated completion.

Limitations

1. Excessive gas or water production difficult to control.
2. Selective fracing or acidizing more difficult.
3. Casing set "in the dark" before the pay is drilled or logged.
4. Requires more rig time on completion.
5. May require frequent cleanout.

Perforated Completion — Casing cemented through producing zone and perforated. See Figure 5-2.

Advantages

1. Excessive gas or water production can be controlled more easily.
2. Can be selectively stimulated.
3. Logs and formations samples available to assist in decision to set casing or abandon.
4. Full diameter opposite pay.
5. Easily deepened.

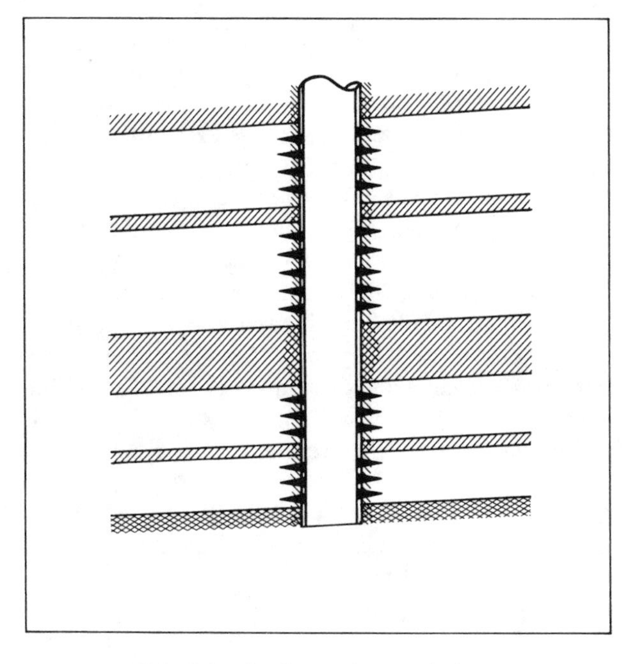

FIG. *5-2—Perforated completion.*

6. Will control most sands, and is adaptable to special sand control techniques.
7. Adaptable to multiple completion techniques.
8. Minimum rig time on completion.

Limitations

1. Cost of perforating long zones may be significant.
2. Not adaptable to special drilling techniques to minimize formation damage.
3. Log interpretation sometimes critical in order not to miss commercial sands, yet avoid perforating submarginal zones.

It should be recognized that a poor primary cement job in effect converts a "perforated casing" completion to an "openhole" completion. Continuous perforating with no "blank" zones between perforated intervals also converts a perforated casing completion to an openhole completion.

CONVENTIONAL TUBULAR CONFIGURATIONS

Single-zone Completion

Factors leading to selection of single-zone "conventional" completions, as opposed to miniaturized completions, multiple inside-casing completions, or multiple tubingless completions are: high producing rates—corrosive well fluids—high pressures—governmental policies—operator tradition.

Various hookups are possible depending on objectives. Basic questions concern use of tubing and packers. Many wells are produced without tubing. This possibility should always be considered.

Valid reasons for tubing may include:

1. Better flow efficiency.
2. Permit circulation of kill fluids, corrosion inhibitors or paraffin solvents.
3. Provide multiple flow paths for artificial lift system.
4. Protect casing from corrosion, abrasion, or pressure.
5. Provide means of monitoring bottom-hole flowing pressure.

Tubing should be run open-ended, and set above highest alternate completion interval to permit thru-tubing wireline surveys and remedial work.

A packer should be run only where it accomplishes a valid objective such as:

—Improve or stabilize flow.

—Protect casing from well fluids or pressure—however, it should be recognized that use of a packer may increase pressure on casing in the event of a tubing leak.
—Contain pressure in conjunction with an artificial lift system or safety shut-in system.
—Hold an annular well-killing fluid.

Where packers are used, landing nipples to permit installation of bottom-hole chokes or safety valves are sometimes desirable. Also a circulating device is sometimes desirable to assist in bringing in or killing the well.

In a high volume, annular flow well, where casing can sustain shut-in well pressure, and a safety shut-in capability is required, it may be desirable to run a tubing string to bottom, but set a packer and surface-controlled safety valve within several hundred feet of the surface. Well fluids then flow through both the annulus and tubing to a point immediately below the packer. Here all flow is brought into the tubing through the safety control valve, and then back into both the annulus and tubing to the surface. Thus, safety valve control is maintained, but pressure restriction is minimized.

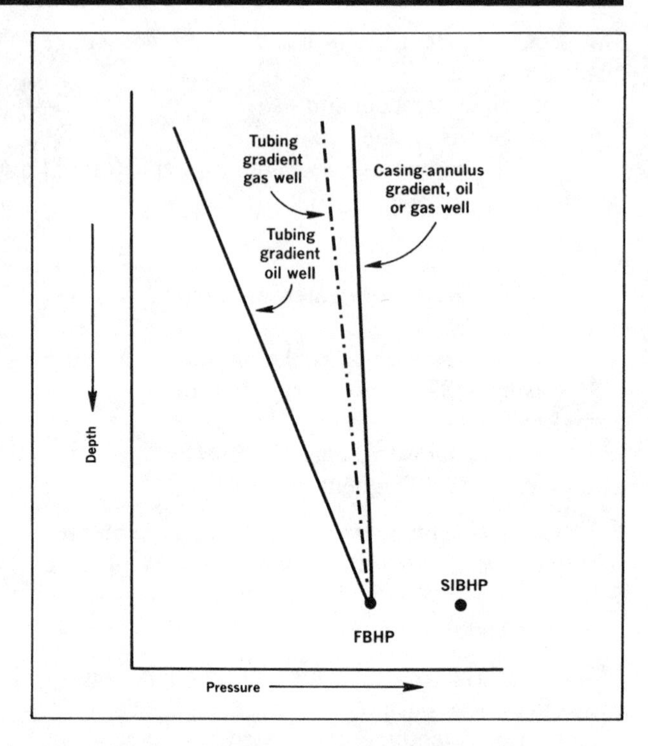

FIG. 5-3—*Pressure gradient for wells with tubing "swung."*

Effect of Tubing and Packer

Effect of tubing, with or without a packer, on well pressure gradients under various conditions should be recognized.

Tubing Without Packer (flowing well)—Figure 5-3 shows pressure gradient situation for an oil well and a gas well with tubing "swung." The annulus acts as separator, thus with a gas gradient in the annulus, annular wellhead pressure is almost equal to bottomhole pressure for the oil or the gas well.

For the gas well, wellhead annular pressure is slightly greater with tubing than without tubing. For the oil well, wellhead annular pressure is considerably greater with tubing installed due to the gas gradient in the annulus.

In the gas well the chances of a tubing leak, with tubing swung, are nil; thus, there is no justification for a premium tubing joint. Chances of a casing leak are essentially the same as if tubing had not been installed.

With the oil well, chances of a tubing leak are maximized. Pressure differential is from annulus to tubing, and in the event of a shallow tubing leak, the wellhead annular pressure will drop as fluid level moves upward in the annulus.

Although a tubing leak is not disastrous, prolonged flow may erode casing and result in a casing leak. Chances of a shallow casing leak are increased with tubing swung due to the pressured annulus.

Tubing With Packer (flowing well)—Figure 5-4 shows tubing and annulus pressure gradients for an oil well and a gas well with tubing set on a packer. The annulus is filled with a liquid providing a slight overburden above shut-in formation pressure.

For the oil well, differential pressure across the tubing is now quite small, and the chance of a tubing leak is nil. The same can be said for the chance of a casing leak, assuming normal formation pressures.

For the gas well, differential pressure across the tubing increases to a maximum near the surface. Chances for a near surface tubing leak are maximized due to an unfavorable situation as regards to tubing load and temperature changes.

Effect of Tubing Leak (flowing well)—Figure 5-5 shows the effect of a tubing leak on the pressure gradient in the annulus in an oil well and a gas

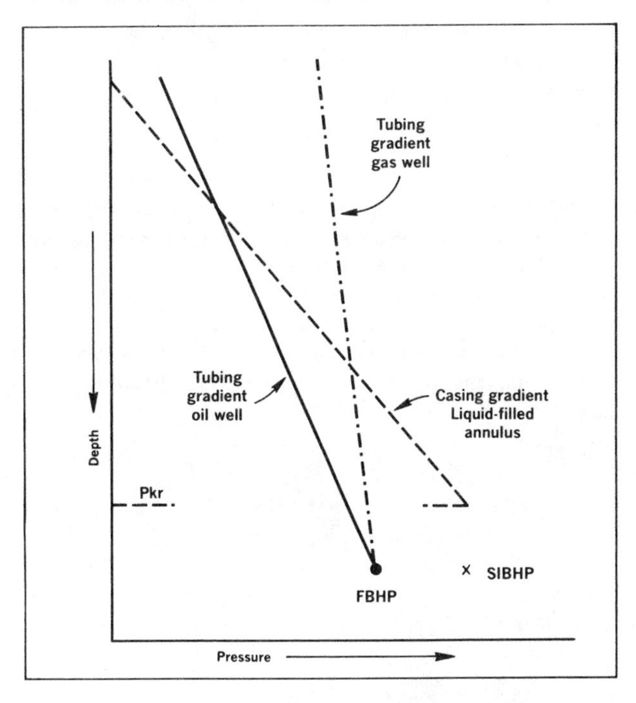

FIG. 5-4—*Pressure gradient for wells with tubing set on a packer.*

well with tubing set on a packer.

In the gas well, chances are that a collar leak will occur at a shallow depth since (1) pressure differential is greatest, (2) tubing tensile load is greatest, and (3) temperature fluctuations are greatest. Tubing pressure added on top of the pressure gradient of a high weight mud column can rupture casing downhole.

One solution to this gas well situation is to use a light liquid (water) in the annulus, then add pressure on top to more nearly match the tubing gradient. This reduces tubing leak probability, provides better retrievability of packers, and permits monitoring of casing pressure to better determine condition of tubing and casing.

In the oil well a near-surface tubing leak imposes additional bottom-hole pressure on the casing, but depending on flowing pressure in the tubing, is not usually a serious downhole threat to the casing.

In either case, oil or gas well, heavy mud is a poor packer fluid from a formation damage standpoint, and in most situations does not provide additional safety over a lighter liquid.

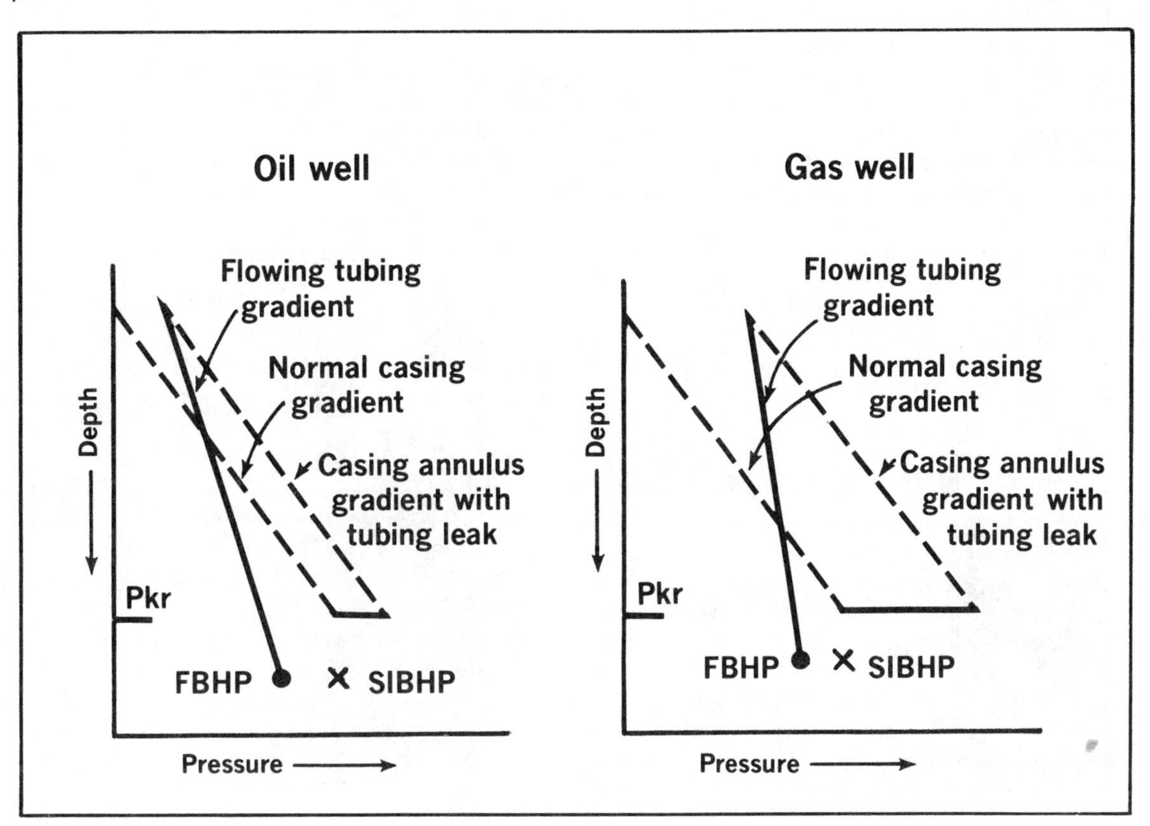

FIG. 5-5—*Effect of tubing leak on pressure gradient in well with tubing set on a packer.*

Multiple Zone Completion

Factors leading to selection of multiple completions are: higher producing rate, faster payout, and multi-reservoir control requirements. Numerous configurations are possible utilizing single or multiple strings of tubing.

Single String—Single Packer—There is both tubing and annulus flow (Figure 5-6). This is the lowest cost conventional dual.

Limitations

1. Upper zone cannot be produced through tubing, unless lower zone is blanked off.
2. Casing subject to pressure and corrosion.
3. Only lower zone can be artificially lifted.
4. Upper zone sand production may stick tubing.
5. Workover of upper zone requires killing lower zone.

Single String—Dual Packer—Again, there is both tubing and annulus flow (Figure 5-7). Advantage is that cross-over choke permits upper zone to be flowed through tubing.

Limitations

1. Casing subjected to pressure and corrosion.
2. Must kill both zones for workover of upper zone.

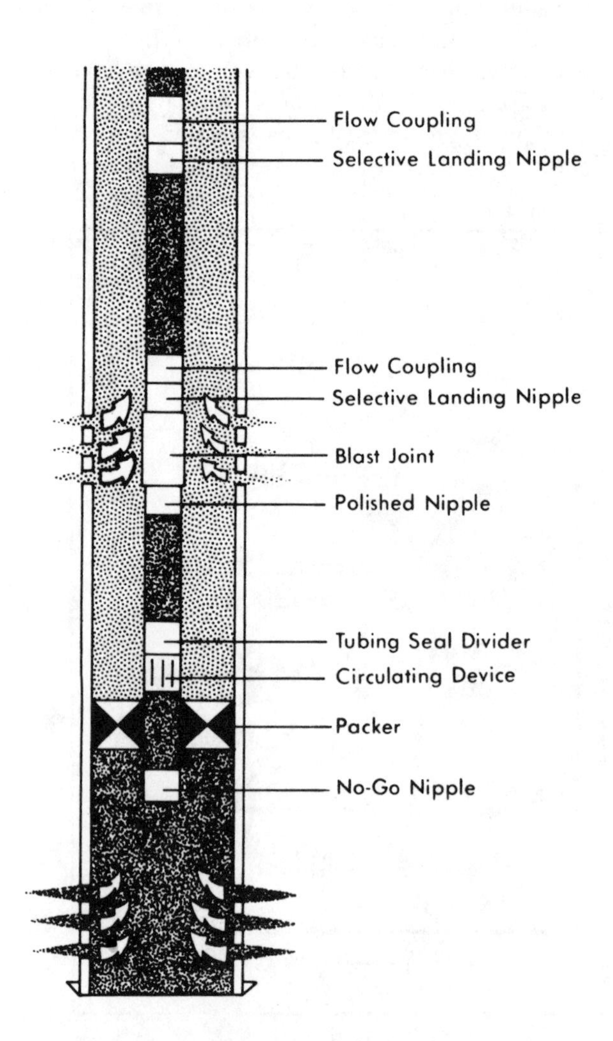

FIG. 5-6—*Single string with single packer.*

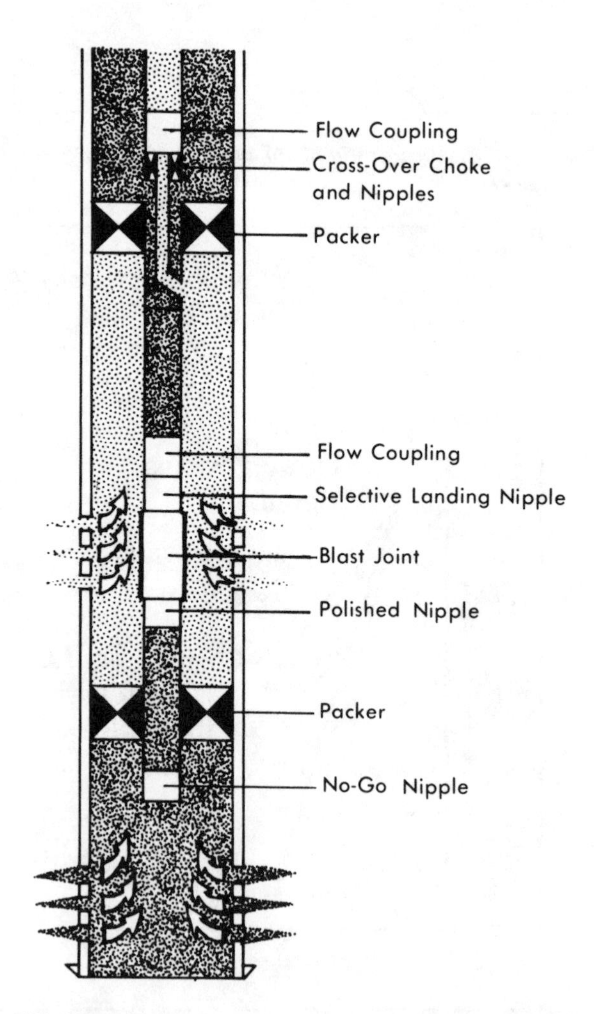

FIG. 5-7—*Single string with dual packers, selective crossover.*

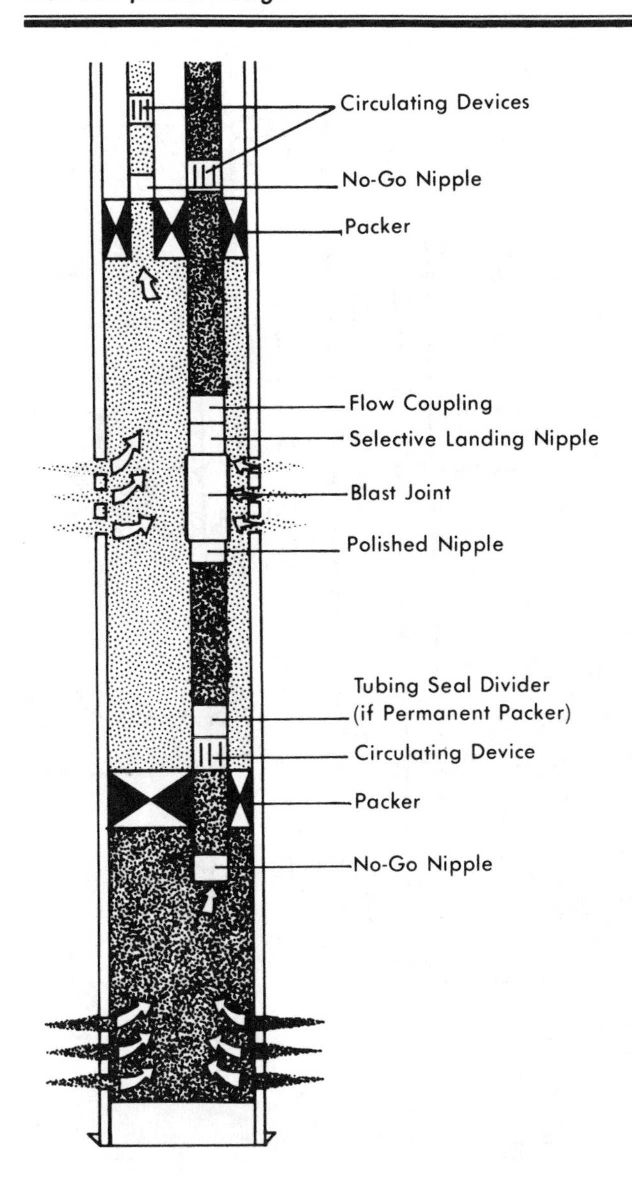

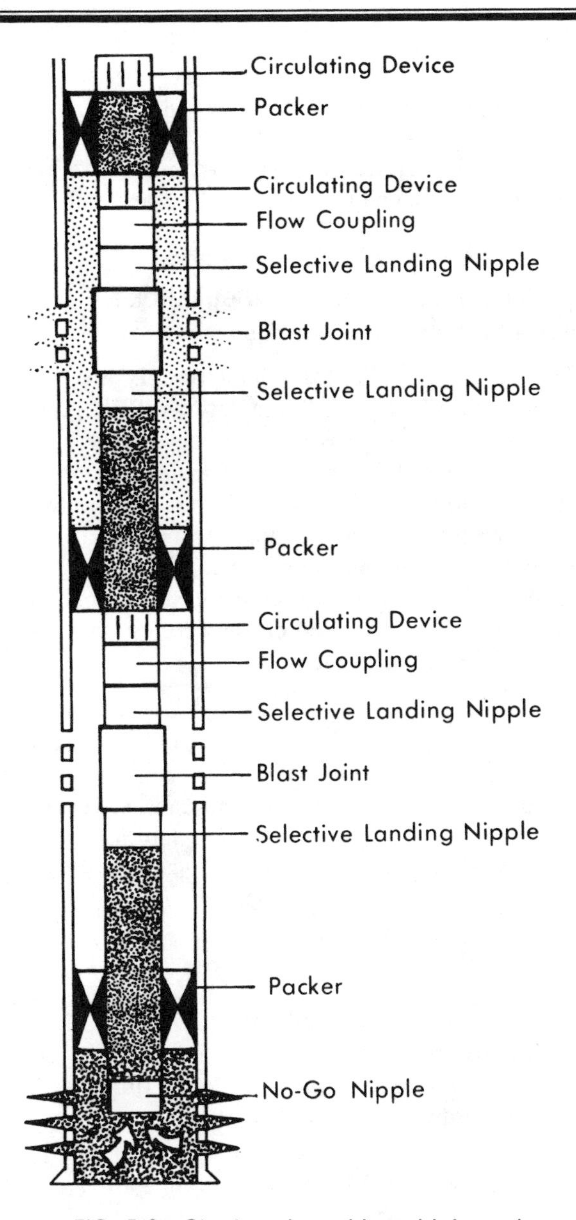

FIG. *5-9—Single string with multiple packers.*

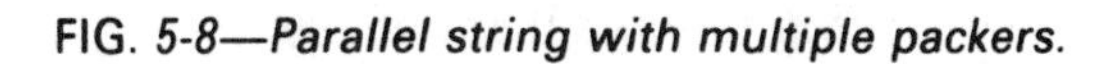

FIG. *5-8—Parallel string with multiple packers.*

Parallel String—Multiple Packer—This is shown in Figure 5-8.

Advantages

1. Can lift several zones simultaneously.
2. Concentric tubing and wireline workovers practical in all zones.

Limitations

1. High cost.
2. Susceptibility to tubing and packer leaks.
3. Hesitation to perform stimulation treatments or workovers of individual zones.

Single String—Multiple Packer—Selective Zone—This is shown in Figure 5-9.

Advantage

1. Producing sections can be opened or closed by use of wireline.

Limitations

1. Difficulty of monitoring flow from individual zones.
2. Difficulty of treating or even reperforating individual zones unless well is killed and tubing is pulled.

UNCONVENTIONAL TUBULAR CONFIGURATIONS

Multiple "Tubingless" Completion

The multiple tubingless completion is an outgrowth of permanent well completions (PWC) and concentric tubing workover technology. It involves cementing several strings of pipe inside one wellbore, as shown in Figure 5-10. Originally this concept was applied to multiple strings of 2⅞-in. pipe; but, currently multiple strings of 3½-in. and even 4½-in. are used. The concept should not be thought of as being limited entirely to low volume producing or injection wells.

Advantages

1. Reduced cost—initial completion and future workover costs are reduced.
2. Each zone is independent and can be worked on without disturbing the other completions.
3. Communication between strings is easily located and eliminated.
4. Procedures are simplified.

Limitations

1. Restricted production rate.
2. Corrosion and paraffin control more critical.
3. Higher risk due to pressured well fluids.
4. High rate stimulation treatments are more difficult.
5. Long-zone sand control more difficult.

The application of multiple tubingless completions, along with procedures for running and cementing multiple strings of casing, are covered in detail in Appendix A of this chapter.

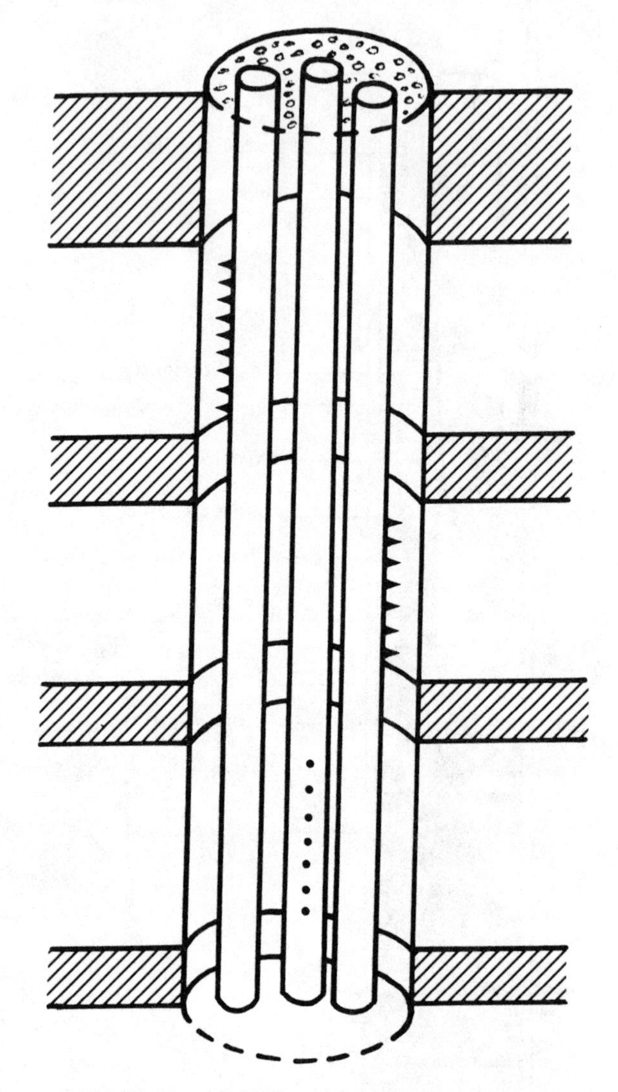

FIG. 5-10—*Multiple tubingless completion.*

SIZING PRODUCTION TUBULARS

The size of the production string casing depends upon the diameter of flow conduit (single or multiple) needed to produce the desired flow stream, the method of artificial lift, if required, or specialized completion problems such as sand control.

Sizing of the production tubing depends primarily on the desired production rate. Maximum production rate in a given well depends upon:

—Static reservoir pressure.
—Inflow performance relation.
—Pressure drop in tubing.
—Pressure drop through the wellhead constrictions.
—Pressure drop through the flow line.
—Pressure level in the surface separating facilities.

Where maximum flow rate is an objective of well completion design all of these factors must be considered.

Well Systems Analysis

Systems Analysis approach (nodal analysis) has been used for many years to analyze systems where various components are interactive. The procedure consists of first selecting an appropriate division point or node. Figure 5-11 shows many possible nodes. All components upstream from node comprise the inflow section, and all components down stream comprise the outflow section.

Next a relation between flow rate and pressure drop must be developed for each component in both sections. Flow rate for the specific system or set of components can then be determined by satisfying the following relationships:

—Flow into node = Flow out of node
—Only one pressure can exist at the node.

At a given time two pressures are fixed in the normal well flow system; i.e., the average reservoir pressure, and the separator pressure.

The basic procedure is to calculate the pressure at the node both ways from the fixed pressure points as follows:

Inflow to node:

$$P_R - \Delta P \text{ (upstream components)} = P_{node}$$

Outflow from node:

$$P_{separator} + \Delta P \text{ (downstream components)} = P_{node}$$

The pressure drop ΔP, in any component varies with flow rate q, therefore, a plot of flow rate vs. node pressure for each section will produce two curves, the intersection of which gives the one flow rate which satisfies the conditions 1 and 2 above. (See Figure 5-12).

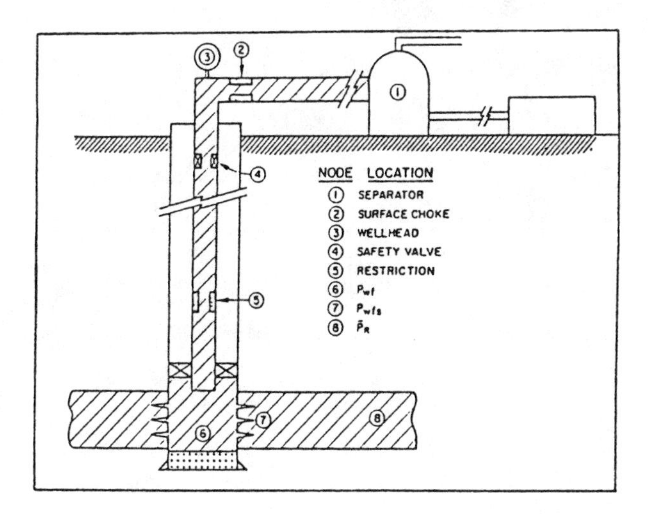

FIG. 5-11—Possible locations of nodes. After H. Dale Beggs.

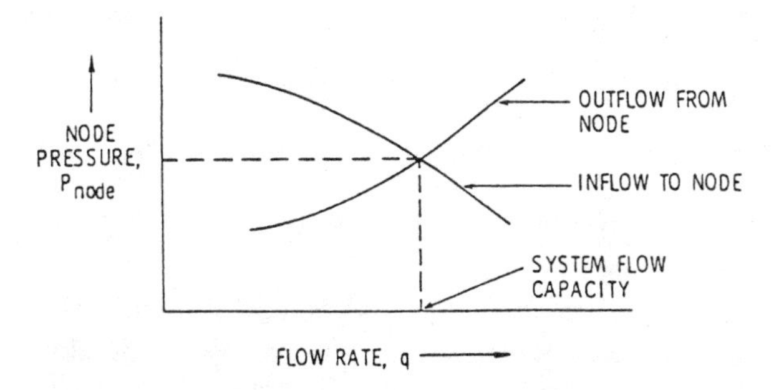

FIG. 5-12—Inflow-Outflow crossplot. After H. Dale Beggs.

A change in the pressure drop across an upstream component (inflow section) will leave the outflow curve unchanged, but the intersection point will change, and thus the flow rate will change. Likewise, a change in the pressure drop across a downstream component will result in an adjustment in the flow rate. Finally, a change in either of the "fixed" pressures (the average reservoir pressure—or the separator pressure) occurring during the life of the well will result in a change in the flow rate.

A frequently used node or division point is inside the casing at the perforations; i.e., between the reservoir and the piping system. Thus, the flow through the rock, the perforations, and the gravel pack (if installed) is one system, and flow up the tubulars, through the well head and through the flow line—and manifold to the separator is the second system.

The total system is optimized by selecting the combination of component characteristics which will maximize production rate for the lowest cost.

The system analysis approach is basically used to optimize flowing well performance, but can also be applied to artificial lift situations if the effect of the artificial lift system on the pressure is a function of the flow rate. Possible applications include:

—Selection of tubing and/or flow line sizing
—Surface choke or subsurface safety valve sizing
—Analyzing effect of perforation density
—Analyzing effect of gravel pack design
—Artificial lift design
—Predicting the effect of depletion on producing capacity

Inflow Performance Relation

The Inflow Performance Relation for a specific well represents the ability of that well to produce fluids against varying bottom hole or "well intake" pressures. Sometimes this relation is assumed to be a straight line (Fig. 5-13). However, except for water drive reservoirs producing above bubble point pressure, flow rate usually drops off significantly from a straight line relation at higher wellbore pressure drawdown.

For a specific well the inflow performance relation often declines with cumulative production from the reservoir. For dissolved gas drive or gas drive reservoirs this decline may be rapid. The occurrence of formation damage or stimulation also affects the inflow performance relation.

Productivity index (bpd / psi pressure drawdown), a popular term used to describe well flow efficiency, represents only one point on the inflow performance curve. Thus, PI usually declines with:

1. Higher wellbore pressure drawdown.
2. Cumulative reservoir fluid withdrawals.
3. Degree of formation damage.

Assuming water-free oil production, P.I. can be estimated from reservoir parameters by:

$$PI = \frac{6 \times 10^{-4} k_o h}{u_o B_o}$$

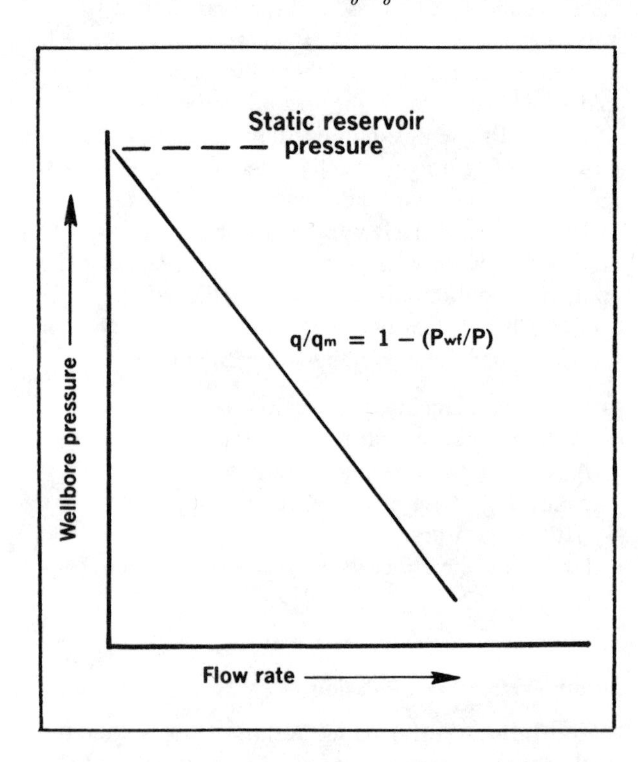

FIG. 5-13—*Generalized inflow performance relation.*

k_o = oil relative permeability, md
h = zone height, feet
u_o = oil viscosity, centipoise
B_o = formation volume factor

The inflow performance relation can best be determined by Isochronal testing of the specific oil (or gas) well. However, simplified techniques are satisfactory for most purposes of well completion design (i.e.: sizing flow conduits). For more detailed study, Kermit Brown[10] is suggested.

Vogel developed the general inflow performance relation of Figure 5-14 through a computer study involving a wide range of parameters applicable to dissolved gas drive reservoirs. Having one well test measuring static reservoir pressure and a flow rate with corresponding bottom hole flowing pressure, Figure 5-14 predicts flow rate at any other BHFP. Future IPR's can be estimated by displacing the curve downward in proportion to the decline of static reservoir pressure.

Strictly the Vogel work applies to a dissolved gas drive reservoir; however, for practical purposes it can be used for any type of reservoir.

Referring to Figure 5-14,

$$\frac{q}{q_m} = 1 - 0.20\frac{p_{wf}}{\bar{p}} - 0.80\left(\frac{p_{wf}}{\bar{p}}\right)^2$$

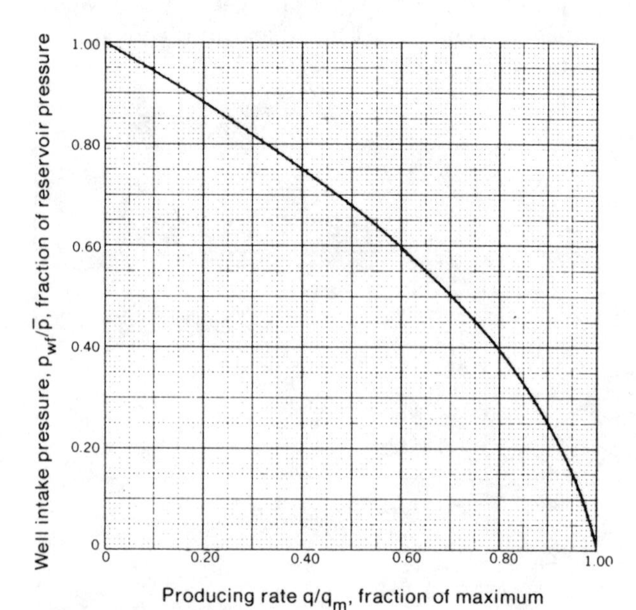

FIG. 5-14—*Inflow performance relation (Vogel).*[11] *Permission to publish by The Society of Petroleum Engineers.*

q = producing rate at given p_{wf}
q_m = producing rate when $p_{wf} = 0$
p_{wf} = well intake pressure
$\bar{p}$ = shut-in reservoir pressure

Example Problem

$\bar{p}$, Static reservoir pressure = 2500 psi
p_{wf}, Well intake pressure = 2000 psi
q_o, Flow rate = 400 bpd

—What would the well make if p_{wf} = 1600 psi?

$$\text{Flow rate} = \frac{.55}{.33} \times 400 = 667 \text{ bpd}$$

In his computer study Vogel did not consider formation damage. Standing[12] extended the usefulness of Vogel's work presenting Figure 5-15 involving flow efficiency. Thus, if FE is known from pressure buildup tests (or can be estimated) the effect of formation damage, or elimination of formation damage can be estimated.

In Figure 5-15,

$$FE = \frac{\text{Ideal drawdown}}{\text{Actual drawdown}}$$

$$= \frac{\bar{p} - p_{wf} - \Delta p_s}{\bar{p} - p_{wf}}$$

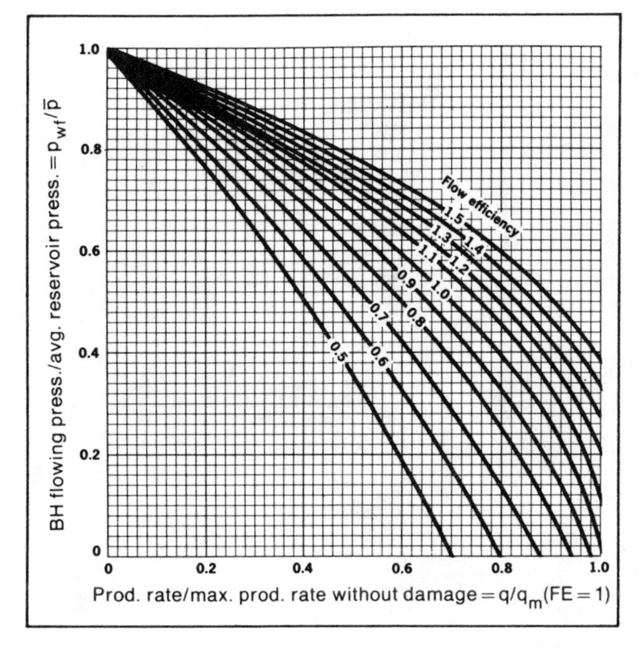

FIG. 5-15—*Inflow performance relation, modified by Standing.*[12]

Example Problem

$\bar{p}$, Static reservoir pressure = 2500 psi
p_{wf}, Well intake pressure = 2000 psi
q_o, Flow rate = 400 bpd
FE, flow efficiency = 0.7

—What would the well make if p_{wf} = 1600 psi?

$$\text{Flow rate} = \frac{.40}{.23} \times 400 = 690 \text{ bpd}$$

—What would the undamaged well make if p_{wf} = 1600 psi?

$$\text{Undamaged flow rate} = \frac{.54}{.23} \times 400 = 940 \text{ bpd}$$

Pressure Drop in Well Tubing

Pressure loss in two-phase or three-phase vertical flow is difficult to calculate since the average density and velocity of the fluid is usually unknown due to gas break-out and fluid slip.

Poettmann and Carpenter[8] developed empirical correlations which can be used to approximate multiphase vertical flow. Generally this correlation applies to 2⅜-in. to 3½-in. od tubing; flow rates greater than 400 bpd; GOR less than 1,500 cu ft/bbl; and viscosity less than 5 cp.

Since this original work, Dun and Ros, Hagedorn and Brown, and Beggs and Brill have developed additional vertical flow correlations aimed at improving the accuracy of pressure loss calculations. Most are applicable to all conditions including annular flow. Also vertical flow relations can be used in holes deviated up to 15° to 20° from vertical.

Figure 5-16 shows typical vertical pressure traverse curves from the Hagedorn and Brown correlation. Use of the curves to obtain flowing pressure drop where surface pressure is known is briefly:

1. Pick proper curve to fit situation, i.e., flow rate, WOR, pipe size, etc.
2. Drop vertical line from surface pressure intersecting gas-liquid ratio to determine "pseudo depth."
3. To this pseudo depth add well depth to determine "pressure depth."
4. Move horizontally from pressure depth to proper GLR and read bottom-hole pressure.

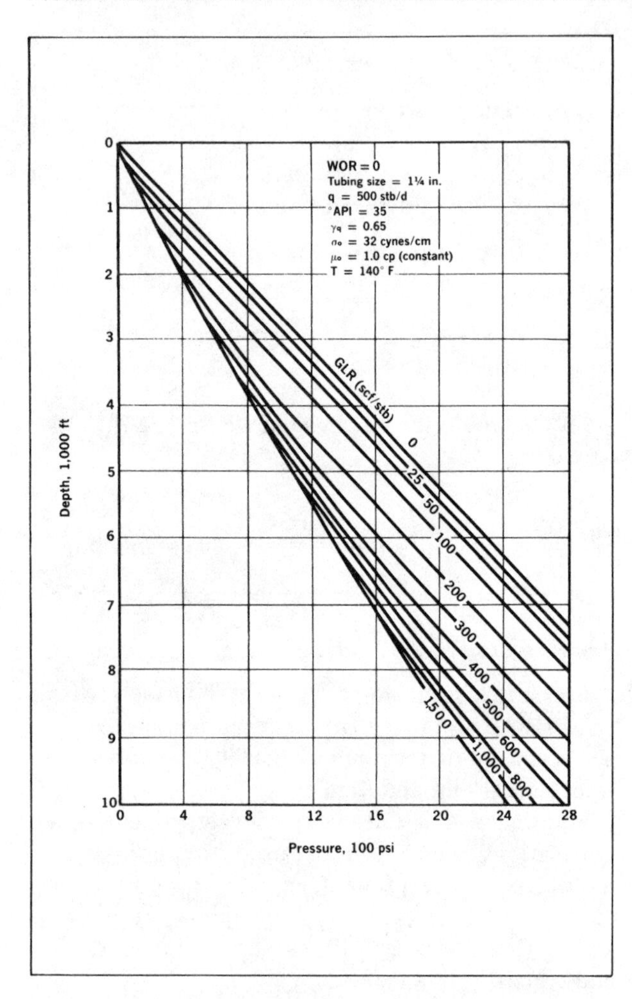

FIG. 5-16—*Typical vertical pressure traverse curves.[10] Permission to publish by The Petroleum Publishing Co.*

5. Subtract surface pressure from BHP to determine pressure drop in tubing.

With gas in the flow stream the effect of increasing surface pressure is to increase pressure loss in the tubing. Thus, back pressure against the formation is increased due to (1) the higher pressure loss in the tubing and (2) the higher surface pressure.

Curves similar to Figure 5-16 are useful for most engineering work where approximate pressure drop calculations are required. Computer solutions of various correlations permit more detailed look at the effect of changing variables.

Brill, Doerr, Hagedorn and Brown studied the effect of certain variables on multiphase vertical flow, and presented the following relationships. These are included here to provide a "feel" for their relative importance.

Effect of tubing size is shown in Figure 5-17 for a flow rate of 200 stb/*d* of 35° API oil. Although larger tubing sizes show an advantage in lowering bottom-hole pressure, the income from the increased cost of the larger tubing. Also as tubing becomes larger, flow velocity decreases and may let gas break through the liquid. Resulting liquid fallback and accumulation may kill the well.

Effect of tubing size on gradient reversal is shown in Figure 5-18. With small diameter tubing producing high gas-liquid ratios (gas lift wells) this gradient reversal effect is common. As the pressure on the mixture flowing up the tubing decreases near the surface, the pressure gradient due to density decreases. However, with reduced pressure the flow velocity increases, and the pressure gradient due to fric-

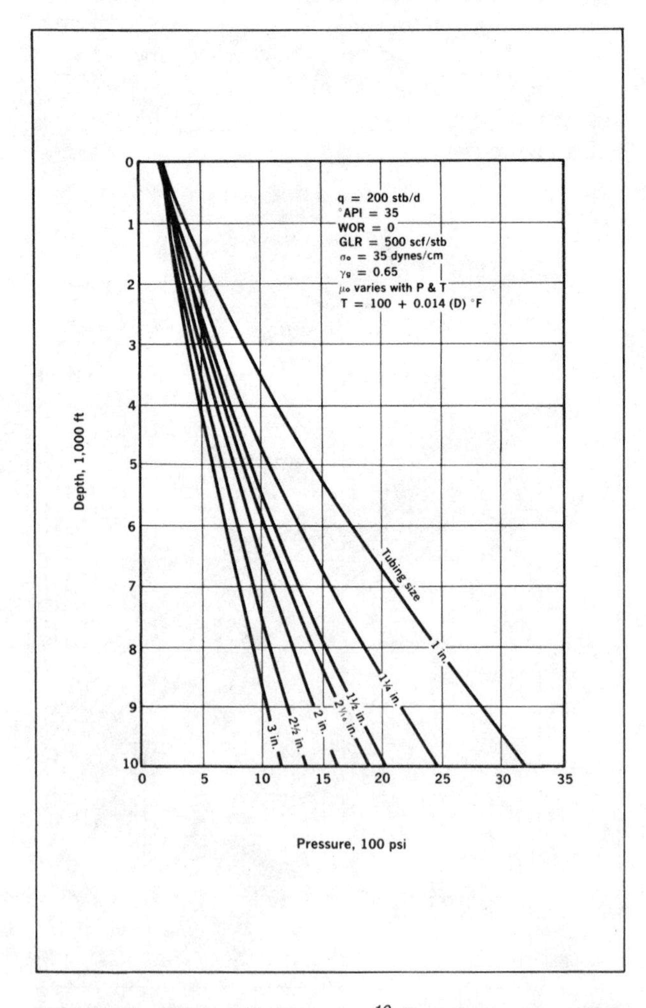

FIG. 5-17—*Effect of tubing size.[10] Permission to publish by The Petroleum Publishing Co.*

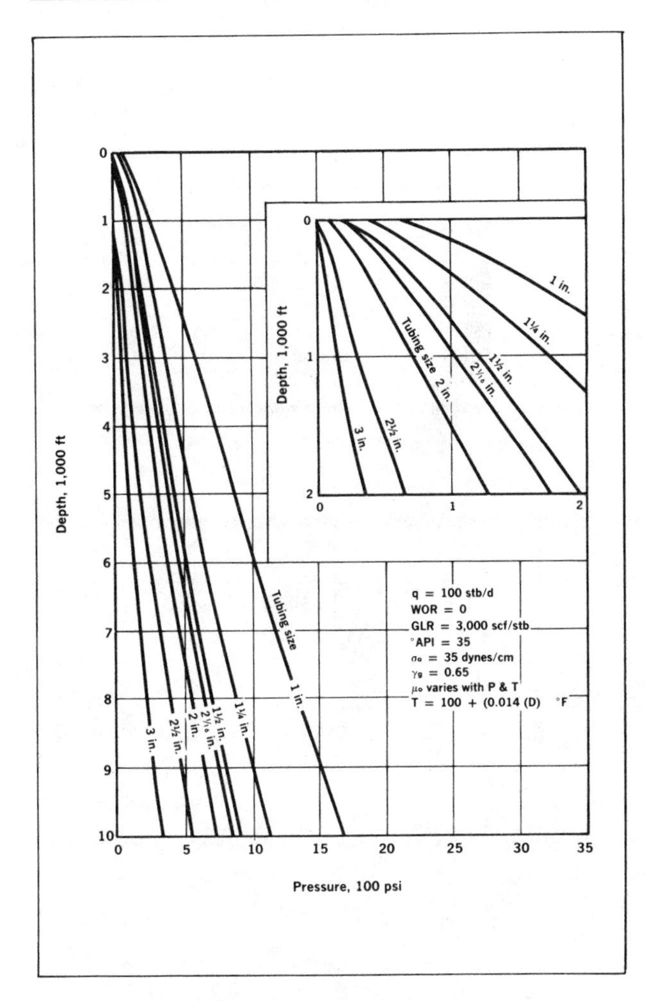

FIG. 5-18—Effect of tubing size on gradient reversal.[10] Permission to publish by The Petroleum Publishing Co.

tion loss increases at a greater rate than the density gradient decreases. The result is an increased pressure loss near the surface.

Effect of surface flow rate is shown in Figure 5-19 for 2-in. tubing. At low rates (50 stb/*d*) these correlation probably break down due to the so-called "heading" effect. Attempts to describe this heading effect mathematically or to predict the rate at which it occurs have not yet been successful.

Effect of gas-liquid ratio is shown in Figure 5-20 for a flow rate of 200 stb/*d* of 35° API oil in $1^1/_4$-in. tubing. As the gas-liquid ratio increases the flowing bottomhole pressure required to produce the rate decreases. However, a point is reached where further increases in gas-liquid ratio actually increase bottomhole pressure.

Effect of liquid density is shown in Figure 5-21, when oil viscosity is held constant at 1 cp. As API gravity increases, flowing pressure decreases. It should be noted that as API gravity increases the amount of solution gas at a given pressure level increases. This increases the liquid holdup factor, however, which in turn increases density and tends to offset the higher gas-liquid ratio.

Effect of liquid viscosity is shown in Figure 5-22 for $1^1/_4$-in. tubing and a flow rate of 200 stb/*d*. Free gas viscosity is assumed to be 0.02 cp while liquid viscosity varies with temperature and solution gas.

Effect of liquid surface tension is shown in Figure 5-23 for $1^1/_4$-in. tubing producing 200 stb/*d* 35° API oil. Larger surface tension results in greater liquid holdup, higher density, and, therefore, high flowing bottom-hole pressures.

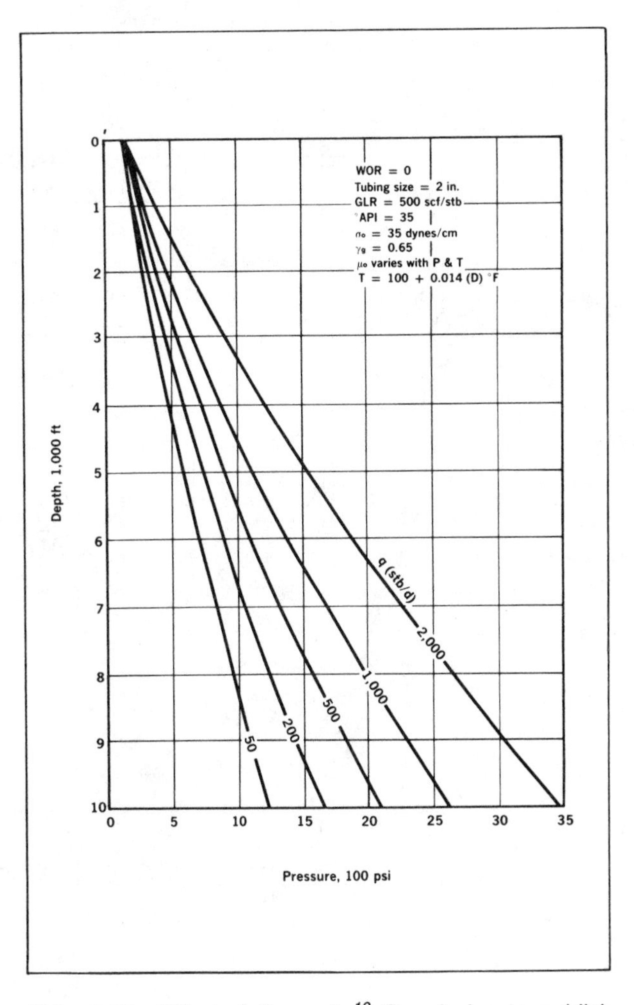

FIG. 5-19—Effect of flow rate.[10] Permission to publish by The Petroleum Publishing Co.

Effect of kinetic energy is often neglected, but can become important with small diameter tubing with high gas-liquid ratios and low pressure levels, Figure 5-24.

Pressure Drop Through Tubing Restrictions and Wellhead

In high rate producing situations pressure drop through the wellhead and through tubing restrictions can be significant. Figure 5-25 shows one typical situation producing 3,600 bopd from 7,700 ft through $3^1/_2$-in. tubing. Tubing restrictions include a sliding side door assembly and a ball-type safety valve.

Pressure Drop in the Flow Line

The problem of horizontal two-phase flow is as complex as that of vertical two-phase flow. A number of correlations have been presented based on empirical data. Probably the best currently is one by Eaton and Brown described in Reference 9. Working curves for this correlation are presented in Reference 10.

A typical example of pressure drops with high flow rates is shown in Figure 5-25.

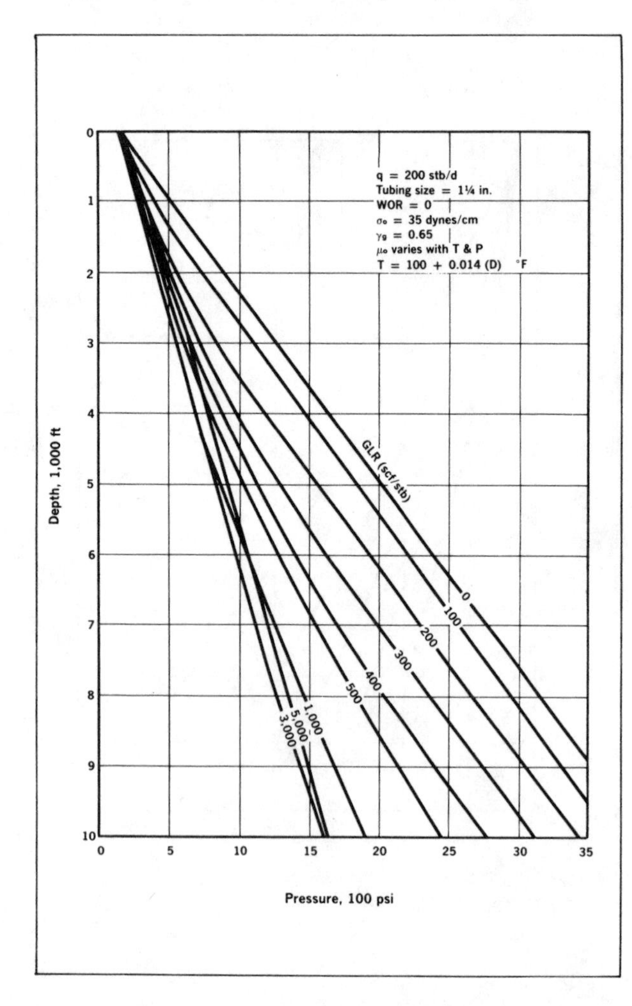

FIG. 5-20—Effect of gas-liquid ratio.[10] Permission to publish by The Petroleum Publishing Co.

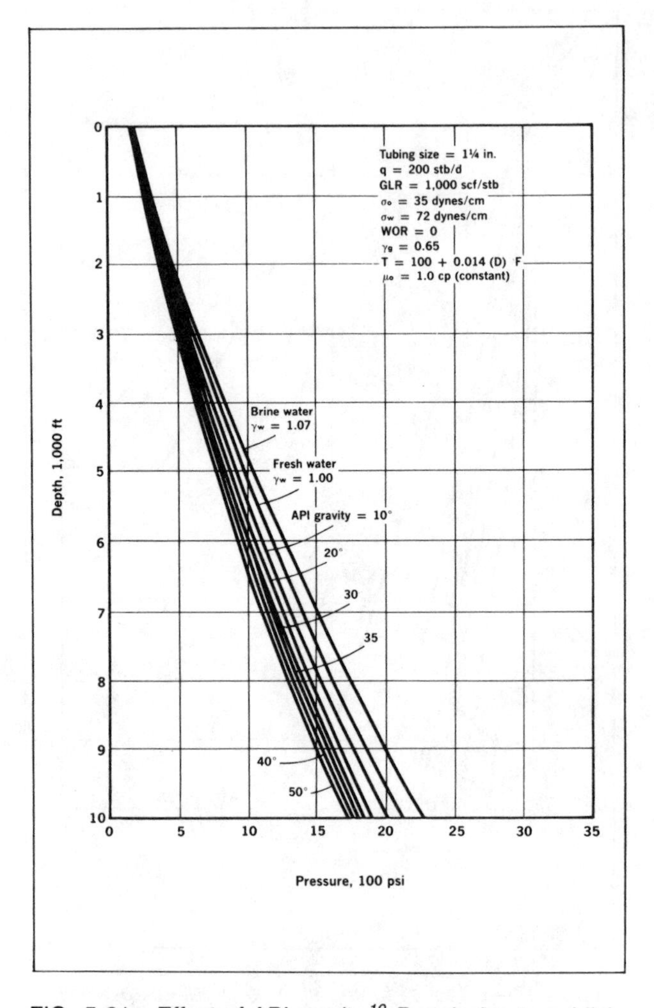

FIG. 5-21—Effect of API gravity.[10] Permission to publish by The Petroleum Publishing Co.

COMPLETION INTERVAL

Selection of the completion interval is dictated by a number of interrelated factors. In thin multi-reservoir fields the choice is usually obvious. In thicker multi-lense zones comprising single reservoirs considerable study of geologic and reservoir fluid flow mechanisms may be required for intelligent interval selection.

Reservoir Drive Mechanism

With reasonable vertical permeability, obviously a well in a water drive reservoir should be completed near the top of the zone, and a well in a gas drive reservoir should be completed near the bottom.

Reservoir Homogeneity

Reservoir homogeneity must be considered in selecting individual well completion intervals. Full advantage should be taken of shale barriers to aid in control of unwanted fluids and to eliminate workover expense.

In sandstone reservoirs depositional environment controls the continuity of shale breaks. Shale or clay breaks in barrier bars can be correlated over a long interval. Clay breaks in fluviatile or tidal channel sediments are difficult to correlate and usually converge. Different types of environments—i.e. channel fills, barrier bars, point bars—have characteristic permeability distribution vertically and laterally.

Similar statements could be made regarding carbonate deposition—even reef type reservoirs (sometimes thought to act like a big tank) can have barriers which totally restrict vertical movement through a portion of the reservoir. By good choice of the completion interval above or beneath a barrier with favorable properties break-through of water or gas into the well may be retarded.

Producing Rate

In a massive reservoir with good permeability and vertical communication, opening a limited in-

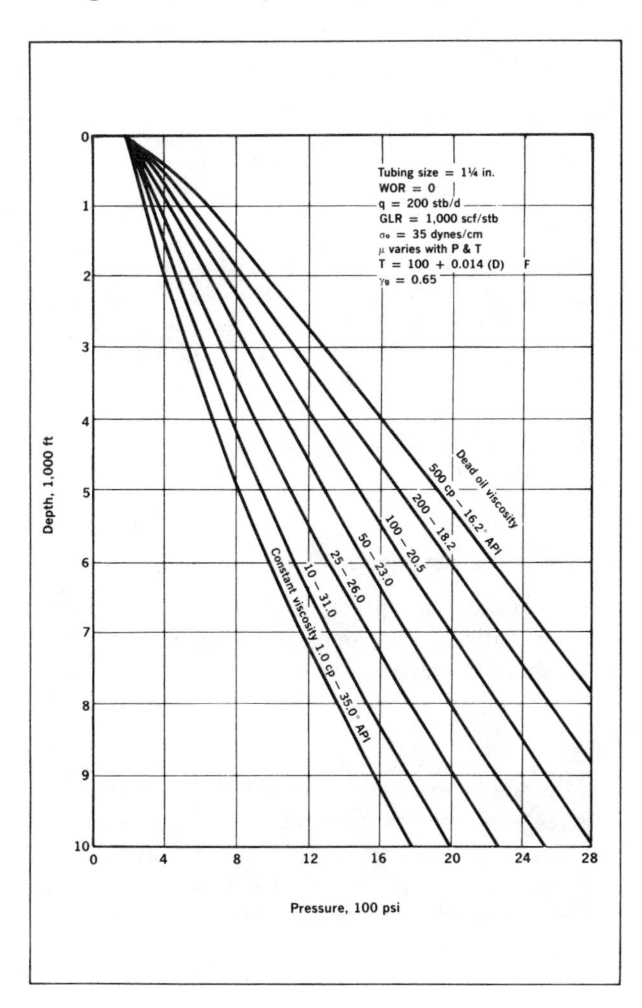

FIG. 5-22—Effect of viscosity.[10] Permission to publish by The Petroleum Publishing Co.

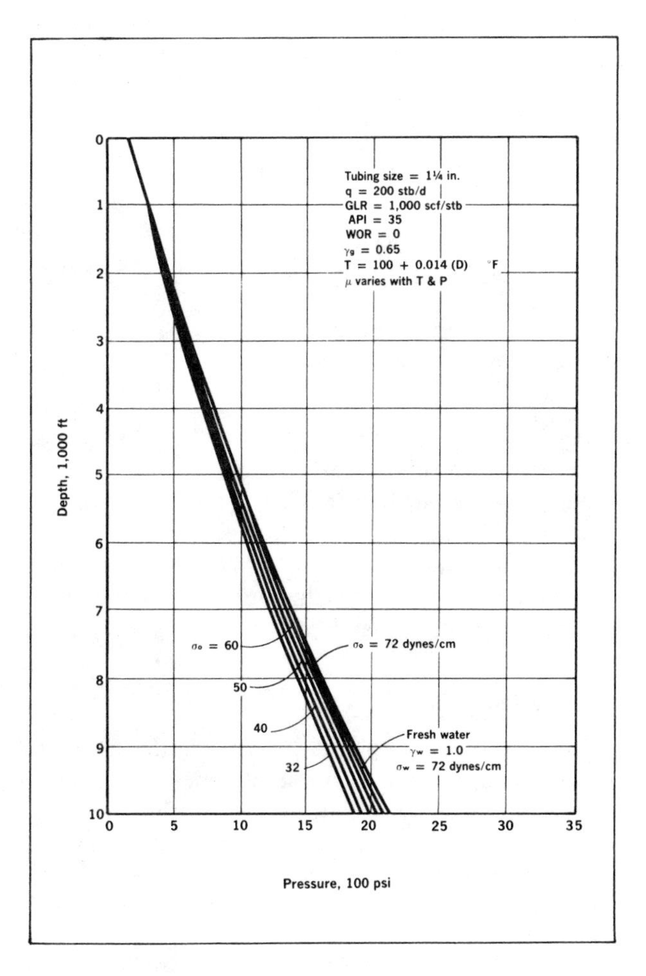

FIG. 5-23—Effect of surface tension.[10] Permission to publish by The Petroleum Publishing Co.

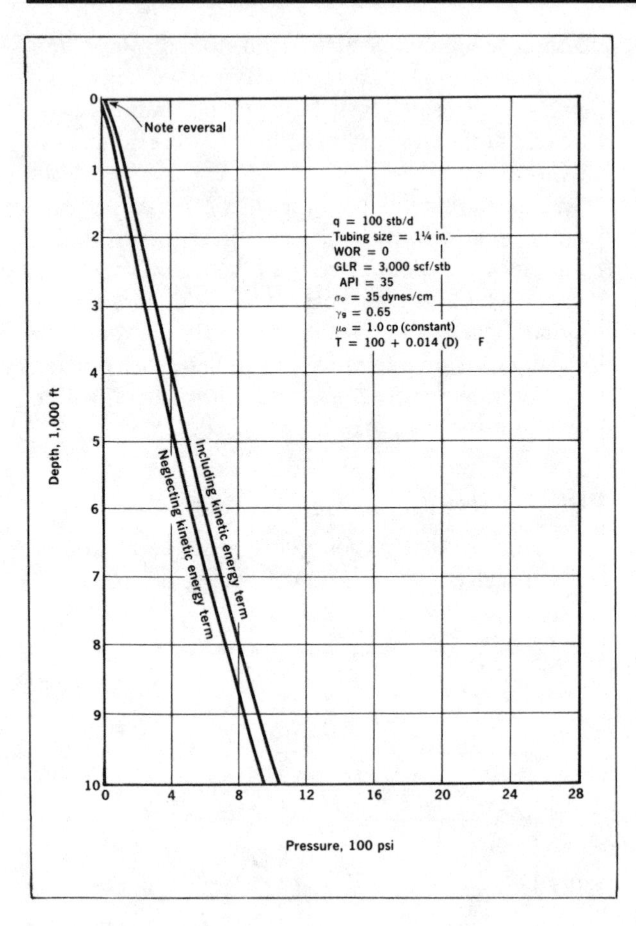

FIG. *5-24—Effect of kinetic energy.[10] Permission to publish by The Petroleum Publishing Co.*

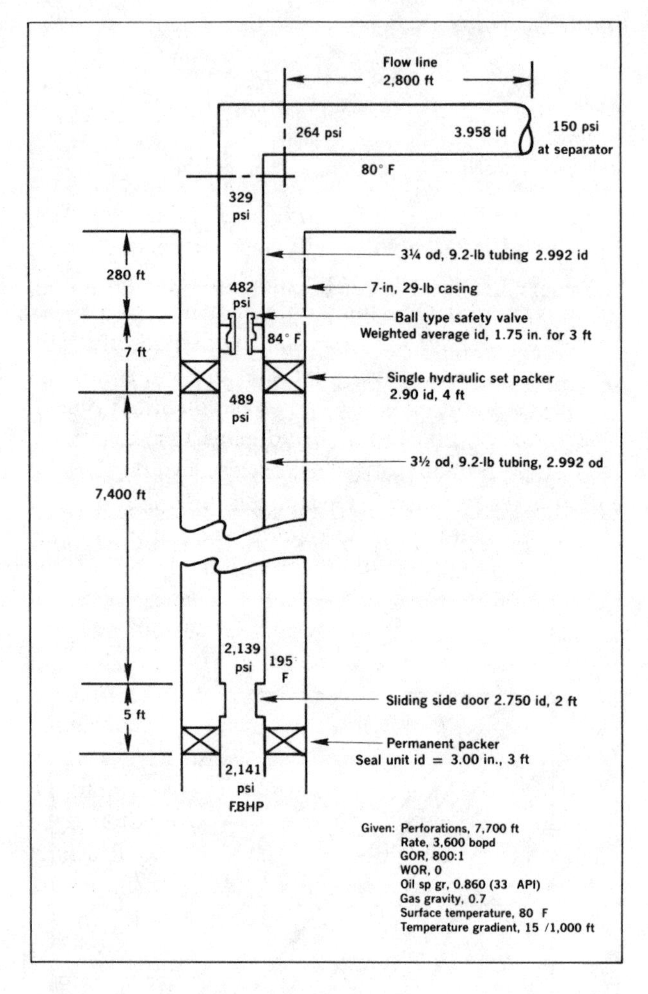

FIG. *5-25—Pressure at various points in high-flow-rate-well system.*

terval will provide high capacity and, at the same time, facilitate future workovers.

In a multi-zone field having marginal stringers, many possibilities are available. These include commingling separate reservoirs. Generally an optimum approach is to open only the number of sands that can be effectively stimulated.

Stimulation Needs

Stimulation is simplified by limiting the number of perforations and the intervals open. If several stringers are to be produced, perforations should be "pinpointed" with as much blank section as possible between zones.

Future Workover Possibilities

It is often sound practice to open only a limited interval initially, and complete in other zones later, if needed, by means of low-cost workovers.

Cost

Minimum perforation density and interval enhances economics through reduced perforating cost, facilitiating limited entry or ball sealer stimulation, and simplifying future squeeze cementing operations to control water or gas entry.

REFERENCES

1. Huber, T. A.; Allen, T. O.; and Abendroth, G. F.: "Well Completion Practices," API, Los Angeles (1950).

2. Huber and Tausch: "Permanent-Type Completions," Trans., AIME (1953) Vol. 198.

3. Althouse, W. S., Jr., and Fisher, H. H.: "The Selection of a Multiple Completion Hookup," *JPT* (Dec. 1958) p. 12.

4. Corey, C. B., Jr., and Rike, J. L.: "Tubingless Completions," Paper 926-4-6, Southwestern District API (March 1959).

5. Willingham, J. E.: "Tubingless Completions in the West Texas Area," Paper 906-4-6, Southwestern District API (March, 1959).

6. Bleakley, W. B.: "What It Takes to Make a Good Well Completion," *Oil and Gas J.* (June 11, 1962).

7. Murphy, L. A.: "Completion Developments and Trends Worth Watching," *Pet. Eng.* (January, 1964).

8. Poettman, F. H., and Carpenter, P. G.: "The Multiphase Flaw of Gas, Oil, and Water Through Vertical Flow Strings with Application to Design of Gas Lift Installations," *Drill. and Prod. Prac.* (1952) p. 257.

9. Eaton, Ben A., and Brown, Kermit E.: "The Prediction of Flow Patterns, Liquid Holdup and Pressure Losses Occurring During Continuous Two-Phase Flaw in Horizontal Pipe Lines," The Univ. of Texas, Pet. Eng. Dept. (Oct. 1965).

10. Brown, Kermit E.: *Gas Lift Theory and Practice,* The Petroleum Publishing Co. (1972).

11. Vogel, J. V.: "Inflow Performance Relationship for Solution-Gas Drive Wells," *JPT* (Jan. 1968).

12. Standing, M. 8.: "Inflow Performance Relationships for Damaged Wells Producing by Solution Gas Drive," *JPT* (Nov. 1970) p. 1399.

13. Beggs, H. D., and Bull, J. P.: "A Study of Two-Phase Flow in Inclined Pipes," *JPT* (May, 1973).

14. Crouch, E. C., and Pack, K. J.: "System Analysis Use for the Design and Evaluation Gas Wells," SPE 9424 (Sept. 1980).

15. Mach, Joe; Proano, E., and Brown, K. E.: "Application of Production Systems Analysis to Determine Completion Sensitivity," ASME #81-PET-13, (Jan. 1982).

16. McLeod, H. O., Jr.: "The Effect of Perforating Conditions on Well Performance," *JPT* (Jan. 1983).

17. Beggs, H. D.: *Production Optimization Training Manual,* Oil & Gas Consultants International, Inc. (1988).

18. Iyoho, A. W., and Lea, J. F.: "Production System Optimization: Model Applications, Capabilities, and Pitfalls," SPE 18220, Houston, (Oct. 1988).

Appendix A

Tubingless Completion Techniques and Equipment

Permanent well completions (PWC), concentric tubing workovers, and tubingless completions should all be considered as a series of well completion developments. Large diameter single completions made without tubing in the Middle East and North Africa are outside the scope of this development.

PERMANENT WELL COMPLETION (PWC)

The overall concepts of PWC had the objective of eliminating the necessity of pulling tubing during the life of a well. Figure 5A-1 illustrates the basic arrangement for permanent well completions with and without a packer. An essential feature of PWC is setting the bottom of the tubing open-ended and above the highest anticipated future completion zone.

Primary developments needed to make the system feasible form the basis of current well completion technology. These developments include:

1. The through tubing perforator, and along with it the concept of underbalanced perforating (differential pressure into the wellbore) to provide debris-free perforations.
2. A concentric tubing extension run and set on a wireline to permit circulating to the desired point in the well. Later the wireline tubing extension was replaced by the use of full string of small diameter tubing which could be run through the normal producing tubing using a small conventional workover rig. Use of this small tubing is termed Concentric Tubing Workover.
3. Low fluid loss, below frac pressure, squeeze cementing, to provide slurry properties so that cement can be placed at the desired point (in the perforations or in a channel behind the casing) and the excess cement subsequently reverse circulated out of the well.
4. Logging devices, gaslift valves, bridge plugs, and other necessary tools designed to be run through tubing on a wireline or electric line.

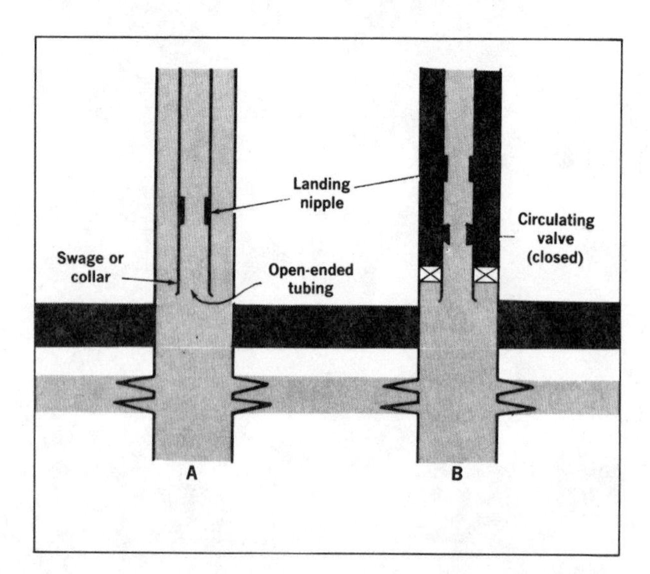

FIG. *5A-1—Permanent-type well completion for single zone, without packer (A), or with packer (B).*

TUBINGLESS COMPLETIONS

This system usually involves the cementing of one or more strings of $2^1/_2$-in. or $3^1/_2$-in., as production casing in a single borehole. Figure 5A-2 shows a comparison between conventional and tubingless completion in a multi-pay field.

The original effort was aimed at reducing initial investment. However, the major economic benefits have been in reducing well servicing and workover costs, with particular application to triple completions in lenticular multi-reservoir fields, and to dual offshore wells.

This type of completion is not necessarily restricted to low-return, low-volume, short-lived wells. Single or multi-pay gas fields are excellent candidates for tubingless completions.

Hole and casing size should be designed to obtain optimum rate of return over the life of the well. Typical casing-hole size combinations for 2-in. or $2^1/_2$-in.

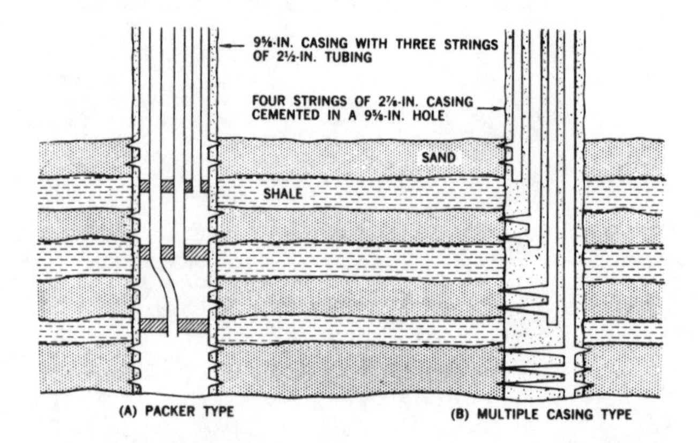

FIG. *5A-2—Standard and tubingless completion in multi-pay field.*[10] *Permission to publish by The Society of Petroleum Engineers.*

TABLE 5A-1
Popular Casing-Hole Size Combinations in the U.S.

Type completion	Surface casing in.	Hole below surface casing in.	Production casing in.
Single	7 to 8⅝	6¼ to 7⅞	2 or 2½
Dual	8⅝	7⅞	2 or 2½
Dual Offshore*	—	11½	3½
Triple	9⅝	8⅝	2 or 2½
Quadruple	10¾	9⅝	2 or 2½

*Deviated wells.

tubingless completions in the U.S. are shown in Table 5A-1.

Casing and Cementing Practices

The following discussion presents details of casing and cementing practices generally used with tubingless or multiple tubingless completions. The mechanical problems of running multiple strings require special casing handling and blowout control equipment. Cementing of multiple tubingless wells is an important concern because the annular cross section with several strings encourages channeling of cement, and ineffective displacement of mud. Good cementing practices discussed under "Primary Cementing" must be adhered to. Pipe centralization, relative pipe movement, and low PV and YP drilling fluid properties are important.

Production Casing—Non-upset *J*-55 casing can be used to about 7,500 feet. Deeper wells should use *J*-55 with a full strength connection such as Buttress or Armco Seallock or a combination of upset *J*-55 and non-upset *J*-55 depending on strength requirements. Both ends of the collars must be beveled. Couplings should be floated on to provide uniform make-up torque on both ends of the collar.

Running and Cementing Practices (based on South Louisiana)—These are:

1. Thoroughly condition mud to optimum properties prior to pulling out to run casing.
2. In case of dual completion, make up and run both strings simultaneously, employing dual slips and elevators. On a triple completion run first two strings together, followed by third string.
3. Use reciprocating wireloop scratchers and bow string centralizers throughout the productive interval on all strings.
4. Use a torque-turn device to assure a tight uniform makeup of tubing joints and to eliminate pressure testing each joint.
5. Make up cementing heads on the handling joints while they are on the pipe rack to allow fast hook-up and initiation of reciprocation as soon as pipe is on bottom. Quick initiation of reciprocation reduces chances of sticking.
6. Maintain reciprocation for at least one full cycle while mud is circulated and conditioned. Employ strain gauges below the cementing heads to assure an even pull on each string and to assure that one string is not trying to stick.
7. Cement slurry should be as "Newtonian" as possible to improve mud displacement. Densified Class H or a mixture of lightweight and Class H cement have desirable flow properties. Batch mix cement for better uniformity.
8. Pump cement down two strings simultaneously (5–6 BPM) while maintaining reciprocation throughout the period cement is being pumped and up to 30 minutes after wiper plugs are bumped.
9. Use slip-type tubing hangers to assure that each tubing string is hung in tension.
10. Test all zones for communication by pressure testing each zone after perforating.

Completing Tubingless Wells

Normal practice is to release the drilling rig after casing is cemented. As a rule, perforating, stimulation, or other well completion operations are carried out without a rig. However, if a rig is required, a small completion or workover rig is moved on the well.

Because of size limitations, careful analysis should be made of available perforators to insure adequate hole size and penetration. Whenever feasible the well is perforated with a differential into the wellbore to avoid swabbing and to improve perforation productivity. Some of the most commonly used perforators are the scallop and strip guns. The scallop-type gun provides less casing splitting.

When perforating multiple tubingless completions, it is necessary to locate adjacent casing strings and rotate the gun by surface control so as to avoid perforating adjacent casing strings. See Figure 5A-3.

For single completions, run radioactivity log and collar log for depth correlation. For multiples, run radioactive collars in each string near perforated intervals. Then run gamma ray—neutron log in one string

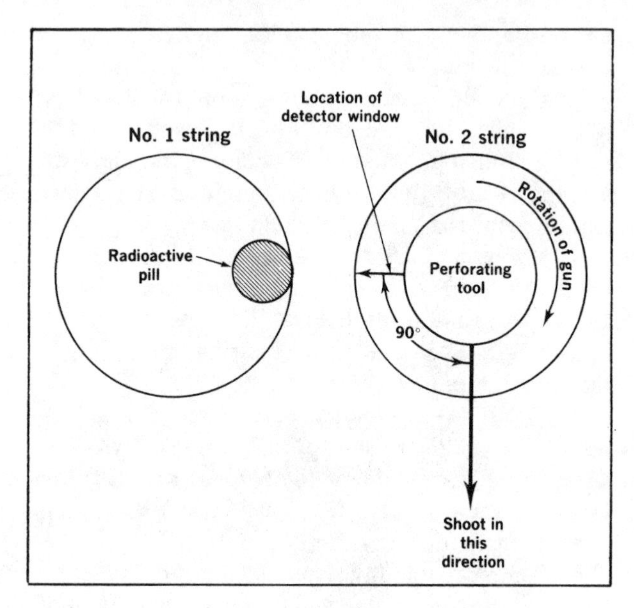

FIG. *5A-3—Orienting perforating technique in dual tubingless completion.*

to locate radioactive collars in all strings. Each casing string can then be perforated with depth correlation with previously located radioactive collars.

Logging tools are available for through casing logging down to the 2-in. size. Open hole logs are available for holes down to about $4^1/_2$ in. in diameter.

Artificial Lift

The artificial lift system should produce fluid volumes required to optimize profits and cash flow rate of return from a tubingless completion program. Five methods are available to artificially lift wells having $2^7/_8$-in. casing:

1. Casing Pumps—Figure 5A-4 depicts a casing pump installation and theoretical volumes produced from various depths from $2^7/_8$-in. od casing.

Factors to consider when planning a casing pump installation for a tubingless completion are: (1) Casing is subject to rod wear; (2) all gas must pass through pump; and (3) if appreciable sand, scale, or paraffin is anticipated, pump sticking may cause expensive workover jobs or possible loss of well.

2. Rod Pumping Inside Macaroni Strings—The most widely used method of pumping miniaturized completions is the insert pump illustrated in Figure 5A-5. A larger volume can be obtained by running a larger pump than an insert-type on the macaroni string. If gas is no particular problem, efficiency can be im-

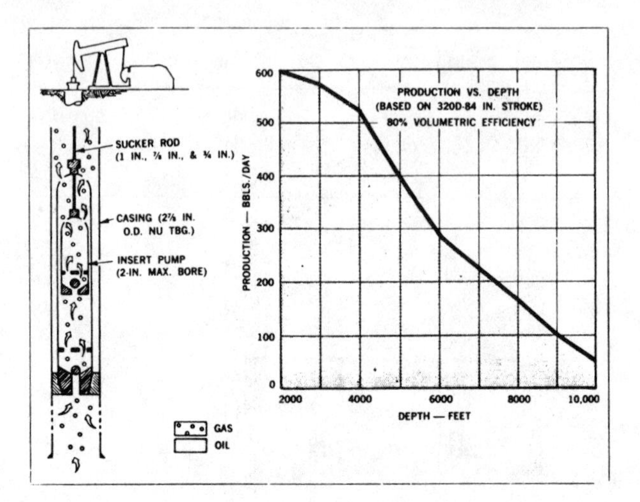

FIG. *5A-4—Casing pumping installation, $2^7/_8$-in. od casing.[10] Permission to publish by The Society of Petroleum Engineers.*

proved by running a bottom hold-down and a top pack-off seal.

Advantages of pumping inside macaroni string over casing pumping: (1) Reduces casing wear, (2) gas can be vented, thereby increasing pump efficiency, (3) corrosion and paraffin inhibitors can be circulated down casing-tubing annulus, and (4) sand can be confined to tubing, thus facilitating fishing if pump becomes stuck.

3. Hollow Sucker Rod Tubing Pumping—Figure 5A-6 illustrates rod pumping of miniaturized completions with hollow rods or sucker rod tubing. This system eliminates the hollow polished rod and flexible hose used in some installations.

The hollow rod system has restricted volume compared with a casing pump. Pumping efficiency is aided by the ability to vent gas. Casing and tubing wear, similar to that with casing pumping, limits the use of hollow rods.

4. Subsurface Hydraulic Pumps—Figure 5A-7 shows a conventional insert pump in which the macaroni string must be pulled to service the pump. The graph in Figure 5A-7 shows pump volumes with a $2^1/_2$-in. standard pump. Standard $1^1/_4$-in. free-type pumps are also applicable in miniaturized completions. High pump rates, 300 to 400 bpd, are being pumped from 10,000 to 12,000 ft in low gas-fluid ratio wells. Corrosion and paraffin can be controlled by additives to power oil.

5. Gas Lift—Gas lift is the most used artificial lift method in South Texas, where there are several thou-

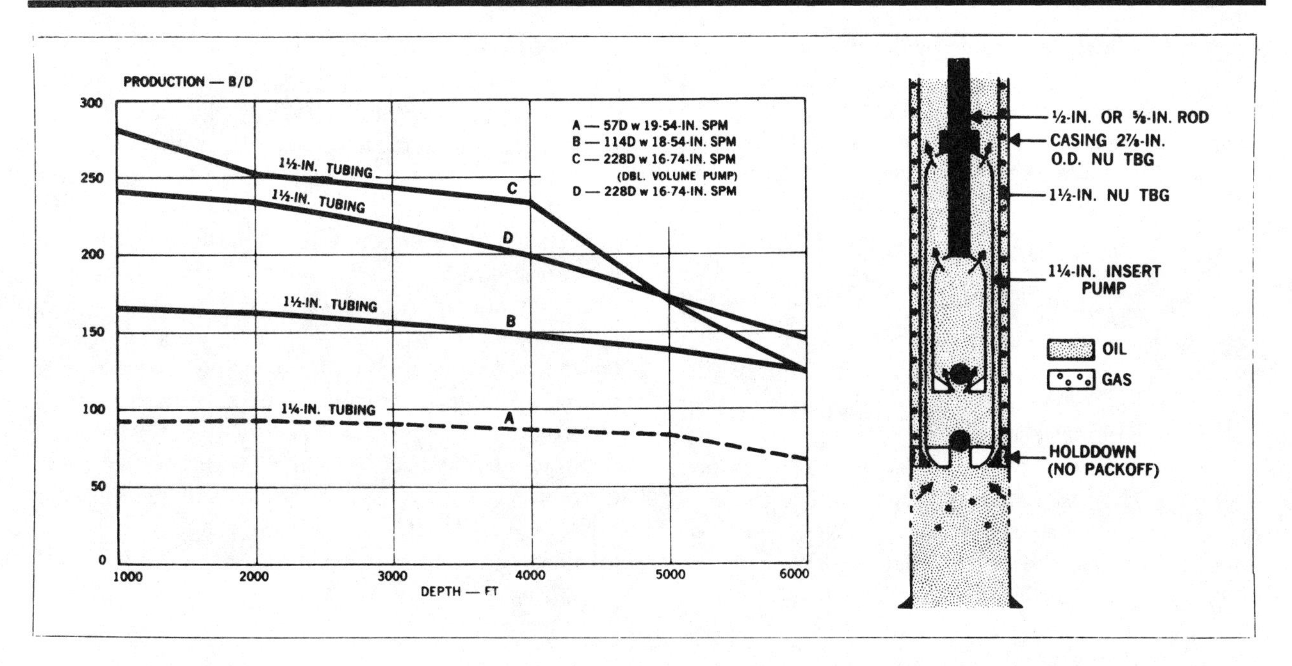

FIG. *5A-5—Insert casing pump inside macaroni string.*[10] *Permission to publish by The Society of Petroleum Engineers.*

sand miniaturized completions and where gas is readily available.

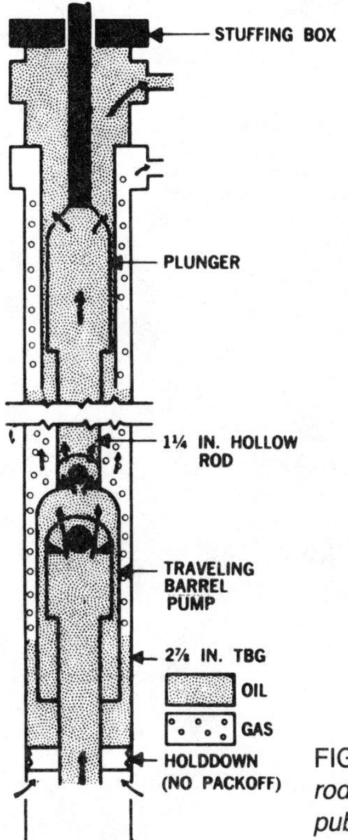

FIG. *5A-6—Use of sucker-rod tubing.*[10] *Permission to publish by The Society of Petroleum Engineers.*

Standard concentric and eccentric gas lift valves are available in tubing sizes down to $1^1/_4$-in. Figure 5A-8 shows theoretical production through gas lift casing pencil valves in an eccentric installation.

Workover of Tubingless Wells

Although the original objective of tubingless or miniaturized completions was to reduce initial well

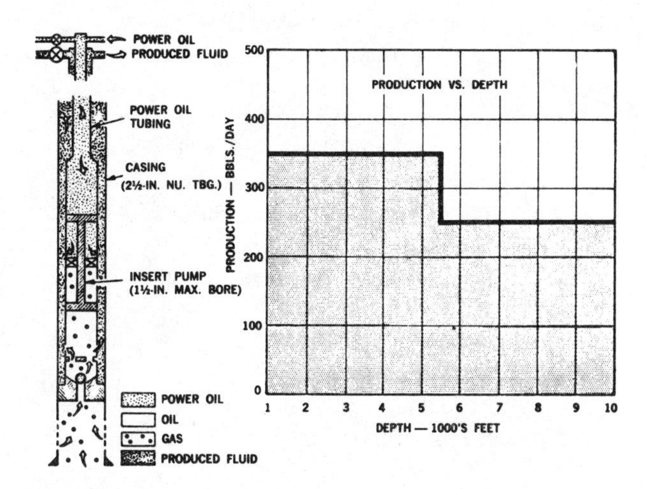

FIG. *5A-7—Hydraulic pumping in miniaturized completion.*[10] *Permission to publish by The Society of Petroleum Engineers.*

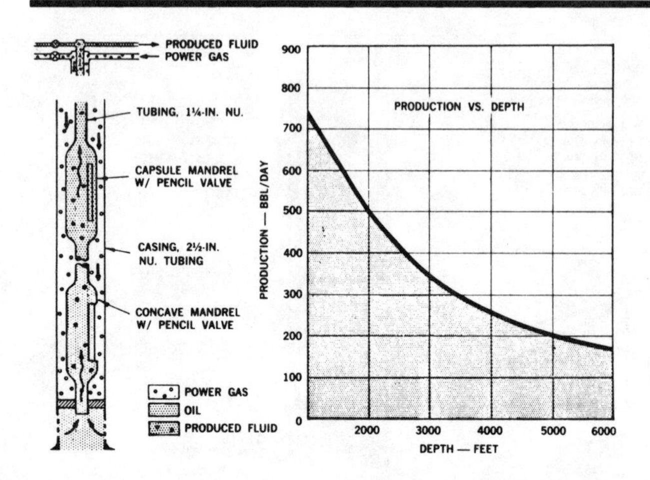

FIG. *5A-8—Gas lift in miniaturized completion.*[10] *Permission to publish by The Society of Petroleum Engineers.*

cost, experience has proved the major long range saving to be in workover of multizone completions. All the necessary operations of well workover can be accomplished (i.e.: squeeze cementing perforating or sand consolidation) using wireline, electric line or concentric tubing techniques.

Multiple tubingless completions have several unique advantages in well servicing and well workover. Communication between zones as a result of tubing and packer leaks is eliminated; plug-back can be accomplished without a rig; by isolating a workover to a single completion, formation damage to other completions in a well and need for subsequent stimulation to overcome damage is eliminated; and during workover or servicing of an individual completion of a multiple well, there is no loss of production from remaining completions.

REFERENCES

1. Allen, T. O., and Atterbury. S. H., Jr.: "Effectiveness of Gun Perforating." *Trans.* AIME (1954) 201, pp. 8–14.

2. Tausch, G. H., and McDonald, Price: "Permanent-Type Completions and Wireline Workovers," *Pet. Eng.* (Sept. 1956) Vol. 28, No. 10, p. 8–39.

3. Corley, C. B., Jr., and Rike, J. L.: "Tubingless Completions," API Paper 926-4-6 (March 1959).

4. Pistole, Harry, and True, Martin E.: "A Challenge to the Operators and Designers of Oil Field Equipment," ASME, 13th Annual Pet. Mech. Eng. Conf., Denver, CO (Sept. 21–24, 1958).

5. Huber, T. A., and Corey. C. B., Jr.: "Permanent-Type Multiple and Tubingless Completions," *Pet. Eng.* (Feb. and March 1961).

6. Lebourg, M. P., and Bell, W. T.: "Perforating of Multiple Tubingless Completions," *JPT* (May 1960) p. 88.

7. Childers, Mark A.: "Primary Cementing of Multiple Casing." *JPT* (July 1968) pp. 751–762.

8. Frank, Wallace J., Jr.: "Improved Concentric Workover Techniques Offshore," *JPT* (April 1969) pp. 401–408.

9. Crosby, George E.: "Miniaturized Completions Can Be Artificially Lifted," *Pet. Eng.* (Feb. 1969) pp. 54–59.

10. Rike, J. L., and McGlamery, R. G.: Recent Innovations in Offshore Completion and Workover Systems," *JPT* (Jan. 1970) pp. 17–24.

11. Clark, Charles R., and Jenkins, Robert C.: "Cementing Practices for Tubingless Completions," SPE 4609 (Sept. 30, 1973).

Chapter 6 Tubing Strings, Packers, Subsurface Control Equipment

Tubing Strings

Proper selection, design, and installation of the tubing string is a critical part of a completion program. The tubing must be sized so that producing operations can be carried out efficiently; it must be designed against failure from tensile forces, internal and external pressures, and corrosive actions; and it must be installed in a pressure-tight and undamaged condition.

A number of grades of steel and types of tubing connections have been developed to meet demands of greater depths and new completion techniques. API Specifications 5CT and API Bulletin 5C2 "Performance Properties of Casing and Tubing" contain detailed specifications for oil well tubular goods. Bulletin 5C1, "Care and Use of Casing and Tubing," contains recommended makeup torque for API connections. See Tables 6-2 through 6-4.

Steel Grades

Standard API steel grades for tubing are H-40, J-55, C-75, L-80, N-80, and P-105. Grades C-75 and L-80 are intended for hydrogen sulfide service where higher strength than J-55 is required. The C grade and L grade steels are heat treated to remove martensitic crystal structure. With the L grade, hardness must not exceed 23 Rockwell C.

Numbers in the grade designations indicate minimum yield strength in 1000 psi. Grade of new pipe can be identified by color bands as follows: J-55 green; K-55 two green; C-75 blue; L-80 red with brown band; N-80 red; C-95 brown, P-105 white.

Tubing Connections

Standard API Coupling Connections—Two standard API coupling tubing connections are available:

The API Non-Upset, Tubing Connection (NU) is a 10-round thread form, wherein the joint has less strength than the pipe body.

The API External Upset Tubing Connection (EUE) is an 8-round thread form wherein the joint has greater strength than the pipe body. For very high pressure service the API EUE connection is available in $2\frac{3}{8}$-, $2\frac{7}{8}$- and $3\frac{1}{2}$-in. sizes having a long thread form (EUE long T & C) wherein the effective thread is 50 percent longer than standard.

Extra Clearance Couplings—Where extra clearance is needed, API couplings can be turned down somewhat without loss of joint strength. Special clearance collars are usually marked with a black ring in the center of the color band indicating steel grade.

Extra clearance coupling-type thread forms have been developed for non-upset tubing which (unlike the API NU connection) have 100 percent joint strength. The National Tube Buttress connection and the Armco Seal Lock connection are examples.

Standard and turned-down diameters of several coupling-type connections are shown in the following table:

Standard and Turned-down Coupling Sizes

Thread Form	Coupling outside diameter—in. Standard	Special clearance
(2⅜ in.)		
API NU—10-round	2.875	2.642
API EUE—8-round	3.063	2.910
National Tube buttress	2.875	2.700
Armco Seal Lock	2.875	2.700
(2⅞ in.)		
API NU—10-round	3.500	3.167
API EUE—8-round	3.688	3.460
National Tube buttress	3.500	3.220
Armco Seal Lock	3.500	3.220

Integral Joint Connections—Several integral-joint thread forms are available from various manufacturers which provide extra clearance. Some can be turned down to provide even greater clearance. These joints usually carry a premium price and must, therefore, be justified by special conditions. An API integral-joint connection (10-round thread form) has recently been adopted for small diameter tubing, sizes 1.315, 1.660, 1.900, 2.063.

Special Connections—Many special tubing connections, both integral joint and threaded and coupled, are available to meet needs involving very high pressures or corrosive service. The World Oil magazine[4] prints annually joint identification tables showing most special connections.

Connection Seals—In order to form a seal with any well-designed connection certain specific make-up requirements must be met. Most connections use a metal-to-metal seal which requires that the mating pin and box surfaces be forced together under sufficient stress to establish a bearing pressure exceeding the differential pressure across the connection. This principle is illustrated in Figure 6-1.

The API Round Thread connection forms several metal-to-metal seals between the tapered portions of pin and box surfaces (Figure 6-2). The small void (0.003-in. clearance) between the crest and root of mating threads must be filled with thread compound solids in order to transmit adequate bearing pressure from one threaded surface to the other.

Buttress and 8-Acme connections are similar to API Round Thread connections in that adequate bearing pressure between mating thread surfaces must be established, and voids must be filled with thread compound solids to transmit bearing loads across void spaces.

So-called "metal-to-metal" connections (Hydril and Extremeline) have large smooth metal-to-metal sealing surfaces. These seals are self-energized by the contained pressure. Threads have relatively large clearance, and do not act as seals. Armco Seal Lock has both a sealing thread and smooth metal sealing surface.

Resilient teflon rings are used in several connections to act as a supplementary seal and to provide corrosion protection. Premium connections are available that have a multiple sealing system, including combinations of sealing threads, metal to metal seals, and/or teflon rings.

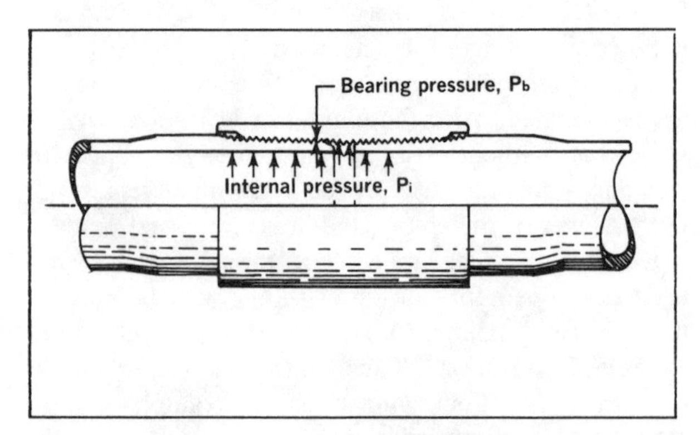

FIG. 6-1—*Bearing pressure must exceed internal pressure to form pressure seal.[2] Permission to publish by The Society of Petroleum Engineers.*

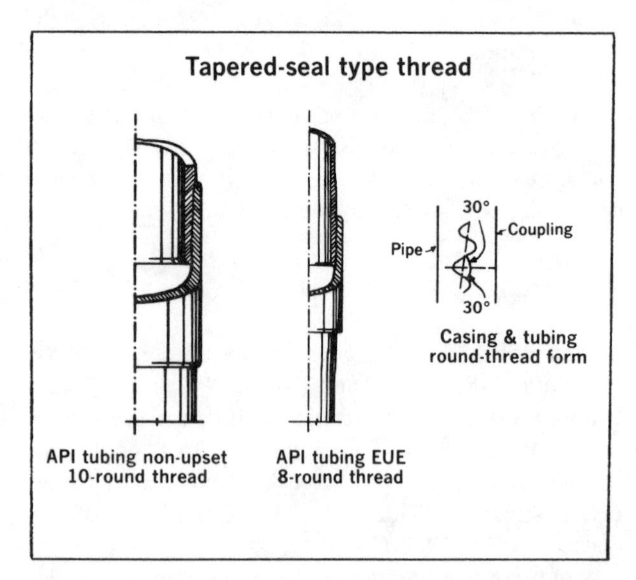

FIG. 6-2—*API round-threaded connections.[2] Permission to publish by The Society of Petroleum Engineers.*

Makeup of API Threaded Connection

The stresses induced in the connection during make-up and subsequent service determine the success of the connection as a sealing element.

Normal Conditions—Traditional field practice often dictates that API tapered thread connections are made up on one of the following bases:

1. Make-up position rules of thumb: "last tool mark on the pin must be 'buried' by the collar"; "two turns past hand tight (50 ft-lbs).
2. Make-up torque recommendations, "Minimum, optimum, or maximum," as shown in API Tubing Table 6-4.

With good thread cleaning and doping practices, these techniques prove satisfactory for "normal" situations.

Critical Situations—To obtain maximum leak resistance with the API tapered thread connection the pin end of connection must be made up to the point of yielding. The problem of make-up is thus to screw the connection up sufficiently to provide the needed seal without permanently damaging the connection. Figure 6-3 shows the relation between "Make-up" (turns past hand tight), Hoop Stress in the coupling, and Radial Pressure between pin and coupling.

Where maximum seal is required, make-up torque is not a good indication of connection stress, because it depends on friction between mating surfaces. Friction, for clean threads, is a function of thread dope lubricating characteristics. This is shown in Figure 6-4. Likewise turns or make-up position is not a reliable indication of pin stress due to size tolerance in machining the threads.

An extensive laboratory testing program by a major operator developed the torque-turn technique of tubing make-up which sets minimum criteria for both torque and make-up position which must be met to insure a seal under high-pressure conditions. For a particular connection and thread dope, the testing program established values for the following parameters:

Reference torque	— Point of intimate contact—start of turns count.
Minimum torque	— Lowest torque (ft. lbs.) needed for seal.
Low turns	— Must get at least this number before reaching minimum torque—or connection is signaled "bad."
Minimum turns	— Turns required with at least minimum torque for good seal.
Maximum turns	— Must not exceed this value before reaching minimum torque or connection is signaled "bad."

Special hydraulic tong equipment is available to measure torque and count turns, to record measure-

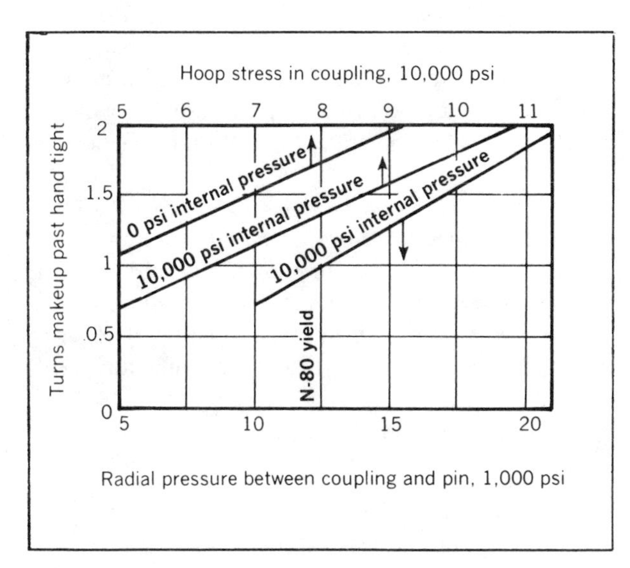

FIG. *6-3—Stresses due to connection makeup and internal pressure.[2] Permission to publish by The Society of Petroleum Engineers.*

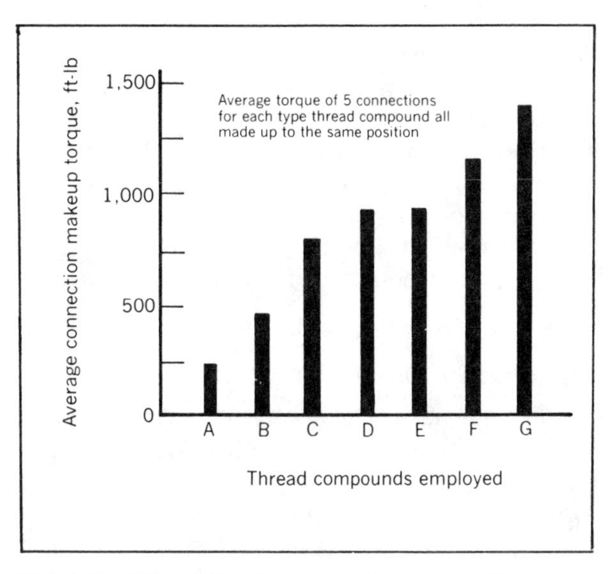

FIG. *6-4—Effect of various thread compounds on connection makeup torque.[2] Permission to publish by The Society of Petroleum Engineers.*

ments, and to signal good or bad connections. For critical situations—high pressure oil or gas—or hostile environment conditions—field results of torque-turn method justifies consideration.

Design of Tubing Strings

Tubing string design is essentially the same as for casing. Tapered strings are becoming more common in deep wells, although uniform strings are desirable (but more expensive) due to difficulty of keeping tubing strings in proper order of weight and grade.

Uniform weight upset tubing strings reach their safe tensile limit in air as follows:

Tensile Setting Depths in Air for Upset Tubing (ft)[(a)(b)]

	Safety Factor		
Grade	*1.50*	*1.60*	*1.75*
J-55	10,200	9,600	8,000
C-75	13,900	13,000	11,900
N-80	14,800	13,900	12,700
P-105	19,500	18,300	16,700

[(a)]Based on minimum yield strength times area of section under root of last perfect thread, or body of pipe whichever is smaller.
[(b)]Values shown apply to normal or special clearance couplings.

A tension design factor of 1.60 is common for uniform tubing strings.

Collapse design factor should not be less than 1.00 based on the pressure differential that may actually be applied; for example, a full annular fluid column and an empty tubing string.

Tubing should not be subjected to burst pressures higher than its rated pressure divided by 1.3, unless it has been previously tested to a higher pressure.

High Strength Tubing

High strength tubing is usually considered to be those grades with a yield strength above 80,000 psi, i.e., C-75, L-80, N-80, and P-105. C-75, L-80, and N-80 are included because their yield strength as manufactured, often exceeds 80,000 psi. High strength tubing, particularly P-105, presents several problems due to decreased ductility and increased sensitivity to sharp notches or cracks.

Physical Properties of Steel

Yield and Tensile Strength—The tensile test provides basic design information on the strength of materials. This test subjects a standard specimen to a gradually increasing load. At relatively low loads (the elastic range) elongation is directly proportional to the load applied and permanent deformation does not occur. As load increases, a point is reached where elongation occurs with no increase in load.

This is the yield point. The load at this point divided by the specimen cross-sectional area is the yield strength. Further increases in load cause permanent deformation (plastic range) and finally the specimen breaks. The load at the breaking point determines the tensile strength or ultimate strength.

The numerical value of tensile strength must be used with care, since it is determined under very restrictive conditions of uniaxial loading, and may not relate closely to the complex conditions of stress and environment encountered in service.

Ductility—Ductility is the ability of material to plastically deform without fracturing. Material with high ductility will permit large deformation in the plastic range before breaking. Ductility is measured by percent elongation of a standard specimen and API standards specify elongation for each grade of tubing.

Toughness and Impact Strength—Material failure is usually classed as ductile fracture or brittle fracture. A ductile fracture occurs with plastic deformation prior to and during propagation of the crack. A brittle fracture occurs with little elongation and at a rapid rate. Toughness refers to the ability of a material to resist brittle failure. Toughness or impact strength is measured by the Charpy impact test, in which a swinging pendulum strikes and fractures a notched specimen.

API Physical Property Specifications—API physical property specifications for tubing cover only basic material properties and are shown in Table 6-1. The maximum yield strength limit and the minimum elongation specification are important factors in ensuring satisfactory mill control of tubing manufacture.

Sensitivities of High-Strength Tubing

Notch Sensitivity—Failures of high-strength tubing are normally caused by: (1) manufacturing defects, (2) damage during transportation, handling and running, or (3) hydrogen embrittlement.

Any sharp-edged notch or crack in the surface of a material is a point of stress concentration which

TABLE 6-1
API Physical Property Specifications for Tubing

Grade	Yield Strength psi		Minimum Tensile Strength psi	Min. Elongation % in 2 Inches[a]	Typical Hardness Rc
	Minimum	Maximum			
H-40	40,000	80,000	60,000	29.5	—
J-55	55,000	80,000	75,000	24.0	14
C-75	75,000	90,000	95,000	19.5	22
L-80	85,000	95,000	95,000	19.5	23[b]
N-80	80,000	110,000	100,000	18.5	24
P-105	105,000	135,000	120,000	16.0	35

[a]Based on $e = 625{,}000 \frac{A^{0.2}}{U^{0.9}}$ (b) Maximum hardness Rc

e = Min. Elongation in 2 inches, %
A = Cross Sectional Area (in.^{2})
U = Tensile Strength psi

tends to extend the crack progressively deeper into the material much like driving a wedge. Low strength materials are soft and ductile and will yield plastically to relieve the stress concentration. High strength materials do not yield to relieve the stress concentration and, thus, fatigue or fail rapidly when subjected to cyclic stresses.

Hydrogen Embrittlement—In the presence of moisture only a trace of H_2S is needed to cause hydrogen embrittlement. The embrittlement process (sometimes called sulfide corrosion cracking), is not fully understood, but apparently results from the penetration of hydrogen atoms into the lattice structure of a high-strength steel. The hydrogen atom is smaller than the lattice structure and can migrate into the steel similarly to fine sand passing through a gravel pack.

When two hydrogen atoms meet within the steel lattice they combine to form a hydrogen molecule with a resulting increase in size. The stress created by this increase in size can part the grains of a low ductility steel structure. Low strength high ductility steels will yield to relieve the stress without failing; thus are more suitable for use in a hydrogen sulfide environment.

It is generally accepted that below a hardness of 22 Rockwell C embrittlement or sulfide cracking does not take place. Figure 6-5 shows correlation between hardness, applied stress (% of yield stress) and time to failure.

P-105 tubing with a hardness of about 35 Rc could be expected to fail due to embrittlement within a few days if loaded to even 50% of its yield stress in the presence of H_2S and water. Recent experience shows that the problem of embrittlement is reduced at temperatures above 180°F. L-80, which is the only API grade that has a maximum hardness limitation—23 Rockwell C, is the best API grade for a hydrogen embrittlement environment.

Embrittlement of high strength tubing has been reported as a result of acid used in mill cleaning op-

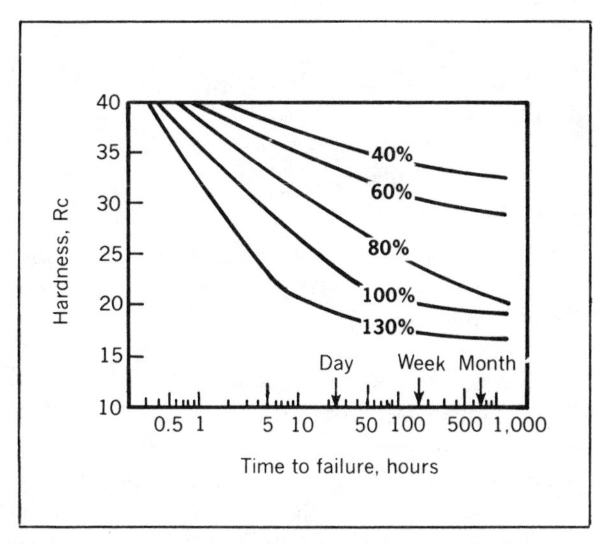

FIG. 6-5—*Approximate correlation of failure time to hardness and applied stress for carbon steel in 5% NaCl solution containing 3,000 ppm H_2S (from CORROSION, 22(8), Aug. 1966).*

erations, improperly inhibited formation acidizing operations, or sulfides formed by packer fluid degradation due to thermal, bacterial, or electrochemical effects.

Tubing Inspection

Methods—Visual inspection of any tubing string regardless of grade should always be mandatory before tubing is run in a well. Defects recognizable by visual inspection include (1) mill defects such as seams, slugs, pits or cracks; (2) poorly machined threads; or (3) shipping or handling damage to the pipe body, coupling and threads. An alert crew can often prevent installation of pipe weakened by excessive slip or tong damage.

Hydrostatic pressure tests of tubing and connections as the string is being run in the hole are often considered good practice. The general procedure is to test only the connection unless there is definite indication of a leak within the joint body. Test pressures are usually 80 percent of internal yield pressure based on 100 percent wall thickness and minimum yield strength values. A successful pressure test is not positive proof of the lack of mill defects, since these may show up only after a number of cycles of pressure or temperature change. Where the Torque Turn Method of make up is used hydrostatic testing is probably not necessary.

Electromagnetic Search Coil inspection methods include two types which will identify defects as follows:

1. Corrosion pits and transverse defects—Tuboscope Company's Sonoscope; Plastic Applicator's Scanograph.
2. Corrosion pits, transverse defects and longitudinal defects—Tuboscope's Amalog; Plastic Applicator's Scanalog.

Since the conductor coils must be in contact with the pipe these electromagnetic inspections are questionable in the upset portion of tubing or in threaded areas.

Magnetic Particle inspection methods induce a transverse magnetic field in the tubing, and magnetic particles dusted on the tubing line up to point out longitudinal defects. Special end-area magnetic particle inspection methods have been developed to evaluate both longitudinal and transverse defects in coupling and upset areas that could not be detected by electromagnetic search coil inspections.

Magnetic particle inspections are considered to be less positive than electromagnetic search coil inspections because: (1) effectiveness is reliant on operator's attitude and efficiency, (2) environmental conditions (wind, rain, and light), (3) inspection is limited to outside of pipe, and (4) transverse defects are not detected (except with the end-area technique).

Inspection Criteria—In spite of mill inspections, new pipe containing serious defects is often received in the field. As a result, one major operator established these inspection criteria:

1. Inspections, other than visual inspection, of new J-55 tubing are generally uneconomical because failures caused by mill defects are uncommon in these grades.
2. New C-75 and higher grade tubing should have Tuboscope's Amalog (or equivalent) inspection. Defects from 5 to 12½ percent of wall thickness should be removed, and joints with defects greater than 12½ percent should be rejected.
3. New C-75 and higher grade couplings should be boxed separately and should have magnetic particle inspection.
4. Used C-75 or higher grade couplings should be visually inspected.
5. Used N-80 or higher grade tubing should have Tuboscope's Sonoscope (or equivalent) inspection when service conditions indicate that corrosion pits, transverse cracks or service-induced defects are present.

Tubing Handling Practices

High-strength tubing can be easily damaged by inadequate shipping and handling practices, or by careless use of slips and tongs. Improper use of tongs and slips is probably responsible for more critical damage than all other things combined. Proper handling practices are absolutely essential for high-strength materials.

Mill Coupling Buck-On Practices—Field experience shows that mills often do not obtain proper thread make-up. One major operator found that 85% of collar leaks were in the mill end of the collar. Some operators order C-75 and higher grade couplings boxed separately to insure proper doping and make-up. Where mills are using the Torque Turn make-up method the practice of separate boxing of collars is probably not necessary.

Loading, Transportation, and Unloading—P-105 and higher grade pipe should be unloaded with a gin-pole truck using a spreader, and nylon or webbing slings which will not scratch the pipe. It must not be unloaded by the common practice of rolling from trucks to pipe racks. Woodsills should be placed between the pipe rack and first row of pipe, and between each row of pipe to minimize pipe contact. Pipe should be racked in a stairstep fashion away from the catwalk so that pipe from the top row never falls more than one pipe diameter as it is rolled to the pipe rack.

Equipment for Use with High-Strength Tubing

Tongs—Use of power tongs is considered a necessity to obtain adequate make-up. Tong dies are available which (1) are contoured to fit the outside diameter of the pipe, (2) have larger surface areas to contact the pipe, and (3) have teeth separated by cross-hatches to minimize slippage and notch depth.

Elevators and Spiders—Collar-type tubing elevators with level bearing surfaces are satisfactory for running high-strength tubing with API non-beveled couplings fully made up. Picking up heavy tubing strings with collar-type elevators where collars are made up only "hand tight" may cause partially engaged threads to fail.

Integral joint or collars with beveled underside must, of course, be handled with slip-type elevators.

Satisfactory hand-set rotary slips are available for running high-strength tubing; however, a worn rotary table or bushings can provide uneven support for rotary slips causing non-uniform contact between tubing and rotary slips which can damage the tubing.

Tubing Running Practices

Adequate thread cleaning to assure removal of all sand, dirt, and dried thread dope is the key to proper connection make-up and pressure-tight strings. The following procedure describes a field-proven technique for proper thread cleaning and running practices for high strength tubing with the couplings removed.

1. Remove the thread protectors.
2. Clean the threads with kerosene and a wire brush; satisfactory cleaning requires complete removal of all dope, dirt, sand, and other foreign material to 100 percent bare, shiny steel. The use of kerosene in a compressed air spray gun operating off the rig air system or from a portable unit is also a satisfactory method and will accelerate job completion. If steam is available, steam cleaning is an excellent method.
3. Dry the threads with clean rags or compressed air.
4. Re-install clean thread protectors on the dry pin ends.
5. Apply thread compound to the male threads at the box end of the tube.
6. Install and make-up tubing couplings manually with about 300 foot-pounds of torque using special friction-type tongs to eliminate notching. (Installing the couplings before picking up the tubing minimizes the possibility of dropping the string, in the event slippage occurs between the pipe and the elevators as the joint is lowered in the derrick.)
7. Wash all dirt off the ramp and catwalk.
8. Roll one joint of pipe at a time from the upper tiers onto wooden sills placed across the catwalk. Roll pipe slowly and maintain control at all times to prevent colliding of joints.
9. Steel thread protectors should remain on the pin ends of the pipe while picking up pipe from the catwalk.
10. While picking up each joint from the catwalk with a plaited pickup cable and air hoist or cathead, use a snub rope attached to the pin end to enable one man to restrain lateral motion of the joint and minimize contact between the pipe and ramp. Minimizing contact between the pipe and ramp by snubbing should permit the use of a clean plastic bucket over the box end as a dirt deflector. A rope hold-back should also be used at the "V" door to catch the lower end of the joint as it swings on to the derrick floor.
11. With the box end of the joint resting on the derrick floor, remove the plastic bucket and use dry compressed air or clean, dry rags to remove sand or other foreign material from the dry box threads.
12. After the traveling block pulls the joint up to a vertical position using a plaited pickup cable, remove the pin-end thread protector. Some new pipe has loose mill scale inside which should be allowed to fall out prior to stabbing to minimize contamination of clean threads. Use dry compressed air or clean, dry rags to remove mill scale or other foreign material which may accumulate on the dry pin-end threads while picking up the joint. Apply a light coat of thread compound to clean pin-end threads and make up the joint.

Torque make-up should be measured by a direct reading Martin-Decker or equivalent power tong hydraulic torque gauge. A pressure gauge on air tongs is unsatisfactory, because the torque exerted for a given pressure is dependent on the condition of the air motor and the amount of lubricant used. The pointer of a hydraulic torque gauge should always return to zero when tong torque is zero to permit accurate measurements at low and high torques. Accurate torque measurement requires that the backup cable connect to the tong arm at a 90-degree angle and in the same horizontal plane; periodic leveling of the tong suspension system may be necessary to permit even die contact with the pipe.

Make-up torque values for API tubing connections are in the "Tubing Tables" at the end of this section. It should be remembered that in the case of API connections, torque values are a guide only and that proper make-up should move the coupling face to or past the "last scratch" (transverse plane passing through the last mark on the pipe OD made by thread-cutting tools). Manufacturers' current recommendations should be used for making up all other connections.

A satisfactory method of establishing the correct torque is to record the torque measurement necessary to place the coupling face as described above on the first 10 or 12 joints and then use this torque reading for the remainder of the string. Should a particular length require make-up torques much higher or lower than other lengths of the same size, weight, grade, and having the same type connection, back out the length and visually inspect for defects, thread damage, and foreign material on the threads and remove the joint from the string if necessary. Couplings should not be hammered to break out joints except as a final resort. Hammered couplings should be replaced before running the pipe.

REFERENCES

1. Hilliard, Harold: "How Corrosive Environments Affect Drill Stem," *Pet. Eng.* (Sept. 1968) p. 84.

2. Weiner, P. D., Sewell, F. D., Jr.: "New Technology for Improved Tubular Connection Performance," *JPT* (March, 1967) p. 337.

3. Weiner, P. D., True, Martin: "A Method of Obtaining Leakproof API Threaded Connections in High Pressure Gas Service," API Paper 926-14-17 (March 5, 1969).

4. 1988 Tubing Tables, World Oil (Jan. 1989).

API Publication List

Specifications

Spec 5CT: Specification for Casing and Tubing.

Std 5B: Specification for Threading, Gaging, and Thread inspection of Casing, Tubing, and Line Pipe Threads.

Recommended Practices

RP 5C1: Recommended Practice for Care and Use of Casing and Tubing.

RP 5A5: Recommended Practice for Field Inspection of New Casing, Tubing, and Plain End Drill Pipe.

Bulletins

Bul 5A2: Bulletin on Thread Compounds.

Bul 5C2: Bulletin on Performance Properties of Casing and Tubing.

TABLE 6-2
Casing Minimum

Dimensions						*Minimumum Collapse Pressure (psi)*				*Minimum Internal Yield Pressure (psi)*			
OD Size (In.)	*Weight Per Foot Nom. (Lbs.)*	*Wall (In.)*	*ID (In.)*	*Drift (In.)*	*Coupling OD (In.)*	*J55*	*N80*	*P110*	*Q125*	*J55*	*N80*	*P110*	*Q125*
4-1/2	9.50	.250	4.090	3.965	5.000	3310	. . .	. . .	. . .	4380	. . .	. . .	. . .
	11.60	.250	4.000	3,875	5.000	4960	6350	7560	. . .	5350	7780	10690	. . .
	13.50	.290	3.920	3.795	5.000	. . .	8540	10670	. . .	. . .	9020	12410	. . .
	15.10	.337	3.826	3.701	5.000	. . .	. . .	14320	15840	. . .	. . .	14420	16380
5	11.50	.220	4.560	4.435	5.563	3060	. . .	. . .	. . .	4240	. . .	. . .	. . .
	13.00	.253	4.494	4.369	5.563	4140	. . .	. . .	. . .	4870	. . .	. . .	. . .
	15.00	.296	4.408	4.283	5.563	5550	7250	8830	. . .	5700	8290	11400	. . .
	18.00	.362	4.276	4.151	5.563	. . .	10490	13450	14830	. . .	10140	13940	15840
5-1/2	14.00	.244	5.012	4.887	6.050	3120	. . .	. . .	. . .	4270	. . .	. . .	. . .
	15.50	.275	4.950	4.825	6.050	4040	. . .	. . .	. . .	4810	. . .	. . .	. . .
	17.00	.304	4.892	4.767	6.050	4910	6280	7460	. . .	5320	7740	10640	. . .
	20.00	.361	4.778	4.653	6.050	. . .	8830	11080	. . .	. . .	9190	12640	. . .
	23.00	.415	4.670	4.545	6.050	. . .	11160	14520	16070	. . .	9880	14520	16510
6-5/8	20.00	.288	6.049	5.924	7.390	2970	. . .	. . .	. . .	4180	. . .	. . .	. . .
	24.00	.352	5.921	5.796	7.390	4560	5760	6710	. . .	5110	7440	10230	. . .
	28.00	.417	5.791	5.666	7.390	. . .	8170	10140	. . .	. . .	8810	12120	. . .
	32.00	.475	5.675	5.550	7.390	. . .	10320	13200	14530	. . .	10040	13800	15680
7	20.00	.272	6.456	6.331	7.656	2270	. . .	. . .	. . .	3740	. . .	. . .	. . .
	23.00	.317	6.366	6.241	7.656	3270	3830	. . .	. . .	4360	6340	. . .	. . .
	26.00	.362	6.276	6.151	7.656	4320	5410	6210	. . .	4980	7240	9960	. . .
	29.00	.408	6.184	6.059	7.656	. . .	7020	8510	. . .	. . .	8160	11220	. . .
	32.00	.453	6.094	5.969	7.656	. . .	8600	10760	. . .	. . .	9060	12460	. . .
	35.00	.498	6.004	5.879	7.656	. . .	10180	13010	14310	. . .	9240	12700	15560
	38.00	.540	5.920	5.795	7.656	. . .	11390	15110	16750	. . .	9240	12700	16880
7-5/8	26.40	.328	6.969	6.844	8.500	2890	3400	. . .	. . .	4140	6020	. . .	. . .
	29.70	.375	6.875	6.750	8.500	. . .	4790	5340	. . .	. . .	6890	9470	. . .
	33.70	.430	6.765	6.640	8.500	. . .	6560	7850	. . .	. . .	7900	10860	. . .
	39.00	.500	6.625	6.500	8.500	. . .	8810	11060	12060	. . .	9180	12620	14340
8-5/8	24.00	.264	8.097	7.972	9.625	1370	. . .	. . .	. . .	2950	. . .	. . .	. . .
	32.00	.352	7.921	7.796	9.625	2530	. . .	. . .	. . .	3930	. . .	. . .	. . .
	36.00	.400	7.825	7.700	9.625	3450	4100	. . .	. . .	4460	6490	. . .	. . .
	40.00	.450	7.725	7.600	9.625	. . .	5520	6380	. . .	. . .	7300	10040	. . .
	44.00	.500	7.625	7.500	9.625	. . .	6950	8400	. . .	. . .	8120	11160	. . .
	49.00	.557	7.511	7.386	9.625	. . .	8570	10720	11660	. . .	9040	12430	14130
9-5/8	36.00	.352	8.921	8.765	10.625	2020	. . .	. . .	. . .	3520	. . .	. . .	. . .
	40.00	.395	8.835	8.679	10.625	2570	3090	. . .	. . .	3950	5750	. . .	. . .
	43.50	.435	8.755	8.599	10.625	. . .	3810	4430	. . .	. . .	6330	8700	. . .
	47.00	.472	8.681	8.525	10.625	. . .	4750	5310	5640	. . .	6870	9440	10730
	53.50	.545	8.535	8.379	10.625	. . .	6620	7930	8440	. . .	7930	10900	12390
10-3/4	40.50	.350	10.050	9.894	11.750	1580	. . .	. . .	. . .	3130	. . .	. . .	. . .
	45.50	.400	8.950	9.794	11.750	2090	. . .	. . .	. . .	3580	. . .	. . .	. . .
	51.00	.450	9.850	9.694	11.750	2700	3220	3670	. . .	4030	5860	8060	. . .
	55.50	.495	9.760	9.604	11.750	. . .	4020	4630	. . .	. . .	6450	8860	. . .
	60.70	.545	9.660	9.504	11.750	. . .	. . .	5860	6080	. . .	. . .	9760	11090
	65.70	.595	9.560	9.404	11.750	. . .	. . .	7490	7920	. . .	. . .	10650	12110
11-3/4	47.00	.375	11.000	10.844	12.750	1510	. . .	. . .	. . .	3070	. . .	. . .	. . .
	54.00	.435	10.880	10.724	12.750	2070	. . .	. . .	. . .	3560	. . .	. . .	. . .
	60.00	.489	10.772	10.616	12.750	2660	3180	3610	3680	4010	5830	8010	9100
13-3/8	54.50	.380	12.615	12.459	14.375	1130	. . .	. . .	. . .	2730	. . .	. . .	. . .
	61.00	.430	12.515	12.359	14.375	1540	. . .	. . .	. . .	3090	. . .	. . .	. . .
	68.00	.480	12.415	12.259	14.375	1950	2260	2330	. . .	3450	5020	6910	. . .
	72.00	.514	12.347	12.191	14.375	. . .	2670	2890	2880	. . .	5380	7400	8410
16	75.00	.438	15.124	14.936	17.000	1020	. . .	. . .	. . .	2630	. . .	. . .	. . .
	84.00	.495	15.010	14.822	17.000	1410	. . .	. . .	. . .	2980	. . .	. . .	. . .
18-5/8	87.50	.435	17.755	17.567	19.625	630	. . .	. . .	. . .	2250	. . .	. . .	. . .
20	94.00	.438	19.124	18.936	21.000	520	. . .	. . .	. . .	2110	. . .	. . .	. . .
	106.50	.500	19.000	18.812	21.000	770	. . .	. . .	. . .	2410	. . .	. . .	. . .
	133.00	.635	18.730	18.542	21.000	1500	. . .	. . .	. . .	3060	. . .	. . .	. . .

Reprinted in part from API Bulletin 5C2, May 1987

Performance Properties

		Joint Yield Strength (1000 Lbs.)											
		Short Thread						Long Thread					
OD Size (In.)	Weight Per Foot Nom. (Lbs.)	J55	K55	N80	L80	P110	Q125	J55	K55	N80	L80	P110	Q125
4-1/2	9.50	101	112	. . .	. . .	. . .	. . .	. . .	. . .	. . .	. . .	. . .	. . .
	11.60	154	170	. . .	. . .	. . .	. . .	162	180	223	212	279	. . .
	13.50	. . .	. . .	. . .	. . .	. . .	. . .	. . .	. . .	270	257	338	. . .
	15.10	. . .	. . .	. . .	. . .	. . .	. . .	. . .	. . .	. . .	. . .	406	438
5	11.50	133	147	. . .	. . .	. . .	. . .	. . .	. . .	. . .	. . .	. . .	. . .
	13.00	169	186	. . .	. . .	. . .	. . .	182	. . .	. . .	. . .	. . .	. . .
	15.00	207	228	. . .	. . .	. . .	. . .	223	201	311	295	388	. . .
	18.00	. . .	. . .	. . .	. . .	. . .	. . .	. . .	246	396	376	495	535
5-1/2	14.00	172	189	. . .	. . .	. . .	. . .	. . .	. . .	. . .	. . .	. . .	. . .
	15.50	202	222	. . .	. . .	. . .	. . .	217	239	. . .	. . .	. . .	. . .
	17.00	229	252	. . .	. . .	. . .	. . .	247	272	348	338	445	. . .
	20.00	. . .	. . .	. . .	. . .	. . .	. . .	. . .	. . .	428	416	548	. . .
	23.00	. . .	. . .	. . .	. . .	. . .	. . .	. . .	. . .	502	489	643	694
6-5/8	20.00	245	267	. . .	. . .	. . .	. . .	266	290	. . .	. . .	. . .	. . .
	24.00	314	342	. . .	. . .	. . .	. . .	340	372	481	473	641	. . .
	28.00	. . .	. . .	. . .	. . .	. . .	. . .	. . .	. . .	586	576	781	. . .
	32.00	. . .	. . .	. . .	. . .	. . .	. . .	. . .	. . .	677	666	904	989
7	20.00	234	254	. . .	. . .	. . .	. . .	. . .	. . .	. . .	. . .	. . .	. . .
	23.00	284	309	. . .	. . .	. . .	. . .	313	341	442	435	. . .	. . .
	26.00	334	364	. . .	. . .	. . .	. . .	367	401	519	511	693	. . .
	29.00	. . .	. . .	. . .	. . .	. . .	. . .	. . .	. . .	597	587	797	. . .
	32.00	. . .	. . .	. . .	. . .	. . .	. . .	. . .	. . .	672	661	897	. . .
	35.00	. . .	. . .	. . .	. . .	. . .	. . .	. . .	. . .	746	734	996	1106
	38.00	. . .	. . .	. . .	. . .	. . .	. . .	. . .	. . .	814	801	1087	1207
7-5/8	26.40	315	342	. . .	. . .	. . .	. . .	346	377	490	482	. . .	. . .
	29.70	. . .	. . .	. . .	. . .	. . .	. . .	. . .	. . .	575	566	769	. . .
	33.70	. . .	. . .	. . .	. . .	. . .	. . .	. . .	. . .	674	664	901	. . .
	39.00	. . .	. . .	. . .	. . .	. . .	. . .	. . .	. . .	798	786	1066	1194
8-5/8	24.00	244	. . .	. . .	. . .	. . .	. . .	. . .	. . .	. . .	. . .	. . .	. . .
	32.00	372	. . .	. . .	. . .	. . .	. . .	417	452	. . .	. . .	. . .	. . .
	36.00	434	. . .	. . .	. . .	. . .	. . .	486	526	688	678	. . .	. . .
	40.00	. . .	. . .	. . .	. . .	. . .	. . .	. . .	. . .	788	776	1055	. . .
	44.00	. . .	. . .	. . .	. . .	. . .	. . .	. . .	. . .	887	874	1186	. . .
	49.00	. . .	. . .	. . .	. . .	. . .	. . .	. . .	. . .	997	983	1335	1496
9-5/8	36.00	394	423	. . .	. . .	. . .	. . .	453	489	. . .	. . .	. . .	. . .
	40.00	452	486	. . .	. . .	. . .	. . .	520	561	737	727	. . .	. . .
	43.50	. . .	. . .	. . .	. . .	. . .	. . .	. . .	. . .	825	813	1106	. . .
	47.00	. . .	. . .	. . .	. . .	. . .	. . .	. . .	. . .	905	893	1213	1360
	53.50	. . .	. . .	. . .	. . .	. . .	. . .	. . .	. . .	1062	1047	1422	1595
10-3/4	40.50	420	450	. . .	. . .	. . .	. . .	. . .	. . .	. . .	. . .	. . .	. . .
	45.50	493	528	. . .	. . .	. . .	. . .	. . .	. . .	. . .	. . .	. . .	. . .
	51.00	565	606	804	794	1080	. . .	. . .	. . .	. . .	. . .	. . .	. . .
	55.50	. . .	. . .	895	884	1203	. . .	. . .	. . .	. . .	. . .	. . .	. . .
	60.70	. . .	. . .	. . .	. . .	1338	1502	. . .	. . .	. . .	. . .	. . .	. . .
	65.70	. . .	. . .	. . .	. . .	1472	1652	. . .	. . .	. . .	. . .	. . .	. . .
11-3/4	47.00	477	509	. . .	. . .	. . .	. . .	. . .	. . .	. . .	. . .	. . .	. . .
	54.00	568	606	. . .	. . .	. . .	. . .	. . .	. . .	. . .	. . .	. . .	. . .
	60.00	649	693	924	913	1242	1395	. . .	. . .	. . .	. . .	. . .	. . .
13-3/8	54.50	514	547	. . .	. . .	. . .	. . .	. . .	. . .	. . .	. . .	. . .	. . .
	61.00	595	633	. . .	. . .	. . .	. . .	. . .	. . .	. . .	. . .	. . .	. . .
	68.00	675	718	963	952	1217	. . .	. . .	. . .	. . .	. . .	. . .	. . .
	72.00	. . .	. . .	1040	1029	1402	1576	. . .	. . .	. . .	. . .	. . .	. . .
16	75.00	710	752	. . .	. . .	. . .	. . .	. . .	. . .	. . .	. . .	. . .	. . .
	84.00	817	865	. . .	. . .	. . .	. . .	. . .	. . .	. . .	. . .	. . .	. . .
18-5/8	87.50	754	794	. . .	. . .	. . .	. . .	. . .	. . .	. . .	. . .	. . .	. . .
20	94.00	784	824	. . .	. . .	. . .	. . .	907	955	. . .	. . .	. . .	. . .
	106.50	913	960	. . .	. . .	. . .	. . .	1057	1113	. . .	. . .	. . .	. . .
	133.00	1192	1253	. . .	. . .	. . .	. . .	1380	1453	. . .	. . .	. . .	. . .

Reprinted in part from API Bulletin 5C2, May 1987

TABLE 6-3
Tubing Minimum Performance Properties

Tubing Size		Nominal Weight						Threaded and Coupled				Integral Joint				Joint Yield Strength		
									Coup. Outside Dia.									
Nom. (in.)	OD (in.)	T & C Non-Upset (lb/ft)	T & C Upset (lb/ft)	Int. Jt. (lb/ft)	Grade	Wall Thick-ness (in.)	Inside Dia. (in.)	Drift Dia. (in.)	Non-Upset (in.)	Upset Reg. (in.)	Upset Spec. (in.)	Drift Dia. (in.)	Box OD (in.)	Collapse Resis-tance (psi)	Inter-nal Yield Pres-sure (psi)	T & C Non-Upset (lb)	T & C Upset (lb)	Int. Jt. (lb)
3/4	1.050	1.14	1.20		H-40	.113	.824	.730	1.313	1.660				7,680	7,530	6,360	13,300	
	1.050	1.14	1.20		J-55	.113	.824	.730	1.313	1.660				10,560	10,360	8,740	18,290	
	1.050	1.14	1.20		C-75	.113	.824	.730	1.313	1.660				14,410	14,120	11,920	24,940	
	1.050	1.14	1.20		L,N-80	.113	.824	.730	1.313	1.660				15,370	15,070	12,710	26,610	
1	1.315	1.70	1.80	1.72	H-40	.133	1.049	.995	1.660	1.900		.955	1.550	7,270	7,080	10,960	19,760	15,970
	1.315	1.70	1.80	1.72	J-55	.133	1.049	.955	1.660	1.900		.955	1.550	10,000	9,730	15,060	27,160	21,960
	1.315	1.70	1.80	1.72	C-75	.133	1.049	.955	1.660	1.900		.955	1.550	13,640	13,270	20,540	37,040	29,940
	1.315	1.70	1.80	1.72	L,N-80	.133	1.049	.955	1.660	1.900		.955	1.550	14,550	14,160	21,910	39,510	31,940
1-1/4	1.660			2.10	H-40	.125	1.410					1.286	1.880	5,570	5,270			22,180
	1.660	2.30	2.40	2.33	H-40	.140	1.380	1.286	2.054	2.200		1.286	1.880	6,180	5,900	15,530	26,740	22,180
	1.660			2.10	J-55	.125	1.410					1.286	1.880	7,660	7,250			30,500
	1.660	2.30	2.40	2.33	J-55	.140	1.380	1.286	2.054	2.200		1.286	1.880	8,500	8,120	21,360	36,770	30,500
	1.660	2.30	2.40	2.33	C-75	.140	1.380	1.286	2.054	2.200		1.286	1.880	11,580	11,070	29,120	50,140	41,600
	1.660	2.30	2.40	2.33	L,N-80	.140	1.380	1.286	2.054	2.200		1.286	1.880	12,360	11,810	31,060	53,480	44,370
1-1/2	1.900			2.40	H-40	.125	1.650					1.516	2.110	4,920	4,610			26,890
	1.900	2.75	2.90	2.76	H-40	.145	1.610	1.516	2.200	2.500		1.516	2.110	5,640	5,340	19,090	31,980	26,890
	1.900			2.40	J-55	.125	1.650					1.516	2.110	6,640	6,330			36,970
	1.900	2.75	2.90	2.76	J-55	.145	1.610	1.516	2.200	2.500		1.516	2.110	7,750	7,350	26,250	43,970	36,970
	1.900	2.75	2.90	2.76	C-75	.145	1.610	1.516	2.200	2.500		1.516	2.110	10,570	10,020	35,800	59,960	50,420
	1.900	2.75	2.90	2.76	L,N-80	.145	1.610	1.516	2.200	2.500		1.516	2.110	11,280	10,680	38,180	63,960	53,780
2-1/16	2.063			3.25	H-40	1.56	1.751					1.657	2.325	5,590	5,290			35,690
	2.063			3.25	J-55	.156	1.751					1.657	2.325	7,690	7,280			49.070
	2.063			3.25	C-75	.156	1.751					1.657	2.325	10,480	9,920			66,910
	2.063			3.25	L,N-80	.156	1.751					1.657	2.325	11,180	10,590			71,370
2-3/8	2.375	4.00			H-40	.167	2.041	1.947	2.875					5,230	4,920	30,130		
	2.375	4.60	4.70		H-40	.190	1.995	1.901	2.875	3.063	2.910			5,890	5,600	35,960	52,170	
	2.375	4.00			J-55	.167	2.041	1.947	2.873					7,190	6,770	41,430		
	2.375	4.60	4.70		J-55	.190	1.995	1.901	2.875	3.063	2.910			8,100	7,700	49,450	71,730	
	2.375	4.00			C-75	.167	2.041	1.947	2.875					9,520	9,230	56,500		
	2.375	4.60	4.70		C-75	.190	1.995	1.901	2.875	3.063	2.910			11,040	10,500	67,430	97,820	
	2.375	5.80	5.95		C-75	.254	1.867	1.773	2.875	3.063	2.910			14,330	14,040	96,560	126,940	
	2.375	4.00			N-80	.167	2.041	1.947	2.875					9,980	9,840	60,260		
	2.375	4.60	4.70		L,N-80	.190	1.995	1.901	2.875	3.063	2.910			11,780	11,200	71,930	104,340	
	2.375	5.80	5.95		L,N-80	.254	1.867	1.773	2.875	3.063	2.910			15,280	14,970	102,990	135,400	
	2.375	4.60	4.70		P-105	.190	1.995	1.901	2.875	3.063	2.910			15,460	14,700	94,410	136,940	
	2.375	5.80	5.95		P-105	.254	1.867	1.773	2.875	3.063	2.910			20,060	19,650	135,180	177,710	
2-7/8	2.875	6.40	6.50		H-40	.217	2.441	2.347	3.500	3.668	3.460			5,580	5,280	52,780	72,480	
	2.875	6.40	6.50		J-55	.217	2.441	2.347	3.500	3.668	3.460			7,680	7,260	72,580	99,660	
	2.875	6.40	6.50		C-75	.217	2.441	2.347	3.500	3.668	3.460			10,470	9,910	98,970	135,900	
	2.875	8.60	8.70		C-75	.308	2.259	2.165	3.500	3.668	3.460			14,350	14,060	149,360	186,290	
	2.875	6.40	6.50		L,N-80	.217	2.441	2.347	3.500	3.668	3.460			11,160	10,570	105,570	144,960	
	2.875	8.60	8.70		L,N-80	.308	2.259	2.165	3.500	3.668	3.460			15,300	15,000	159,310	198,710	
	2.875	6.40	6.50		P-105	.217	2.441	2.347	3.500	3.668	3.460			14,010	13,870	138,560	190,260	
	2.875	8.60	8.70		P-105	.308	2.259	2.165	3.500	3.668	3.460			20,090	19,690	209,100	260,810	
3-1/2	3.500	7.70			H-40	.216	3.068	2.943	4.250					4,630	4,320	65,070		
	3.500	9.20	9.30		H-40	.254	2.992	2.867	4.250	4.500	4.180			5,380	5,080	79,540	103,610	
	3.500	10.20			H-40	.289	2.922	2.797	4.250					6,060	5,780	92,550		
	3.500	7.70			J-55	.216	3.068	2.943	4.250					5,970	5,940	89,470		
	3.500	9.20	9.30		J-55	.254	2.992	2.867	4.250	4.500	4.180			7,400	6,980	109,370	142,460	
	3.500	10.20			J-55	.289	2.922	2.797	4.250					8,330	7,950	127,250		
	3.500	7.70			C-75	.216	3.068	2.943	4.250					7,540	8,100	122,010		
	3.500	9.20	9.30		C-75	.254	2.992	2.867	4.250	4.500	4.180			10,040	9,520	149,140	194,260	
	3.500	10.20			C-75	.289	2.922	2.797	4.250					11,360	10,840	173,530		
	3.500	12.70	12.95		C-75	.375	2.750	2.625	4.250	4.500	4.180			14,350	14,060	230,990	276,120	
	3.500	7.70			L,N-80	.216	3.068	2.943	4.250					7,870	8,640	130,140		
	3.500	9.20	9.30		L,N-80	.254	2.992	2.867	4.250	4.500	4.180			10,530	10,160	159,090	207,220	
	3.500	10.20			L,N-80	.289	2.922	2.797	4.250					12,120	11,560	185,100		
	3.500	12.70	12.95		L,N-80	.375	2.750	2.625	4.250	4.500	4.180			15,310	15,000	246,390	294,530	
	3.500	9.20	9.30		P-105	.254	2.992	2.867	4.250	4.500	4.180			13,050	13,330	208,800	271,970	
	3.500	12.70	12.95		P-105	.375	2.750	2.625	4.250	4.500	4.180			20,090	19,690	323,390	386,570	
4	4.000	9.50			H-40	.226	3.548	3.423	4.750					4,060	3,960	72,000		
	4.000		11.00		H-40	.262	3.476	3.351		5.000				4,900	4,580		123,070	
	4.000	9.50			J-55	.226	3.548	3.423	4.750					5,110	5,440	99,010		
	4.000		11.00		J-55	.262	3.476	3.351		5.000				6,590	6,300		169,220	
	4.000	9.50			C-75	.226	3.548	3.423	4.750					6,350	7,420	135,010		
	4.000		11.00		C-75	.262	3.476	3.351		5.000				8,410	8,600		230,750	
	4.000	9.50			L,N-80	.226	3.548	3.423	4.750					6,590	7,910	144,010		
	4.000		11.00		L,N-80	.262	3.376	3.351		5.000				8,800	9,170		246,140	
4-1/2	4.500	12.60	12.75		H-40	.271	3.958	3.833	5.200	5.563				4,500	4,220	104,360	144,020	
	4.500	12.60	12.75		J-55	.271	3.958	3.833	5.200	5.563				5,720	5,800	143,500	198,030	
	4.500	12.60	12.75		C-75	.271	3.958	3.833	5.200	5.563				7,200	7,900	195,680	270,040	
	4.500	12.60	12.75		L,N-80	.271	3.958	3.833	5.200	5.563				7,500	8,430	208,730	288,040	

Reprinted in part from API Bulletin 5C2, May 1987

TABLE 6-4
Recommended Tubing Makeup Torque, API Connections[1]

1	2	3	4	5	6	7	8	9	10	11	12	13	14
	Nominal Weight lb. per ft.				*Torque, ft-lb*								
	Threads and Coupling				*Non-Upset*			*Upset*			*Integral Joint*		
Size: Outside Diameter in.	*Non-Upset*	*Upset*	*Integral Joint*	*Grade*	*Opt.*	*Min.*	*Max.*	*Opt.*	*Min.*	*Max.*	*Opt.*	*Min.*	*Max.*
1.050	1.14	1.20	—	H-40	140	110	180	460	350	580	—	—	—
	1.14	1.20	—	J-55	180	140	230	600	450	750	—	—	—
	1.14	1.20	—	C-75	230	170	290	780	590	980	—	—	—
	1.14	1.20	—	L-80	240	180	300	810	610	1010	—	—	—
	1.14	1.20	—	N-80	250	190	310	830	620	1040	—	—	—
1.315	1.70	180	1.72	H-40	210	160	260	440	330	550	310	230	390
	1.70	1.80	1.72	J-55	270	200	340	570	430	710	400	300	500
	1.70	1.80	1.72	C-75	360	270	450	740	560	930	520	390	650
	1.70	1.80	1.72	L-80	370	280	460	760	570	950	540	400	680
	1.70	1.80	1.72	N-80	380	290	480	790	590	990	550	410	690
1.660	—	—	2.10	H-40	—	—	—	—	—	—	380	280	480
	2.30	2.40	2.33	H-40	270	200	340	530	400	660	380	280	480
	—	—	2.10	J-55	—	—	—	—	—	—	500	380	630
	2.30	2.40	2.33	J-55	350	260	440	690	520	860	500	380	630
	2.30	2.40	2.33	C-75	460	350	580	910	680	1140	650	490	810
	2.30	2.40	2.33	L-80	470	350	590	940	710	1180	670	500	850
	2.30	2.40	2.33	N-80	490	370	610	960	720	1200	690	520	860
1.900	—	—	2.40	H-40	—	—	—	—	—	—	450	340	560
	2.75	2.90	2.76	H-40	320	240	400	670	500	840	450	340	560
	—	—	2.40	J-55	—	—	—	—	—	—	580	440	730
	2.75	2.90	2.76	J-55	410	310	510	880	660	1100	580	440	730
	2.75	2.90	2.76	C-75	540	410	680	1150	860	1440	760	570	950
	2.75	2.90	2.76	L-80	560	420	700	1190	890	1490	790	590	990
	2.75	2.90	2.76	N-80	570	430	710	1220	920	1530	810	610	1010
2.063	—	—	3.25	H-40	—	—	—	—	—	—	570	430	710
	—	—	3.25	J-66	—	—	—	—	—	—	740	560	920
	—	—	3.25	C-75	—	—	—	—	—	—	970	730	1210
	—	—	3.25	L-80	—	—	—	—	—	—	1010	760	1260
	—	—	3.25	N-80	—	—	—	—	—	—	1030	770	1290
2⅜	4.00	—	—	H-40	470	350	590	—	—	—	—	—	—
	4.60	4.70	—	H-40	560	420	700	990	740	1240	—	—	—
	4.00	—	—	J-55	610	460	760	—	—	—	—	—	—
	4.60	4.70	—	J-55	730	550	910	1290	970	1610	—	—	—
	4.00	—	—	C-75	800	600	1000	—	—	—	—	—	—
	4.60	4.70	—	C-75	960	720	1200	1700	1280	2130	—	—	—
	5.80	5.95	—	C-75	1380	1040	1730	2120	1590	2650	—	—	—
	4.00	—	—	L-80	830	620	1040	—	—	—	—	—	—
	4.60	4.70	—	L-80	990	740	1240	1760	1320	2200	—	—	—
	5.80	5.95	—	L-80	1420	1070	1780	2190	1640	2740	—	—	—
	4.00	—	—	N-80	850	640	1060	—	—	—	—	—	—
	4.60	4.70	—	N-80	1020	770	1280	1800	1350	2250	—	—	—
	5.80	5.95	—	N-80	1460	1100	1830	2240	1680	2800	—	—	—
	4.60	4.70	—	P-105	1280	960	1600	2270	1700	2840	—	—	—
	5.80	5.95	—	P-105	1840	1380	2300	2830	2120	3540	—	—	—

[1]For non-API connections, see manufacturers' literature
Reprinted from API Bulletin 5C1, March 1981

TABLE 6-4 (cont.)
Recommended Tubing Makeup Torque, API Connections

1	2	3	4	5	6	7	8	9	10	11	12	13	14
	Nominal Weight lb. per ft.				Torque, ft-lb								
	Threads and Coupling				Non-Upset			Upset			Integral Joint		
Size: Outside Diameter in.	Non-Upset	Upset	Integral Joint	Grade	Opt.	Min.	Max.	Opt.	Min.	Max.	Opt.	Min.	Max.
$2\frac{7}{8}$	6.40	6.50	—	H-40	800	600	1000	1250	940	1560	—	—	—
	6.40	6.50	—	J-55	1050	790	1310	1650	1240	2060	—	—	—
	6.40	6.50	—	C-75	1380	1040	1730	2170	1630	2710	—	—	—
	8.60	8.70	—	C-75	2090	1570	2610	2850	2140	3560	—	—	—
	6.40	6.50	—	L-80	1430	1070	1790	2250	1690	2810	—	—	—
	8.60	8.70	—	L-80	2160	1620	2700	2950	2210	3690	—	—	—
	6.40	6.50	—	N-80	1470	1100	1840	2300	1730	2880	—	—	—
	8.60	8.70	—	N-80	2210	1660	2760	3020	2270	3780	—	—	—
	6.40	6.50	—	P-105	1850	1390	2310	2910	2180	3640	—	—	—
	8.60	8.70	—	P-105	2790	2090	3490	3810	2860	4760	—	—	—
$3\frac{1}{2}$	7.70	—	—	H-40	920	690	1150	—	—	—	—	—	—
	9.20	9.30	—	H-40	1120	840	1400	1730	1300	2160	—	—	—
	10.20	—	—	H-40	1310	980	1640	—	—	—	—	—	—
	7.70	—	—	J-55	1210	910	1510	—	—	—	—	—	—
	9.20	9.30	—	J-55	1480	1110	1850	2280	1710	2850	—	—	—
	10.20	—	—	J-55	1720	1290	2150	—	—	—	—	—	—
	7.70	—	—	C-75	1600	1200	2000	—	—	—	—	—	—
	9.20	9.30	—	C-75	1950	1460	2440	3010	2260	3760	—	—	—
	10.20	—	—	C-75	2270	1700	2840	—	—	—	—	—	—
	12.70	12.95	—	C-75	3030	2270	3790	4040	3030	5050	—	—	—
	7.70	—	—	L-80	1660	1250	2080	—	—	—	—	—	—
	9.20	9.30	—	L-80	2030	1520	2540	3130	2350	3910	—	—	—
	10.20	—	—	L-80	2360	1770	2950	—	—	—	—	—	—
	12.70	12.95	—	L-80	3140	2360	3930	4200	3150	5250	—	—	—
	7.70	—	—	N-80	1700	1280	2130	—	—	—	—	—	—
	9.20	9.30	—	N-80	2070	1550	2590	3200	2400	4000	—	—	—
	10.20	—	—	N-80	2410	1810	3010	—	—	—	—	—	—
	12.70	12.95	—	N-80	3210	2410	4010	4290	3220	5360	—	—	—
	9.20	9.30	—	P-105	2620	1970	3280	4050	3040	5060	—	—	—
	12.70	12.95	—	P-105	4060	3050	5080	5430	4070	6790	—	—	—
4	9.50	—	—	H-40	940	710	1180	—	—	—	—	—	—
	—	11.00	—	H-40	—	—	—	1940	1460	2430	—	—	—
	9.50	—	—	J-55	1240	930	1550	—	—	—	—	—	—
	—	11.00	—	J-55	—	—	—	2560	1920	3200	—	—	—
	9.50	—	—	C-75	1640	1230	2050	—	—	—	—	—	—
	—	11.00	—	C-75	—	—	—	3390	2540	4240	—	—	—
	9.50	—	—	L-80	1710	1280	2140	—	—	—	—	—	—
	—	11.00	—	L-80	—	—	—	3530	2650	4410	—	—	—
	9.50	—	—	N-80	1740	1310	2180	—	—	—	—	—	—
	—	11.00	—	N-80	—	—	—	3600	2700	4500	—	—	—
$4\frac{1}{2}$	12.60	12.75	—	H-40	1320	990	1650	2160	1620	2700	—	—	—
	12.60	12.75	—	J-55	1740	1310	2180	2860	2150	3580	—	—	—
	12.60	12.75	—	C-75	2300	1730	2880	3780	2840	4730	—	—	—
	12.60	12.75	—	L-80	2400	1800	3000	3940	2960	4930	—	—	—
	12.60	12.75	—	N-80	2440	1830	3050	4020	3020	5030	—	—	—

Packers and Subsurface Control Equipment

PRODUCTION PACKERS

Production packers are generally classified as either retrievable-type or permanent-type. Recent packer innovations include the retrievable seal nipple packers or permanent retrievable type, packer bore receptable and "cement packer."

Packers are sometimes run when they serve no useful purpose, resulting in unnecessary initial investment and the possibility of high future removal cost. Routine use of packers should be limited to these situations:

1. Protection of casing from pressure (including both well and stimulation pressures) and corrosive fluids,
2. Isolation of casing leaks, squeezed perforations, or multiple producing intervals.
3. Elimination of inefficient "heading" or "surging".
4. Some artificial lift installations.
5. In conjunction with subsurface safety valves.
6. To hold kill fluids or treating fluids in casing annulus.

General Considerations in Packer Selection

Packer selection involves an analysis of packer objectives in anticipated well operations, such as initial completions, production stimulation, and workover procedures. The packer with the minimum overall cost that will accomplish the objective, considering both current and future well conditions, should be selected. Initial investment and installation costs should not be the only criterion. Overall packer cost is directly related to retrievability and failure rate and to such diverse factors as formation damage during subsequent well operations.

Retrievability will be greatly enhanced by utilizing oil or saltwater rather than mud for the packer fluid. Frequency of packer failures may be minimized by utilizing the proper packer for the well condition and by anticipating future conditions when setting the packer. The permanent packer is by far the most reliable and, properly equipped and set, is excellent for the high pressure differentials imposed during stimulation, or when reservoir pressures vary significantly between zones in multicompletions.

Weight-set and tension types of retrievable packers will perform satisfactorily when the force on the packer is in one direction only and is not excessive.

Purchase Price—Table 6-5 presents a range of packer costs. The most economical types are weight-set and tension packers. However, inclusion of a hydraulic hold-down with a weight-set packer will increase the initial cost 20 to 100%. Multistring hydraulic-set packers are usually the most expensive and also require many accessories.

Packer Mechanics—The end result of most packer setting mechanisms is to (1) drive a cone behind a tapered slip to force the slip into the casing wall and prevent packer movement, and (2) compress a packing element to effect a seal. Although the end result is relatively simple, the means of accomplishing it and subsequent packer retrieval varies markedly between the several types of packers.

Some packers involve two or more round trips, some wireline time and some eliminate trips by hydraulic setting. The time cost should be examined carefully, especially on deep wells using high cost rigs. In some cases higher first cost of the packer may be more than offset by saving in rig time, especially on high cost offshore rigs.

Sealing Element—The ability of a seal to hold differential pressure is a function of the rubber pressure or stress developed in the seal, i.e., the stress must exceed the differential pressure. In a packer sealing element, the stress developed depends on the packer setting force, and the back-up provided to limit extrusion.

The sealing element may consist of one piece or may be composed of multiple elements of different hardness. In a three-element packer, the upper and lowermost elements are usually harder than the center element.

TABLE 6-5
Cost Comparison of Production Packer Types

Packer type	Tubing-casing size, in.	Typical cost
Weight set	2 × 5½	$ 1,800
Tension set	2 × 5½	1,800
Mechanical set	2 × 5½	3,000
Hydraulic set	2 × 5½	4,500
	2 × 2 × 7	11,000
Permanent[a]	2 × 5½	3,600
Semi permanent[a]	2 × 5½	5,000

[a]Electric line setting charge not included.

The lower durometer center seals off against imperfections in the casing, while the harder outside elements restrict extrusion and seal with high temperature and pressure differentials. Many packers also include metallic back-up rings to impede extrusion.

Where H_2S or CO_2 are present, seal materials and conditions must be carefully considered. Temperature is a primary factor. Below 250°F, Nitrile rubber can be used with metallic backup for static seals. Viton becomes marginal at 300°F. Recent tests show that a tubing-to-packer seal consisting of V-type rings of Kalrez, Teflon, and Rylon in sequence with metallic backup is satisfactory under limited movement to 300°F and 10,000 psi differential pressure. Seal sticking is minimal.

Teflon resists H_2S or chemical attack up to 450°F, but extrusion can be a problem. With controlled clearance and suitable metallic backup to prevent extrusion, glass-filled Teflon has performed satisfactorily to 450°F and 15,000 psi differential pressure. Due to seal rigidity it may not perform well below 300°F, however.

Corrosive Well Fluids—Materials used in packer construction must be considered where well fluids contain CO_2 or H_2S in the presence of water.

Sweet corrosion: CO_2 and water cause iron carbonate corrosion, resulting in deep pitting. For ferrous materials low strength steels or cast iron are desirable to resist stress concentrations from pitting. Depending on economics, corrosion inhibitors may be required to protect exposed surfaces. Critical parts of production equipment can be made of stainless steel with 12% or higher chromium.

Sour corrosion: Even small amounts of H_2S with water produce iron sulfide corrosion and hydrogen embrittlement. NACE specifies materials for H_2S conditions be heat-treated to a maximum hardness of 22 Rockwell C to prevent embrittlement. AISI 4140 steel heat-treated to 25 Rockwell C has more usable strength than 22 Rockwell C and compares favorably with L-80 steel for H_2S service. Hardness has no effect on iron sulfide corrosion however. For critical parts where high strength is required K-Monel is resistant to both embrittlement and iron sulfide corrosion.

Bimetallic or galvanic corrosion resulting from contact of dissimilar metals should be considered. Usually this is not a problem, however, since steel is the anode or sacrificial member, and resulting damage is negligible due to the massive area of the steel compared to the more noble stainless or K-Monel.

Retrievability—Retrievability is a combination of several factors, related to packer design and packer use. Retrievable packers are released by either straight pull or by rotation. In deviated hole torque usually develops more downhole releasing force than pull, although sometimes it is necessary to manipulate the tubing to transmit the torque to bottom.

The packer sealing element should prevent solids from settling around the slips. Usually the bypass opens before the seal is released, to permit circulation to remove sand or foreign material.

High setting force is needed to provide a reliable seal under high differential pressures, but it should be recognized that seal extrusion can contribute to the retrieval problem. A jar stroke between release and pickup positions is an aid in packer removal.

The method of retaining slip segments is a factor in retrievability. Bypass area is also important. Where external clearance is minimized to promote sealing, internal bypass area must be sufficient to prevent swabbing off the sealing element in pulling out of the hole.

Fishing Characteristics—A permanent packer must be drilled out to effect removal. This usually presents little problem because all material is drillable. Recent expensive variations of permanent packers provide for retrieval but retain the removable seal tube feature. Removal of "stuck" retrievable packers usually results in an expensive fishing operation because components are "non-drillable." In comparing packers consider the volume of metal that must be removed and the presence of rings or hold-down buttons that may act as ball bearings to milling tools.

Through-Tubing Operations—Packers with internal diameters equal to that of the tubing should be utilized to facilitate through-tubing operations. Also tubing should be set so as to minimize buckling where through-tubing operations are anticipated.

Surface Equipment–Downhole Correlation—Setting a packer always requires surface action and in most cases either vertical or rotational movement of the tubing. Thus selection of the packer must be related to wellhead equipment. The well completion must be considered as a coordinated operation and the surface and downhole equipment selected to work together to insure a safe completion, especially in high pressure wells.

For example, a wrap-around tubing hanger with a smooth joint designed to allow vertical and rotational tubing movement under pressure may be used with J-tool or rotational set compression packers. However, if a well is to be washed-in and cleaned before the packer is set, a hydraulic-set packer may be needed for proper well control.

Effect of Pressure, Temperature Changes

Changes in temperature, and in pressure inside and outside of tubing sealed in a packer, depending on the type of packer and how it is set will:

1. Increase or decrease the length of the tubing with a packer permitting free motion of the tubing.
2. Induce tensile or compressive forces in the tubing and packer if free motion is not permitted within a packer which is rigidly attached to the casing.
3. Unseat a packer not rigidly attached to the casing.
4. Unseal a permanent-type packer where the seal nipple section is not long enough and the tubing is not latched into the packer.

Several effects must be evaluated to accurately determine the tubing movement or stress situation as shown in Figure 6-6. Lubinski[1] and Hammerlindl[7] present an excellent discussion of these effects, show calculation procedures, and give examples based on field situations. The following simplified discussion concerns vertical uniform completions. For a combination completion that contains more than one size of tubing or casing, or more than one fluid in the tubing or casing refer to the Hammerlindl paper.[7]

Hooke's Law Effect—The pressures inside the tubing and in the annulus above the packer act on the differential area presented by the tubing and the packer mandrel to change the tubing length according to Hooke's Law. If tubing motion is limited by the packer, the force of the packer on the tubing also affects tubing length according to Hooke's Law:

$$\Delta L_1 = \frac{L\Delta F}{E\,A_s} \tag{1}$$

With a free motion seal where only pressure induced forces are acting:

$$\Delta F = (A_p - A_i)\Delta P_i - (A_p - A_o)\Delta P_o$$

Where:

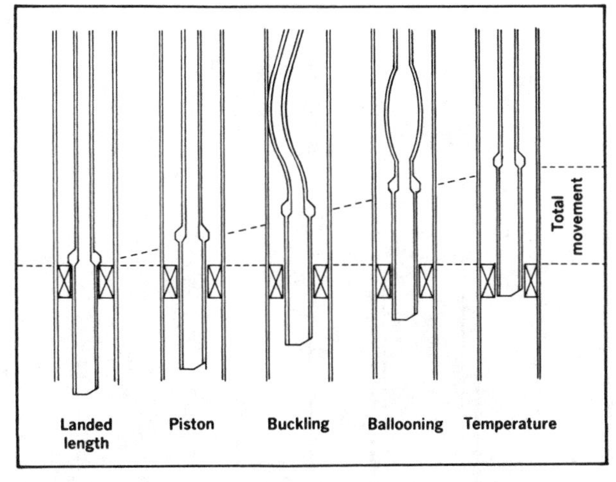

FIG. 6-6—*Effect of various forces on tubing movement.*

ΔL_1 = change in length due to Hooke's Law effect, in.
L = length of tubing—in.
F = force acting on bottom of tubing, lb.
E = modulus of elasticity (30×10^6 psi for steel)
A_s = cross-sectional area of tubing, in^2
A_i = area based on inside diameter of tubing, in^2
A_o = area based on outside diameter of tubing, in^2
A_p = area based on diameter of packer seal, in^2
P_o = pressure at packer seal in annulus, psi
P_i = pressure at packer seal in tubing, psi

See Fig. 6-7.
Notes:

1. ΔL, ΔF, ΔP_i or ΔP_o indicate change from initial packer setting conditions. It is assumed $P_i = P_o$ when packer is initially set.
2. For values of dimensional tubing and packer constants, see Table 6-7.

Helical Buckling—Buckling or "corkscrewing" of the tubing above the packer may shorten the tubing. With a packer permitting free motion of the tubing, buckling is a result of the differential between the pressure inside and that outside the tubing acting on the full cross-sectional area of packer bore at the tubing seal. Where the packer limits tubing motion, the tubing weight set on the packer must also be considered.

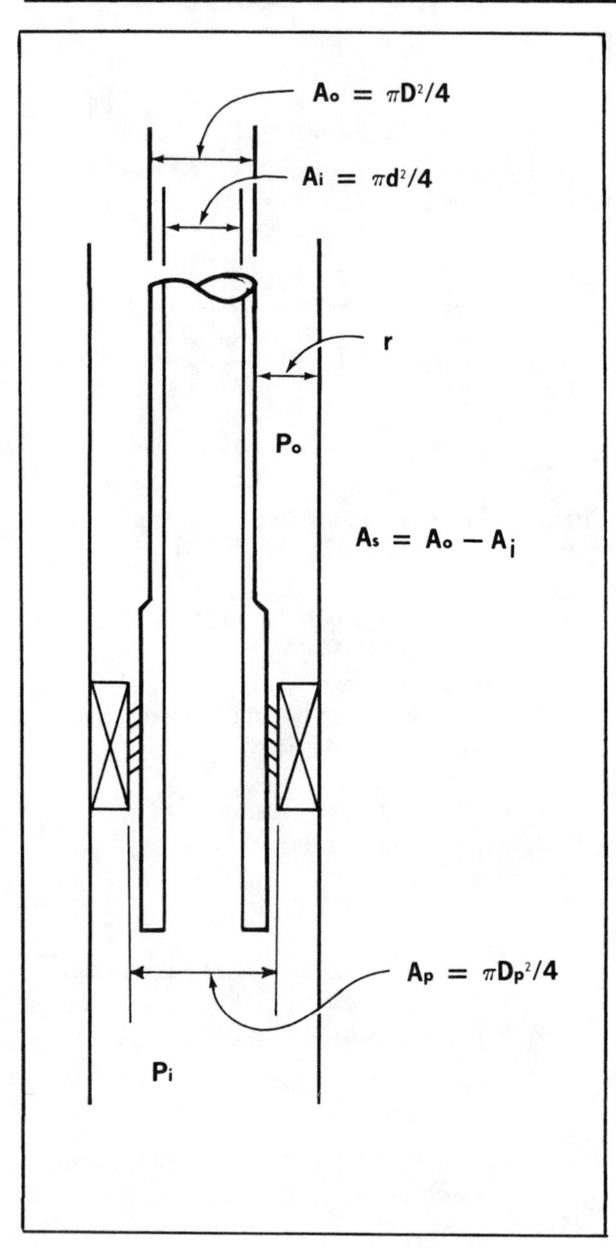

FIG. 6-7—*Definition of terms.*

Shortening due to helical buckling with a free motion seal and where the buckling affects only a portion of the length of the tubing, may be calculated by the following equation:

$$\Delta L_2 = \frac{r^2 A_p^2 (\Delta P_i - \Delta P_o)^2}{8\,E\,I\,(w_s + w_i - w_o)} \qquad (2)$$

$$I = \frac{\pi}{64}(D^4 - d^4)$$

Force causing buckling: $F_f = A_p\,(\Delta P_i = \Delta P_o)$
(If F_f is negative, there is no buckling)

Length of tubing buckled: $n = \dfrac{F_f}{w}$

Where:

ΔL_2 = change in length due to buckling—in.
r = radial clearance between tubing and casing—in.
w = $w_s + w_i - w_o$ (See Table 6-7 for values)
w_s = weight of tubing—lb/in.
w_i = weight of fluid contained inside tubing—lb/in. (based on id of tubing).
w_o = weight of annulus fluid displaced by bulk volume of tubing—lb/in. (based on od of tubing).
D = tubing outside diameter—in.
d = tubing inside diameter—in.

Buckling is increased by higher pressure inside the tubing, by a larger ratio of casing to tubing diameter, by lower density fluid in the tubing, by a larger packer bore, and by an upward acting packer force.

With a free motion seal, buckling cannot occur if the pressure at the seal in the annulus (P_o) is greater than the pressure at the seal inside the tubing (P_i). Again it is assumed that $P_i = P_o$ when the packer was set initially.

Ballooning Effect—Radial pressure inside the tubing tends to increase tubing diameter and thereby shorten the tubing. Greater pressure outside the tubing lengthens the string due to "reverse ballooning."

Change in length due to ballooning or reverse ballooning can be calculated according to the following equation:

$$\Delta L_3 = \frac{\mu L^2}{E}\left(\frac{\Delta\rho_i - R^2\Delta\rho_o - \dfrac{1+2\mu}{2\mu}\,\delta}{R^2 - 1}\right)$$

(Density effect)

$$+\ \frac{2\mu L}{E}\left(\frac{\Delta p_i - R^2\Delta p_o}{R^2 - 1}\right) \qquad (3)$$

(Surface pressure effect)

ΔL_3 = change in length due to ballooning: in.
μ = Poisson's ratio (0.3 for steel)
R = tubing od/tubing id

$\Delta\rho_i$ = change in density of fluid inside tubing: lb/cu in.
$\Delta\rho_o$ = change in density of fluid outside tubing: lb/cu in.
Δp_i = change in *surface* pressure inside tubing
Δp_o = change in *surface* pressure outside tubing
δ = pressure drop in tubing due to flow psi/in. (usually considered as $\delta = o$)

Temperature Effect—A change in temperature due to producing hot fluids or injecting cold fluid changes the tubing length as follows:

$$\Delta L_4 = LC\Delta T \qquad (4)$$

ΔL_4 = change in length: in.
L = length of tubing string: in.
C = 6.9×10^{-6}, coefficient of expansion of steel per °F
ΔT = temperature change, °F

Packer Setting Force—The initial compressive force or tensile force used in setting the packer has a direct bearing on the subsequent situation. Proper choice of the initial setting force must anticipate future conditions resulting from production, stimulation or remedial operations.

To convert tubing length change due to slack-off to compressive force, the following equation may be used:

$$\Delta L = \frac{LF}{E A_s} + \frac{r^2F^2}{8\,E\,I\,(w_s + w_i - w_o)} \qquad (5)$$

Permanent Buckling or "Corkscrewing"

1. *Due to Initial Slack Off:*—Excessive initial set-down weight or "slack off' on a packer may cause permanent buckling of the tubing even before pressures and temperatures are changed by completion or production operations. To investigate this possibility, the following equation may be used. If S_o exceeds the yield strength of the tubing grade, then permanent damage will result.

$$S_o = \frac{F}{A_s} + \frac{DrF}{4I} \qquad (6)$$

F = set-down force

2. *Due to Subsequent Operations:*—If stress induced in the tubing by pressure or temperature changes or by movements exceed the yield stress of the tubing material, then permanent deformation will result. Stress level should be checked at both the inner and outer wall of the pipe using the following equations. These stresses, S_i and S_o, should not be allowed to exceed the yield stress of the tubing.

$$S_i = \sqrt{3\left[\frac{R^2(P_i - P_o)}{R^2 - 1}\right]^2 + \left[\frac{P_i - R^2P_o}{R^2 - 1} + \sigma_a \pm \frac{\sigma_b}{R}\right]^2} \qquad (7)$$

$$S_o = \sqrt{3\left[\frac{P_i - P_o}{R^2 - 1}\right]^2 + \left[\frac{P_i - R^2P_o}{R^2 - 1} + \sigma_a \pm \sigma_b\right]^2} \qquad (8)$$

$$\sigma_a = \frac{F_a}{A_s}$$

For Free Motion Pkr.:

$$F_a = (A_p - A_i)P_i - (A_p - A_o)P_o$$

$$\sigma_b = \frac{Dr}{4I}F_f$$

For Free Motion Pkr.:

$$F_f = A_p(\Delta P_i - \Delta P_o)$$

Where packer exerts force on tubing, this force must be added to F_a or F_f.

Sign of σ_b in Equations 7 and 8 should be chosen to maximize S_i or S_o.

Table 6-6 shows examples of tubing movement in a permanent-type packer (3¼-in. bore) set at 10,000 ft in 7-in. casing. 2⅞-in. tubing with no locator sub was landed without latching into the packer; thus there is free motion of the tubing within the packer.

With the described conditions the permanent packer setup would not be suitable for fracing or high pressure squeeze cementing unless an unusually long seal nipple section had been provided.

Considering the high pressure squeeze cementing operation, if the tubing were latched into the permanent packer (with zero tubing pickup or slack-off) the forces on the tubing would result in an upward pull on the packer latch-in section of 70,000 lb. At the surface tensile force in the tubing would be 87,000 lb as compared to its weight in air of 64,000 lb.

Under these conditions the tubing would be buckled

TABLE 6-6
Tubing Movement in Permanent Packer

	Type of Operation			
Conditions:	*Swab*	*Production*	*Frac*	*Squeeze Cement*
Initial Fluid	10 lb/gal Mud	10 lb/gal Mud	9.2 lb/gal Saltwater	30° Oil
Final Fluid:				
Tubing	45° Oil to 5000 ft	45° Oil	9.0 lb/gal frac fluid	15 lb/gal cmt
Annulus	10 lb/gal Mud	10 lb/gal Mud	9.2 lb/gal Saltwater	30°Oil
Final Pressure:				
Tubing	0	1000 psi	3000 psi	5000 psi
Annulus	0 psi	0 psi	1000 psi	1000 psi
Temp. Change	+10°F	+20°F	−50°F	−20°F
Tbg. Movement due to:				
Hooke's Law	+27.6 in.	+5.7 in.	−19.2 in.	−67.9 in.
Buckling	0	0	−3.6	−46.1
Ballooning	+13.4	−0.9	−9.7	−34.6
Temperature	+8.3	+16.6	−41.5	−16.6
Total Movement	+49.3 in.	+21.4 in.	−74.0 in.	−165.2 in.

Note: (−) indicates shortening, (+) indicates lengthening.

slightly; an initial pickup of 18,000 lb would be required to eliminate buckling. Depending on the tubing grade, the string might be permanently deformed by the latch-in procedure.

As a better alternative, if a locator sub were run, sufficient tubing weight could be slacked off on the packer to restrict tubing movement. Tubing slack-off of 72,000 lb would prevent tubing movement under the conditions of the high pressure squeeze job. However, excessive tubing slack-off is undesirable due to buckling and probably hindrance to use of through-tubing tools.

In deep wells, pressure fluctuations in normal producing operations can result in continual small movements of the seals and eventual failure. Best practice in this situation is to slack off sufficient tubing weight to prevent these small seal movements, but install sufficient sliding seal section to provide for upward movement of tubing during well treatments.

For a comprehensive understanding of pressure-temperature effects on packer hookups the Lubinski et al paper and the Hammerlindl papers, References 1 and 7, should be studied. References 13 and 14 (Hammerlindl and Durham, respectively) consider dual packer hookups.

Unseating of Weight Set Packer

To determine whether or not a weight-set packer will be unseated by injecting in the tubing a simple force balance calculation, as shown in Figure 6-8, is sufficient. In this example, the well is equipped with 5½-in. casing, the packer is set with 7,000 lb tubing weight at 6,000 ft on 2⅜-in. od tubing. The annulus contains salt water. Acid is displaced down the tubing with crude oil. The resulting surface pressure of 1,000 psi imposes a 3,800-lb upward force on the bottom of the packer which will unseat it.

Unseating may be avoided by (1) applying pressure to the casing-tubing annulus or (2) utilizing a hydraulic hold-down. Figure 6-9 pictures two types of hydraulic hold-downs.

Retrievable Packers

The following discussion describes briefly the operation and application of various "families" of Retrievable packers. The details of specific packers are best studied by reference to manufacturers literature and setting instructions.

Weight-set Packers—The weight-set packer is economical and ideally suited to low pressure situations

TABLE 6-7
Tubing and Packer Constants

Tubing Constants

OD IN (Inches)	WT. IN (Lbs/Ft)	A_o IN (Sq. In.)	A_i IN (Sq. In.)	A_s IN (Sq. In.)	I IN ($In.^4$)	R^2
1.660	2.40	2.164	1.496	.668	.195	1.448
1.900	2.90	2.835	2.036	.799	.310	1.393
2.000	3.40	3.142	2.190	.952	.404	1.434
2-1/16	3.40	3.341	2.405	.936	.428	1.389
2-3/8	4.70	4.430	3.126	1.304	.784	1.417
2-7/8	6.50	6.492	4.680	1.812	1.611	1.387
3-1/2	9.20	9.621	7.031	2.590	3.434	1.368

$w_s + w_i - w_o$

Tubing OD (Inches)	Weight (Lbs/In.)	w_i and w_o (Lbs/In.)	7.0 52.3	8.0 59.8	9.0 67.3	10.0 74.8	11.0 82.3	12.0 89.8	13.0 97.2	14.0 104.7	15.0 112.2	16.0 119.7	17.0 127.2	18.0 134.6	Lbs/Gal. Lbs/Cu. Ft.
1.660	w_s = .200	w_i	.045	.052	.058	.065	.071	.078	.084	.091	.097	.104	.110	.116	
		w_o	.065	.075	.084	.094	.103	.112	.122	.131	.140	.150	.159	.169	
1.900	w_s = .242	w_i	.062	.070	.079	.088	.097	.106	.115	.123	.132	.141	.150	.159	
		w_o	.086	.098	.110	.123	.135	.147	.159	.172	.184	.196	.209	.221	
2.000	w_s = .283	w_i	.066	.076	.085	.095	.104	.114	.123	.133	.142	.152	.161	.171	
		w_o	.095	.109	.122	.136	.150	.163	.177	.190	.204	.218	.231	.245	
2-1/16	w_s = .283	w_i	.073	.083	.094	.104	.114	.125	.135	.146	.156	.167	.177	.187	
		w_o	.101	.116	.130	.145	.159	.174	.188	.202	.217	.231	.246	.260	
2-3/8	w_s = .392	w_i	.095	.108	.122	.135	.149	.162	.176	.189	.203	.217	.230	.243	
		w_o	.134	.153	.172	.192	.211	.230	.249	.268	.288	.307	.326	.345	
2-7/8	w_s = .542	w_i	.142	.162	.182	.203	.223	.243	.263	.284	.304	.324	.344	.364	
		w_o	.196	.225	.253	.281	.309	.337	.365	.393	.421	.450	.478	.506	
3-1/2	w_s = .767	w_i	.213	.243	.274	.304	.335	.365	.395	.426	.456	.487	.517	.548	
		w_o	.291	.333	365	.416	.458	.500	.541	.583	.625	.666	.708	.749	

Area of Packer Bores

Bore (Inches)	Area (Sq. In.)	Bore (Inches)	Area (Sq. In.)
6.00	28.26	2.50	4.91
5.24	21.55	2.42	4.60
4.75	17.71	2.28	4.08
4.40	15.20	2.06	3.33
4.00	12.56	1.96	3.00
3.87	11.76	1.87	2.75
3.62	10.29	1.68	2.22
3.25	8.30	1.53	1.84
3.00	7.07	1.43	1.61
2.68	5.67	1.25	1.23

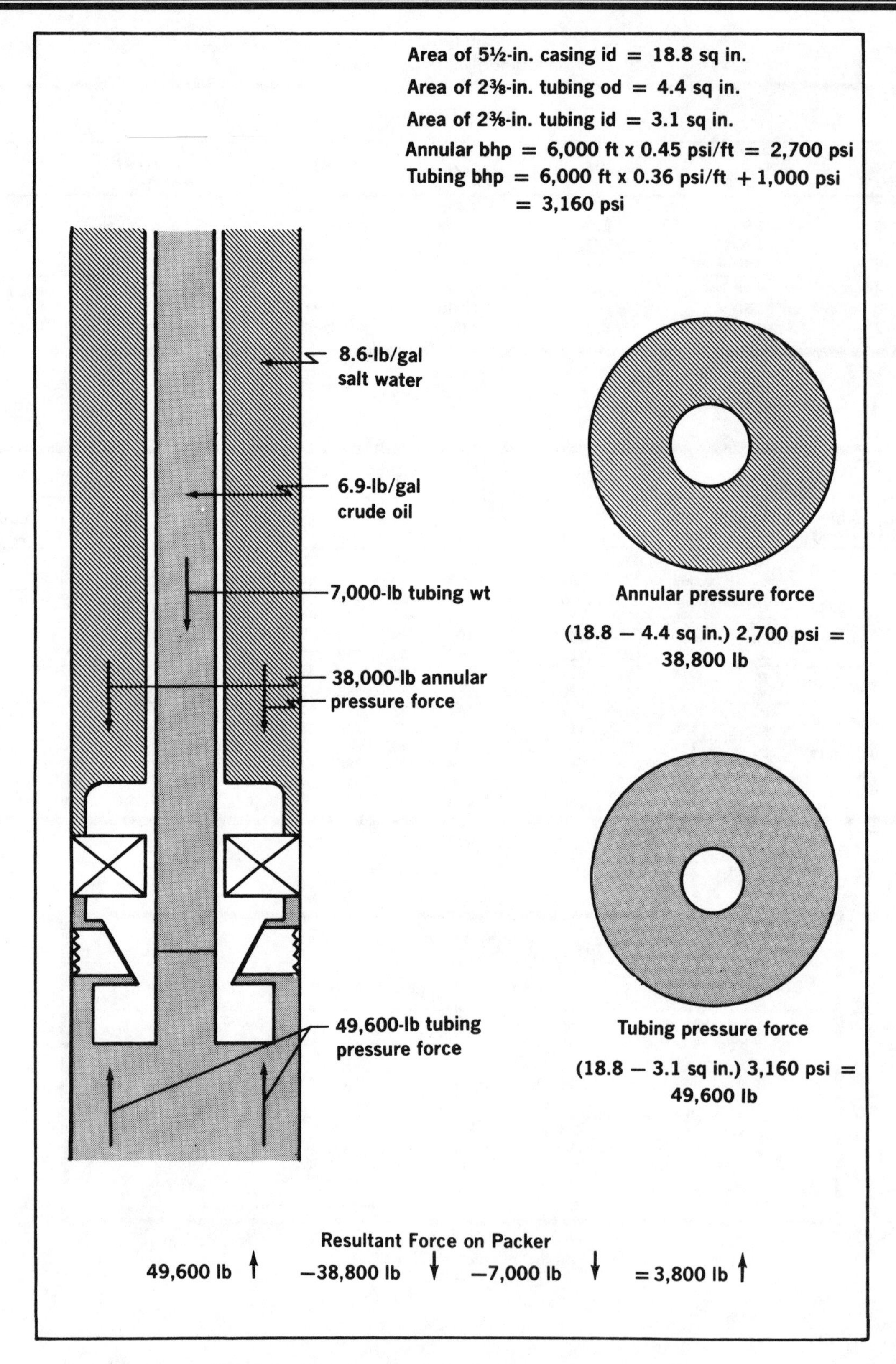

FIG. *6-8—Weight-set packer force-balance calculation.*

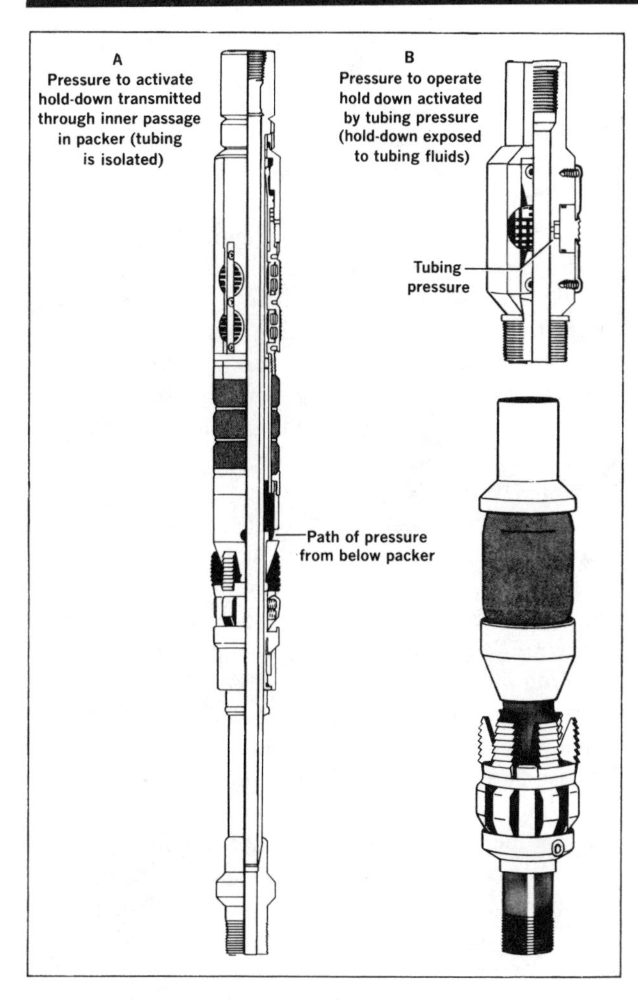

FIG. *6-9—Types of hydraulic hold-downs.*

where the annulus pressure above the packer always exceeds the tubing pressure below the packer. A hydraulic holddown is often included where pressure differentials from both directions are anticipated.

Weight-set packers (Figure 6-10) employ a slip and cone arrangement with the slips attached to a friction device such as drag springs or drag blocks. The friction device engages the casing and holds the slips stationary with respect to the remainder of the packer.

A "J" slot device permits vertical movement of the tubing and causes the cone to move behind the slips and anchor the packer in the casing. Tubing weight is then applied to expand the packing element. Frictional drag limits tubing weight and thus the setting force that can be applied. Release is effected by picking up tubing weight to pull the cone from behind the slips.

Tension-set Packers—Tension packers are essentially weight-set packers run upside down and set by pulling tension on the tubing. After a tension packer is set, a differential pressure from below increases the setting force on the packer and holds it in place automatically. This feature renders the tension packer particularly suitable for water injection wells or stimulation work.

Tension packers are frequently used in shallow wells where insufficient tubing weight is available to seal a weight-set packer. Being short and compact, they require a minimum of space for setting and little maintenance. Temperature should be taken into account in determining the initial setting force on the packer; injection or production of hot fluids can cause the tubing to elongate and release the packer.

Mechanical-set Packers—Tubing rotation plays a major role in setting and retrieving mechanical-set

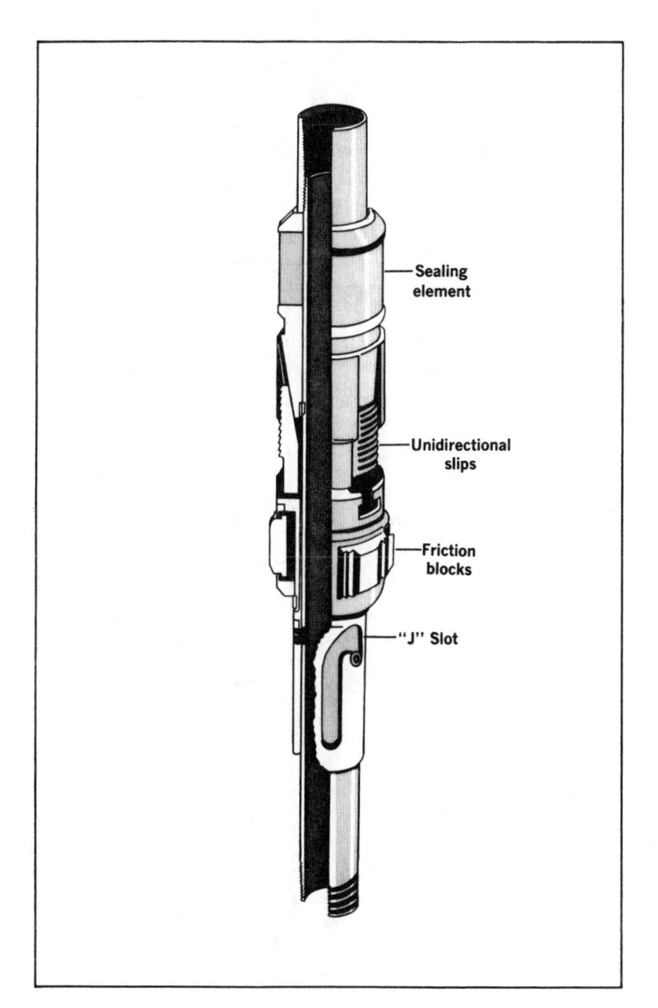

FIG. *6-10—Weight-set packer.*

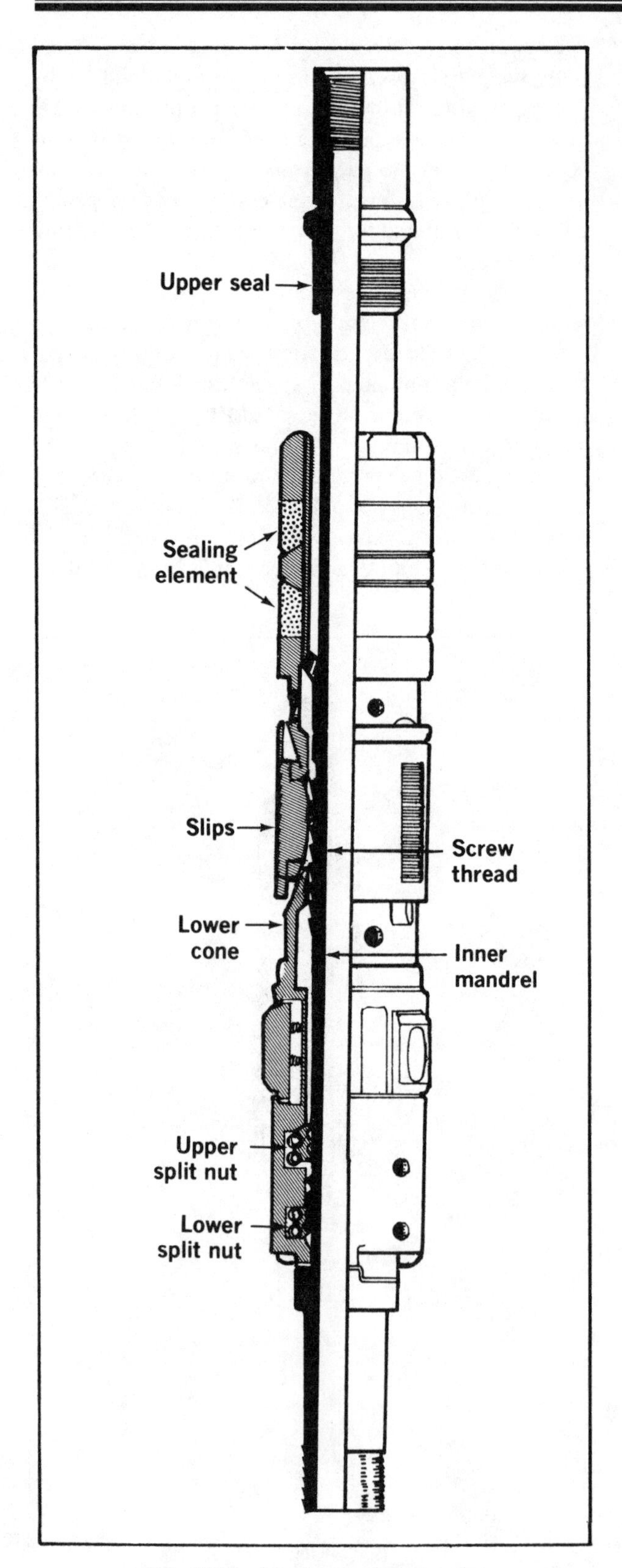

FIG. *6-11—Mechanical-set packer.*

packers. Tubing rotation may either (1) simultaneously set the seals and slips in one continuous motion with a screw-thread and cone arrangement which forces the cones behind the slips and compresses the seals or (2) release the inner mandrel and allow tubing weight to drive the cones behind the slips and compress the sealing element. See Figure 6-11.

These packers generally incorporate a non-directional slip system which prevents movement from either direction thus eliminating the necessity for a hydraulic holddown. After the packer is set, tension may be applied to the tubing to reduce buckling and thus to facilitate passage of through-tubing tools.

Release is effected by righthand pipe rotation. The necessity to release the packer by rotating the tubing is the principal disadvantage of mechanical-set packers; the screwthread may be inoperative after extended periods of time, or solids settling on top of the packer may make it impossible to rotate the tubing.

Hydraulic-set Packers—Hydraulic-set packers utilize fluid pressure acting on a piston-cylinder arrangement to drive the cone behind the slips. The packer remains set by a pressure actuated mechanical lock. Release is accomplished by picking up the weight of the tubing string. A hydraulic holddown is required because the slips are unidirectional. Figure 6-12 pictures a single hydraulic-set packer and a schematic of the setting and releasing mechanism. Multi-string hydraulic packers are set and retrived by essentially the same process.

A primary use of hydraulic-set packers is in multistring conventional wells. Principal advantages are: (1) the tubing can be landed, christmas tree installed, and well circulated with a light fluid or gas before setting the packer to initiate production without swabbing, (2) all strings can be landed in tension to enhance the passage of wireline tools and concentric tubing, (3) since tubing motion is not required, all strings in a multiple completion may be run and landed simultaneously with multiple slips and elevators before setting the packers, and (4) high setting force can be applied to hold large differential pressures.

Despite their advantages, the expense of hydraulic-set packers limits their use to situations where other packers are not applicable, such as in dual, triple, and quadruple-string completions, deviated holes, or ocean floor completions.

Hydraulic Expansion Seal Packers—An interesting version of the hydraulic-set principle is utilized to effect a seal in open hole. An inflatable rubber ele-

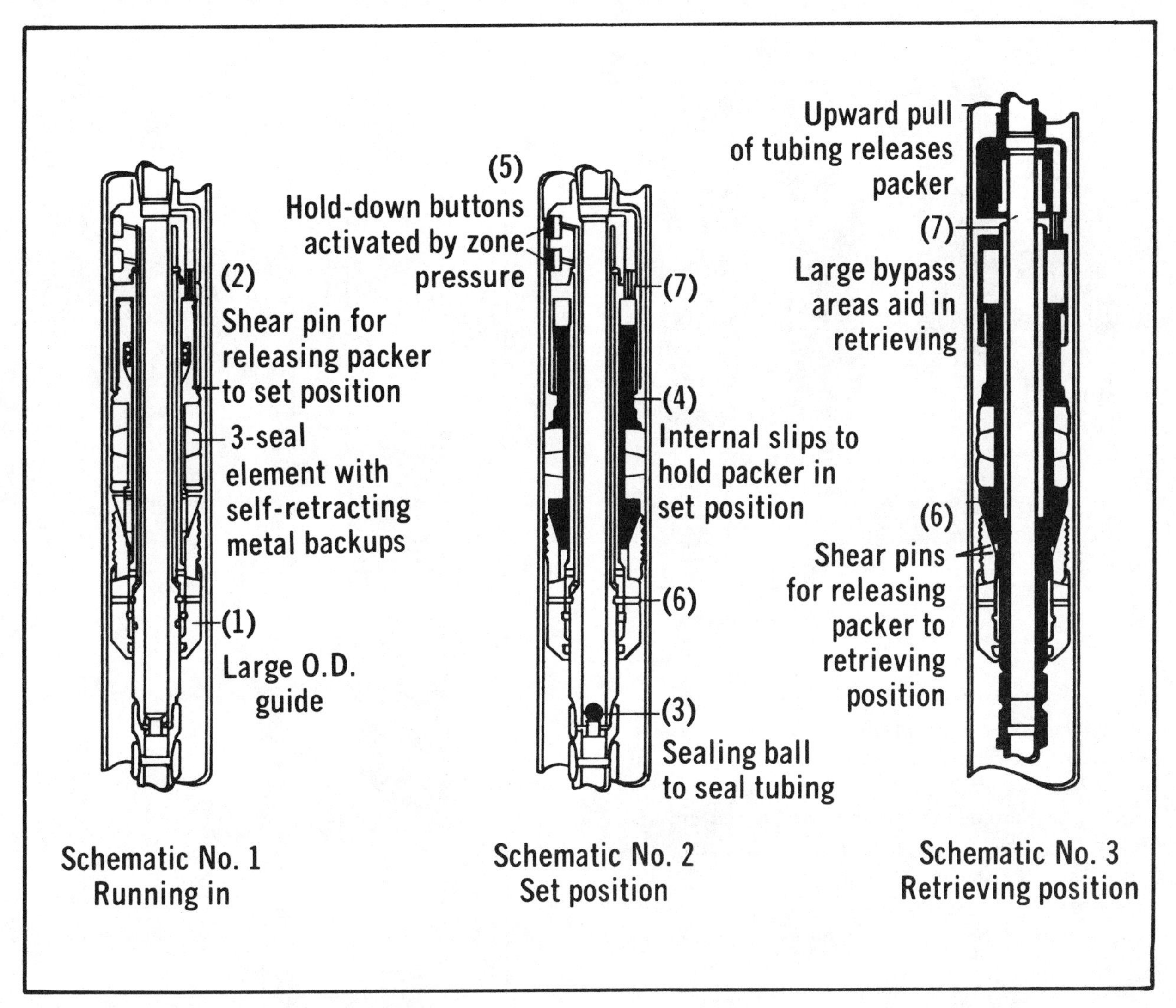

FIG. *6-12—Single hydraulic-set packer, schematic of setting and releasing mechanism.*

ment packer has been successfully employed to shut off bottom water and isolate high gas-oil ratio zones. Figure 6-13 shows the operating mechanics of the packer when utilized conventionally. Bull plugging the bottom of the mandrel and including an entry sub above the packer permits fluid production from above the packer only.

The packer is available with several different length seals. Selecting a long sealing element and setting the packer in a relatively gauge portion of the hole will increase the chances of obtaining an effective barrier. Due to the large diametral expansion the inflatable packer also has application inside casing where it may be necessary to set a packer below an obstruction.

Permanent Packers

Permanent packers utilize opposed slips with a compressible sealing element between the slips. They may be run on tubing or electric conductor cable. Wireline setting is a valuable asset where precise packer location is necessary. Since the tubing may be run separately from the packer, trip time is faster and replacement of the packing in the seal nipples is the only dressing required. The metallic back-up ring for the sealing element and the opposed-slip principle of this type packer is outstanding in applications involving high pressure differentials.

Figure 6-14A pictures a permanent packer. The

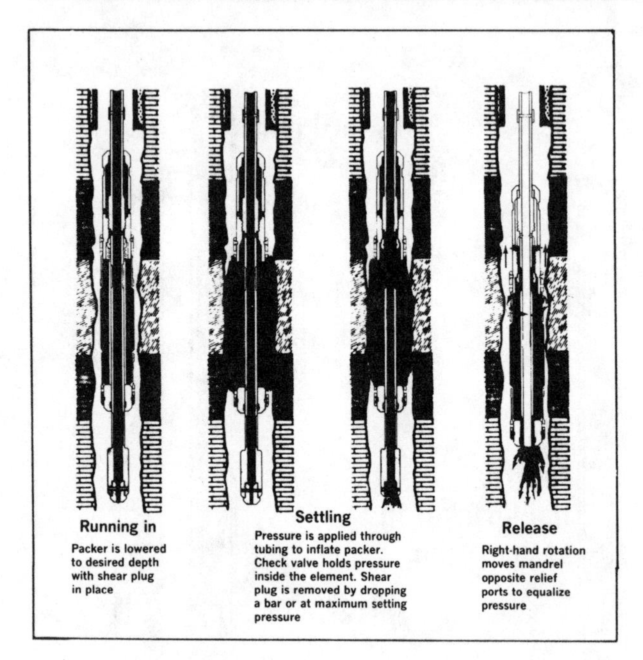

FIG. *6-13—Open-hole inflatable packer.*

tubing is sealed off in the bore of the packer by V-type chevron packing, attached to the seal nipples. Downward movement of the tubing is impeded by either a locator or anchor sub (Figure 6-14B). The anchor type permits the tubing to be latched into the packer, thus avoiding contraction. Right-hand rotation releases the anchor assembly from the packer. As an alternative, any number of seals may be run to compensate for tubing movement with the locator sub.

Accessories are available to convert the packer to a temporary bridge plug to test, squeeze cement, or frac above the packer.

To set the packer by wireline methods, a setting tool and collar locator are attached to the packer and the entire assembly is run to the desired setting depth. An electrically detonated powder charge builds up gas pressure that is transmitted hydraulically through a piston arrangement to mechanical forces which set the packer. A release stud then shears and the setting tool is retrieved from the well. Seal nipples to pack off in the bore of the packer are run on the tubing.

A combination of right-hand pipe rotation, tension, and weight is utilized to set the packer on tubing. The seal nipple assembly is normally run with the packer, in this instance, to avoid an extra round trip.

The primary objection to permanent packers, their permanence, has been partially overcome by the de-

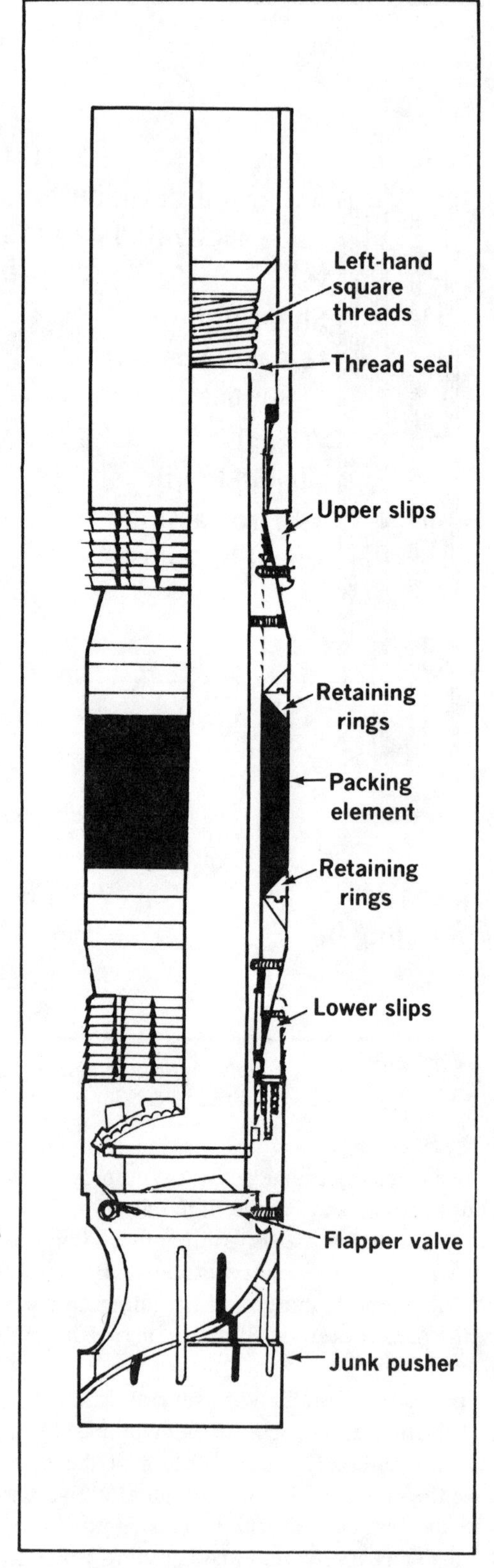

FIG. *6-14A—Permanent packer.*

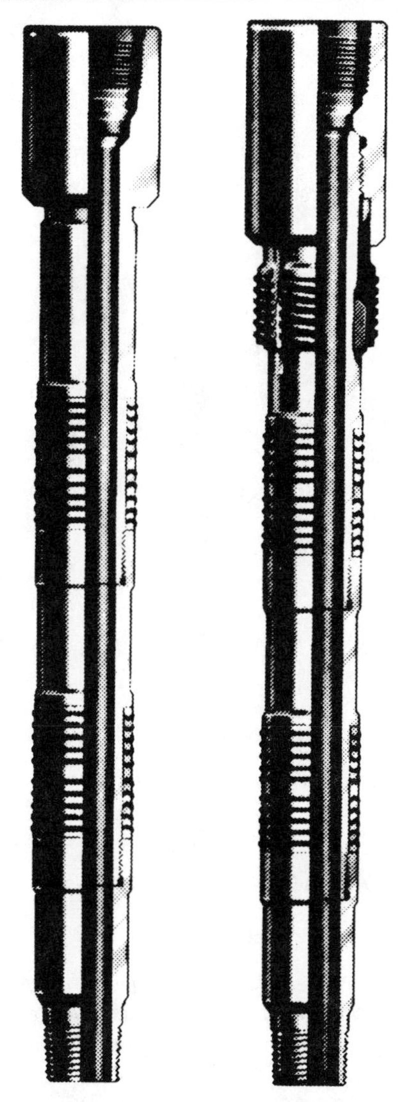

FIG. 6-14B—*Seal assemblies, locator-type (left), and anchor-type (right).*

velopment of through-tubing workover methods. Low-pressure squeeze cementing, dump bailers, through-tubing perforators, and concentric tubing techniques usually eliminate the necessity for packer removal.

It is usually possible to set a permanent packer above several alternate completion intervals and successively plugback without pulling the tubing string. If a desired completion interval is above the packer, the permanent packer serves as an excellent cement retainer to squeeze cement the zone below the packer.

Omission of the locator or latch-in sub permits running seals to pack off in any number of permanent packers set in the same wellbore. In this manner, permanent packers are adapatable as isolation packers.

Permanent packers are readily removed in two to three hours drilling time using a flat-bottom packer mill or about six hours with a rock bit. By comparison, removal of a stuck retrievable packer may require two or three days and considerable tool expense. Packer milling and retrieving tools are also available to recover the permanent packer by cutting the upper slips and catching the remainder of the packer.

Initial cost of permanent packers is somewhat higher than weight-set or tension packers, but less than hydraulic-set models. If accuracy within a few feet is not necessary, permanent packers are set most economically on tubing, thus avoiding per-foot depth and service charges. However, where a wireline truck is necessary for other services, such as perforating, the service charge is absorbed in the perforating operation.

Permanent-Retrievable Packers—(Figure 6-15) Permanent-retrievable packers are designed to incorporate the advantages of the permanent packers, i.e., opposed slips to hold high pressure differential in either direction, V-type Chevron packing within a polished bore, and wireline setting, in a packer that can be retrieved. This type of packer, which is made of steel instead of cast iron and is longer than regular retrievable packers, is removed by straight upward pull from a retrieving tool run on tubing or drill pipe. The setting procedure is the same as for permanent packers, i.e.: by electric wireline, by rotation or by hydraulics.

The retrieving tool carries cams that engage collets in the inner packer mandrel sleeve, or as shown in Figure 6-15 engage a pin in the upper portion of the packer mandrel. Upward pull by retrieving tool shears pins or shear rings that allow the upper slips to move off the expander cone and the lower cone to move out of the slips. In actual practice, care must be taken to prevent solids from settling around the outside of packer above the rubber sealing elements. Because of the large O.D. of these packers, such solids cannot be removed by washing over. A stuck packer can only be removed by an extensive milling and fishing operation.

Inner seal assemblies for these packers are the same as shown in Figure 6-14B for permanent packers.

Packer Bore Receptacle

In deep gas wells or other situations where casing or liner diameter is limited and maximum packer bore

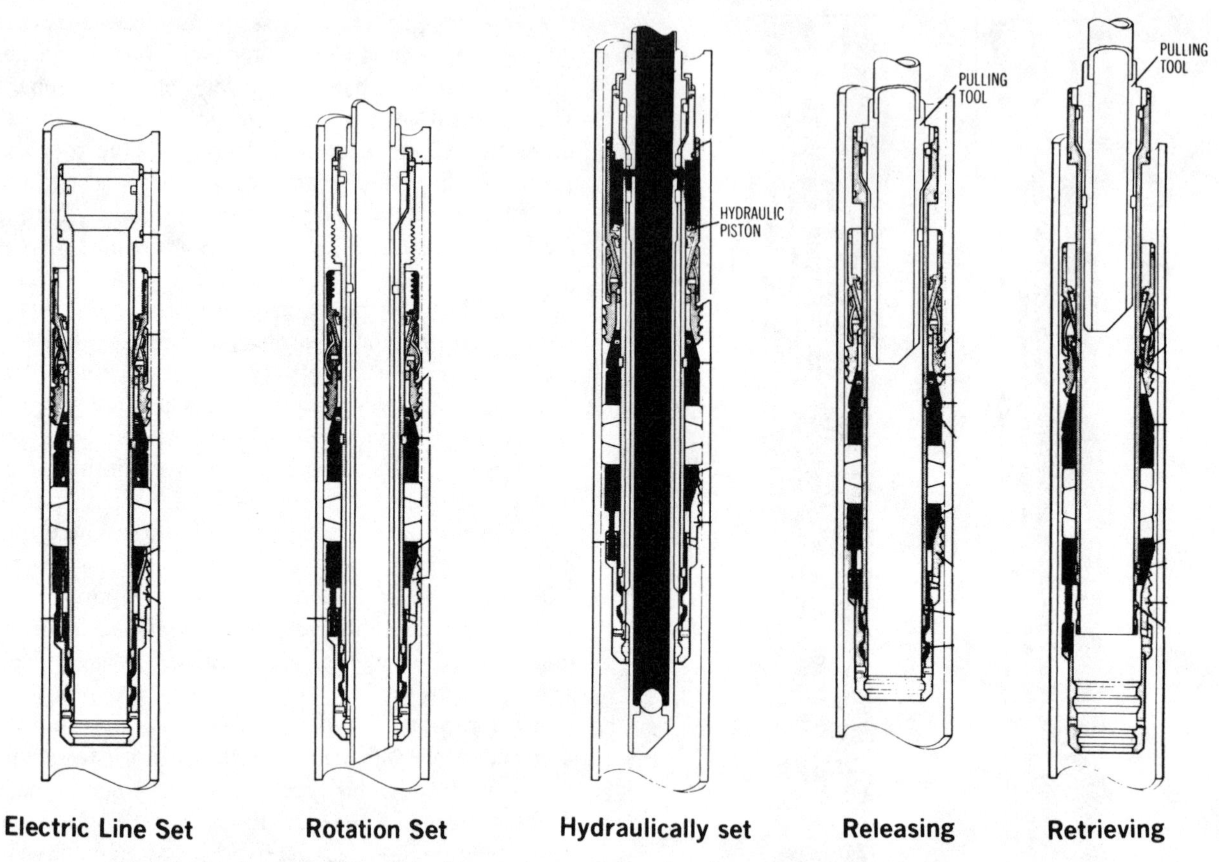

FIG. *6-15—Permanent-retrievable packer, schematic of setting and releasing mechanism.*

is desired, the Packer Bore Receptacle, Figure 6-16, may have application.

The seal arrangement of Figure 6-16 allows sufficient free upward tubing movement during stimulation treatments but permits tubing weight slack-off to eliminate seal movements during the producing life of the well. Produced fluids are not in contact with the intermediate casing or the polished sealing surface and the liner top is not exposed to pressure.

Cement Packers

The term ''cement packer'' has become accepted for tubing cemented inside conventional casing, which essentially creates a tubingless well. Cement is circulated into the tubing-casing annulus just as in cementing tubing in the open hole for tubingless wells. The cement packer replaces the conventional packer by sealing against vertical flow in the tubing-casing annulus.

The technique has these advantages:

1. Isolating leaking squeezed perforations and casing failures without remedial squeeze cementing.
2. Avoiding setting liners during deepening operations.
3. Minimizing the need for wireline completion equipment in multiple wells.
4. Eliminating communication repair jobs due to tubing and packer leaks.
5. Permitting application of tubingless completion techniques during future servicing and workovers.

Figures 6-17, 6-18, and 6-19 illustrate typical applications. The workover objective in the first example was to deepen the well several hundred feet and complete for water injection in both the existing and deepened interval. Conventional methods would dictate setting and cementing a liner, re-running tubing on a packer and perforating. Utilizing the cement packer concept, tubing was merely cemented in the wellbore and the formation perforated through both strings of casing.

In the second example, a casing leak prohibited

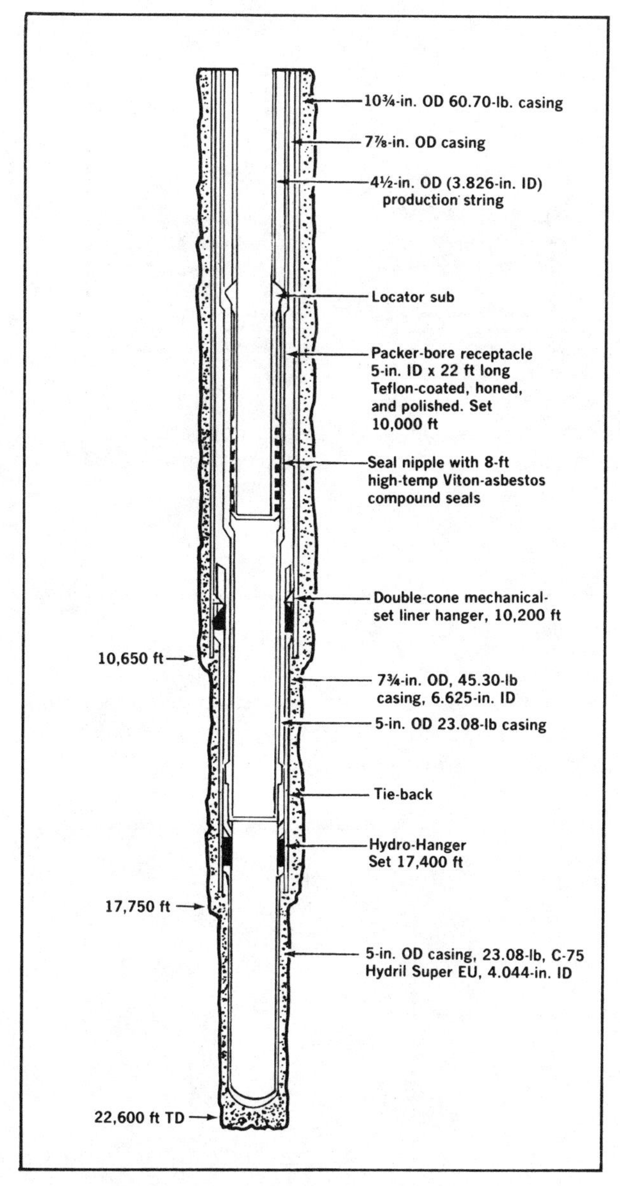

FIG. *6-16—Deep gas well completion with maximum bore from surface to TD.*[6] *Permission to publish by The Society of Petroleum Engineers.*

routine single-string dual completion of a conventionally cased well. Running and cementing two strings of tubing accomplished the desired results at minimum cost and eliminated the possibility of future problems with tubing or packer leaks should it be desired to stimulate.

Figure 6-19 illustrates the use of the cement packer idea to repair a leaking packer, or to recomplete above the upper packer in a dual packer well. Briefly the procedure is to wireline plug the short tubing string and perforate the short string with one or two holes immediately above the upper packer.

A concentric workstring is then used to circulate low fluid loss cement into the annulus above the upper packer. Excess cement can be reversed back, and the annulus cement held in place by pressure or by balancing fluids.

SUBSURFACE CONTROL EQUIPMENT

Subsurface control equipment includes: (1) safety valves which plug the tubing at some point below the wellhead should the surface controls become damaged or completely removed; (2) bottomhole chokes and regulators which reduce the wellhead flowing pressure and prevent the freezing of surface controls and lines by taking a pressure drop, downhole; and (3) check valves that prevent backflow of injection wells. The essential working elements of each of these devices can be installed or removed with a wireline.

Since all these tools are susceptible to erosion damage, the well should be brought in and thoroughly cleaned prior to installing a subsurface control device.

Safety Systems

Surface Safety Systems—The surface safety system is the first line of protection against minor mishaps in surface treating facilities. The surface system generally consists of normally closed valves held open by low pressure gas acting on a piston. If gas pressure is bled off, internal spring action closes valve against line pressure. Valves and controls are readily accessible for maintenance.

Low pressure gas can be tied into a network of sensors to detect abnormal conditions.

Catastrophe Systems—The catastrophe system is the downhole shut-in system activated (except for testing) only in the event of imminent disaster. In its most rigorous form this system consists of a near-surface packer or hanger and master valve, Figure 6-20, supplementing the surface christmas tree and master valve. The downhole valve may be direct controlled (self controlled) or remote controlled from the surface. Remote control application is becoming almost universal on new installations, because of more effective control of all wells on the platform.

Direct Controlled or self-controlled valves (storm chokes) are preset to close when conditions at the valve reach a particular criteria. Two basic types are available—one operated by differential pressure across the valve—Figure 6-21, the other operated by am-

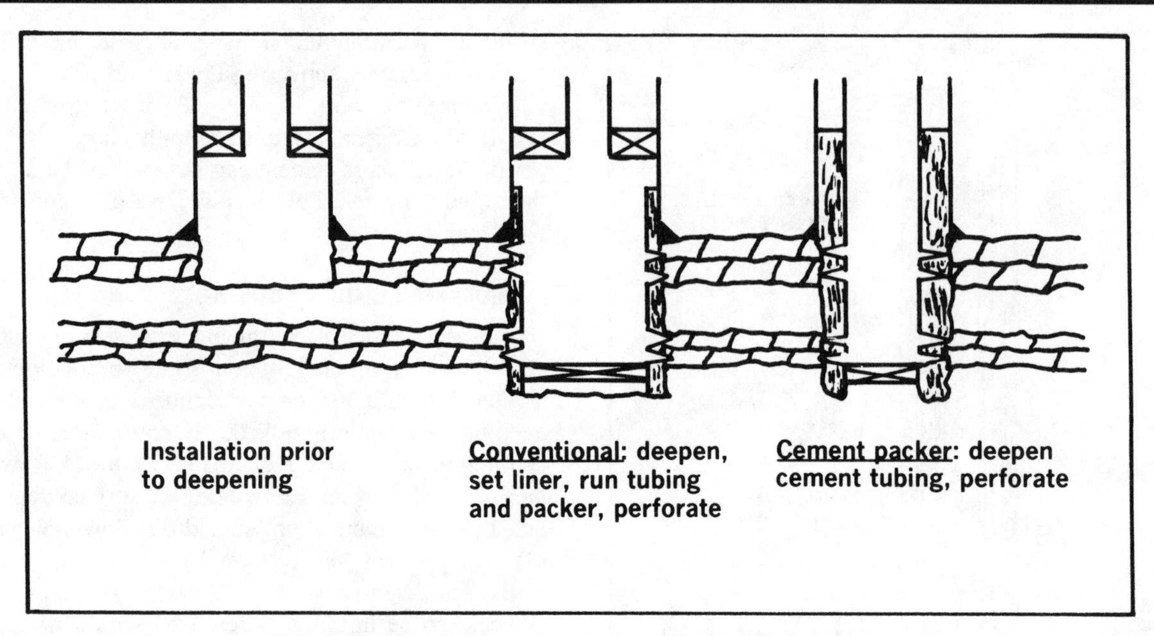

FIG. 6-17—Application of cement packer for deepening.

bient pressure in a precharged bellows. Both are normally wireline installed in a tubing string landing nipple.

Differential Pressure Valves—Flow of well fluids through an orifice creates a pressure drop related to flow rate. The valve is held open by a preset spring tension, but as flow-rate increases pressure drop across the orifice eventually causes the valve to close. Various types of closure are available as shown in Figure 6-22.

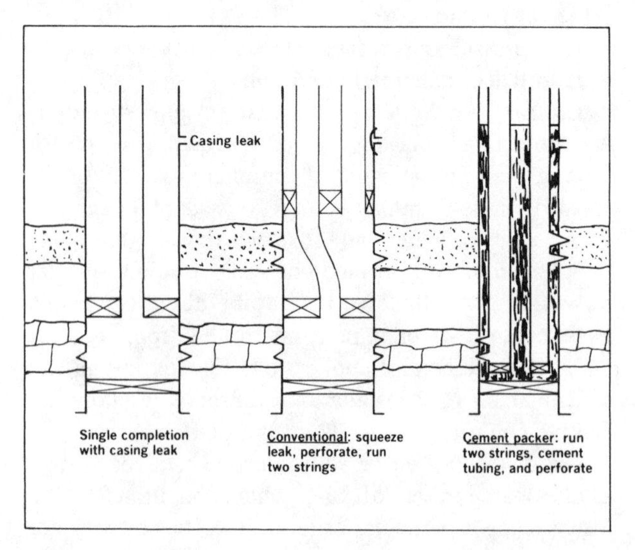

FIG. 6-18—Application of cement packer for dual completion.

If a valve closes, it can be reopened by applying pressure to the tubing string or actuating the equalizing sub by running an equalizing prong on a wire-

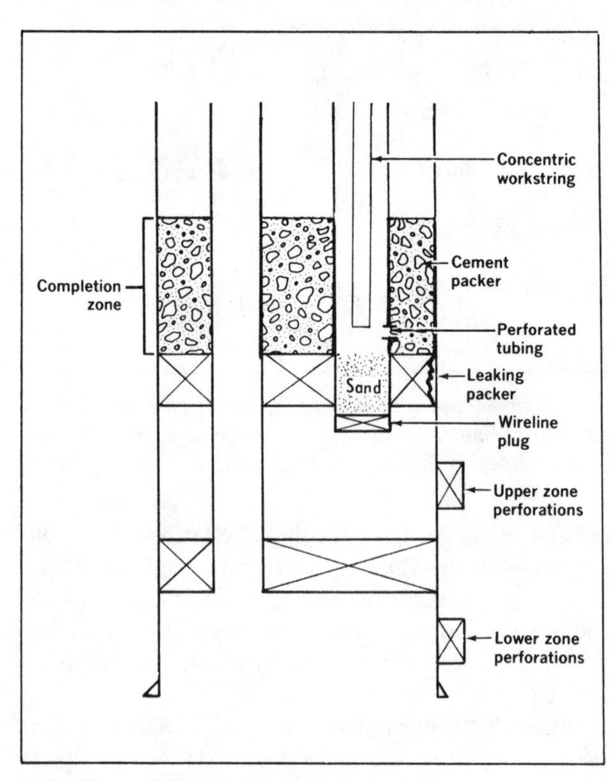

FIG. 6-19—Application of cement packer for top packer repair or recompletion above top packer.

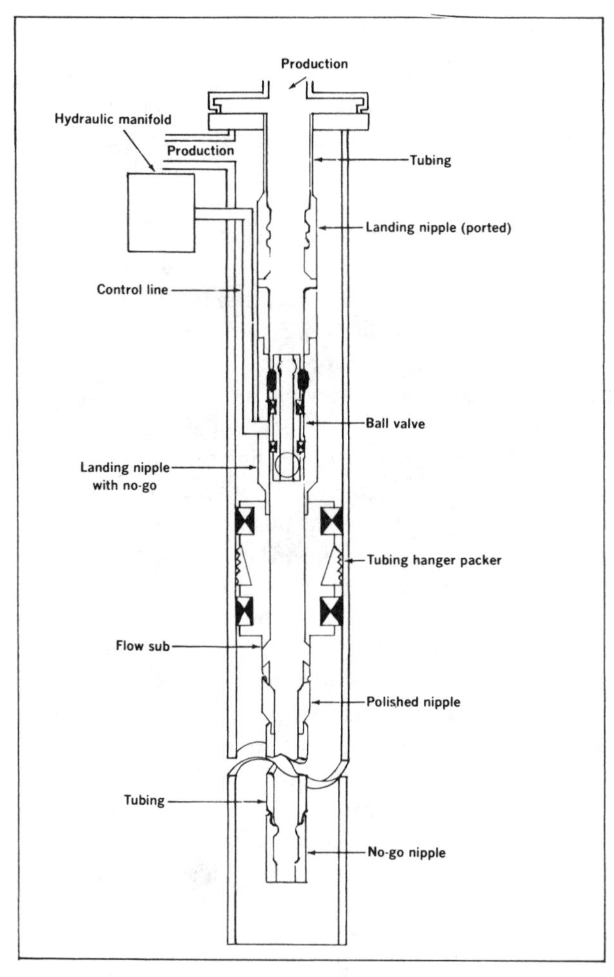

FIG. 6-20—*Ideal downhole shut-in system, annular production.*

line. When the pressure is equalized across the safety valve, the compressed spring causes the valve to open automatically.

Ambient Pressure Valve—Valve bellows is precharged to a pressure less than the pressure in the tubing under normal flowing conditions at the valve.

Under abnormally high flow rates the tubing pressure at the valve drops to a point permitting bellows charge pressure to close the valve. Shown in Figure 6-23, the ambient valve does not require a choke bean; thus, pressure drop across the valve is minimized.

Closing Calculations—Provided sufficiently accurate information is available on well flowing characteristics closing pressures can be calculated for direct controlled valves with reasonable accuracy. However, actual tests must be run under actual stabilized

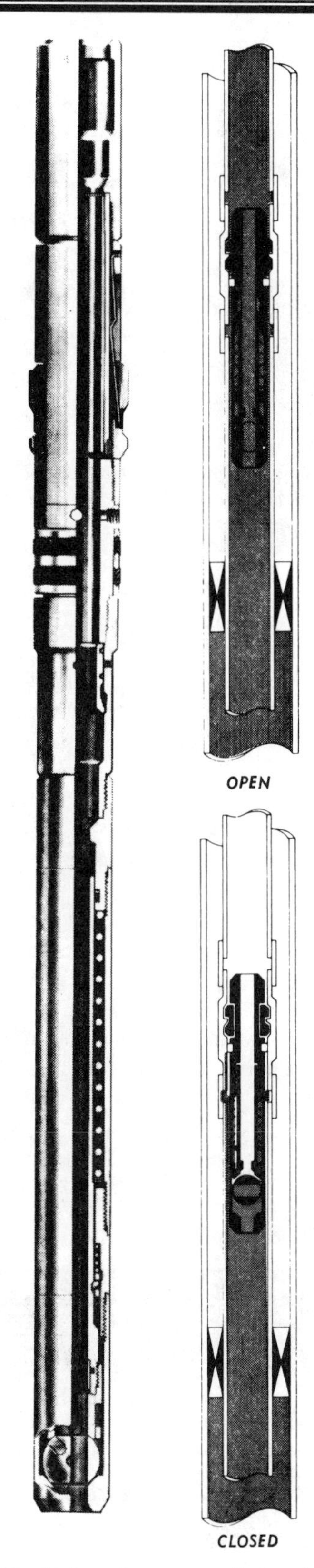

FIG. 6-21—*Subsurface controlled safety valves.*

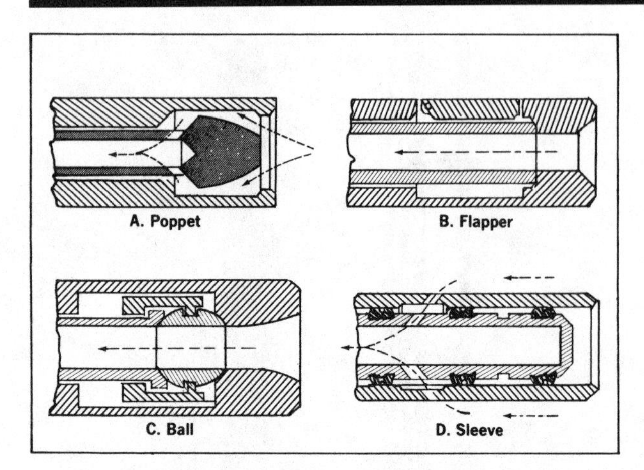

FIG. *6-22—Types of closures for differential-pressure valves.*[5] *Permission to publish by The Society of Petroleum Engineers.*

flowing conditions to ensure proper operation.

Several conditions might occur that could prevent a direct controlled valve from closing when desired even though it was correctly set when initially installed in the well:

1. Maximum flow rate capability of well may decline to less than original value.
2. Paraffin or sand accumulation may restrict flow rate.
3. A catastrophe on the platform may result in a small wellhead leak—disastrous perhaps, but not causing sufficient flow rate to close valve.

For these reasons, many operators and regulatory bodies are reexaming the application of direct controlled valves.

Surface-Controlled Safety Valves

Surface-controlled safety valve systems are normally positioned slightly below the ocean floor or at 200 to 300 ft on land locations. Major items of equipment include: (1) a special landing nipple with an external ⅛-in. id (0.409-in. od) control line which is made up and run into the well with the tubing string, (2) the safety valve, (3) an exit assembly for the external line at the christmas tree, and (4) a surface control unit and related lines and pressure pilots.

Principal advantages of surface-controlled safety valves over subsurface models are (1) larger internal diameters for tubing-retrievable types which permit higher flow rates and the ability to lower wireline tools through the valve, (2) insensitivity to pressure and fluid surges, (3) more positive control because operation does not depend on an orifice which may be damaged by sand, and (4) simplified testing. Disadvantages include more complicated design and higher cost.

Figure 6-24 pictures a wireline retrievable valve which is typical of most designs. It operates on a piston principle and is controlled from the surface by the aforementioned control manifold. Hydraulic pressure is exerted down the ¼-in. line from the surface, enters the landing nipple and safety valve, and acts against a piston within the safety valve. This holds a spring in compression and maintains the flow tube in the down position, holding the closing mechanism open.

If surface controls fail, pressure exhausts from the hydraulic line at the surface, causing well pressure acting with the coil spring to move the flow tube upward and allow the valve to close. The valve is reopened by again applying pressure through the external line. It is also possible to pump through the valve

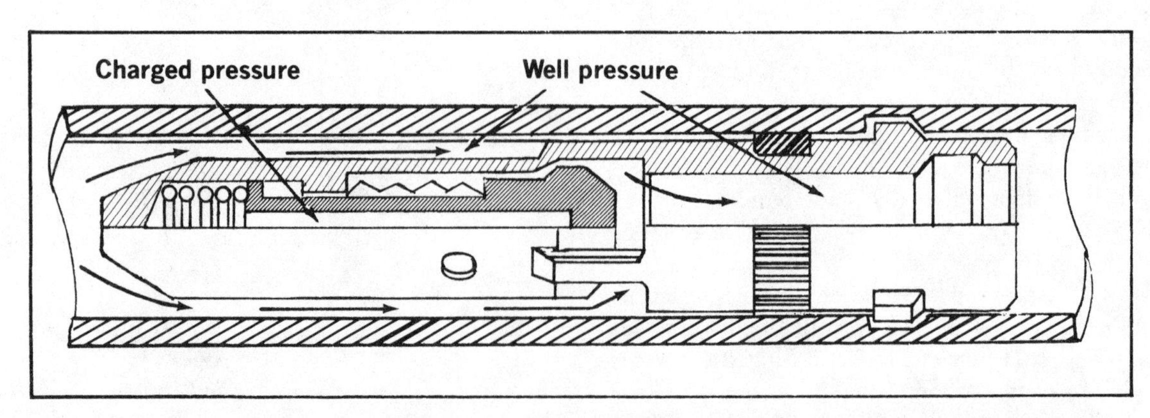

FIG. *6-23—Ambient-pressure valve.*[5] *Permission to publish by The Society of Petroleum Engineers.*

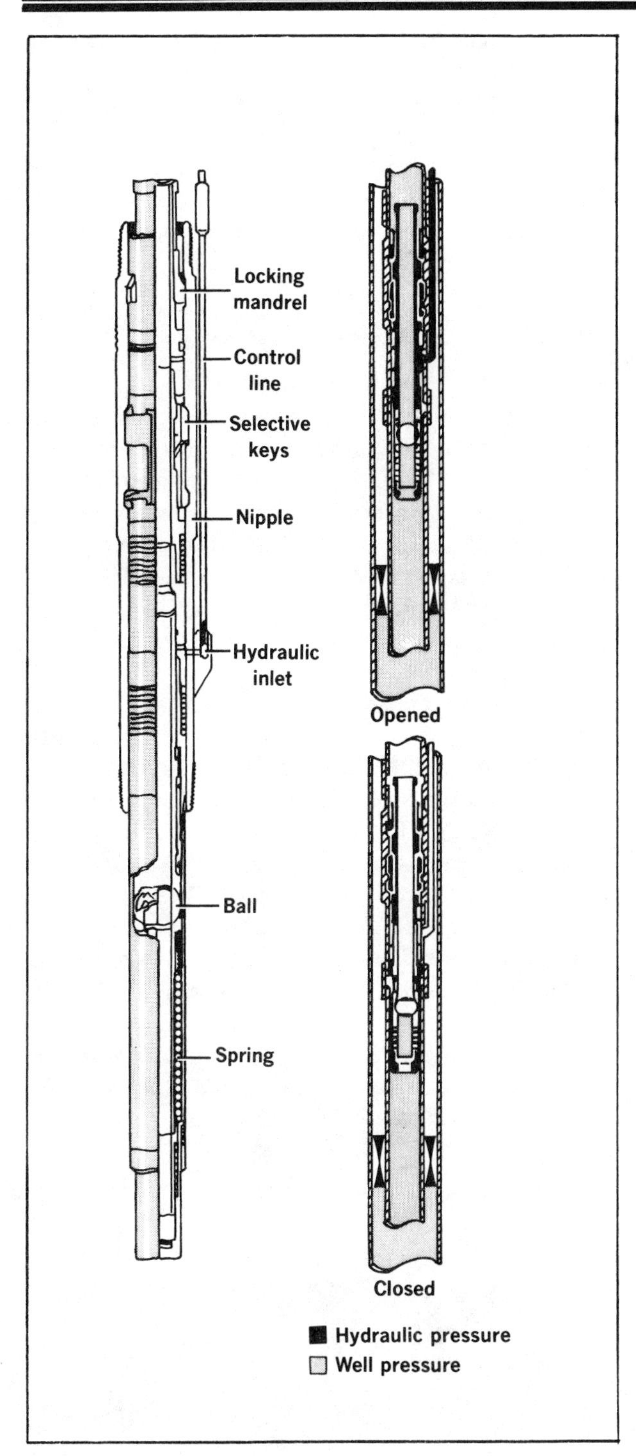

FIG. *6-24—Wireline-removable surface-controlled safety valve.*

when it is in the closed position to kill the well if necessary.

Figure 6-25A pictures a typical wellhead adapter necessary to accommodate the external line. A new style wellhead adapter permits retrieving the external line without pulling the tubing. Figure 6-25B compares a conventional control line system with an annular pressure control system eliminating the control line and utilizing a tapered tubing string to provide a large diameter through the safety device. The safety device can be either wireline or tubing retrievable.

The control unit is normally placed at or near the wellhead. The main component of the unit is a pump which maintains a constant, pre-set pressure on the control line to hold the safety valve open. Figure 6-26 pictures one type of control unit and shows the pump, fluid reservoir, and miscellaneous equipment.

Operation Considerations

Excessive Setting Depth—To maintain surface control, hydrostatic head of fluid in the control line must present less valve opening force than the spring closing force available at the valve. Otherwise if tubing

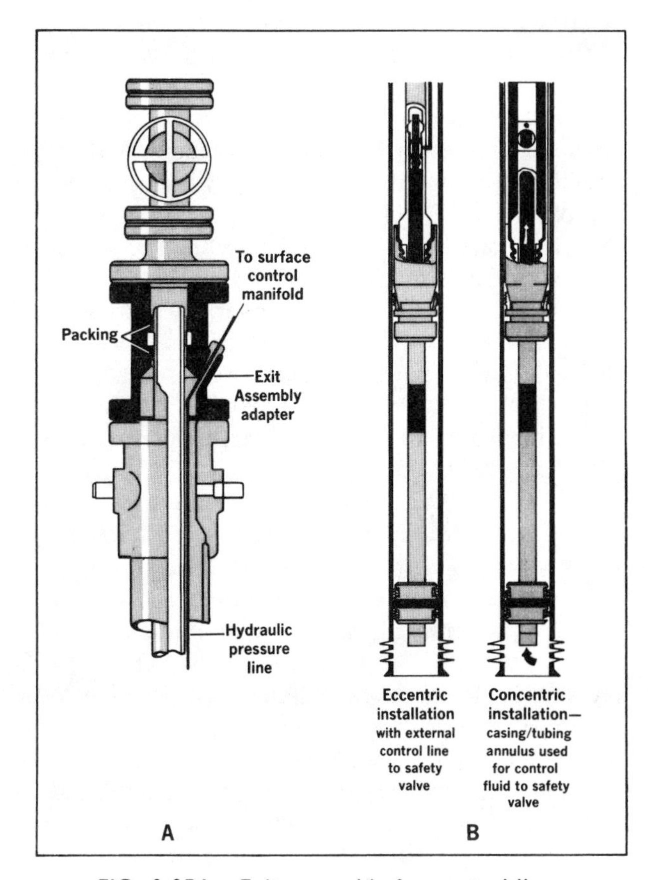

FIG. *6-25A—Exit assembly for control line.*
FIG. *6-25B—Pressure-control systems.*

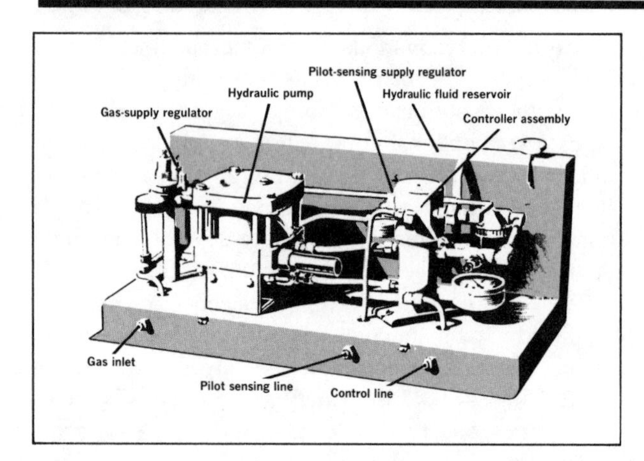

FIG. 6-26—*Surface control unit.*

pressure at the valve dropped to zero hydrostatic control line pressure would keep the valve open even with no positive surface control line pressure.

If it is necessary to set the valve at a depth where this effect becomes a problem, response to surface control must be assured by using additional spring force, adjustment of piston areas or by providing a second small diameter "balancing" line from the surface.

Control Fluids—The following control fluids can be used:—treated water—water containing a water-soluble oil to provide lubrication and corrosion protection; diesel oil—add light oil to ensure lubricity; and light-weight oil—good lubricity and flow characteristics unless temperature drops below 50°F.

Gas should probably not be used since a control line failure near the valve would expose the valve to the hydrostatic head of the annular fluid rather than gas.

It is impossible to completely negate the possibility of solids settling out in parts of the valve. Periodic testing not only determines if the valve is working properly but also helps in reducing seal "striction" and keeping operating surfaces free of deposits.

Surface Sensing Systems—The primary problem in setting up a Sensing System is to design the system so that it actually "sees" a proper malfunction. The pressure pilot, the most common sensor, is intended to activate safety valves when pressure at some point in the system exceeds or falls below pre-set limits.

It must be remembered that a small leak can be "disastrous" but not actually lower flow line pressure from the wells. Also the high-low pressure malfunction can often be handled by surface safety shut-in equipment rather than downhole shut-in. Routine problems should not activate the primary downhole system.

Fire detectors include fusible plugs, or ultraviolet-light detectors. Proper placement is critical with either. Obviously fire requires downhole shut-in. Collision or storm damage is usually detected with fragile or brittle control lines. Downhole shut-in is warranted.

Bottom-Hole Chokes and Regulators

Bottom-hole chokes and regulators are utilized to eliminate the freezing of surface controls and lines due to the formation of hydrates by moving the point of pressure reduction and attendant temperature decrease to the lower portion of the wellbore. This allows the higher temperature at depth to reheat the flow stream before it reaches the surface. The lower surface flowing pressure is also advantageous in reducing the continuous surface pressure on the tubing and christmas tree.

Bottom-hole choke beans (Figure 6-27) are attached to the lower end of mandrels designed to be set in a landing nipple or anchored to the tubing wall. Type A bean is a spring-loaded ground seat bean that is recommended for high pressure or heavy fluid wells. The bean is mounted inside a cage and seats against the lower end of the mandrel. The bean is held on seat by the spring inside the bean cage. Type B bean incorporates a positive orifice and will accommodate any size choke up to the internal diameter of the mandrel.

A bottom-hole choke is susceptible to erosion and is also rate sensitive; however, if a well produces without sand and at a fairly constant rate, this positive

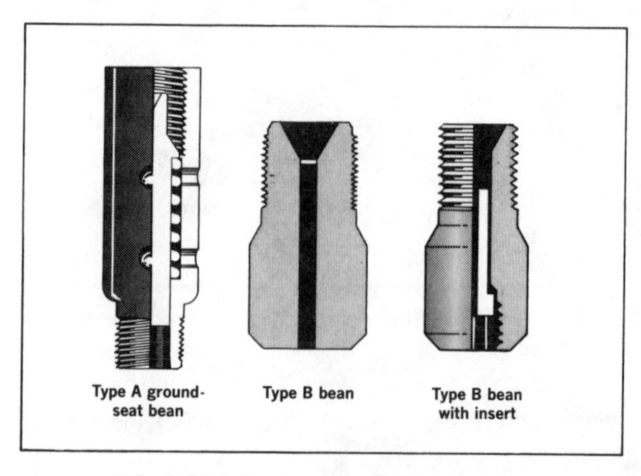

FIG. 6-27—*Bottom-hole choke beans.*

choke assembly is superior to the bottom-hole regulator due to economy, simplicity, and minimum maintenance.

Bottom-hole Regulator (Figure 6-28)—This consists of a valve and a spring-loaded valve seat. A predetermined spring tension determines the pressure differential across the regulator. When this predetermined differential is reached, the valve seat moves off the valve to allow the well to flow.

The regulator differs from a bottom-hole choke in that it maintains a constant pressure differential across the valve, regardless of the flow rate. Well production is adjusted at the surface by a conventional choke. Theoretically, a regulator will also reduce the surface shut-in pressure. However, the sealing mechanism is not sufficiently reliable or durable for this purpose, particularly after the regulator has been installed for extended periods of time.

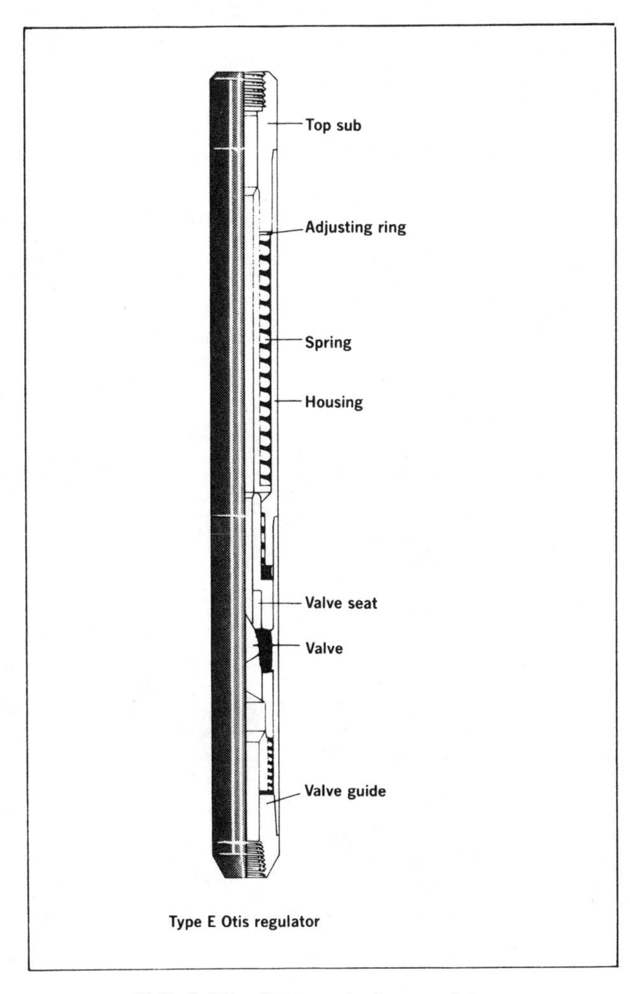

FIG. 6-28—*Bottom-hole regulator.*

Regulators can be installed in all sizes of tubing, either in a landing nipple, slip-type mandrel, or collar lock mandrel. Pressure differentials up to 1,500 psi can be taken across a single regulator. If a larger reduction in surface pressure is desired, two regulators can be connected in tandem, or several separate regulators can be set at various depths in the tubing string.

Data required to properly calculate bottom-hole choke and regulator sizes include: (1) tubing size, (2) setting depth of valve, (3) surface shut-in pressure, (4) desired rate of production, (5) flowing pressure and temperature at valve depth, (6) specific gravity of produced fluids, and (7) desired pressure drop across the choke or regulator.

Bottom-hole chokes and regulators should be set in the lower portion of the wellbore. Increasing the setting depth reduces the length of tubing exposed to full flowing pressure and permits additional time and temperature for heating the fluids before they reach the surface.

Inclusion of flow coupling in the tubing string above and below landing nipples contemplated to house bottom-hole chokes and regulators should be considered. Field experience has shown that the flow restrictions from turbulence are sufficient to seriously erode the pipe. A 3-ft length has proven satisfactory in most circumstances.

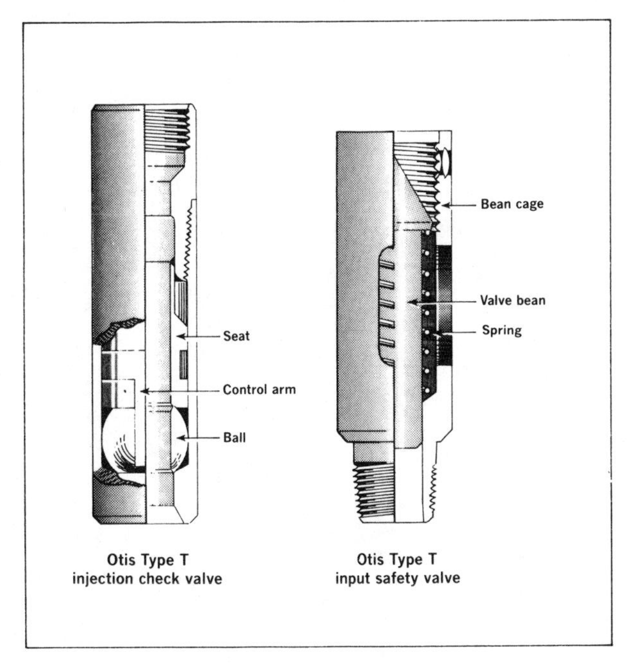

FIG. 6-29—*Injection safety valves.*

Subsurface Injection Safety Valves (Figure 6-29)

There are two types of injection safety valves which will shut the well if flow is reversed. The Otis Type T injection check valve is designed to be either made up on the end of the tubing string and run into the well with the tubing or set in a landing nipple by wireline means. The Type T valve is a ball-and-seat type valve and is designed so that injection or static pressure will hold the ball in the open position for fluid passage in the upstream direction.

Should the pressure reverse, the ball in the valve will rotate to the closed position and shut the well in. The valve will remain closed until the pressure differential across it is equalized by resuming injection. The Type T Otis input safety valve is a simple, spring-loaded valve and seat mechanism. Injection pressure forces the valve open for fluid passage.

If injection ceases or reverses for any reason, the spring tension and the flow pressure act to force the valve to the closed position. This valve may also be seated in a landing nipple or may be set on a collar stop at any point in the tubing.

REFERENCES

1. Lubinski, Arthur; Althouse, W. S.; and Logan, J. L: "Helical Buckling of Tubing Sealed in Packers," *JPT* (June, 1962) p. 655.

2. Leutwyler, Kurt: "Completion Design for Corrosive Environment," *Pet. Eng.* (Feb. 1970).

3. Krause, W. E., Jr., and Sizer, P. S.: "Selective Criteria for Subsurface Safety Equipment for Offshore Completion," *JPT* (July, 1970) p. 793.

4. Logan, J. L.: "How to Keep Tubing Sealed in Packers," *J. Canadian Pet. Tech.* (Summer, 1963).

5. Raulins, G. M.: "Platform Safety by Downole Well Control," *JPT* (March, 1972) p. 263.

6. Lindsey, H. E., Jr.: "Deep Gas Well Completions Practices," SPE 3908, Amarillo, TX (Sept. 11, 1972).

7. Hammerlindl, D. J.: "Movement, Forces and Stresses Associated with Combination Tubing Strings Sealed in Packers," *JPT* (Feb. 1977).

8. Baker Oil Tools-Packer Calculations handbook.

9. API Specification 14A Subsurface Safety Valve.

10. API Recommended Practice 148, Design, Installation and Operation of Subsurface Safety Systems.

11. Purser, P. E.: "Review of Reliability and Performance of Subsurface Safety Valves," OTC 2770 (May, 1977).

12. Burley, J. D., and Holland, W. E.: "Recent Developments in Packer Seal Systems for Sour Oil and Gas Wells," SPE 6762 (Oct. 1977).

13. Hammerlindl, D. J.: "Packer-to-Tubing Forces for Intermediate Packers," SPE 7552 (Oct. 1978).

14. Durham, Kenneth S.: "Tubing Movement, Forces, and Stresses in Dual Flow Assembly Installations," SPE 9265 (Sept. 1980).

Chapter 7 Perforating Oil and Gas Wells

Perforating methods
Perforator performance evaluation
Factors affecting perforating results
Fluids, pressure differential, formation properties
Operational considerations
Optimum perforating practices
API RP-43 tests

INTRODUCTION

Perforating is probably the most important of all completion functions in cased holes. Adequate communication between the wellbore and all desired zones, as well as isolation between zones, is essential to evaluate and to optimize production rate as well as oil and gas recovery from each zone.

Research[2,3,6] by Exxon (Humble) developed the significance of (1) plugging of perforations with mud or shaped charge debris, (2) perforating with a pressure differential into wellbore in clean fluid, (3) the necessity of maintaining this differential without interruption into the wellbore until all perforations are cleaned, and (4) the effect of formation and cement compressive strength on perforation hole size and penetration. This work led to the development of nonplugging shaped charges, through-tubing perforators, improved bullet guns, the through-tubing flow meter, and an API standard, Section 2, API RP 43, to evaluate perforators under simulated downhole flow conditions. The development of more effective jet perforators has aided penetration where a high compressive strength formation, high compressive strength cement, and thick-walled, high-strength casing is present.

Although technology is available to insure good perforating in most wells, unsatisfactory perforating tends to be the rule in many areas of the world. The four most prevalent causes for poor perforating results probably are (1) a lack of understanding of the requirements for optimum perforating, (2) after perforating with a pressure differential into the wellbore, the practice of shutting in a well soon after perforating to pull a through-tubing or tubing-conveyed gun, (3) inadequate control of gun clearance, usually requiring decentralization with through-tubing perforators, (4) selecting perforating guns or charges on the basis of surface penetration tests in cement instead of flow efficiency, based on laboratory flow tests, Section 2, API RP 43, and (5) the rather widespread practice of awarding perforating jobs on the basis of price, rather than job quality.

TYPES OF PERFORATORS

Bullet Perforators

Bullet guns, 3 $^1/_4$ in. OD or larger, are applicable in formations with compressive strength less than about 6,000 psi. Bullet perforators in the 3 $^1/_4$ in. or larger size range may provide deeper penetration than many jet guns in formations with less than about 2,000 psi compressive strength. However, tests of specific jet and bullet guns should be made in various compressive strength reservoir rocks, using tests described in Section 2, API RP 43 to validate these conclusions on a current basis. The Western Atlas "E" bullet gun has been a very satisfactory bullet gun for 40 years.

Muzzle velocity of bullet guns is about 3,300 ft/sec. The bullet loses velocity and energy when the gun clearance exceeds 0.5 in., the clearance at which most comparative tests have been made. At zero gun clearance penetration increases about 15% over 0.5

in. clearance, along with a deburring effect. Loss in penetration with one-inch clearance is about 25% of the penetration at 0.5 in. clearance, and at 2 in. clearance the loss is 30%.

Deburring of bullet holes is not dependent on decentralization if the bullet carries a deburring device. In addition, this device is more effective in deburring than using zero gun clearance. Bullet guns can be designed to fire either selectively or simultaneously.

Jet Perforators

The jet perforating process is illustrated in Figure 7-1. An electrically-fired detonator starts a chain re-

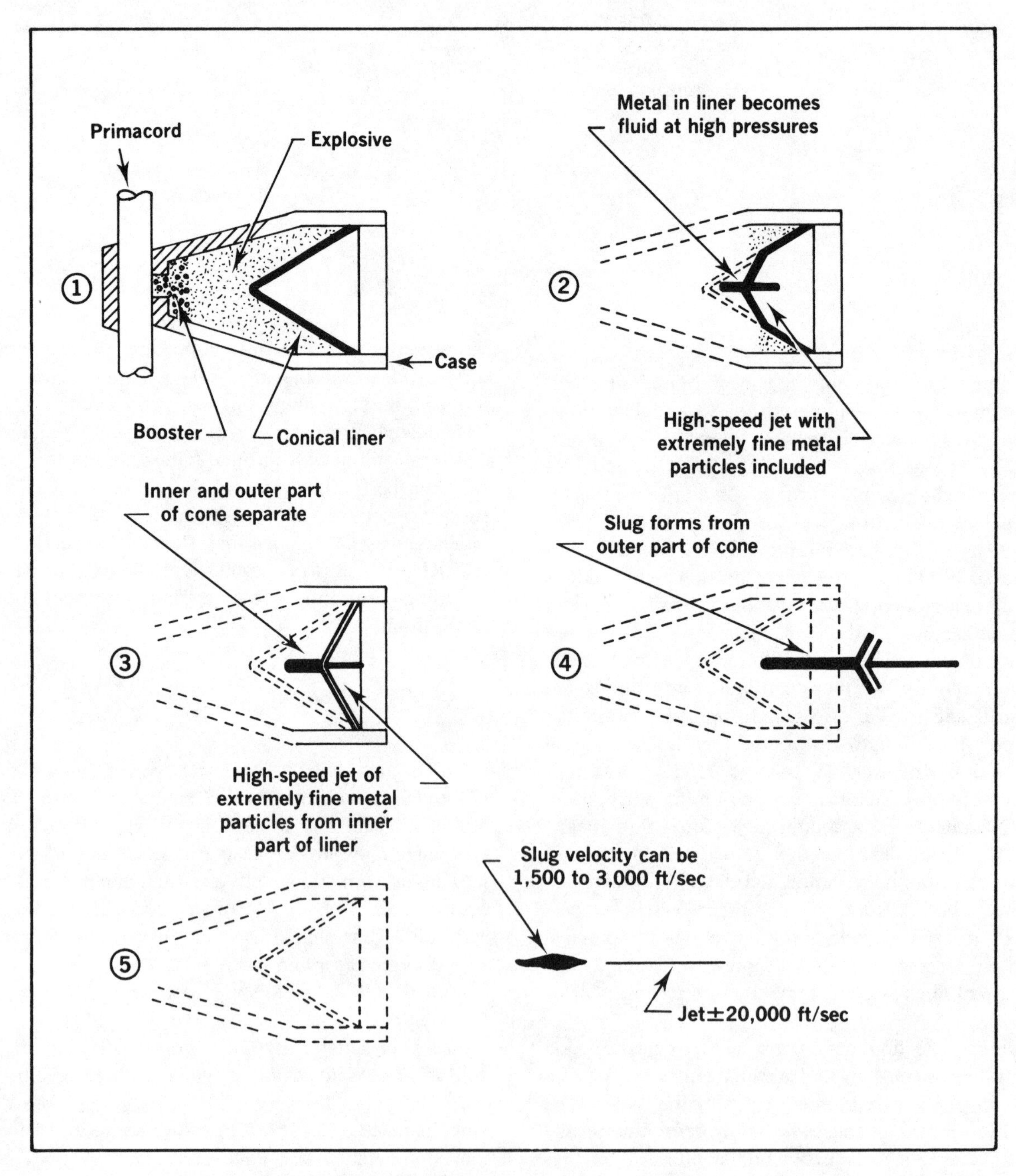

FIG. 7-1—*Jet perforating process using a solid metal liner.*

action which successively detonates the primacord, the high velocity booster in the charge, and finally, the main explosive. High pressure generated by the explosive causes the metal in the charge liner to flow, separating the inner and outer layers of the liner. Continued pressure buildup on the liner causes a needle-like high speed jet of fine particles to spew from the cone at a speed of about 20,000 ft/sec at its tip with a pressure at this point estimated to be 10 million psi.

The outer shell of the charge liner collapses to form a slower moving metal stream, with a velocity between 1,500 and 3,000 ft/sec. This outer liner residue may be in the form of a single slug sometimes called a "carrot" or a stream of metal particles. Premium-priced charges usually are relatively carrot-free, with residue being sand-sized or smaller particles.

Because of the sensitivity of the jet perforating process to a near perfect sequence of events from the firing of the detonator through the formation of the jet, any fault in the system can cause malfunction. This can result in an irregular or inadequate hole size, poor penetration, or no hole at all. Some of the causes of malfunction are: insufficient electric current or voltage to detonator; poor quality or defective detonator; over-age, mashed, or twisted primacord; over-age, poorly packed or poor quality booster and main explosive; and liner incorrectly positioned or not in effective contact with the explosive.

Water or dampness in the gun, primacord, or charge may cause malfunction or low order detonation. High temperature aging of explosive in primacord or charge may reduce charge effectiveness or cause low order detonation. Procedures for loading, running, and firing the gun must be carefully followed to insure good job performance and a safe job.

Conventional and Through-Tubing Jet Perforators—Conventional retrievable hollow carrier steel guns normally provide adequate penetration without damaging the casing. Guns are run in the hole with atmospheric pressure inside the gun carrier. Most of the explosive energy not used in producing the jet is absorbed by the gun carrier rather than the casing.

Through-tubing jet perforators, including capsule-type guns, Swing-Jet guns, wire and strip carrier guns, and thin walled or expendable hollow carrier guns are available. Their primary advantage is being able to run these guns through the tubing to perforate and then retrieve while maintaining a pressure differential into the wellbore.

Many of the through-tubing guns give inadequate penetration and/or hole size. The swing jet provides relatively large hole size and sufficient penetration for most wells. Its major disadvantage is the mechanical manipulation required and the large amount of debris remaining after shooting. Guns with charges exposed to wellbore, such as the capsule gun, will swell the casing. Also, large charge loads in exposed-charge guns may split the casing or liner.

Also it is difficult to obtain proper gun clearance with through-tubing guns to provide needed hole size and penetration. The thin-wall expendable hollow carrier gun eliminates casing-splitting and much of the debris left inside the casing. It also overcomes the gun clearance problem if the gun is properly positioned, but sacrifices some penetration and hole size. To prevent severe swelling of the gun body, most thin-wall hollow carrier guns should normally be fired under a fluid head of at least 500 psi.

Many special-purpose jet perforators are available.

- —Selectively-fired guns are usually available for both conventional and through-tubing completions.
- —Four-way or five-way directional single-plane jet perforators are usually not recommended because of either inadequate penetration or casing-splitting.
- —Perforators are available to penetrate only the tubing. Tubing should be centralized to prevent casing damage; however, it is acceptable to perforate just above the packer or possibly immediately above or below a tubing collar.
- —Jet-type cutters are also available to cut various sizes of tubing, casing, and large-diameter steel piling.
- —Open-hole perforators have primary application in penetrating scale and other damage near the open-hole wellbore. Some premium-type hollow carrier guns provide satisfactory penetration in open hole and reduce open-hole cavings due to perforating.
- —The Jet Vac perforator is designed to clean out perforations with a high pressure differential into the gun carrier immediately after perforating.

Other Perforating Methods

Hydraulic Perforators—Cutting action is obtained by jetting sand-laden fluid through an orifice against the casing. Penetration is greatly reduced as wellbore pressure is increased from zero to 300 psi. Penetration (see Figure 7-7) can be increased appreciably by the addition of nitrogen to fluid stream.

Mechanical Cutters—Knives and milling tools have been used to open slots or windows to provide communication between the wellbore and the formation.

Milling a window in the casing, underreaming, and gravel packing are the standard procedures for sand control in some areas.

EVALUATION OF PERFORATOR PERFORMANCE

Prior to 1952, essentially all perforator evaluation was done in actual downhole tests in wells, or in surface tests at atmospheric temperature and pressure in casing cemented inside steel oil drums, similar to current testing in Sec. 1, API RP-43.

Comparative downhole testing of perforators in oil and gas wells proved to be generally impractical because of the difficulties in obtaining comparable downhole test conditions when perforating various wells in the same or different formations.

Surface tests at atmospheric pressure proved to be misleading for several reasons. Liner "slugs" from the shaped charge which otherwise would plug a perforation downhole tended to be deflected away from the perforation when shooting at atmospheric pressure. Surface tests, made in sand-cement targets such as those used in Section 1, API RP-43, cannot be used to predict downhole flow through perforations. Because fluid flow from or injection into a downhole formation is the objective of all downhole perforating, surface penetration tests in sand-cement targets can not be regarded as adequate for selecting perforators or perforating conditions.

Development of Flow Index System

In 1952, Exxon developed the first reliable testing procedure to simulate perforating and cleaning perforations under downhole conditions[3] and thereby to optimize well productivity. Perforating in a clean fluid with the differential pressure into the wellbore and maintaining this differential pressure into the wellbore continuously until the well is cleaned was proved to be the best well completion method for all sandstone and most carbonate reservoirs. This system initially was called the "Productivity Method of Perforator Testing" or "Well Flow Index" system. The test apparatus, illustrated in Figure 7-2, is very similar to that currently used on API RP 43, Section 2 tests.

The test plan designed to simulate actual downhole conditions include: (1) use of large diameter formation cores conditioned to specific hydrocarbon and interstitial-water saturations; (2) isolation of the formation from the wellbore by casing and a suitable cementing material; (3) perforating the casing, cementing material, and formation with various perforating fluids in the well; (4) determination of comparative flow through the formation core prior to perforating, after perforating, and after cleaning the well; (5) maintenance of well and reservoir temperature, and reservoir and wellbore pressure during and after perforating; (6) testing perforators with both a differential pressure into the formation and with a differential pressure into wellbore; (7) continuous backflowing the well after perforating without interruption to simulate producing a well to clean perforations; (8) evaluation of test results; and (9) proving the validity of simulated well tests in hundreds of wells through field production tests and thru-tubing flow meter tests.

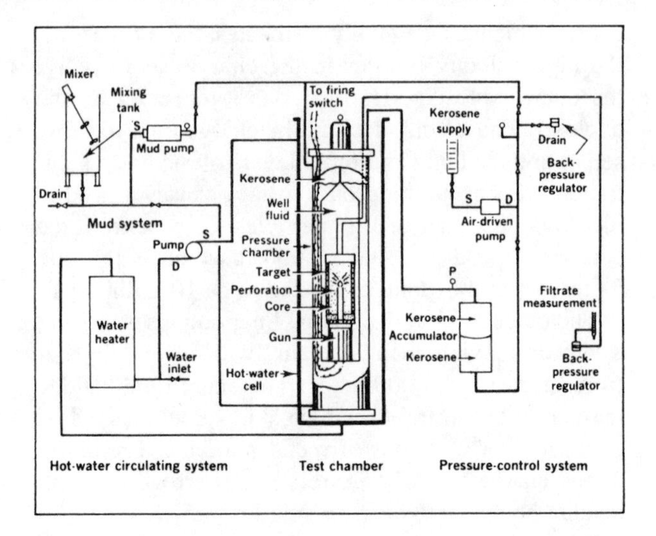

FIG. 7-2—Schematic of simulated downhole gun-perforator testing.[3] Permission to publish by API Production Department.

Current Methods of Evaluating Perforators—API RP 43, Sections 1 and 2 presented in Appendix A, are the primary methods of evaluating both shaped charge and bullet guns. Section 1 covers surface perforating tests at atmospheric temperature and pressure where casing is cemented inside a light steel drum. Cementing material is two parts sand and one part cement.

Section 2 tests provide comparative penetration and flow from the formation through perforations and should be used for all charge and gun selection. When tests are made with a differential pressure into the wellbore, flow into well should be maintained without interruption from the time of perforating until all crushed and compacted material in and around the formation perforation is removed.

Effect of Perforating in Various Fluids

Table 7-1[3] shows the results of Well Flow Index tests of jet and bullets in various fluids. Two types of tests were made with jet perforators in salt water, one with a 200 psi differential pressure into the wellbore, and a second with a 500 psi differential pressure into the formation. Tests of bullet perforators were run with a 500 psi pressure differential into the formation. Test temperature was 180°F. All tests were conducted with the Well Flow Index system illustrated in Figure 7-2. Essentially the same test equipment and procedure is used for tests under Section 2, API RP 43.

Figure 7-3 shows a mud-plugged jet perforation; perforating was performed in 10 lb/gal mud with a pressure differential of 500 psi into formation and cleaned by backflowing. After many hours of backflowing with clean fluids, cleanout of perforation into the formation was only about 1/8-inch. On one test perforated in 16 lb/gal lime-based mud with a pressure differential into the formation, a drawdown pressure into wellbore of 430 psi was required to initiate flow through a single perforation.

Clean perforations, as shown in Figure 7-4, result from perforating in clean oil, clean saltwater, or nitrogen with carrot-free shaped charges with a differential pressure into the wellbore.

Examination of these sandstone targets perforated during these tests also showed that flow patterns and perforation geometry prevent appreciable clean out of most mud-plugged or silt-plugged perforations by swabbing or producing a well perforated with a differential pressure into the formation.

Plugging of perforations may result from killing a well with mud or dirty fluid during well completion, servicing, or workover; however, cleanout can usually be predicted if a well was perforated in clean fluids and if perforating and cleanout of perforation was carried out as a continuous sequence with a differential pressure into the wellbore until all crushed sand and debris is cleaned from the well.

Effect of Formation Strength on Perforator Performance

Simulated downhole tests[3,6] using test apparatus shown in Figure 7-2 showed that perforator penetration varies with formation compressive strength as measured by ASTM C-190 tests of cores. The ASTM test is a crushing test of an unsupported column at atmospheric pressure. The crushing strength or compressive strength of a rock, as shown by this test, may not be representative of downhole conditions, particularly in the case of relatively unconsolidated sandstones. Crushing strength measured on the surface is less than the rock crushing strength downhole in the zone to be perforated. However, with the higher

TABLE 7-1
Comparative Results of Perforating in Various Fluids[3]

Type perforator, well fluid, and differential pressure	Perforation data: Penetration in.	Perforation data: Hole size in.	Average differential to initiate flow, psi	Avg. flow index	Condition of perforation after backflowing
Jet perforator					
10-lb saltwater					
200 psi into wellbore	6½–8	¼–½	0	1.00	Hole clean to total depth.
500 psi into formation for 3 to 10 hours	6½–8	¼–½	0	0.61	Hole clean or partially filled with charge debris and sand.
10-lb caustic-quebracho mud					
500 psi into formation	6½–8	¼–½	30	0.55	Partially or completely filled with mud and charge debris.
16-lb lime-base mud					
500 psi into formation	6½–8	¼–½	100	0.41	Completely filled with mud, sand, and charge debris.
Bullet perforator					
10-lb saltwater					
500 psi into formation	3–3½	½	0	0.61	Cleaned out or partially filled with sand.
10-lb caustic-quebracho mud					
500 psi into formation	3–3½	½	30	0.53	Filled with mud and sand.

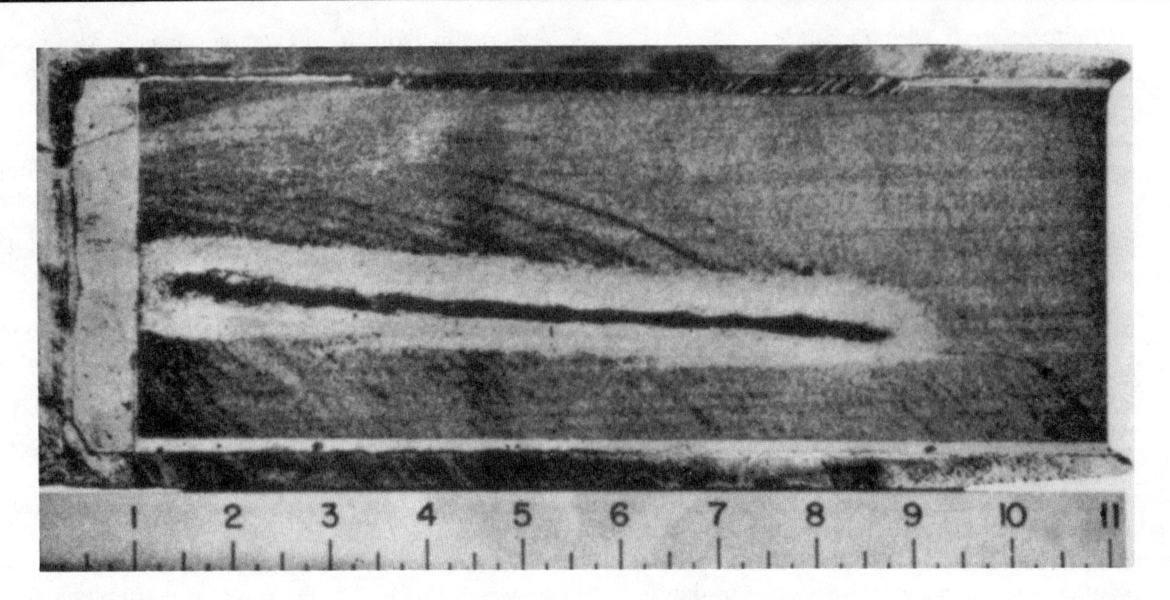

FIG. 7-3—Jet perforation made in mud with a differential pressure into formation.[3] Permission to publish by API Production Department.

strength rocks such as dolomite, crushing strength measured under surface conditions appears to approach downhole crushing strength.

As shown in Figure 7-5, jets penetrate deeper than bullets in hard formation. However, some premium bullet guns may penetrate deeper than some jets in low compressive strength formations, particularly if guns are fired at zero clearance.

Figure 7-6 shows that penetration of jets, bullets, and hydraulic perforators was reduced with increased compressive strength of formation penetrated.

When API RP 43, Section 2, tests were established, the EXXON test system used to obtain data for Figures 7-5 and 7-6 was adopted for the API standard. Unless stated otherwise, all perforating tests for API RP 43, Section 2, are run in Berea sandstone,

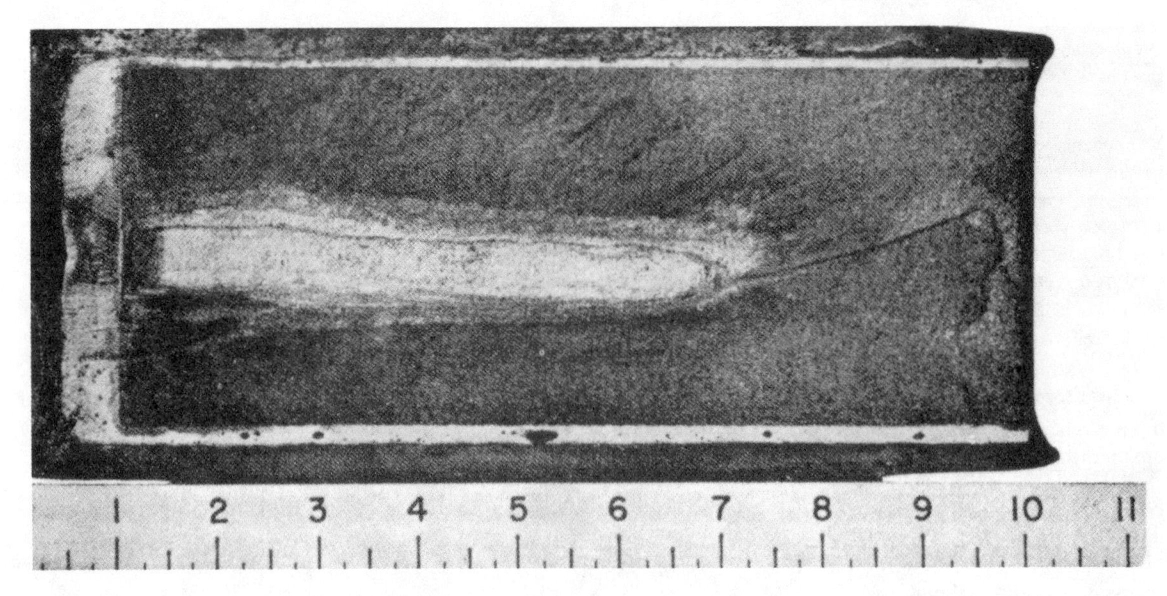

FIG. 7-4—Jet perforation made in saltwater with differential pressure into wellbore.[3] Permission to publish by API Production Department.

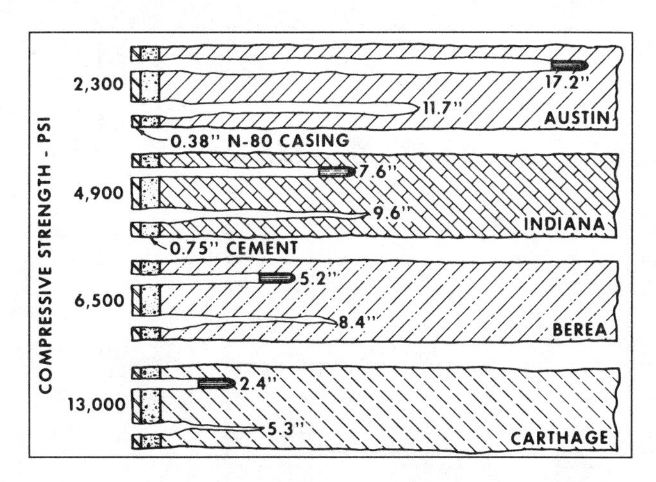

FIG. *7-5—Effect of formation compressive strength on penetrating efficiency of bullet and jet perforators.[6] Permission to publish by API Production Department.*

having a crushing or compressive strength of 6500 psi. Therefore, all penetration test data from API RP 43, Section 2, must be corrected for the difference in compressive strength of Berea and the downhole rock to be perforated.

To illustrate the procedure for estimating penetration of a specific jet gun in a given downhole formation, assume a San Andres dolomite well is to be perforated with a newly designed jet perforator. This gun penetrated 14 in. in a Berea sandstone target employed in standard API RP 43, Section 2 tests. To convert this test data to penetration in the San Andres dolomite, the following steps must be taken:

1. Obtain compressive strengths on San Andres cores from the well in question or from nearby wells, using ASTM C-190 crushing tests. To continue this example calculation, assume the ASTM test showed the dolomite to have a 16,000 psi compressive strength.
2. Next, refer to the 4 in. jet perforator performance envelope of curves shown in Figure 7-6. This envelope of curves for original 4 in. jet guns tested can be used to estimate penetration of any size, type, and design of jet perforator to be employed in various strength formations.
3. Because all API, RP-43, Section 2 tests are run in Berea sandstone with a strength of 6,500 psi, draw a vertical line from top to bottom of the chart in Figure 7-6 at 6,500 psi. All test data from API RP 43, Section 2 tests will fall on this line.
4. Plot the 14 in. penetration from API tests for the proposed jet perforation on this 6500 psi vertical line. Then measure the distance on this 6,500 psi line from the 14 in. penetration down to the top of the envelope of curves for jet guns. In this instance the distance is 4 in.
5. Since the penetration correction is for 16,000 psi rock, take the distance (4 in.) measured on the 6,500 psi line, and plot the 4 in. upward from the top of the jet penetration envelope on the 16,000 psi line. This provides an estimated penetration of about 8 $^3/_8$ in. for the specific 4-in. jet gun in casing cemented inside 16,000 psi dolomite.
6. To obtain penetration into the formation rock, subtract 1 $^1/_8$ in., the thickness of the steel and cement penetrated on the API Berea sandstone tests. For this example, core penetration would be about 7 $^1/_4$ in. into the 16,000 psi dolomite.

For relatively unconsolidated sandstones, it is usually impossible to obtain meaningful compressive strength needed to correct perforator penetration. However, drillability, as obtained from laboratory and field tests, with a sharp rock bit is roughly inversely proportionate to rock compressive strength as measured on ASTM C-190 tests. Based on previous ASTM tests on cores and drillability data from relatively unconsolidated sands, it can be roughly estimated that rock crushing strength is 2,000 psi or less, if the producing formation can be drilled at a rate of 1,000 ft per day or more. On this basis, API RP 43, Section 2, test results can be corrected to 2,000 psi or less rock strength, using the correction chart, Figure 7-6.

Table 7-2 shows permeability, porosity, and compressive strength of representative formations in Texas and Louisiana, U.S.A. Data from Table 7-2 can be used where applicable to estimate penetration using perforator data from Section 2, API RP 43 and curves shown in Figure 7-6.

Figure 7-7 shows semilog plot of rock compressive strength vs. penetration with a hydraulic perforator with and without nitrogen in the hydraulic jet streams.[6]

Downhole Evaluation of Perforations

The thru-tubing flowmeter provides the best method of determining flow rate from each one or two feet of perforations. If the flowmeter is to be used as a basic downhole method of evaluating perforators, it should be ascertained that a good cement-to-formation bond exists in the well prior to perforating.

Another method of evaluating actual perforations and perforation plugging downhole involves running

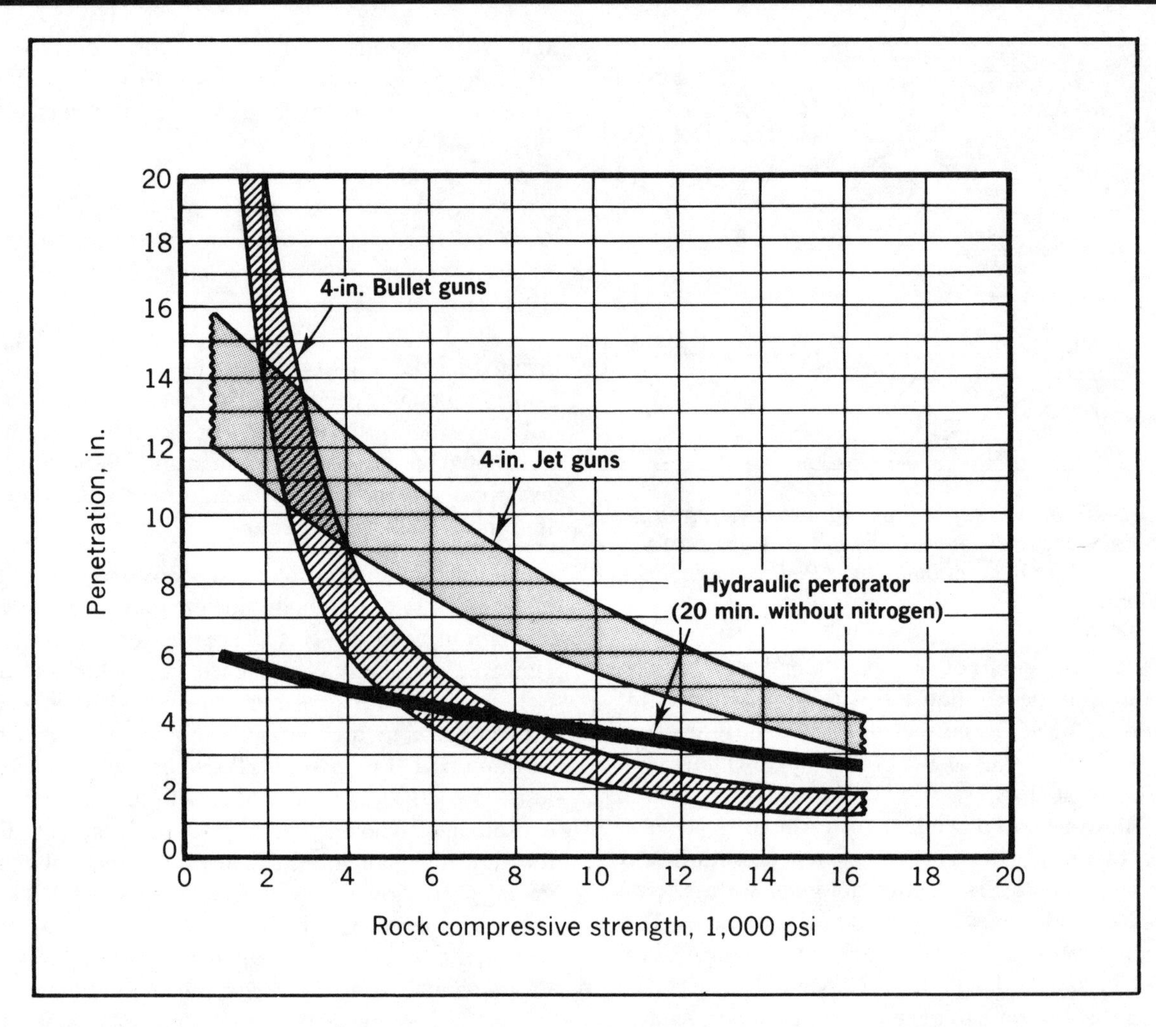

FIG. 7-6—Bullet, jet, and hydraulic perforator performance in formations of various compressive strength.[6] Permission to publish by API Production Department.

a soft rubber impression packer in the well opposite casing perforations. Then the soft rubber packer is hydraulically expanded. If a perforation is open, the soft rubber will extrude into the perforation for about $^1/_4$-in. If any perforation is completely sealed, no extrusion will take place.

FACTORS AFFECTING GUN PERFORATING RESULTS

Perforation Plugging—Plugging of perforations with charge liner residue slugs can be very severe. These slugs may completely plug the perforation in the casing and/or cement. Through the use of sintered or powdered metal charge liners, large residue slugs have been eliminated in most of the premium-priced charges. The slug or charge liner residue is still formed but is carried to the end of the perforation as sand-sized or smaller particles. Surface tests, Section 1, API RP 43, at atmospheric pressure are not reliable to evaluate this type of perforation plugging, because "slugs" are frequently deflected from the perforation when perforating at atmospheric pressure.

If perforations are made in mud or other solids-containing fluids, the perforations are filled with crushed formation rock, mud solids, and charge debris. These plugs are not readily removed by backflowing. Also, crushed and compacted rock around the perforation plugs has essentially zero permeability and further reduces the probability of perforation cleanout. Perforations made in high weight muds, mixed from high density solids such as Barite are very difficult to clean out.

Differential pressure from the formation into the

TABLE 7-2
Physical Characteristics of Some Oil-producing Formations[6]

Formation	Field*	Approx. depth, ft	Avg. perm., md	Average porosity, %	Avg. comp.** strength, psi
Woodbine	Van	2,700	—	—	370
Frio	Tom O'Connor	5,500	1,725	30	245
Frio	Seeligson	5,900	6	18	1,850
Frio	Trull	12,000	50	20	3,620
		12,000	20	18	6,615
Cockfield	Conroe	5,000	—	—	60
Yegua	South Dayton	8,600	—	—	2,400
		8,600	1	13	6,600
Wilcox	Barbston	6,300	500	32	6,000
Paluxy	Merit	10,500	57	9	8,040
		10,500	2	7	13,860
Rob-7	Bayou Sale	13,300	129	23	1,550
		13,300	63	19	2,490
		13,300	12	16	4,940
Grayburg	—	4,000	—	—	12,600
San Andres	—	4,200	—	—	15,800
Wolfcamp	—	8,500	—	—	11,400
Pennsylvanian	—	8,950	—	—	12,400
Devonian	—	10,450	—	—	8,215
Ellenburger	—	12,220	—	—	16,000

*Fields located in the U.S.
**Crushing strength using ASTM C-190 tests.

wellbore, required to initiate flow through each plugged perforation, varies. When a few perforations open up, flow through these perforations and concurrent pressure buildup in the wellbore, make it difficult to create the higher pressure drawdown needed to open additional perforations. This situation is usually most significant in gas wells where pressure buildup due to flow from a few perforations can be very rapid. Under these conditions, a large number of plugged gas well perforations may remain permanently sealed. The result may be failure to drain many specific zones in stratified formations such as sand-shale sequences. When all perforations in one or more zones in a layered multizone well are completely plugged, drill stem tests, production tests, and pressure buildup tests, including skin damage calculations may provide an erroneous evaluation of well damage, potential well productivity, and oil or gas recovery.

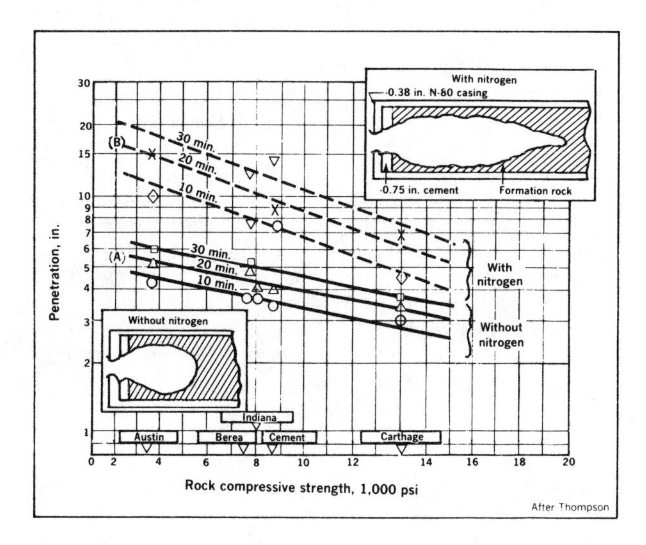

FIG. 7-7—*Effect of rock compressive strength on penetration with hydraulic perforator.[6] Permission to publish by API Production Department.*

Cleanout of Plugged Perforations—For unconsolidated or very permeable sands, the use of perforation wash tools with one foot spacing between packers is probably the best approach. Backsurge tools have been used. However, they tend to break down relatively unconsolidated sands and cause severe caving. If perforations in sandstone wells cannot be cleaned out with perforation washers or backsurge tools, the next approach should usually be to break down or fracture each perforation with clean water or oil using ball sealers. Normally these fractures will heal within about 1/2 hour after frac pressure is removed. Another approach is to reperforate in gas, nitrogen, or in clean oil or water with a pressure differential into the wellbore.

If acid is used as a breakdown or fracture fluid to clean out sandstone perforations, the cement-formation bond above and below perforations is usually permanently ruptured. Acidizing mud-plugged carbonate perforations is usually successful because acid entering a few perforations can usually dissolve enough carbonate rock to open some of the adjacent perforated interval without using breakdown or fracture pressure.

Limestone or dolomite wells are frequently perforated in HCl or Acetic acid with a low differential pressure into formation. Then acid is slowly pumped into the formation. Because of the slow reaction of Acetic acid on limestone, it is usually desirable to leave the Acetic acid in the formation for at least twelve hours after perforating. This procedure usually results in much lower breakdown pressure and pump horsepower requirement if the well is to be hydraulically fractured. Perforating in clean water or clean oil with a pressure differential into the wellbore, however, is quite satisfactory.

If a section of perforated casing is poorly cemented, providing vertical communication behind the pipe between perforations, the resulting condition is similar to a screened open hole. If any flow occurs from the formation, all casing perforations will be cleaned out. However, the perforations in the formation may or may not clean out.

Plugging of perforations during production with paraffin, asphaltenes, or scale is a major problem in many areas of the world. Twenty-four hour solvent soaking will remove paraffin. Seventy-two hour soaking with Zylene or Toluene is required to remove asphaltenes. If perforations are plugged with acid-soluble or acid-insoluble scale, it is often advisable to reperforate and then to treat with acid if required.

Effect of Pressure Differential—When perforating in mud with a pressure differential into the formation, perforations are filled with mud solids, charge debris, and formation particles. Mud plugs are very difficult to remove, often resulting in permanent perforation plugging, reduced well productivity, and reduced oil and gas recovery. These problems may be acute in shale or silt laminated sandstone reservoirs.

When perforating in water or oil with pressure differential into the formation, particles of charge debris, cement, mill scale, rust, clay, or other fines may cause perforation plugging and deep formation damage. Permeabilities of about 250 md or greater will allow clay size particles to be carried deep into formation pores.

For carbonate formations it is often possible to obtain high well productivities and low perforation breakdown pressure when perforating in HCl or Acetic acid with a small differential pressure into the formation.

Perforating with a differential pressure into the wellbore in clean fluids aids in perforation cleanout. This is the preferred method of perforating sandstone wells and many carbonate wells. Recommendations in Table 7-3, are a modification of 1986 field studies by King, Anderson, and Bingham.[24]

Effect of Clean Fluids—If a specific charge or perforator provides adequate hole size and penetration under given well conditions, well productivities in all sandstone formations and many carbonates will be maximized by perforating in clean oil or saltwater with pressure differential into the wellbore during perforating and well clean-up period. If the well is shut in to pull the gun prior to completely cleaning all perforations, many perforations may remain plugged, or become plugged due to settling of solids in the well during the shut-in period.

Effect of Compressive Strength[6,15]—Penetration and hole size made by jet perforators are reduced as compressive strength of casing, cement, and formation rock is increased. Penetration of bullets is severely de-

TABLE 7-3
Recommended Differential Pressure Into Wellbore When Perforating

Reservoir Conditions	*Differential Pressure psi* *Oil Wells*	*Gas Wells*
Weakly Consolidated Sandstone	300–500	300–500
Consolidated Sandstone		
Formation permeability		
Greater than 100 md	500	1000
From 100 down to 10 md	500–1000	2000
Less than 10 md	1000–2000	2000
Carbonates		
Formation permeability		
250 md or above	500	500
100 to 250 md	750	1000
Below 100 md	1000	2000
Below 10 md	2000	2000

creased with increases in strength of the casing, cement, and formation.

Perforation Density—Shot density usually depends on required production rate, which is also dependent on diameter of perforation, formation permeability, and length of perforated interval. For high volume oil or gas wells, shot density should permit the desired flow rate with reasonable pressure drawdown. Four 0.5-in. or larger holes per foot are usually adequate for many wells. Two large diameter holes per foot may be satisfactory for many low-volume wells. In wells to be fractured, four 0.75 to 0.8 in. holes per foot may be the optimum to decrease pressure drop through perforations during fracturing and thereby reduce hydraulic horsepower required for the frac job, including fracture breakdown. For wells requiring sand consolidation jobs, four shots/ft of large-diameter deep perforations is usually preferred. For gravel packed wells, four to twelve shots/ft of very large-diameter perforations, 0.75 to 0.8 in. should be standard procedure. For gravel packing, large-diameter holes penetrating through the casing, cement, and a few inches into the sandstone is usually adequate.

Perforating with four or more shots/ft in low-strength, small-diameter casing, with an exposed-charge perforator such as a strip or capsule gun, may cause severe casing splitting. Also, cement behind the pipe may be badly cracked for several feet above and below the perforated interval. Under these conditions it may be difficult to squeeze off undesired water or gas migration through the cracked cement. High-strength casing collars may be split by multiple perforations in a collar.

Cost—Perforating prices vary from area to area; however, reduced perforation density usually results in lower job costs. Selectively-fired guns can often save appreciable rig time where productive zones are separated by long nonproductive intervals.

The use of through-tubing guns can frequently save rig time if tubing is open-ended and set above all zones to be perforated. With careful control of cement setting time on new wells, tubing may be run within a few hours after the primary cement is in place. Then through-tubing perforating may be carried out without a rig on the well, often resulting in little or no rig time being charged to well completion.

Pressure and Temperature Limitations—Pressure and temperature ratings are available on all perforator charges. Bottom hole pressure may impose limitations on some exposed charge guns. However, there are few wells being perforated where pressure is a problem with most conventional hollow carrier casing-type guns.

Figure 7-8 shows self-detonation time-temperature curves for cyclonite (RDX) powder used in conventional charges. High temperature explosives are only used in very hot wells.

As a general rule, high temperature charges should not be employed even in wells in the 300-340°F range. This recommendation is based on the following: (1) most high temperature charges provide less penetration; (2) high temperature powder is less sensitive, resulting in increased misfires; (3) high temperature charges are more expensive; and (4) there is much less choice in charge selection.

When operating near the upper temperature limit of low temperature charges, these approaches may be used:

1. Wells can be circulated with low temperature fluids to lower the bottom-hole and surrounding formation temperature. This is especially applicable for through-tubing guns which can be run to bottom soon after fluid circulation has been stopped.

2. If there is some question as to whether the temperature limit of the gun will be reached prior to firing the gun, high temperature detonators may be employed in guns equipped with conventional low temperature cyclonite charges. This approach will pre-

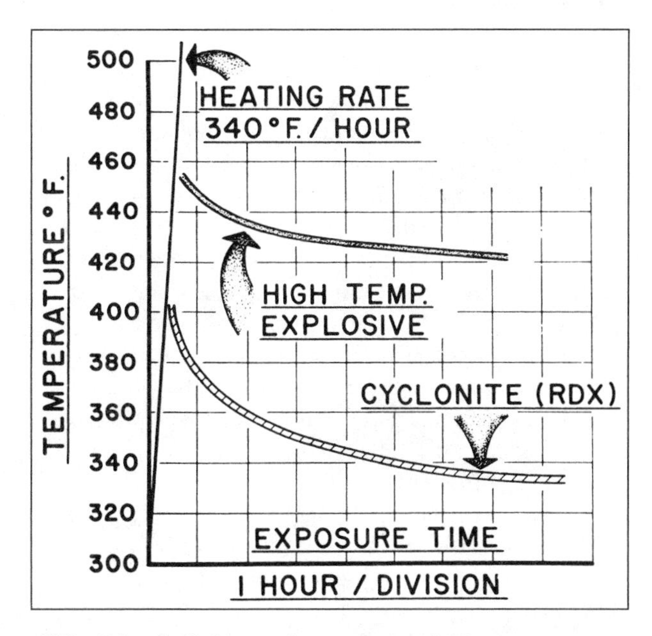

FIG. *7-8—Self-detonation curves of 26-gm encapsulated charges.*[17] *Permission to publish by The Society of Petroleum Engineers.*

TABLE 7-4
Pressure and Temperature Ratings on Selected Perforations and Changes

Gun types	Powder designation	OD in	Pressure ratings psi	Temp. ratings °F
Low temperature powder				
Bullet	...	3 1/4, 3 3/8	15,000	250
Jet (Thru-tubing capsule)	RDX	1 1/16	10,000, 15,000	300
Jet (Thru-tubing Semi-expendable)	RDX	2	20,000	340
Jet (Thru-tubing, hollow carrier)	RDX	2 1/8	15,000, 20,000	340, 325
Jet (Std hollow carrier)	RDX	3 3/8	20,000	325
	RDX	4	20,000	330, 340
	RDX	5	15,000, 20,000	340
	RDX	6	15,000, 20,000	340, 325
Jet (hollow carrier)*	RDX	7 1/4	15,000	340
High temperature powder				
Jet (Thru-tubing, hollow carrier)	PSF	1 11/16	25,000	470
Jet (Thru-tubing, hollow carrier)	HMX	2 1/8	15,000	400
Jet (Casing, hollow carrier)	HMX, PSF	3 3/8	20,000, 25,000	400, 470
	HNS, PYX			500, 550
	HMX, PSF	4	20,000, 25,000	400, 470
	HMX, PSF	4	15,000	400, 470
Jet (Casing, hollow carrier)	HMX	6	15,000	400

*Tubing Conveyed Perforator

vent accidental perforating in the wrong interval due to high downhole temperature. Jet charges exposed to excessive temperature will "fuse off" or burn with no resulting casing perforation unless fired with the perforating gun detonator.

For very high temperature wells there may be no alternative but to run the entire high temperature perforating package. This includes the detonator, the primacord, the booster charge, and the main powder charge. The detonator is the key to the system. Unless the detonator is fired, the shaped charge will not function as intended to perforate. Table 7-4 presents temperature and pressure ratings on selected perforators.

Well Control—Normal pressure onshore oil wells can be perforated with oil or water in the hole with thru-tubing guns using conventional wellhead control and the labyrinth type of "flow-tube" or "blow-by" packing gland. It is always good practice, however, to use a wireline BOP.

A grease seal lubricator should be used on all gas wells, all offshore wells, and on oil wells if greater than 1,000 psi surface pressure is anticipated on the lubricator after perforating. Abnormally high pressure wells can be safely perforated with salt water in the hole with through-tubing guns using high pressure wellhead control equipment. Lubricators for use during perforating are available with working pressure ratings up to 25,000 psi.

Casing and Cement Damage[7,11]—Hollow carrier jet guns absorb unused energy, about 93% of total available energy, from jet charges. This prevents casing splitting and virtually eliminates cement cracking. Little casing damage occurs with conventional bullet guns. Shooting with zero degree charge phasing and zero gun clearance tends to eliminate burrs inside the casing.

Jet guns with exposed charges, such as strip or capsule-type guns, can cause deformation, splitting, and rupture of the casing, and appreciable cracking of cement. Explosive weight, degree of casing support with cement, perforation density, casing diameter, and casing "mass-strength" are factors in casing splitting with exposed jet charges. Casing "mass-strength" has been defined as the product of wt/ft and yield strength. Tests shown in Figure 7-9 indicate that unsupported and semi-supported 5 1/2" J-55 casing can be safely perforated with exposed-charge guns using 20 grams or less of cyclonite powder if the casing is in good condition.

Need for Control of Gun Clearance—Excessive gun clearance with any jet gun, and especially with some through-tubing perforators as illustrated in Figure 7-10, can result in inadequate penetration, inadequate

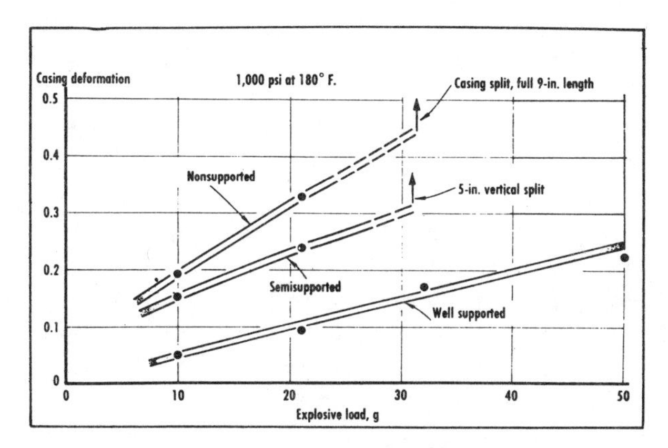

FIG. *7-9—Casing damaged by large explosive loads.*[7] *Permission to publish by API Production Department.*

hole size, and irregularly shaped or "keyed" holes. Bullet guns should usually be fired at zero or 0.5 in. gun clearance to avoid appreciable loss of penetration. There is usually little problem with most large diameter conventional hollow carrier jet guns, except when perforating in 9 $^5/_8$-in. od or larger casing.

Clearance control can be achieved through spring-type deflectors, magnets, and other methods. If magnets are to be used, it is good operating practice to place magnets in a strong magnetic field for a period of time immediately prior to running the gun to insure adequate magnetic strength. If zero degree charge phasing is desired, two magnets, one located at the top and at the bottom of a thru-tubing gun, are needed. Depending on charge and gun design, 0 or $^1/_2$ in.

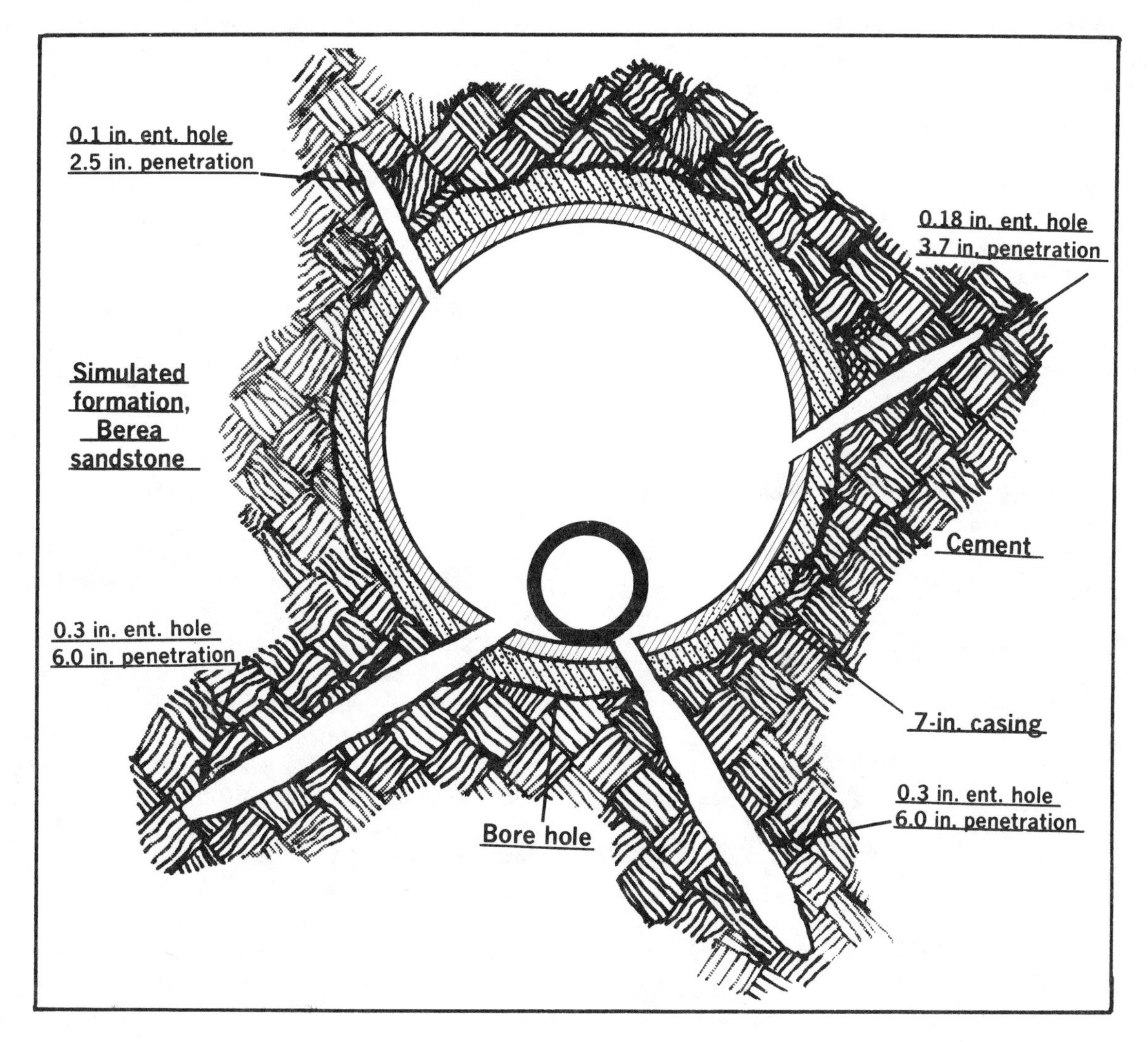

FIG. *7-10—Typical performance variations with a 90° phased $1^{11}/_{16}$-in. perforator inside 7-in. casing.*[17] *Permission to publish by The Society of Petroleum Engineers.*

clearance for jet guns usually provides maximum penetration and hole size. In some hollow carrier casing-type guns, there is significant reduction in hole size as gun clearance is increased from 0 to 2 in. or more. However, in others, centralization of the gun may produce a larger and more consistent hole size. With casing gun clearance greater than about 2 in., it is usually desirable to decentralize, or to use zero degree charge phasing, and to orient the direction of fire from the gun into the casing toward the near side.

Centralizing is not a valid approach for most through-tubing jet guns which are usually designed to be shot with zero gun clearance. Swing-Jets may alleviate the clearance problem for through-tubing guns. However, debris and mechanical problems can be quite severe with swing jets.

Depth Measurements—The accepted method of ensuring accurate perforation depth control is to run a collar locator with the perforator, and to measure from casing collars which have been previously located with respect to the formations using radioactive logs. Radioactive tags can be inserted inside selected shaped charges to assist in locating exact depth of perforations with a Gamma ray after perforating. Collar logs may show location of old or new perforations made with exposed charges. In this case the collar log records permanent swelling or bulging of the casing, due to detonation of exposed charges inside the casing.

Oriented Perforating—Oriented perforating is required for through- tubing perforating in multiple completions where adjacent tubing strings are present, as illustrated in Figure 7-11. Mechanical, radioactive, and electromagnetic[13] gun orientation devices are available. When using oriented through-tubing

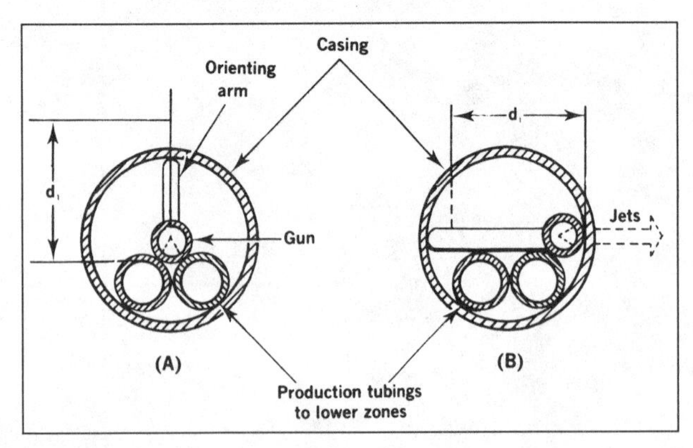

FIG. 7-11—*Oriented perforating of zones between packers. In A, gun cannot be fired, as orienting arm does not extend beyond d-1 arming distance. In B, firing position: gun armed.*

perforators in multiple completions, as shown in Figure 7-11 a thin wall hollow carrier gun should always be used. Capsule-type guns or other exposed charge perforators can cause collapse of an adjacent tubing string.

To avoid perforating adjacent strings of casing cemented in the same borehole, the most prevalent practice is to run the radioactive source and detector on the same electric cable as the perforating gun. The gun is then rotated to avoid perforating adjacent strings of casing. If there is doubt in interpretation, a radioactive pill is run in adjacent casing to aid in location of adjacent casing.

Penetration vs. Hole Size—In the design of any shaped charge, greater penetration can be achieved by sacrificing hole size. Because maximum penetration appears to be important on the basis of theoretical flow calculation,[1] the petroleum industry has often requested and received greater penetration at the sacrifice of hole size. When perforating thick-walled high-strength casing, high-strength cement, and dense high-strength formations, maximum penetration is probably required even if hole size is reduced to 0.4 in.

However, for most situations, because of the difficulties in removing mud, shaped charge debris, sand, carbonate particles, formation fines, asphaltenes and paraffin from a long small-diameter perforation[2,3], the perforation should normally have an entrance hole diameter of 0.5 to 0.8 in. with a uniform smooth bore to maximum depth.

For specific perforating situations the following points should be considered:

1. For gravel packing, perforations should be large and round with a minimum diameter of 0.75 in. and a shot density of 4 to 12 per ft. When perforations are to be cleaned with a perforation washer prior to gravel packing, penetration into the formation of 4 to 6 in. should be adequate.
2. In plastic sand consolidation, perforations should have a minimum diameter of 0.5 in. along with maximum obtainable depth. Perforate with a differential pressure into wellbore and/or clean perforations with a perforation washer, if needed.
3. When ball sealers are to be used as a diverting device in acid fracturing or hydraulic fracturing, entrance holes should be large, round and smooth. A hole size of about 0.75 in. is desired where practical.
4. When perforating carbonates in acid, hole size should be about 0.75 in. in diameter. Four to 6 in. of

penetration into formation is probably adequate because the acid will dissolve carbonates around and ahead of the perforation.

5. If paraffin or asphalt plugging is a problem, in an oil field, large smooth-bore perforations are desired to facilitate cleanout.

6. If scale plugging is anticipated, hole size should be as large as practical, probably 0.75 in. or larger to reduce plugging tendencies. To aid in perforation cleanout, hole should be large, uniform, and smooth to maximum depth.

The length of time required to plug a perforation with scale in producing, shut-in, or injection wells appears to be related to hole diameter, smoothness of perforation tunnel in the formation, the concentration of scale crystals in the produced or injection water in contact with the perforations, and velocity of water moving through the perforations. Increasing pressure drop through small diameter perforations increases scaling of $CaCO_3$, $BaSO_4$ and $CaSO_4$. Temperature drop through small diameter perforations in gas wells will increase scaling of $BaSO_4$.

Limitations in the Use of Exposed-Charge Jet Perforators

Most exposed-charge guns are more difficult to decentralize or centralize for proper gun clearance than are hollow carrier through-tubing or conventional perforators.

When employing exposed charge guns, there is usually no way to determine whether all exposed charges produced perforations. With top-fired exposed charge guns, charges may drop off the gun after one or more charges have been detonated. To minimize this problem, exposed charge perforators should be bottom-fired.

With jet perforating, about 7% of the explosive energy from the charge is utilized to produce the perforation. Therefore, with an exposed charge gun, 93% of the total energy resulting from charge detonation must be absorbed by the casing and the well fluid system; thus, the chances of splitting the casing are great, particularly if there is a channel in the cement between the casing and formation. With the hollow carrier gun, this excess energy is absorbed by the steel gun carrier.

Charges in exposed charge guns are in direct contact with well fluids and well pressure. Many instances of leaky or cracked charge cases have been reported when exposed charge guns were used in water, or gas-filled casing. Penetration has been reduced as much as 50% when exposed charges, having aluminum- or ceramic-cases, are run in gas-filled wells. For this reason, exposed charge guns should not be run in gas filled wells. When perforating in HCl acid, exposed aluminum charges should not be employed because of the solubility of aluminum in HCl acid. Acetic acid does not dissolve aluminum.

PERFORATING IN A CLEAN FLUID WITH A DIFFERENTIAL PRESSURE INTO THE WELLBORE

For most formations, perforating in a clean solids-free fluid with a differential pressure into wellbore results in the highest well productivity. The ideal method would be to perforate with a through-tubing gun with zero gun clearance in clean fluids with a differential pressure into the wellbore. In many cases, however, because of diameter limitations of tubing, landing nipples, and "No-Go" nipples, through-tubing guns frequently do not provide adequate hole size and penetration. Also, some decentralizing devices required for zero gun clearance may not be effective.

If available through-tubing guns have insufficient power for a particular perforating job, there are several other possible means of optimizing perforating results.

1. One approach requires the use of a large diameter lubricator on the wellhead and employs the following procedures:

Perforate with a full-sized casing gun in clean fluid with a differential pressure into wellbore and then retrieve the gun under well pressure. For relatively high permeability, normal-pressure wells, a differential pressure of 500 psi into the wellbore during perforating is usually adequate. Required differential is usually obtained by partially or completely filling the casing with salt water, fresh water, or oil. Gas or nitrogen may be used to provide desired pressure in the casing in place of water or oil. After the perforating gun is removed from the well, a permanent type packer body is then run on an electric cable under well pressure and set at the desired depth. After setting the packer, equipped with a positive seal on bottom, pressure is bled off the casing above the packer. One of several available systems may then be employed to complete the well, depending on packer design.

With one type of packer having a retrievable inside blanking plug on the lower end of the packer body, tubing is run and set with tubing seals positioned inside the packer body. After the christmas tree is in place, the blanking plug in the lower part of the packer unit is retrieved through the tubing with a wireline, allowing full well-pressure communication between the wellbore and the tubing.

With another type of packer, the lower part of the packer body is equipped with a frangible seal. After the tubing has been run and tubing seals are positioned inside the packer body, continued weight on the tubing is employed to push out the frangible disc, thus opening the tubing to wellbore pressure. After the christmas tree is installed, a blanking plug inside the tubing landing nipple is pulled with a wireline.

This system is limited by the ability of the stuffing box on the electric cable lubricator to seal around the cable. Therefore, completions of this type have usually been limited to cases where not more than about 3,000 psi pressure is anticipated on the wellhead after perforating or where formation permeability is low enough to allow the well pressure to be bled down after perforating.

2. A second approach involves relatively low pressure wells, where the reservoir pressure gradient is less than about 0.465 psi/ft of depth. These wells, which are partially filled with oil, fresh water, salt water, KCl, or $CaCl_2$, are swabbed down, gas-lifted, or circulated with nitrogen using reelable tubing to provide the desired pressure differential into the wellbore. Then the well is perforated with a conventional casing perforator and the gun is retrieved. If the well is an oil producer, entry of oil into the wellbore will gradually increase the pressure head on the formation. If the well pressure cannot be bled down prior to running tubing, it may be necessary to pump in several barrels of clean NaCl, $CaCl_2$ or KCl water or oil to kill the well. Perforations should be cleaned thoroughly prior to running the tubing. If all loose sand and debris has been cleaned from perforations, killing a well with clean fluid should do little permanent damage.

After perforating a potential oil or injection well with an adequate differential pressure into the wellbore in clean fluids, the perforation debris and surrounding crushed rock will be removed if the differential pressure is maintained continuously into the wellbore until an average of 0.5 to 2.0 barrels of fluid are flowed, pumped or swabbed per perforation. Gas wells also require continuous differential pressure into the wellbore after perforating and until the well is cleaned to remove perforation debris and surrounding crushed rock. If a well is shut in soon after perforating to pull the gun, most perforations will remain plugged unless communication exists between perforations behind the casing because of channels in the cement. If perforation debris and surrounding crushed rock are cleaned from all perforations, killing a well with clean water or oil prior to pulling the gun and running tubing, should not cause appreciable permanent perforation plugging. The validity of this procedure can be verified on any well with a through-tubing flowmeter which will show volume of flow from each perforated interval. This procedure is applicable to most lower pressure wells in the U.S., Canada, and many other parts of the world.

3. A third approach is to perforate casing with a conventional casing gun in clean fluid (water, oil, or nitrogen) with a pressure differential into wellbore, to pull the gun under pressure with the well flowing at a low rate, and to produce the well without tubing. Producing oil and gas wells without tubing has been the usual procedure for a number of years on many wells in the U.S., and in the Middle East. This technique has also been a part of the tubingless completion extensively used in the U.S. for more than 30 years.

4. A fourth approach is to perforate with a conventional casing gun, with a pressure differential into wellbore to allow perforations to be totally cleaned. Then pull the gun under full well pressure and snub tubing into the well.

5. In some high volume wells equipped with 5 $^1/_2$ to 7-in. diameter tubing, a conventional hollow carrier gun may be run through a large diameter lubricator and through the tubing instead of employing smaller diameter through-tubing type guns. These wells can be perforated with a differential pressure into the wellbore in clean fluids.

6. A sixth method of perforating employs a tubing conveyed perforator (TCP) which is shot in clean fluid with a differential pressure into the wellbore. This system was developed and patented in the early 1950's by Exxon as a tubing-conveyed gun called a "gun-screen," because the gun assembly also included a screen. This system was developed to overcome perforation and formation damage problems experienced when perforating with conventional casing perforators in mud or other fluids with a differential pressure into the formation. Other perforating methods, including the through-tubing guns, were developed to overcome these mud-damage problems; however, some perforating problems remained, especially those due to in-

adequate power of many through-tubing guns.

Although most of the potential advantages of perforating with a differential pressure into the wellbore have been available for more than 35 years, recent publicity and marketing of tubing conveyed perforating (TCP) has made the petroleum industry more aware of the benefits of perforating in a clean fluid with a differential pressure into the wellbore.

The tubing-conveyed perforating (TCP) involves running a standard casing gun on tubing or drill pipe below a tension-set or compression-set packer. A sliding sleeve mandrel is installed in the tubing between the packer and the perforating gun and a shear disk is located in tubing at a predetermined point above the packer. Prior to setting the packer, the gun must be accurately located in relation to the casing interval to be perforated. To aid in this operation, a tubing sub equipped with a radioactive tag in the bottom collar is often run one joint above the packer. When the gun has been run to point near the correct perforating depth, the tubing is hung from the slips. Then, a Gamma-ray or neutron logging tool is run to locate the radioactive-tagged collar in the tubing and to tie in this depth with the formation zone to be perforated. The tubing is spaced to locate the gun correctly opposite the desired interval to be perforated. As part of this spacing-out operation, the packer is set and the flow ports are opened in the sliding sleeve mandrel below the packer.

In preparation for firing the gun, the tubing is partially filled with fluid to provide a predetermined fluid head above the shear disk, the location of the disk and the fluid head above it being the major factors in controlling the pressure head on the formation at the time of perforating.

With this system, the gun is fired in one of two ways: (1) by dropping a detonating bar down the tubing until it strikes the firing pin on the gun, or (2) by applying pressure to a detonating piston assembly on top of the gun.

If the detonating bar system is employed, the bar is dropped in the tubing or drill pipe and falls freely to the shear disk, breaks the disk, and then falls to the top of the gun where it strikes the detonator and fires the gun. Under normal situations, with a low pressure fluid head in the well tubing and a differential pressure into the wellbore, the well begins to flow immediately.

With pressure-controlled firing, the gun is fired by building up pressure on the firing head at the top of the perforator.

The gun may either be equipped to remain in place on the tubing, the tubing and gun may be pulled from the hole, or the gun can be released so that it can drop to the bottom of the hole. The release mechanism on the gun usually involves jarring in an upward direction on the latching mechanism.

A COMPARISON BETWEEN TUBING CONVEYED PERFORATORS (TCP) AND OTHER PERFORATING METHODS IN WELLS EQUIPPED WITH PACKERS

I. Comparison with Through-Tubing Perforating
   A. Advantages of the Tubing Conveyed Perforating (TCP) over Through-Tubing Perforating
      1. The larger diameter tubing conveyed gun can provide deeper penetration and larger hole size.
      2. Through-tubing gun problems related to getting through landing nipples in tubing are eliminated.
      3. In long zone completions, a much longer (600 ft or more) tubing conveyed gun can be run and fired with a differential pressure into the wellbore than with through-tubing guns (limited to about 40 ft).
   B. Disadvantages of Tubing Conveyed Perforator(TCP) when compared to Through-Tubing Gun
      1. Normally, perforating with the Tubing Conveyed gun is more expensive. Also a Gamma-ray neutron log must be run to locate the gun at the correct depth to perforate desired zones.
      2. The large diameter tubing conveyed gun is either left in place, dropped to bottom, or pulled from the well.The necessity of leaving the gun in place or dropping it to bottom are both undesirable options, especially if the well should become sanded-up, scaled-up, or filled with wax or other debris.
      3. Because mud or other debris accumulation on the mechanical detonator of a Tubing-Conveyed gun may prevent the detonating bar from reaching the firing cap, the gun is usually run with a shear disk in the tubing above the gun. This minimizes the potential for mud solids or other debris from settling out on the gun detonator and tends to maximize pressure differential into the wellbore at the time of perforating. If

a high permeability relatively unconsolidated sandstone well is perforated with a high differential pressure, greater than about 500 psi, into the wellbore, fines in the formation may surge toward the wellbore jamming and blocking formation pores. Also, the high differential pressure may cause the formation to collapse.

If desired, a column of oil, water or nitrogen may be placed above the shear disk to reduce the differential pressure into the wellbore. An appreciable fluid column below the shear disk will increase the probability that silt, mill scale, rust, or other debris will fall on top of the gun's mechanical detonator, and prevent the gun from being mechanically fired. However, many TCP's are equipped with an alternate hydraulically-controlled firing system.

II. Comparison of Tubing-Conveyed perforating and conventional casing perforating where a Wireline-Set Packer is run in the well through a large-diameter lubricator at the surface under full well pressure after perforating.

A. Probable advantages of the Tubing-Conveyed gun (TCP) over Conventional Perforating.

1. TCP may be preferred when very high pressures, probably in excess of 3,000 psi, are anticipated at the wellhead soon after perforating with a differential pressure into the wellbore.
2. A Tubing-Conveyed gun longer than a conventional perforator can be run, thus allowing all zones to be perforated at the same time with a differential pressure into the wellbore in clean fluid.
3. The Tubing-Conveyed gun is available in sizes up to 7 $^1/_4$ in., perforating as many as 12 or more holes per foot.

B. Disadvantages of Tubing-Conveyed gun when compared with conventional casing perforating

1. Perforating costs are higher. (Overall completion costs may be more or less with the Tubing-Conveyed gun than with conventional perforating, depending on the entire completion system.)
2. A Gamma-ray or Neutron log must be run inside tubing to insure perforating at correct depth with the Tubing-Conveyed gun.
3. Perforator is left in the hole unless the tubing is pulled to retrieve the gun. (The gun is often pulled before all perforations are cleaned to save rig time, thereby, nullifying part of the benefits of TCP.)
4. When perforating high permeability sandstone with a high differential pressure into the wellbore, the result may be reduced well productivity due to bridging of fines in formation pores or collapse of sand and shale behind the casing.

III. Comparative Summary of Gun Application

To take advantage of the benefits of conventional jet perforating, where packer completions are planned, more extensive availability of large diameter lubricators is needed so that wireline-set packers and large diameter conventional perforators can be run routinely through the lubricator. For most relatively low pressure wells, perforating with a conventional casing perforator with a differential into the wellbore should be the preferred method rather than the Tubing-Conveyed perforator (TCP). Through-tubing perforating with a pressure differential into the wellbore remains a very safe and effective system of perforating if hole size, penetration, and flow through perforations is adequate to provide anticipated producing rates.

SUMMARY OF OPTIMUM PERFORATING PRACTICES

1. Select a perforator on the basis of test data from Section 2, API RP 43, Fourth Edition, August, 1985. (API RP 43, Section 2 test results should be corrected for compressive strength of actual formation to be perforated).The only value of surface tests made under Section 1, API RP 43 is entrance hole determination; penetration of cement is misleading and of no practical value. Also this test provides no basis for estimating plugged perforations or comparing anticipated flow resulting from competitive perforators.
2. Perforating Sandstone Wells
   a. Perforate all wells in clean, non-damaging fluids, normally NaCl, KCl, oil, or Nitrogen with a differential pressure into wellbore.
   b. Guns and completion system should be designed to match given well conditions.
   c. Essentially all sandstone wells should be perforated with a standard casing gun, a through-tubing gun, or a tubing-conveyed gun with differential pressure into wellbore.
   d. Sandstone wells should not be perforated in acid because of severe emulsion problems and other formation damage.
   e. Selection of perforators on the basis of pen-

etration alone in cement and/or sandstone has led to the development of many poor quality deep-penetrating charges. Flow through the formation and perforation tunnel, including ability to readily clean out, must be the basic criteria for perforator selection with penetration being the controlling factor only in very high strength formations.

3. Perforating Carbonate Wells--Alternate systems may include:

a. Perforating in clean fluid, water or oil,with a pressure differential into wellbore.

b. In limestone, chalk, or dolomite wells, it may be desirable to perforate in HCl or Acetic acid with a 200 to 500 psi differential pressure into the formation. If a carbonate well is perforated in Acetic acid, this acid should remain under some differential pressure into the formation for 8 to 12 hours to allow time for dissolution of carbonates with HAc acid. Perforating in acid is especially beneficial if the well is to be fractured.

4. Perforating in oil, water, or acid below a higher weight mud column is not satisfactory and is not recommended.

5. When perforating with a differential pressure into the formation in mud or other fluids containing fines, it should be recognized that:

—It is virtually impossible to remove mud or silt plugs from all perforations by backflowing or swabbing.
—Mud or silt plugs are not readily removed from perforations with acid or other chemicals unless each perforation is "broken down" or fractured with ball sealers.
—Backsurge tools and perforation wash tools have been effective in removing plugs from some perforations in sandstone wells. However, the use of backsurge tools should be avoided in any well to be gravel packed.

6. Drilling mud and completion fluid containing appreciable fines should not be allowed to enter perforations throughout the life of the well. Water containing fines or dirty oil may be very damaging due to perforation plugging and plugging of formation pores with solids for a great distance from the wellbore. After drilling the producing formation with high filtration water-base muds, an 18-foot lateral log is useful to determine the depth of mud filtrate invasion, which has been found to be greater than 18-feet in some high permeability sandstones.

7. Mud, clay, or silt-plugged perforations contribute to the following problems:

—Well productivity can be appreciably reduced.
—Depending on type of formation, reservoir drive and completion practices, oil or gas recovery can be appreciably reduced.
—Efficiency of waterflooding or improved recovery methods can be greatly reduced.
—Exploratory wells may be abandoned as a result of erroneously indicated poor well productivities during drill stem tests or production tests. Skin damage tests cannot indicate the magnitude of damage if one or more zones in a laminated formation is 100% plugged.
—Effectiveness of squeeze cementing, gravel packing, and sand consolidation can be appreciably lowered. Well sanding problems often result from creating high rates of flow through a few perforations when most perforations are plugged.
—Blast joints in dual-completed or triple-completed gas wells usually cut out because most perforations are plugged. This is very apparent when perforations have been made in mud or fluids containing appreciable undissolved solids.
—Screens will cut out very quickly in oil or gas wells flowing at relatively high rates if most casing perforations are plugged.
—The probability of gas or water coning or fingering is increased if a high percentage of perforations are plugged.

REFERENCES

1. McDowell, J. M., and Muskat, M,: "The Effect on Well Productivity of Formation Penetration Beyond Perforated Casing," Trans., AIME (1950).

2. Allen T. O., and Atterbury, J. H.: "Effectiveness of Gun Perforating," Trans., AIME (1954) Vol. 201.

3. Allen, T. O., and Worzel, H. C.: "Productivity Method of Evaluating Gun Perforating," Drill. and Prod. Practice, API (1956).

4. Delacour, J.; Lebourg, M. P.; and Bell, W. T.: "A New Approach Toward Elimination of Slug in Shaped Charge Perforating," JPT (March 1958) p. 15.

5. Lebourg, M. P., and Bell, W. T.: "Perforating of Multiple Tubingless Completions," JPT (May 1960) p. 88.

6. Thompson, G. D.: "Effects of Formation Compressive Strength on Perforator Performance," Drill. and Prod. Practice, API (1962) 225, pp. 191-197.

7. Bell, W. T., and Shore, J. B.: "Preliminary Studies of Casing Damage From Gun Perforators," API Paper 906-8-He, Fort Worth (March 1963).

8. Harris, M. H.: "The Effect of Perforating on Well Productivity," Trans., AIME (1966) Vol. 237.

9. Carter, L. Gregory; Slagle, Knox A.; and Smith, Dwight K.: "Resilient Cement Decreases Perforating Damage," API Div. of Prod. meeting, Paper No. 851-42-E (April 1968).

10. Bell, W. T.: "Recent Developments in Perforating Techniques," Seventh World Petroleum Congress, Mexico City (1967).

11. Godfrey, W. K., and Methven, N. E.: "Casing Damage Caused By Jet Perforating," SPE 3043 (Oct. 1970).

12. "Standard Procedure for Evaluation of Well Perforators," API RP 43, Fourth Edition (Aug. 1985).

13. Stroud, S. G., and DeGough, K. G.: "An Electromagnetic Method of Orienting a Gun Perforator in Multiple Tubingless Completions," SPE 3446 (Oct. 1971).

14. Bell, W. T.; Brieger, E. F.; and Harrigan, Jr.: "Laboratory Flow Characteristics of Gun Perforators," JPT (Sept. 1972) p. 1095.

15. Weeks, S. G.: "Formation Damage of Limited Perforating Penetration? Test-Well Shooting May Give a Clue," JPT (Sept. 1974) p. 979.

16. Keese, J. A., and Oden, A. L.: "A Comparison of Jet Perforating Services, Kern River Field," SPE 5690 (Jan. 1976).

17. Bell, W. T., and Auberlinder, G.A.: "Perforating High Temperature Wells," JPT (March 1961) p. 211.

18. Wade, R. T.; Pohoriles, E. M.; and Bell, W. T.: "Field Tests Indicate New Perforating Devices Improve Efficiency in Casing Completion Operations," JPT (Oct. 1962) p. 1069.

19. Saucier, R. J., and Lands, J. F.: "A Laboratory Study of Perforations in Stressed Formation Rocks," SPE 6758 (Oct. 1977).

20. Cole, Ed. A., Jr.: "A Complete Technology for Overcoming Formation Damage," SPE 7009 (Feb. 1978).

21. Blacker, Laura, K.: "An Analysis of Rate-Sensitive Skin in Oil Wells," SPE 11187 (Sept. 1982).

22. Diver, C. J.; Hart, J. W.; and Graham, G. A.: "Performance of the Hemlock Reservoir-McArthur River Field," SPE 5530 (Sept. 1975).

23. Locke, Stanley: "An Advance Method in Predicting the Productivity Ratio of a Perforated Well," SPE 8804 (Jan. 1980).

24. King, G. E.; Anderson, A. R.; and Bingham, M. D.: "A Field Study of Underbalance Pressures Necessary to Obtain Clean Perforations Using Tubing-Conveyed Perforating," JPT (June 1986) p. 662.

25. Carter, L. Gregory; Slagle, Knox A.; and Smith, Dwight K.: "Resilient Cement Decreases Perforating Damage," API Div. of Prod. meeting, Paper No. 851-42-E (April 1968).

26. Krueger, Roland F.: "An Overview of Formation Damage and Well Productivity in Oilfield Operations," JPT (Feb. 1986) p. 131.

Appendix A

API Tests for Evaluation of Perforators

Section 1, API RP 43, Fourth Edition, Aug. 1985[12], is a revision of surface tests system developed in the early 1940's. These surface tests were designed to evaluate penetration of multi-shot bullet and jet perforators at surface conditions in a concrete test target.

To prepare the target, casing is cemented inside a steel form using one part, or 94 lb (1 sack), of Class A cement, and two parts, or 188 lb, dry sand, mixed with 0.45 part, or 43.3 lb (5.1 gal), of potable water. At the time of perforating, cement-sand mixture must have set a minimum of 28 days and have a minimum tensile strength of 400 psi.

Bullet guns are shot in air and jets in water. The outside steel form may either be left on the target or removed when firing shaped charges; it must remain in place when testing bullet guns. Hole size, burr height, and penetration are measured and recorded on API Form 43D.

Value of Section 1, API RP 43—These surface tests may be beneficial to evaluate changes in hole size, due to changes in gun clearance, and consistency of charge performance. It may be useful outside the U.S. to determine whether the shaped charges have been damaged in transit. Results are not indicative of flow potential through perforations or susceptibility of perforations to be plugged or cleaned out. Also, penetration tests in cement are not comparable to penetration tests in formation sandstones or carbonates.

These surface tests do not evaluate plugging with liner residue slugs, or "carrots," because the slugs tend to tumble and be diverted away from perforations during surface tests at atmospheric pressure. Also debris is not flushed out of perforations as in the flow test. Data from tests under Sections 1 and 2, API RP 43, Fourth Edition, Aug. 1985, are reported on API Form 43D. Appendix I provides examples of test data reported on 43D for specific jet guns.

Tests under Section 2, API RP 43, are designed to evaluate, under simulated downhole wellbore and reservoir conditions, the hole size, penetration, and ability of specific perforators to provide perforations with high flow efficiency. The test apparatus and procedure described earlier in this chapter, are used in tests under Section 2, API RP 43. However, a change was made in method of calculating flow efficiency in the 1971 revised API Standards. The current evaluation procedure is described here.

Evaluation of Flow From Perforations—Original effective permeability (k_o), to kerosene of a "restored state" Berea sandstone core is measured prior to preparing the target.

Perforated effective permeability (k_p) to kerosene of the Berea sandstone target is measured after perforating and cleaning out by backflowing.

(k_p/k_o) the ratio of the perforated effective permeability to original effective core permeability, is then calculated. This experimental permeability ratio was originally called WFI or Well Flow Index.

K_i, is the ideal perforated permeability of an ideal clean undamaged hole with the same depth as the perforated hole used to obtain k_p and is based on a 0.4-in. diameter perforation. k_i, for any given depth perforation in a given core length has been developed from a computer program and is available in Appendix 1, API RP 43, Fourth Edition, Aug. 1985.

(K_i/k_o) is the ratio of the ideal perforated effective permeability to original core permeability.

Core Flow Efficiency, CFE = $(k_p/k_o)/(k_i/k_o)$ CFE represents the relative effectiveness of the perforation for conducting fluid compared to an ideal undamaged 0.4-in. diameter perforation.

Total core penetration in inches is designated as TCP. It is always 1 $^1/_8$-in. less than TTP, Total Target Penetration.

Effective Core Penetration, ECP, is core flow efficiency, CFE, multiplied by total *core* penetration, TCP. ECP = (CFE)(TCP). This provides a means of comparing fluid flow capacity of perforations with different penetration (TCP) and different core flow efficiency (CFE). The perforation with the highest ECP, effective core penetration. in the same length core target, should have the highest flow rate under clean downhole conditions at a given differential pressure.

All perforator tests reported by service companies in Section 2, API Form 43D, are made in Berea sandstone with an average compressive strength of about 6,500 psi, unless a different compressive strength is noted. Therefore, total target penetration (TTP) for each gun should be corrected for compressive strength of the particular downhole formation if the API test data is to be used for perforator selection for a specific well.

For example, Section 2, API RP 43 test data reported on Form 43D for a specific 1 $^{11}/_{16}$ in. through-tubing perforator showed the TTP to be 5.03 in. To illustrate the effect of formation compressive strength on perforator selection, assume the compressive strength of one formation to be 2,000 psi and a second formation to be 14,000 psi. Referring to Figure 7-6, perforator performance on formations of various compressive strengths, the corrected TTP for the 2,000 psi formation is about 8.4 in.; and for the 14,000 psi formation about 1.4 in.

To obtain the correct TCP, subtract the 1 $^1/_8$-in. thickness of the casing and cement from the TTP. leaving about 7.3 in. of penetration in 2,000 psi rock and about 0.3 in. of penetration in 14,000 compressive strength rock. Penetration in the low compressive strength formation would be adequate, whereas penetration in the 14,000 psi formation is completely unsatisfactory.

Recommended Use for API RP 43 Test Data

Concrete Test Target. Section 1, API RP 43—The entry hole diameter obtained in Section 1. Concrete Target Tests, is a most essential part of this test because entry hole diameter is measured in casing. a curved surface, supported by cement.

Section 1, Concrete Target Data, should not be used for evaluating penetration because the sand-cement target is not representative of the downhole formation. The consistency of set cement is such that penetration into cement cannot be reliably compared with penetration of downhole formations.

Berea Sandstone Core Target, Section 2, API RP 43—The Berea target data will provide more reliable and useful information relative to penetration, flow characteristics of perforations, well productivity. and perforation plugging. Entry hole diameter in Berea sandstone targets will usually be slightly larger than those in cemented tests made on the surface tests because Berea targets have a mild steel face plate.

Use of Data on Form API 43D--Data from both Section 1 and Section 2 is reported on Form 43D.

Figure 7A-1 shows test data on the Western Atlas Wireline Services 2 $^1/_8$ in. Silver Jet Capsule Retrievable Strip gun. Data on Section 2 API Form 43D shows average penetration of 10.55 in. and average hole size of 0.39 in. in the Berea sandstone flow target. This through-tubing perforator is designed to be run through 2 $^7/_8$-in. OD and larger tubing.

Figure 7A-2 shows test data on the Western Atlas Wireline Services 4-in. NCF VI JUMBO JET II charge used in an Atlas 4-in. Kaneshot gun. The average penetration in the Berea sandstone flow target is 15.07 in., with average entrance hole size of 0.43 in.

Figure 7A-3 shows test data on the Vann systems 6-in. DP charge used in the Halliburton Wireline Services Vann TCP Hollow Carrier, scalloped, expendable Jet Gun. Average penetration in the Berea flow target is 15.90 in. with an average hole size of 0.45 in. This charge is representative of many competitive charges used in tubing conveyed perforators. This type of charge should be a preferred charge when penetration is a problem in tubing conveyed perforators.

Figure 7A-4 shows test data on the Western Atlas Wireline Services 4-in. JUMBO JET BIG HOLE Charge for use in the Atlas 4-in. KONESHOT BIG HOLE retrievable hollow carrier gun. Data from Section 2 API Form 43D shows average penetration of 9.84 in. and average hole size of .72 in. in the Berea sandstone flow target. The large hole produced by this gun is particularly applicable to sandstone wells to be gravel-packed. Also, the large hole perforator is beneficial in any well where appreciable paraffin, asphaltenes, scale, or other types of perforation plugging are anticipated.

API FORM 43D

CERTIFICATION DATA SHEET

Service Company: ATLAS WIRELINE SERVICES
Gun OD and Designation: 2-1/8 IN. SILVER JET
Charge Name: 2-1/8 IN. SILVER JET
Type Perforator: CAPSULE RETRIEVABLE STRIP
Recommended Minimum Pipe OD: 2-7/8 in.
Standard Gun Length: 10 ft
Maximum Gun One Run: 120 shots
Spacing Available: 1-6 shots/ft
Phasings Available: 0° (ZERO) degree
Firing Order: ✓ Top down, ✓ Bottom up
Available Firing: ✓ Selective, ✓ Simultaneous

Charge Weight: 16 gm RDX powder
Maximum 1 hour Temperature: 300° F
Maximum Pressure: 15,000 psi
Charge Case Material: STEEL
Carrier Material: STEEL (ALLOY)
Bullet OD: ___ in., Length ___ in.
Debris Weight: 160 gm/charge
Debris Description: SMALL STEEL FRAGMENTS .16 IN. FILL UP PER CHARGE IN 5-1/2 IN. CASING.
Remarks: THE CHARGE LID AND LOCK NUT RETRIEVED WITH STRIP.

SECTION 1 — CONCRETE TARGET

Casing Data: 5-1/2 IN. OD, 17 lb/ft, J-55 API Grade
Target Data: 36 IN. OD, Tensile Strength 420 psi, Compressive Strength 4200 psi
Date of Test: JANUARY 8, 1985 Age of Target 28 days

Shot No	No. 1	No. 2	No. 3	No. 4	No. 5	No. 6	Average
Clearance, in.	0	0	0	0	0	0	0
Casing Hole Diameter, Short Axis, in.	.39	.38	.35	.34	.36	.36	.36
Casing Hole Diameter, Long Axis, in.	.37	.38	.38	.36	.36	.37	.37
Average Casing Hole Diameter, in.	.38	.38	.37	.35	.36	.37	.37
Total Depth, in.	17.10	17.68	17.05	18.13	17.80	17.62	17.56
Burr Height, in.	.03	.03	.02	.03	.03	.02	.03
Depth to Jet Debris or Bullet, in.							

Remarks

SECTION 2 — BEREA SANDSTONE CORE TARGET

CORE DATA						SHOOTING CONDITIONS			FLOW CONDITIONS				RESULTS				
Test No.	Test Date	Porosity, %	S_w, %	Core Length, in.	k_o, md	Pressure, psi Well	Pressure, psi Core Target	Clearance, in.	Pressure, psi Well	Pressure, psi Core Target	Total Flow, liters	k_p, md	Entrance Hole Diam. in.	CFE	TTP, in.	TCP, in.	ECP, in.
1348	12/13/84	20.8	72.3	18	293	1000	1200	0	1000	1200	30	426	.40	.79	10.52	9.40	7.43
1349	12/17/84	18.6	81.5	18	184	1000	1200	0	1000	1200	28	324	.40	.90	11.13	10.00	9.00
1350	12/18/84	20.3	77.4	21	256	1000	1200	0	1000	1200	26	342	.38	.91	10.00	8.87	8.07
											Average →		.39	.87	10.55	9.42	8.17

CERTIFICATION

I certify that these tests were made according to the procedures as outlined in *API RP 43: Standard Procedure for Evaluation of Well Perforators*, Second Edition, November 1971. All of the equipment used in these tests, such as the guns, jet charges, detonator cord, etc., was standard equipment with our company for use in the gun being tested, and was not changed in any manner for the test. Furthermore, the equipment was chosen at random from stock and therefore will be substantially the same as the equipment which would be furnished to perforate a well for any operator.

CERTIFIED BY [signature] 5/20/85 Date — Executive Vice President Title — ATLAS WIRELINE SERVICES, 10201 WESTHEIMER, HOUSTON, TEXAS 77042 Address

RECERTIFIED BY ______ Date ______ Title ______ Address ______

(See reverse side)

4 1/2-IN., 16.60-LB/FT DRILL PIPE
HYDROMITE-CM*
MILD-STEEL FACE PLATE

*Registered trade name, U. S. Gypsum Company, Chicago, Illinois.

FIG. 7A-1—Test data on Western Atlas Wireline Services 2 1/8 in. Silver Jet Capsule Retrievable Strip gun.

API FORM 43D

CERTIFICATION DATA SHEET

Service Company ATLAS WIRELINE SERVICES
Gun OD and Designation 4 IN. KONESHOT
Charge Name 4 NCF VI JUMBO JET II
Type Perforator RETRIEVABLE HOLLOW CARRIER
Recommended Minimum Pipe OD 5-1/2 in.
Standard Gun Length 6 & 10 ft
Maximum Gun One Run 168 shots
Spacing Available 1,2,3,4 shots/ft
Phasings Available 120 degree
Firing Order ____ Top down, √ Bottom up
Available Firing SEMI Selective, √ Simultaneous

Charge Weight 22.5 gm RDX powder
Maximum 1-hour Temperature 325° F
Maximum Pressure 20,000 psi
Charge Case Material RUBBER AND STEEL
Carrier Material H.T. STEEL
Bullet OD ____ in., Length ____ in.
Debris Weight ____ gm/charge
Debris Description DEBRIS CONTAINED IN GUN BODY.

Remarks NON-CARROT FORMING PERFORATOR

SECTION 1 — CONCRETE TARGET

Casing Data 5-1/2 IN. OD, 17 lb/ft, J-55 API Grade
Target Data 60 IN. OD, Tensile Strength 445 psi, Compressive Strength 6125 psi
Date of Test AUGUST 30, 1979 Age of Target 37 days

Shot No	No. 1	No. 2	No. 3	No. 4	No. 5	No. 6	Average
Clearance, in.	0	.68	.68	0	.68	.68	.45
Casing Hole Diameter, Short Axis, in.	.35	.39	.37	.35	.39	.36	.37
Casing Hole Diameter, Long Axis, in.	.38	.43	.39	.39	.40	.39	.40
Average Casing Hole Diameter, in.	.37	.41	.38	.37	.40	.38	.39
Total Depth, in.	23.86	23.56	23.61	22.86	23.36	24.86	23.68
Burr Height, in.	.04	.07	.08	.06	.07	.06	.06

Depth to Jet Debris or Bullet, in. THE LINER USED IN THIS CHARGE IS NON-CARROT FORMING.
Remarks

SECTION 2 — BEREA SANDSTONE CORE TARGET

CORE DATA						SHOOTING CONDITIONS			FLOW CONDITIONS				RESULTS				
Test No.	Test Date	Porosity %	S_o, %	Core Length, in.	k_o, md	Pressure, psi: Well	Pressure, psi: Core Target	Clearance, in.	Pressure, psi: Well	Pressure, psi: Core Target	Total Flow, liters	k_p, md	Entrance Hole Diam. in.	CFE	TTP, in.	TCP, in.	ECP, in.
1190	4-03-79	19.3	73.3	21	277.8	1500	1000	.50	1000	1200	28	608.9	.43	.84	15.30	14.17	11.90
1191	4-06-79	19.5	71.8	21	282.3	1500	1000	.50	1000	1200	28	594.2	.44	.86	14.80	13.67	11.75
1192	4-12-79	19.2	73.5	21	265.8	1500	1000	.50	1000	1200	28	560.7	.43	.83	15.10	13.97	11.56
												Average →	.43	.84	15.07	13.93	11.74

CERTIFICATION

I certify that these tests were made according to the procedures as outlined in *API RP 43 Standard Procedure for Evaluation of Well Perforators*, Second Edition, November 1971. All of the equipment used in these tests, such as the guns, jet charges, detonator cord, etc., was standard equipment with our company for use in the gun being tested, and was not changed in any manner for the test. Furthermore, the equipment was chosen at random from stock and therefore will be substantially the same as the equipment which would be furnished to perforate a well for any operator.

CERTIFIED BY [signature] 5/20/88 Date — Executive Vice President Title — 10201 WESTHEIMER, DC-8 HOUSTON, TEXAS 77042 Address

RECERTIFIED BY ____ Date ____ Title (See reverse side) ____ Address

4 1/2-IN., 16.60-LB/FT DRILL PIPE
HYDROMITE-CM*
MILD-STEEL FACE PLATE

*Registered trade name, U. S. Gypsum Company, Chicago, Illinois.

FIG. 7A-2—*Test data on Western Atlas Wireline Services 4–in. NCF VI JUMBO JET II charge used in an Atlas 4–in. Kaneshot gun.*

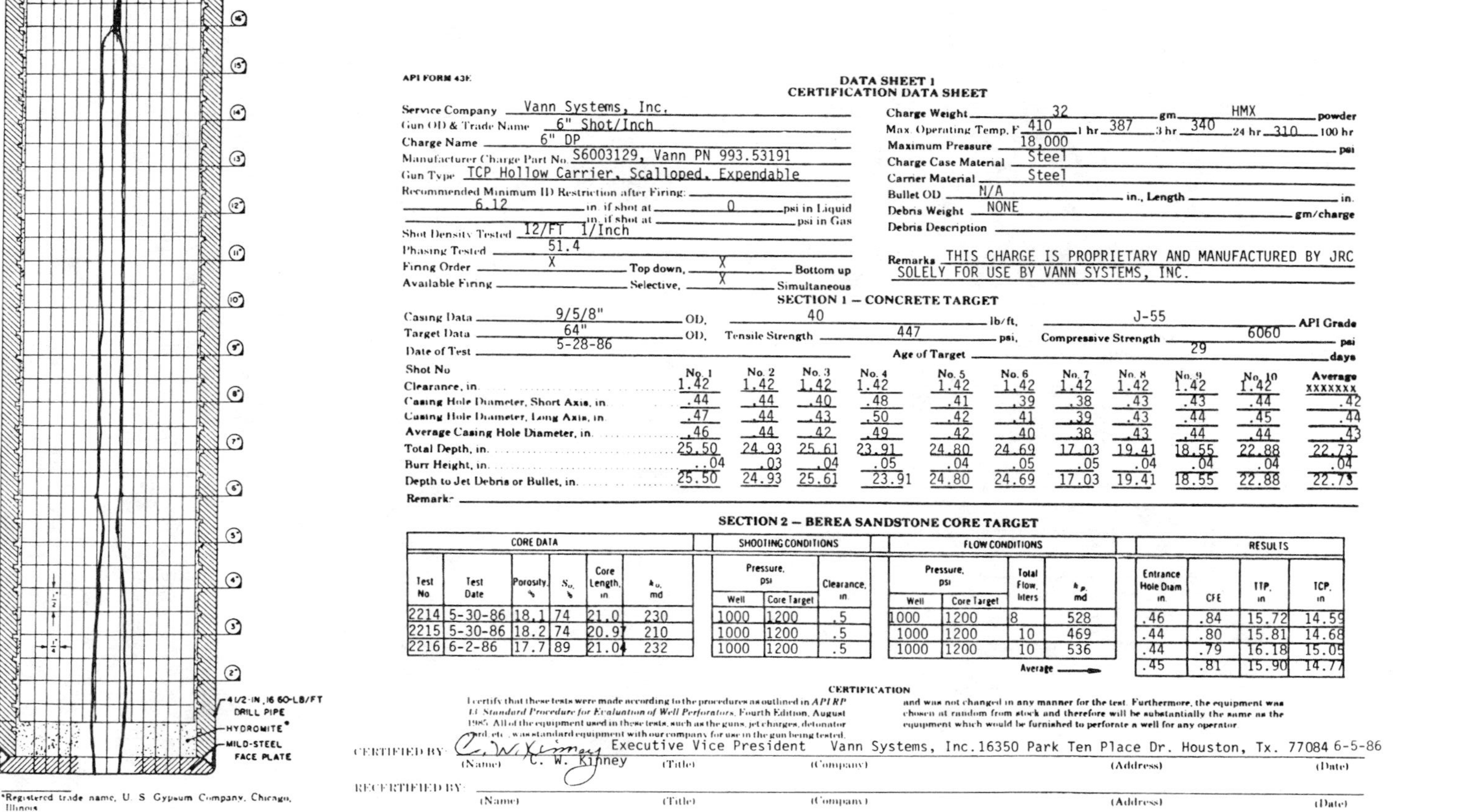

*Registered trade name, U. S. Gypsum Company, Chicago, Illinois

API FORM 43E

DATA SHEET 1
CERTIFICATION DATA SHEET

Service Company: Vann Systems, Inc.
Gun OD & Trade Name: 6" Shot/Inch
Charge Name: 6" DP
Manufacturer Charge Part No. S6003129, Vann PN 993.53191
Gun Type: TCP Hollow Carrier, Scalloped, Expendable
Recommended Minimum ID Restriction after Firing: 6.12 in. if shot at 0 psi in Liquid; ___ in. if shot at ___ psi in Gas
Shot Density Tested: 12/FT 1/Inch
Phasing Tested: 51.4
Firing Order: X Top down, X Bottom up
Available Firing: Selective, X Simultaneous

Charge Weight: 32 gm. HMX powder
Max. Operating Temp. F: 410 1 hr 387 3 hr 340 24 hr 310 100 hr
Maximum Pressure: 18,000 psi
Charge Case Material: Steel
Carrier Material: Steel
Bullet OD: N/A in., Length ___ in.
Debris Weight: NONE gm/charge
Debris Description:

Remarks: THIS CHARGE IS PROPRIETARY AND MANUFACTURED BY JRC SOLELY FOR USE BY VANN SYSTEMS, INC.

SECTION 1 — CONCRETE TARGET

Casing Data: 9/5/8" OD, 40 lb/ft, J-55 API Grade
Target Data: 64" OD, Tensile Strength 447 psi, Compressive Strength 6060 psi
Date of Test: 5-28-86, Age of Target 29 days

Shot No	No. 1	No. 2	No. 3	No. 4	No. 5	No. 6	No. 7	No. 8	No. 9	No. 10	Average
Clearance, in.	1.42	1.42	1.42	1.42	1.42	1.42	1.42	1.42	1.42	1.42	xxxxxxx
Casing Hole Diameter, Short Axis, in.	.44	.44	.40	.48	.41	.39	.38	.43	.43	.44	.42
Casing Hole Diameter, Long Axis, in.	.47	.44	.43	.50	.42	.41	.39	.43	.44	.45	.44
Average Casing Hole Diameter, in.	.46	.44	.42	.49	.42	.40	.38	.43	.44	.44	.43
Total Depth, in.	25.50	24.93	25.61	23.91	24.80	24.69	17.03	19.41	18.55	22.88	22.73
Burr Height, in.	.04	.03	.04	.05	.04	.05	.05	.04	.04	.04	.04
Depth to Jet Debris or Bullet, in.	25.50	24.93	25.61	23.91	24.80	24.69	17.03	19.41	18.55	22.88	22.73

Remarks:

SECTION 2 — BEREA SANDSTONE CORE TARGET

CORE DATA						SHOOTING CONDITIONS			FLOW CONDITIONS				RESULTS			
Test No	Test Date	Porosity, %	S_w, %	Core Length, in	k_a, md	Pressure, psi Well	Pressure, psi Core Target	Clearance, in	Pressure, psi Well	Pressure, psi Core Target	Total Flow, liters	k_p, md	Entrance Hole Diam in	CFE	TTP, in	TCP, in
2214	5-30-86	18.1	74	21.0	230	1000	1200	.5	1000	1200	8	528	.46	.84	15.72	14.59
2215	5-30-86	18.2	74	20.97	210	1000	1200	.5	1000	1200	10	469	.44	.80	15.81	14.68
2216	6-2-86	17.7	89	21.04	232	1000	1200	.5	1000	1200	10	536	.44	.79	16.18	15.05
												Average →	.45	.81	15.90	14.77

CERTIFICATION

I certify that these tests were made according to the procedures as outlined in *API RP 43 Standard Procedure for Evaluation of Well Perforators*, Fourth Edition, August 1985. All of the equipment used in these tests, such as the guns, jet charges, detonator cord, etc., was standard equipment with our company for use in the gun being tested, and was not changed in any manner for the test. Furthermore, the equipment was chosen at random from stock and therefore will be substantially the same as the equipment which would be furnished to perforate a well for any operator.

CERTIFIED BY: C. W. Kinney (Name) Executive Vice President (Title) Vann Systems, Inc. (Company) 16350 Park Ten Place Dr. Houston, Tx. 77084 (Address) 6-5-86 (Date)

RECERTIFIED BY: (Name) (Title) (Company) (Address) (Date)

FIG. 7A-3—Test data on Halliburton's Vann systems 6–in. DP charge used in Vann TCP Hallow Carrier, scalloped, expendable jet gun.

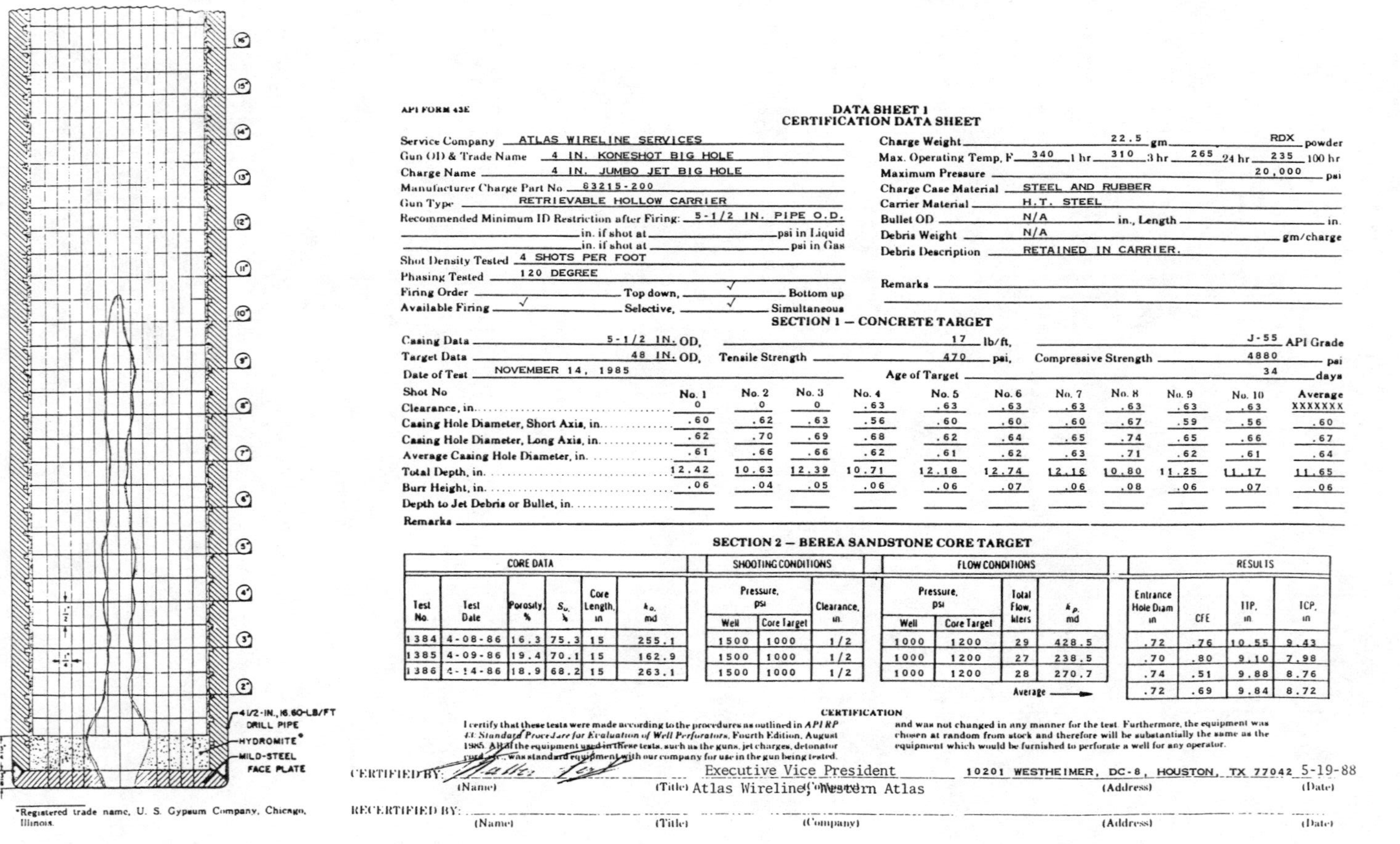

API FORM 43E

DATA SHEET 1
CERTIFICATION DATA SHEET

Service Company: ATLAS WIRELINE SERVICES
Gun OD & Trade Name: 4 IN. KONESHOT BIG HOLE
Charge Name: 4 IN. JUMBO JET BIG HOLE
Manufacturer Charge Part No: 63215-200
Gun Type: RETRIEVABLE HOLLOW CARRIER
Recommended Minimum ID Restriction after Firing: 5-1/2 IN. PIPE O.D.
______ in. if shot at ______ psi in Liquid
______ in. if shot at ______ psi in Gas
Shot Density Tested: 4 SHOTS PER FOOT
Phasing Tested: 120 DEGREE
Firing Order: ______ Top down, ✓ Bottom up
Available Firing: ✓ Selective, ✓ Simultaneous

Charge Weight: 22.5 gm RDX powder
Max. Operating Temp, F: 340 1 hr 310 3 hr 265 24 hr 235 100 hr
Maximum Pressure: 20,000 psi
Charge Case Material: STEEL AND RUBBER
Carrier Material: H.T. STEEL
Bullet OD: N/A in., Length ______ in.
Debris Weight: N/A gm/charge
Debris Description: RETAINED IN CARRIER.
Remarks: ______

SECTION 1 – CONCRETE TARGET

Casing Data: 5-1/2 IN. OD, 17 lb/ft, J-55 API Grade
Target Data: 48 IN. OD, Tensile Strength 470 psi, Compressive Strength 4880 psi
Date of Test: NOVEMBER 14, 1985 Age of Target: 34 days

Shot No	No. 1	No. 2	No. 3	No. 4	No. 5	No. 6	No. 7	No. 8	No. 9	No. 10	Average
Clearance, in.	0	0	0	.63	.63	.63	.63	.63	.63	.63	XXXXXXX
Casing Hole Diameter, Short Axis, in.	.60	.62	.63	.56	.60	.60	.60	.67	.59	.56	.60
Casing Hole Diameter, Long Axis, in.	.62	.70	.69	.68	.62	.64	.65	.74	.65	.66	.67
Average Casing Hole Diameter, in.	.61	.66	.66	.62	.61	.62	.63	.71	.62	.61	.64
Total Depth, in.	12.42	10.63	12.39	10.71	12.18	12.74	12.16	10.80	11.25	11.17	11.65
Burr Height, in.	.06	.04	.05	.06	.06	.07	.06	.08	.06	.07	.06
Depth to Jet Debris or Bullet, in.											

Remarks ______

SECTION 2 – BEREA SANDSTONE CORE TARGET

Core Data: Test No.	Test Date	Porosity, %	S_w, %	Core Length, in	k_o, md	Shooting Conditions: Pressure, psi: Well	Core Target	Clearance, in	Flow Conditions: Pressure, psi: Well	Core Target	Total Flow, liters	k_p, md	Results: Entrance Hole Diam in	CFE	TTP, in	TCP, in
1384	4-08-86	16.3	75.3	15	255.1	1500	1000	1/2	1000	1200	29	428.5	.72	.76	10.55	9.43
1385	4-09-86	19.4	70.1	15	162.9	1500	1000	1/2	1000	1200	27	238.5	.70	.80	9.10	7.98
1386	4-14-86	18.9	68.2	15	263.1	1500	1000	1/2	1000	1200	28	270.7	.74	.51	9.88	8.76
												Average →	.72	.69	9.84	8.72

CERTIFICATION

I certify that these tests were made according to the procedures as outlined in *API RP 43: Standard Procedure for Evaluation of Well Perforators*, Fourth Edition, August 1985. All of the equipment used in these tests, such as the guns, jet charges, detonator cord, etc., was standard equipment with our company for use in the gun being tested, and was not changed in any manner for the test. Furthermore, the equipment was chosen at random from stock and therefore will be substantially the same as the equipment which would be furnished to perforate a well for any operator.

CERTIFIED BY: [signature] (Name) Executive Vice President (Title) Atlas Wireline, Western Atlas (Company) 10201 WESTHEIMER, DC-8, HOUSTON, TX 77042 (Address) 5-19-88 (Date)

RECERTIFIED BY: ______ (Name) ______ (Title) ______ (Company) ______ (Address) ______ (Date)

FIG. 7A-4—Test data on Western Atlas Wireline Services 4–in. JUMBO JET BIG HOLE charge for use in the Atlas 4–in. KONESHOT BIG HOLE retrievable hollow carrier gun.

Chapter 8 Completion and Workover Fluids

Functions, requirements, selection criteria
Formation damage considerations
Oil fluids, practical applications
Clear-water fluids, practical applications
Clay, emulsion, and wettability problems
Viscosity, fluid loss, and density control
Maintenance of clean fluids
Conventional drilling fluids
Aerated fluids
Perforating fluids, packer fluids

FUNCTIONS—REQUIREMENTS—SELECTION CRITERIA

By definition a completion or workover fluid is a fluid placed against the producing formation while conducting such operations as well killing, cleaning out, drilling in, plugging back, controlling sand, or perforating. Frac fluids are sometimes considered to be workover fluids. Similar materials are used to build viscosity and provide fluid loss control; however, frac fluid requirements are significantly different than for workover fluids. See Chapter 8 Volume 2 for a discussion of frac fluids. Packer fluids remain in the well for an extended period; thus stability and corrosion are important concerns. Packer fluids are discussed in this chapter.

Basic completion and workover fluid functions are to facilitate movement of treating fluids to a particular point downhole, to remove solids from the well, and to control formation pressures. Required fluid properties vary depending on the operation—but the possibility of formation damage should always be an important concern. In certain operations, such as sand consolidation or gravel packing, sand-face or perforation plugging is a prime concern. In recent years many new fluid systems have appeared, most due to the recognition of the high risk of reducing the productivity, or completely plugging certain sections of the producing zone, through contact with a foreign fluid.

These points should be considered in selecting a workover or completion fluid:

Fluid Density—Fluid density should be no higher than needed to control formation pressure. With reasonable precautions a hydrostatic pressure of 100–200 psi over formation pressure should be adequate. Balanced pressure workovers are ideal from the standpoint of formation damage and, with proper equipment to contain the surface pressure, are practical for some operations.

Solids Content—Ideally, the fluid should contain no solids to avoid formation and perforation plugging. Figure 8-1 shows plugging of Cypress sandstone (450-md brine permeability) with salt water fluids containing various sizes and concentrations of solids. Particles up to 5 micron size caused significantly more plugging than particles less than 2 micron size. In both cases plugging occurred within the core pore channels.

Particles larger than about one-half the average pore diameter should bridge at the entrance to the pore. These larger particles are probably not detrimental if they are removed by backflow or degraded by acid or crude oil.

Particles which plate out to plug the face of the formation or a perforation obviously obstruct operations, such as sand consolidation, gravel packing, or squeeze cementing.

Filtrate Characteristics—Characteristics of the fil-

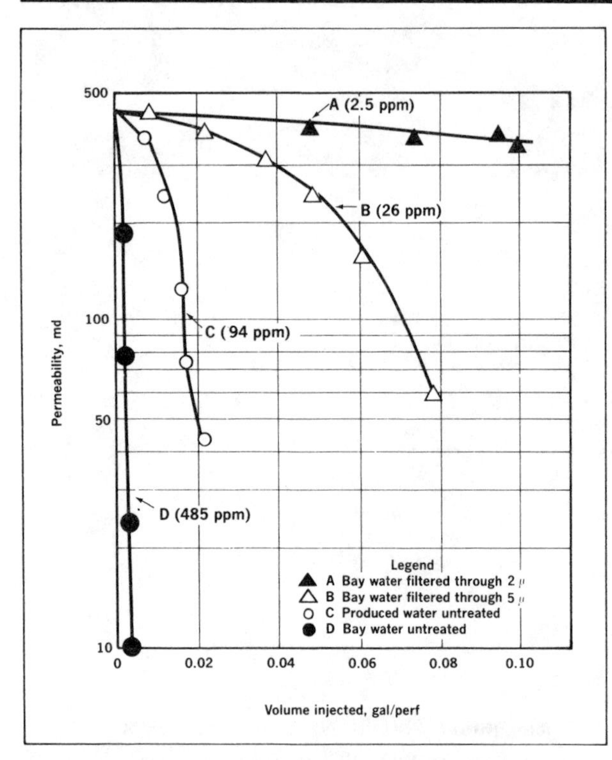

FIG. 8-1—Permeability reduction in Cypress sandstone cores.[7] Permission to publish by The Society of Petroleum Engineers.

trate should be tailored to minimize formation damage considering swelling or dispersion of clays, wettability changes, and emulsion stabilization. Many times this means that the fluid should contain the proper surfactant as well as the proper electrolyte.

Fluid Loss—Fluid loss characteristics may have to be tailored to prevent loss of excessive quantities of fluid to the formation, or to permit application of "hydraulic stress" to an unconsolidated sand formation. Bridging at the formation face by properly sized acid-soluble particles (calcium carbonate) is a desirable approach to fluid loss control. Where limitations permit, oil soluble resin particles may substitute for the calcium carbonate. In either case colloidal particles are also required for an effective seal.

Viscosity-Related Characteristics—Viscosity-related characteristics, such as yield point, plastic viscosity, and gel strength, may have to be tailored to provide fluid lifting capacity required to bring sand or cuttings to the surface at reasonable circulating rates. Lab tests show that many viscosity builders cause permanent reduction in permeability. This can be minimized by careful polymer selection along with adequate fluid loss control to limit invasion.

Corrosion Products—The fluid should be chemically stable so that reaction of free oxygen with tubular steels is minimized, and that iron in solution is sequestered and not permitted to precipitate in the formation. A reasonable upper limit on corrosivity for a completion or workover fluid is 0.05 lb/ft^2 (about 1 mil) per workover. For a packer fluid, the corrosivity target should be about 1 mil per year, but 5 mils per year is considered to be an acceptable upper limit.

Mechanical Considerations—Rig equipment available for mixing, storage, solids removal, and circulating is often a factor in fluid selection.

Economics—The most economical fluid commensurate with the well's susceptibility to damage should be selected.

FORMATION DAMAGE RELATED TO SOLIDS

There are two basic approaches to minimize formation damage due to solids entrained in the completion fluid.

Complete Solids Removal—To be effective, fluid in contact with the formation must not contain any solids larger than 2 micron size. These points are involved:

1. Eliminate all solids to the greatest extent possible through the use of 2-micron surface filters, backed up by other measures to minimize solids pickup downstream from the filter: i.e., control of oxygen to minimize iron oxide, careful use of thread dope, removal of rust, scale, etc. from down-hole tubulars using HCl, IPA, or sand scouring techniques.
2. Accept loss of fluid to the formation and, as a practical matter, movement of very small fines into the formation. The quantity of fines is limited by minimizing differential pressure into the formation. Removal of fines from the pore system after the job is maximized by returning the well to production through gradual increases in flow rate.
3. Accept possible difficulties in removing large particles from the hole due to low viscosity and carrying capacity. Many times velocity can be substituted for viscosity in lifting particles. A rising velocity of 150 ft/min should be sufficient to remove formation cuttings or sand even with 1.0 cp viscosity clear salt water.

This velocity is within the limits of many workover rig pumping systems: i.e., a circulating rate of 5 bbl/min gives 150 ft/min annular velocity with 2⅞ in. tubing inside 7-in. casing or 2⅜ in. tubing inside 5½-in. casing. A no-solids viscosity greater than 1.0 cp may be possible with filtered crude oil or perhaps brine viscosifiers having no residue.

Complete Fluid Loss Control—To be effective, particles must not be allowed to move past the face of the formation into the pore system. These points are involved:

1. Stop all solids at the formation face by carrying in the workover fluid solid particles properly sized to bridge quickly, and colloids to maximize the effectiveness of the seal.
2. Removal of formation-face plug after the job by backflow—and/or by degradation of the calcium carbonate solid particles and colloids with acid. With certain limitations resin solids dissolvable in crude oil could be used rather than calcium carbonate.
3. Accept the possibility that after the job pressure differential needed to unplug all sections of the zone will not be available and that it will not be possible to contact all the plugged section of the formation with acid due to bypassing tendencies.

The optimum approach depends on specific well conditions and operations. For certain critical operations such as sand consolidation and gravel packing through perforations, the "complete-solids-removal" approach often provides highest productivity and minimum cost.

OIL FLUIDS—PRACTICAL APPLICATION

Crude Oil—Availability makes crude oil a logical choice where its density is sufficient. Density considerations may make it particularly desirable in low-pressure formations. A low-viscosity crude has limited carrying capacity and no gel strength and thus should drop out non-hydrocarbon solids in surface pits. See Figure 2-10A of this volume for correlation of oil viscosity vs. API gravity and temperature. Oil is an excellent packer fluid from the standpoint of minimizing corrosion, and gel strength can be provided to limit solids settling. Crude obtained from the stock tank has usually weathered enough to reduce (but not eliminate) the fire hazard.

Loss of oil to the formation is usually not harmful from the standpoint of clay disturbance or from saturation effects, as might be the case with salt water in a low pressure formation. It has no fluid loss control; thus any entrained fine solids could be carried into the pore system.

Crude oil should always be checked for the presence of asphaltenes or paraffins that could plug the formation. This can be done in the field using API fluid loss test equipment to observe the quantity of solids collected on the filter paper.

Also, crude oil should be checked for possibility of emulsions with formation water. Techniques of the API RP 42 test (visual wettability and emulsion breakout) are suitable for field use. If stable emulsions are formed, a suitable surfactant should be added.

Diesel Oil—This may be ideal where an especially clean fluid is required for operations such as sand consolidation. It may even be advantageous to work under pressure at the surface where the density of diesel oil is not sufficient to overcome formation pressure.

Depending on hauling and handling practices, diesel oil should also be checked for solids. Emulsion and wettability problems should be nonexistent if the diesel is obtained at the refinery before certain motor fuel additives are included.

CLEAR WATER FLUIDS—PRACTICAL APPLICATION

Source of Water

Formation Salt Water—When available, formation salt water is a common workover fluid since the cost is low. If it is clean, formation salt water is ideal from the standpoint of minimizing formation damage due to swelling or dispersion of clays in sandstone formations.

Although "gun barrel" salt water is frequently considered to be natural water from the formation, it often contains treating chemicals, fine particles of oil, clay, silt, paraffin, asphalt, or scale, and it therefore may cause appreciable formation damage. (Fig. 8-1).

Even filtered formation salt water may contain oil treating surfactants (cationic emulsion breakers) which may cause wettability or emulsion problems. Field checks can be run using API RP 42 procedures.

Seawater or Bay Water—Due to availability, it is often used in coastal areas. Again, it frequently contains clays and other fines that cause plugging.

As shown in Figure 8-1, untreated bay water caused serious plugging of Cypress sandstone cores. Depending on the salinity of bay water, it may be necessary to add NaCl or KCl to prevent clay disturbance.

Prepared Salt Water—Fresh water is often desirable as a basic fluid due to the difficulty of obtaining clean sea or formation water. Desired type and amount of salt is then added. Where clean brine is available at low cost, it may be preferable to purchase brine rather than mix it on location.

Salt Type, Concentration for Prepared Salt Water

Practicalities—From the standpoint of preventing formation damage in sandstones due to disturbance of smectite or mixed-layer clays, the prepared salt water should, theoretically, match the formation water in cation type and concentration.

It is difficult to match formation brine, however, and laboratory results show that 3% to 5% sodium chloride, 1% calcium chloride, or 1% potassium chloride will limit swelling of clays in most formations. In practice these concentrations are often doubled.

Limitations of $CaCl_2$—In certain formations sodium smectite can be flocculated (shrunk) by contact with calcium ions even in low concentrations. Thus, the clay may become mobile and could cause permeability reduction.

Where this is the case, 1% or 2% potassium chloride should be used rather than calcium chloride since the potassium ion will prevent swelling; in addition, low concentrations will not flocculate the sodium smectite.

Additional objections to the use of calcium chloride result from the observation that field mixed solutions of $CaCl_2$ usually exhibit pH of 10–10.5 which may disperse formation clays. $CaCl_2$ is also incompatible with some viscosifiers and many formation saltwaters.

Extreme Water Sensitivity—In some very water sensitive formations, 2% ammonium chloride brine (while quite expensive) seems to stabilize formation clays. Some small number of sand formations should not be contacted by water of any ionic characteristics.

Emulsion—Wettability Problems

When the brine fluid base is clean fresh water, wettability and emulsion problems theoretically should not be a concern. However, even here contamination from any one of many sources often occurs.

Field Checks—Best practice dictates that the actual workover fluid be checked on location to insure that it does not form a stable emulsion with the reservoir oil or that it does not oil wet the reservoir rock.

This is particularly true where formation salt water is used, or where corrosion inhibitors or biocides are used. Field checks can be run using the simple techniques of the API RP 42 Visual Wettability and Emulsion Breakout tests.

Prevention is the Key—Usually an unsatisfactory emulsifying or wettability situation can be corrected by the addition of a small amount (0.1%) of the *proper* surfactant.

As a general rule, workover fluids for sandstone formations where productivity is important should contain the *proper* surfactant to prevent any possibility of emulsion in the formation and to leave the formation around wellbore strongly water wet.

Viscosity Control—Fluid Loss Control

A number of additives are available to provide "viscosity," thereby increasing the lifting, carrying, and suspending capacity of the fluid. In the Bingham plastic representation of viscosity, "plastic viscosity" relates to flow resistance between particles as well as the viscosity of the continuous fluid phase; and "yield point" relates to suspending capability when the fluid is at rest.

Completion fluid viscosity builders are all long chain polymers or colloids. They also provide fluid loss control by an indepth plugging mechanism which extends some distance back within the radial pore system. Plugging is subsequently reduced by backflow and degradation, but this process is usually not complete, and formation damage remains.

Ideally, fluid loss control should be obtained strictly by a bridging mechanism at the face of the formation. This can be done effectively by use of properly sized particles. Particles larger than one-half the pore size should bridge at the pore entrance.

However, a range of particle sizes is required to reduce bridge permeability. Colloids or "plastic particles" are needed to complete the plug and further reduce permeability. The bridge should form and stabilize quickly to minimize movement of fines into the pore system.

"Viscosity" Builders

Both natural and processed polymers are used in completion fluid formulations. Among them are: guar gum, starch, hydroxyethyl cellulose (HEC), carboxymethyl cellulose (CMC), and biopolymer (xanthan gum). Appendix H, Chapter 8 of Volume 2 contains a detailed discussion of oil field polymers.

Natural Polymers—Guar gum is a hydrocolloid that swells on contact with water to provide viscosity and fluid loss control. Typical fluid properties are shown in Table 8-1. A filter cake is deposited, which may interfere even with squeeze cementing.

Figure 8-2 shows plugging resulting from injection

TABLE 8-1
Properties, Costs of Completion Fluids

	Polymer	*Starch*	*Guar*
Plastic viscosity, cp	27	20	25
Yield point, lb/100 ft²	31	10	65
10-sec gel, lb/100 ft²	2	1	6
10-min gel, lb/100 ft²	5	2	12
API fluid loss, cc/30 min	14.6	10.0	42.0
Cost, $/bbl (relative value)	4.50	3.00	3.90

Composition of the fluid is:

Polymer: 2 lb polysaccaride, 1 lb polyanonic cellulose per bbl salt water.
Starch: 10 lb starch, 1 lb polyanonic cellulos per bbl salt water.
Guar gum: 2.5 lb of guar gum per bbl salt water.

of about twenty pore volumes of unbroken guar gum into a sandstone core. Permeability regain after backflow was only 25%.

In a radial flow system, reduction in productivity would depend on the depth of the reduced permeability zone back from the wellbore. If, for example, the depth of damage was 12 in., productivity would be reduced to about 65% of the undamaged productivity.

Guar gum usually contains from 5–15% impurities. Guar gum stability is affected by changes in pH. It forms an insoluble floc in contact with isopropyl alcohol; thus guar should not be used where sand consolidation with Eposand will be carried out. Industry trend is away from guar gum for workover fluids.

Starch primarily is used to provide fluid loss control. Other polymers may be needed for carrying capacity. Table 8-1 shows typical properties. Overall cost of starch fluids is significantly lower than guar gum or other polymers, but higher concentrations of starch are required. Starch has no inherent bacterial control. Permeability loss due to plugging is significant; thus starch is losing popularity.

Bio-polymer (xanthan gum) provides good carrying capacity and fluid loss control. Gel strength properties provide stable suspensions of calcium carbonate bridging (or weighting) particles, but may make removal of undesirable fine solids more difficult. Biopolymer is not completely removed by HCl acid.

Processed Natural Polymers—HEC, (hydroxyethyl cellulose) sometimes combined with calcium lignosulfonate, has many desirable properties and is currently the preferred viscosifier. It provides:

—Good carrying capacity for hole cleaning.
—Good fluid loss control (in combination with bridging solids)
—Low gel strength to drop out undesirable solids in surface pits.
—Is degradable in HCl—or with enzyme breakers.

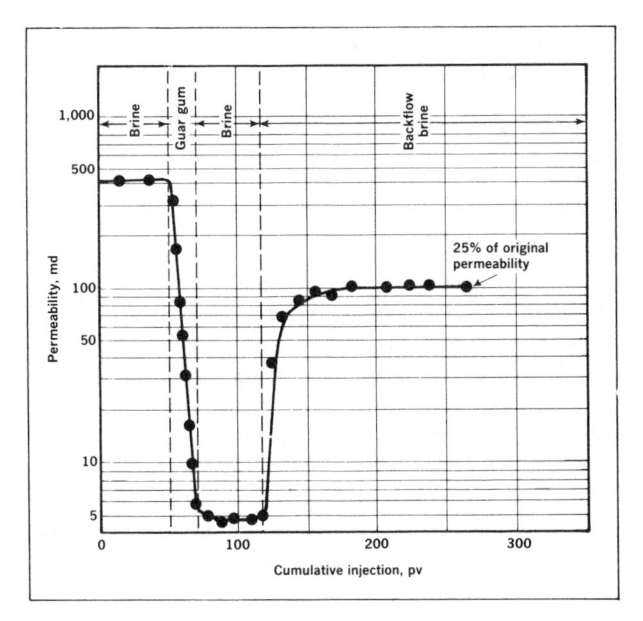

FIG. 8-2—Unbroken Guar gum.[8] Permission to publish by The Society of Petroleum Engineers.

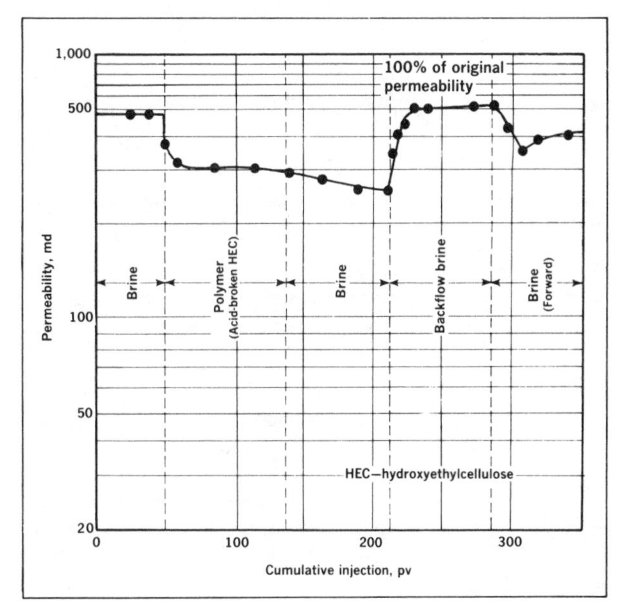

FIG. 8-3—Acid-broken HEC.[8] Permission to publish by The Society of Petroleum Engineers.

Unbroken HEC (without bridging particles) causes significant permeability reduction even after backflow. Damage is not as great as with unbroken guar. Acid-broken HEC (Fig. 8-3) shows little fluid loss control (without bridging solids) but complete permeability regain.

CMC (carboxymethyl cellulose) of the commercial grade used in drilling fluids should never be used in contact with a producing zone, due to irreparable formation damage. Insoluble products are formed in contact with trivalent ions.

Comparison of Solids-Free Fluid Properties—Typical fluid properties and relative costs of three solids-free completion fluids are shown in Table 8-1.

Fluid Loss Control

Due to unregained permeability loss, currently available viscosity builders should not be used without proper bridging particles to prevent movement of the viscosity colloids into the formation pore system. Bridging particles must meet two criteria:

—Form a stable, low-permeability bridge quickly.
—Be removable by degradation or backflow.

Calcium Carbonate—This material is available in several size ranges as shown in Table 8-2. For most formation pore sizes "200 mesh" particle range should be used. $CaCO_3$ is completely soluble in HCl.

Used in conjunction with HEC to provide colloids for carrying capacity and to further reduce the permeability of the "bridge," excellent fluid loss control is provided with almost perfect permeability regain if contacted with HCl acid.

Figure 8-4 shows results of lab tests using an unfiltered by water (136 ppm fine solids) with 5% NaCl, 1.0 lb/bbl HEC and 10 lb/bbl $CaCO_3$. Bridging was almost immediate, fluid loss control excellent, and after contact with HCl acid regain was 93%.

It should be noted that in a field situation acid con-

TABLE 8-2
Calcium Carbonate Particle Sizes

Designation	Diameter—Microns Mean	Maximum
Micro (400 mesh)	3.2	18
Fine (200 mesh)	60	160
Medium (70 mesh)	213	420

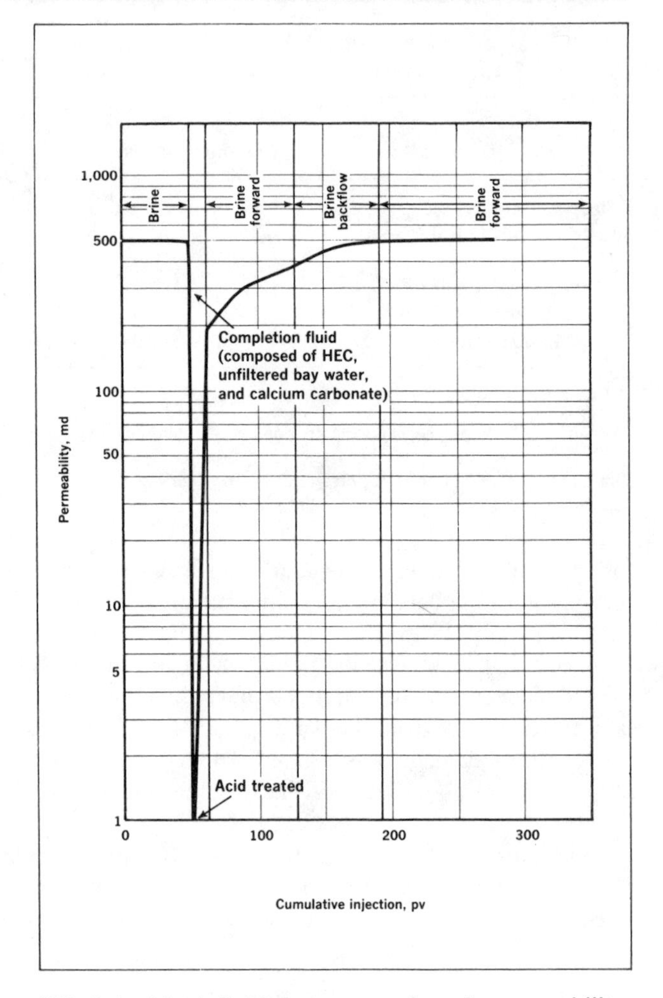

FIG. *8-4—Ideal fluid loss control and permeability regain in lab test.*[8] *Permission to publish by The Society of Petroleum Engineers.*

tact with the bridged $CaCO_3$ cannot always be assured, and some permeability loss may remain.

In gravel-packing at least partial backflow of "200 mesh" bridging solids through the pack is possible with 10–20 mesh gravel. Backflow probably would not occur through a tight pack of 20–40 mesh gravel.

Oil-Soluble Resins—These are available in graded size ranges needed for effective bridging action. While quite effective in brine water fluids they are quickly removed by low oil concentrations (2%). Also, temperature stability is a concern since some resins tend to melt even at relatively low temperature.

Within these limitations the use of oil-soluble resins appears to be an excellent approach to the problems of effective fluid loss control with effective permeability regain.

Field Applications

To facilitate field use, several suppliers combine polymer viscosifiers and graded calcium carbonate into one-package completion fluid systems. Polybrine and Solukleen are typical of these systems. Solukleen contains oil soluble resin rather than $CaCO_3$. Usually 1–4 lbs/bbl of these products are required to prepare a satisfactory fluid. Commercially available products are shown in Appendix A.

Polybrine (Magcobar) is a prepared mixture of polymers with a low concentration of 200-mesh calcium carbonate particles. All components are acid soluble. In 8.5–10.0 lb/gal brine (KCl, NaCl or $CaCl_2$) 1.0 to 4.0 lb/bbl Polybrine provides Marsh Funnel viscosity of 33 to 50 sec/qt and API fluid loss of 13–18 cc. Aluminum stearate (0.25 to 0.50 lb/bbl) must be added to brine prior to mixing Polybrine to prevent foaming.

High shear is needed to effectively hydrate the polymer. Low pressure centrifugal mixing systems are marginal. Dry salt added to Polybrine mixture causes severe foaming. Polybrine must be degraded with HCl before being contacted with isopropyl alcohol (IPA) to prevent formation of insoluble flocs. Above 250°F Polybrine becomes unstable.

A satisfactory calcium carbonate fluid can be prepared by adding 5 to 15 lb/bbl of 200-mesh calcium carbonate to solids-free salt solution containing 0.25 to 0.50 lb/bbl biopolymer. This should provide effective seepage control, and sufficient viscosity to circulate out sand or silt. Higher concentrations of polymer may be required to lift large cuttings or shale if the rig circulating capacity is limited.

In well killing, an effective technique is to circulate a "pill" (10–15 bbl) of fluid containing a high concentration of polymer and calcium carbonate particles. This establishes an initial bridge. Polymer and particle concentration can then be reduced. When and if additional cleaning or chip lifting capacity is needed, another pill can be circulated.

Loss of Circulation problems can usually be solved by the pill technique using additional polymer for carrying capacity with a coarser grade of calcium carbonate to bridge.

Salt Solutions Where Increased Density is Required

Table 8-3 shows the approximate density range of solids-free salt solutions:

Potassium chloride can be mixed to provide densities up to about 9.7 lb/gal at 85°F, as shown in Table 8-4.

TABLE 8-3
Density Range of Salt Solutions

Density (lb/gal)	*Salt solutions*
8.3– 9.7	Potassium chloride
8.3– 9.8	Sodium chloride
9.8–11.0	Sodium chloride—calcium chloride
11.0–11.7	Calcium chloride
11.7–15.1	Calcium chloride-calcium bromide
15.2–19.2	Calcium chloride-calcium bromide, zinc bromide

Sodium chloride can be mixed to provide densities up to 9.8 lb/gal. Figure 8-5 shows quantities of water and salt required for 100 barrels of solution. Sodium chloride-calcium chloride mixtures can provide densities from 10.0 to 11.0 lb/gal. Calcium chloride could be used alone, but addition of sodium chloride reduces cost. Material requirements are shown in Figure 8-6.

Calcium chloride can be used for weights up to 11.7 lb/gal with material requirements as shown in Figure 8-7. Dry calcium chloride is available in two grades, 77% and 95%. The 95% minimum is preferred since fewer unidentified solids are included.

Formulations of calcium chloride and calcium bromide can provide solids free fluid densities up to 15.5 lb/gal. Use of zinc bromide can increase solids free fluid density to 19.2 lb/gal. Mixing tables for high density brines are shown in Appendix B.

Formation Damage—Laboratory core flow tests apparently do not show unfavorable fluid/rock interactions with heavy brines—$CaCl_2$, $CaBr_2/ZnBr_2$ or

TABLE 8-4
Density of KCl Fluids

% KCl	*Lb KCl per bbl*	*Sp Gr of solution*	*Density lb/gal*
1	3.52	1.0046	8.37
2	7.03	1.0110	8.42
3	10.56	1.0175	8.48
4	14.08	1.0239	8.53
6	21.12	1.0369	8.64
8	28.16	1.0500	8.75
10	35.20	1.0633	8.86
15	52.79	1.0986	9.15
20	70.39	1.1328	9.44
25	87.98	1.1670	9.72

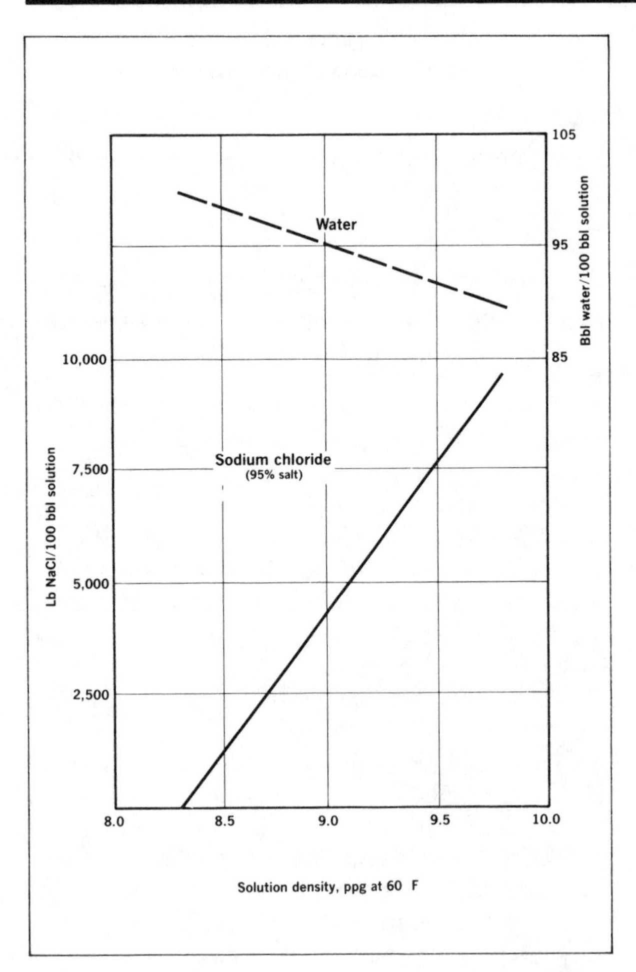

FIG. *8-5—Material requirements for preparing sodium chloride solutions.*

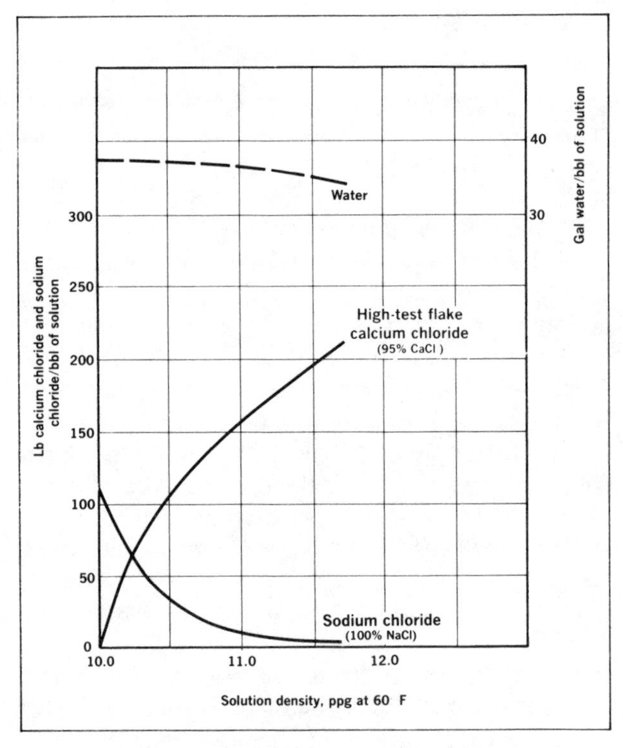

FIG. *8-6—Material requirements for preparing calcium chloride-sodium chloride solutions.*

$CaCl_2/CaBr_2/ZnBr_2$. Recovery of heavy brines can be slow due to high viscosity. Even though reaction with formation materials does not appear to be a serious problem, incompatibility between $CaCl_2$ completion fluids and formation brine can result in damaging precipitates. This problem is more severe if the formation brine contains significant sulfate or bicarbonate concentration. $Ca(OH)_2$ and $CaCO_3$ might be precipitated due to a temperature increase. These salts are acid soluble, thus, reduction of pH tends to prevent precipitation. Mixing of high density brines with seawater (high in SO_4) should be avoided.

Crystalization—Since salt solubility varies with temperature, each brine formulation will have a minimum temperature at which it can be used. This crystalization temperature (Fig. 8-8) is the temperature at which salt crystals first form as the brine is cooled. Lower crystalization temperatures can be obtained by using higher percentages of the more soluble salt. Most brines can be cooled to temperatures well below the crystalization point before the quantity of precipitated salt interferes with routine handling. Crystals redissolve when the brine is heated above the crystalization temperature.

Corrosivity—Toxicity—Safety—When mixing high concentrations of $CaCl_2$, $CaBr_2$, or $ZnBr_2$, precautions should be taken to keep the dry chemical dust out of eyes and lungs. Rubber protective clothing should be worn to prevent skin damage. Considerable heat may be generated thus precautions should be taken to prevent burns. $CaCl_2$–$CaBr_2$ brine toxicity is low enough to allow use of these solutions in marine waters. $ZnBr_2$ can be toxic to fish, which limits its use in offshore areas. Onshore, precautions must be taken to avoid contamination of water supplies. $CaCl_2$–$CaBr_2$ brines are alkaline, whereas $ZnBr_2$ brines are slightly acidic and therefore more corrosive. Figure 8-9 shows corrosion rates of high weight brines inhibited with 0.4% Corban 333 (Dowell). Oxygen solubility decreases as brine density increases; thus oxygen scavengers are not needed in heavy brines.

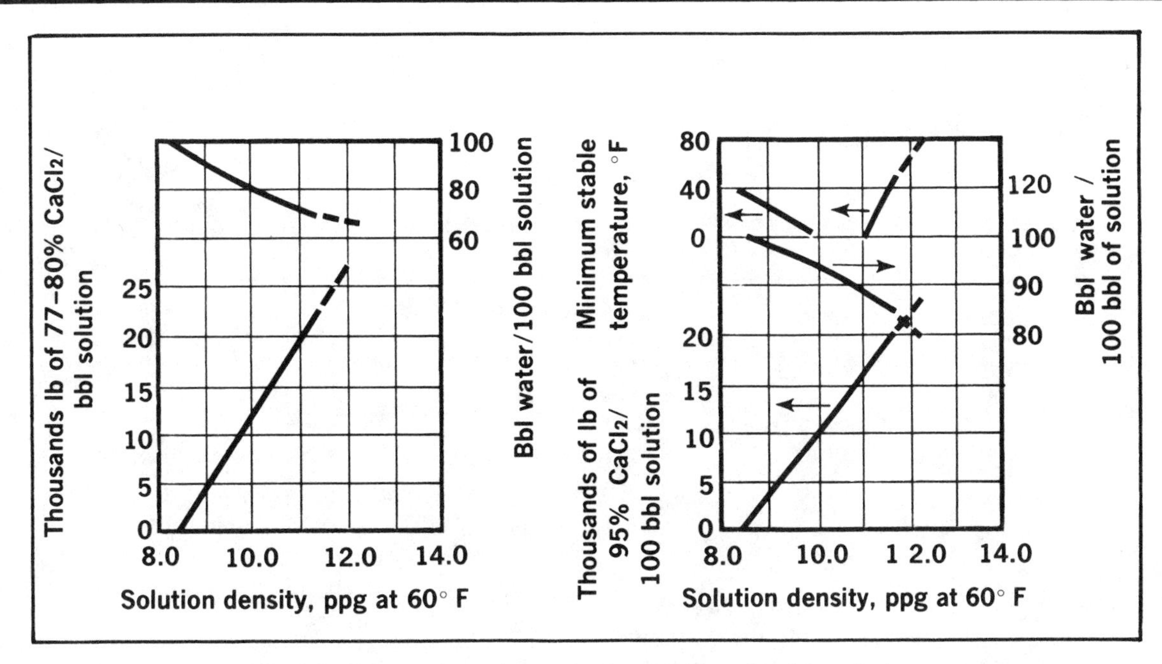

FIG. 8-7—*Material requirements for calcium chloride solutions.*

Viscosity—Heavy brines have inherent viscosity (3–20 cps). If more viscosity is needed HEC can be added, usually in concentrations of 0.25 to 1.0 lb/bbl. Polymer hydration is slower in high density brines. Rapid agitation is important and heat may be needed. Too rapid addition will cause clumping. All polymer should be added over one circulation since dispersion in previously thickened brine is more difficult. Brines thickened with HEC can be thinned by addition of an oxidizing breaker such as calcium hypochlorite.

Cost—Heavy brines are very expensive. A 15.1 lb/gal $CaCl_2$–$CaBr_2$ brine costs about 25 times more than a 10.0 lb/gal $CaCl_2$ brine. An 18.0 lb/gal

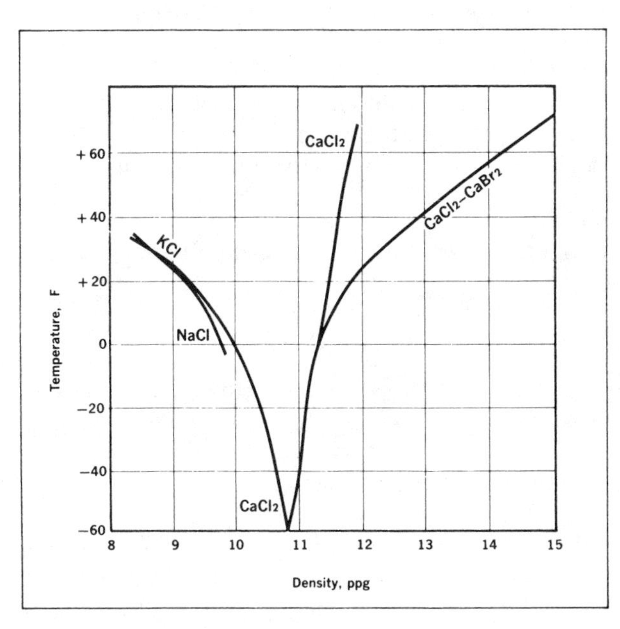

FIG. *8-8—Crystalization temperature of salt solutions.*[9] *Permission to publish by Gulf Publishing Co.*

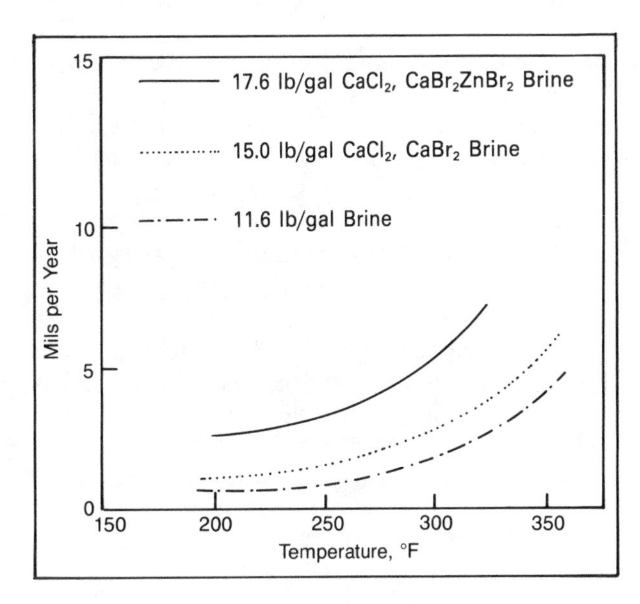

FIG. *8-9—Corrosion rates of weighted brines inhibited with 0.4% Corban 333, based on 30 day exposure of N80 coupons. Courtesy Dowell.*

TABLE 8-5
Fluid Density Adjustment For Downhole Temperature Effect

Surface-measured density		*Loss in density per 100°F rise in average circulating temperature above surface-measured temperature*	
lb/gal	*sp gr*	*lb/gal*	*sp gr*
8.5	1.020	0.35	0.042
9	1.080	0.29	0.035
10	1.201	0.26	0.031
11	1.321	0.23	0.028
12	1.441	0.20	0.024
13	1.561	0.16	0.019
14	1.681	0.13	0.016
15	1.801	0.12	0.014

$CaCl_2$–$CaBr_2$–$ZnBr_2$ brine costs over 80 times more than a 10.0 lb/gal $CaCl_2$ brine. Fluid recovery, reuse, and resale must be considered. Measures to avoid accidental contamination or loss of brines in transport must be taken. Closed tanks may be needed to prevent hygroscopic absorption of moisture from the air with attendant loss of density. Fluid loss to the formation must be considered.

Temperature Effects on Brine Density—As the temperature of any water solution is raised, the volume is increased, and density is reduced. Where minimum formation pressure overload is maintained this effect can be important. Table 8–5 can be used to make the needed adjustment. The decrease in density is more important in the lower weight fluids.

Use of Solid Particles to Provide Density—Calcium carbonate (200 mesh) can be used to obtain higher density fluids. Although viscosity and gel strength will be needed to suspend the solid material, this approach may be desirable where densities exceed 11.0 lb/gal.

Iron carbonate, barium carbonate, and ferric oxide (all acid soluble) also can be used as weighting solids. The following equations estimate the weight of solids needed to obtain a certain fluid density and the resulting increase in fluid volume.

Weighting material required:

$$W = \frac{K(\rho_f - \rho_i)}{C - \rho_f}$$

Increase in volume: $\Delta V_i = W/K$

W = weighting material needed, lb/bbl of initial fluid.

TABLE 8-6
Solid Particles to Provide Density Increase

Weighting material	*Specific Gravity sp gr*	*Density increase obtainable, lb/gal*	*Equation constants* *K*	*C*
Calcium carbonate	2.7	3.5	945	22.5
Iron carbonate	3.85	6.5	1348	32.1
Barium carbonate	4.43	8.0	1551	37.0
Ferric oxide	5.24	10.0	1834	43.7

ρ_f = fluid density desired, lb/gal.
ρ_i = density of available brine, lb/gal.
ΔV_i = volume increase, bbl/bbl initial fluid.
K, C = constants for weighting material (from Table 8–6).

Based on the above relations, 11.5 lb/gal $CaCl_2$ brine would require 150 lb/bbl of calcium carbonate to prepare a 13.0 lb/gal fluid. Volume increase would be 16 bbl/100 bbl of initial brine. Adequate suspension qualities would probably require 0.5 to 1.0 lb/bbl polymer. Settling should be checked before putting fluid in the hole. If settling occurs, more polymer is required.

For effective fluid loss control some of the calcium carbonate should be larger than the 200-mesh size range. Suspension improves, however, as particle size decreases. As an alternate approach initial fluid loss control could be established with a "pill" of graded calcium carbonate, after which 200-mesh size range could be used for density requirements.

Care and Maintenance of Clean Salt Water Fluids

These points are important in the care and maintenance of clean salt-water fluids:

1. Dirty mixing and storage tanks or dirty vacuum tanks are a common source of contamination to a clean workover fluid system. Tanks must be thoroughly cleaned before use.
2. Workover rig tanks should be equipped with sumps and bottom baffles to contain settlings. Suction should be about 18 in. off bottom. Tanks should have easily accessible clean-out plates. Rounded corners aid cleaning.
3. Settlings in workover tanks should be checked hourly and cleaned as needed. Samples from the

pump discharge are helpful in checking for undesirable solids.

4. Dirty tubing strings are often a source of rust, scale, pipe dope, etc. They can best be cleaned in the well by setting a wireline plug at the bottom, running a string of 1-in. and circulating HCl, IPA or simply water with about 1 lb/gal frac sand for scouring. If a sliding sleeve is available above the packer, HCl can be circulated to the bottom of the tubing, then reversed out; however, a bottom tubing plug is desirable.

5. A 4-in. cone desilter properly operated should take out a high percentage of solids down to 10–20 microns, with decreasing percentage removal of solids down to 2–3 microns.

6. For situations where clean no-solids fluids are critical (i.e., sand control), filters should be used. One type of filter frequently used is the Peco filter having 2–5 micron sock-type cartridge. Skid-mounted unit with two cartridge containers will filter water with 2-micron cartridge filter at 72 gal/min through one filter unit, or 144 gal/min through both. With 10 micron cartridge, capacity is five times as great.

Polypropylene sock-type filter cartridges appear to be the most practical. Filter life can be quite short (30 min.) depending on the solids in the unfiltered fluid. Series filtration with a 50 to 75 micron lead filter, and one or two 2 to 5 micro polish filters, increases filter life and efficiency. For continuous filtration, two skids, each having two or three filter pots, are needed in order that one skid can be serviced while the other is in use. Careful supervision is the key to successful operation.

Sand filters similar to those used with waterfloods have also been used. Periodic back washing is required but is sometimes easier to accomplish than changing sock-type filters.

7. Recent Shell work has shown that surface filtration alone is not enough to insure a no-solids fluid at the bottom of the hole. Fluid previously filtered through a 2-micron unit was circulated through a tubing string having additional filters at the top and bottom. The top tubing filter remained clean, but the bottom filter consistently plugged with chemically-formed iron oxide, and particulate matter from the tubing, i.e., pipe dope, scale, and rust.

8. Laboratory work by Union has emphasized the point that iron oxide particles are a serious plugging material. The reaction between oxygen and iron can be prevented by addition of sodium sulfite (and cobalt sulfate as a catalyst) to scavenge the oxygen and sodium citrate to sequester the iron.

CONVENTIONAL WATER-BASE MUD

Economics and availability sometimes suggest use of water-base mud rather than weighted salt water where weights above 10.5 to 11.0 lb/gal are required. However, the filtrate contains clay thinners and dispersants and a high concentration of fine solids certain to cause irreparable formation damage both inside and on the face of the formation.

Thus, water mud should never be used except in zones to be abandoned.

In new wells the presence of mud can be avoided economically by pumping the primary cement plug down with salt water or oil. Mud can be circulated out before perforating using production tubing string; however, once mud solids are inside the production casing complete removal may be difficult.

OIL BASE OR INVERT-EMULSION MUDS

These muds are usually less damaging from the standpoint of clay problems than conventional water-base muds since filtrate is oil and very low filtration rates can be obtained. Most oil-base systems contain strong emulsifiers which may oil-wet the formation, and blown asphalt which can plug the formation as well as present an oil-wet surface. Thus, emulsion blockage could be severe.

Some recent invert-emulsion systems utilize emulsifiers chosen to minimize this problem. Perforating in weighted oil muds could form mud plugs which would not be removed by backflowing. Although lab tests show flow index is not reduced, the mud plugs are detrimental to subsequent squeeze cement jobs.

The cost of oil-base and invert-emulsion muds is relatively high, and usually can be justified only in cases where formation clays would be seriously damaged by conventional water-base mud.

Oil-base or invert-emulsion muds are better packer fluids than water-base muds from the standpoint of corrosion and settling of solids. Temperature stability is very good. Solids do tend to settle over long periods unless the mud is properly treated with oil-dispersible clay to develop gel strength.

FOAM

In low fluid level wells where circulation of solids-free oil or water-base fluids would not be possible, foams can be used for certain workover operations such as washing out sand, drilling in or deepening. Depending on the ratio of air to foam water circulated, flow gradients as low as 0.1 to 0.2 psi/ft are possible.

Foam is a simple mechanical mixture of air or gas dispersed in clean fresh water or field brine containing a small amount of surfactant. Surfactant type and concentration should be selected to develop a stable foam with the specific well fluids encountered.

Equipment requirements include an air compressor or source of measured gas, mixing tanks for foamer solution, a liquid pump, metering facilities for air and liquid volumes, and a foam generator to provide good dispersion of the air in the foam solution.

Equipment needed to handle foam returns includes a tubing rotating head or stripper assembly at the wellhead to divert the foam returns into a blooie line and to a disposal pit. At the pit, a water spray system may be required to break the foam. Aluminum Stearate acts as a good defoamer.

Typical air compressor requirements are 500 to 1000 cfm at a pressure of about 500 psi. Water and foaming agent are mixed and injected into the air stream at a rate of 10 to 20 gal/min. Foaming agent concentrations of 0.5 to 1.0% are typical. Bentonite or polymers are added to the water to produce a ''stiff foam'' having greater carrying capacity.

The prime advantage of foam is the combination of low density and high lifting capacity at moderate flow velocities. Bottom-hole pressures as low as 50 psi have been measured at 2,900 ft while circulating foam. Use of foam in sand washing is justified on the basis of a much faster operation and more complete sand removal.

Foam generated with natural gas or nitrogen has been used in connection with small diameter reelable tubing or snubbing equipment to clean out higher pressure wells without killing them. Foam returns in these cases are directed through the normal flowline system to production separation facilities.

Being a compressible two-phase fluid, the rheology of a foam system is complicated. One major company has developed a computer program to determine injection pressures, bottomhole circulating pressure, annular velocity, and foam-lifting ability at various gas and foamer solution rates. The program considers liquid and gas entry from the formation, temperature gradients, hole deviation, etc.

PERFORATING FLUIDS

Perforating fluids are not necessarily a distinct type of fluid, but are distinguished here to emphasize the importance of perforating in a no-solids fluid.

Salt Water or Oil—When clean, these do not cause mud plugging of perforations, but if the pressure differential is into the formation, fine particles of charge debris will be carried into the perforation.

Acetic Acid—This is an excellent perforating fluid under most conditions. In the absence of H_2S, acetic acid can be inhibited against any type of steel corrosion for long periods at high temperatures. Normally a ten percent solution is used. Acetic acid plus H_2S is very difficult to inhibit against embrittlement. Acetic acid will put iron sulfide and mineral carbonate in solution. These may result in added corrosion problems.

Nitrogen—This has advantages as a perforating fluid in low pressure formations, or where rig time or swabbing costs are very high, or where special test programs make it imperative that formation contamination be avoided.

Gas Wells—These can be completed economically in ''clean fluid'' by perforating one or two holes, bringing the well in and cleaning to remove as much wellbore fluid as possible, then perforating the remaining zones as desired.

PACKER FLUIDS

Criteria—Water-base drilling muds as used today are generally not good packer muds. An acceptable packer fluid must meet two major criteria:

1. Limit settling of mud solids and/or development of high gelation characteristics.
2. Provide protection from corrosion or embrittlement.

Corrosion Protection.

Laboratory tests have shown that many drilling mud additives degrade upon prolonged exposure to high temperatures to form carbon dioxide and hydrogen sulfide. Both are corrosive in water-base mud. Bacterial activity in water-base mud can cause breakdown of organic additives to form organic acids, carbon dioxide, and soluble sulfide. Lignosulfonate solutions can react electrochemically at a metal surface to form sulfide even at moderate temperatures.

Thermal degradation of drilling fluid additives and the action of sulfate-reducing bacteria have been recognized for some time. Normally, bacterial action does not occur above temperatures of 150° to 175°F or in fluids treated with sufficient biocide, or where pH is maintained above 12.

Normally serious thermal degradation of drilling mud chemicals will not occur at temperatures below 300°F. Previously, recommendations for packer fluids were based on avoiding these conditions. In special cases low temperature degradation may be important and would limit the use of some chemicals in packer fluid environments.

Case histories have shown that, even when thermal degradation and bacterial activity were eliminated, serious corrosion and embrittlement of high strength tubing still occurred. Laboratory evidence shows that sulfide can be formed under the influence of electrochemical action where even small amounts of sulfur-containing organics are present.

If a corrosion cell is formed on tubing or casing with enough potential to release hydrogen (0.4 volts in lab experiments), then a sulfur-containing organic mud additive could be reduced to sulfide even at room temperature and with no bacterial activity. These conditions can exist downhole since the potential between mill scale and pipe wall has been reported to be 0.5 volts and under some conditions as high as 0.9 volts.

A number of detailed case histories are available from deep wells in South Louisiana which indicate electrochemical reactions may have been the primary cause of failures of N-80 and harder tubing and casing. Even where chemically-treated drilling muds are replaced by a CMC/Bentonite weighted fluid corrosion occurred since the chemically-treated fluid was not successfully removed by circulation.

Where oil muds were used in drilling and as the packer fluids, local cell electrochemical actions apparently did not occur. Casing potential surveys have shown that current is not entering or leaving the pipe, indicating corrosion cell action is not occurring.

Although there is still much to be learned about corrosion and embrittlement due to packer fluids, the following recommendations suggested by Baroid should be considered:

PACKER FLUID RECOMMENDATIONS

Condition A

No high strength pipe involved in completion (N-80 is borderline case).

Packer fluid density of less than 11.5 ppg required.

Recommendation:

1. Use diesel oil or sweet crude treated with an inhibitor where density requirements permit.
2. Use clear water or brine with an inhibitor and a biocide. Inhibitor and biocide must be compatible.

Condition B

No high strength pipe involved in completion.

Fluid density greater than 11.5 ppg required.

Bottom-hole temperature does *not* exceed 300°F.

Recommendation:

1. Economics of workover must be considered. Where workovers are inexpensive, a water-base mud treated with a biocide might be economical. Tests should be made to ascertain that mud does not contain soluble sulfide: pH should be maintained at 11.5 for a few days prior to completion if possible. Solids should be kept to a minimum to avoid gelation with high pH.
2. In remote locations where workovers are expensive or where workover frequency has been found to be high with water-base muds, use a properly formulated oil mud.

Condition C

No high strength pipe involved in completion.

Density of more than 11.5 ppg required.

Bottom-hole temperature exceeds 300°F.

Recommendation:

1. Use properly formulated oil mud.

Condition D

High strength pipe to be used under any condition of fluid density or bottom-hole temperature.

Recommendation:

1. Where fluid density requirements permit, use oil treated with both an oil-soluble and a brine-dispersible corrosion inhibitor.
2. Use oil mud formulated to meet density and temperature requirements.

In a conductive water-base packer fluid no sulfide can be tolerated without risk of stress cracking of high strength pipe. If the well has been drilled with water-base muds as presently formulated, some of this mud would remain and electrochemical reactions might result in sulfide being formed unless corrosion inhibition were 100% effective.

Amine compounds in the oil-mud packer fluid should oil-wet the surface of the pipe and protect against this type of stress corrosion cracking. Maximum protection would be obtained by using an oil mud for drilling and then gelling the oil mud to be left as the packer fluid.

WELL KILLING

Circulation rather than bullheading is the preferable way to kill conventional completions.

An adjustable choke should be used to hold casing back pressure on the formation when killing a well by circulation. For a high pressure well, a Swaco well control choke may be desirable.

For single completions on a packer, the recommended procedure is as follows:

1. Fill the annulus.
2. Open circulating port in tubing or punch hole in tubing above packer.
3. Pump slowly down casing-tubing annulus ($\frac{1}{4}-\frac{1}{2}$ BPM) as wireline tools are retrieved to build up a back pressure on formation.
4. After wireline tools are retrieved, pump at a constant rate of 2–3 BPM to build up 200–300 psi on tubing.
5. Maintain a constant pump rate and manipulate the adjustable choke, controlling tubing returns to keep casing pressure constant.

For a tubingless completion—or where circulation is not possible—bullheading a non-damaging fluid is best if formation will take fluid without "breakdown" or fracture. Here are four important points.

1. Breaking down the formation may cause difficult squeeze cementing and producing problems.
2. For "bullhead" well killing the surface pressure plus fluid gradient times depth should be less than formation breakdown pressures.
3. It may be necessary to have a surface pressure regulator to prevent over-pressuring.
4. If it is necessary to break down the formation, the size of the resulting fracture can be minimized by low injection rates and high fluid loss.

REFERENCES

1. Glenn, E. E., and Slusser, M. L.: "Factors Affecting Well Productivity: I, Drilling Fluid Filtration; and II, Drilling Fluid Particle Invasion into Porous Media," *Trans.,* AIME (1957) 210, 126 and 132.

2. Monaghan, P. H.; Salathiel, R. A.; Morgan, B. E.; and Kaiser, A. D., Jr.: "Laboratory Studies of Formation Damage in Sands Containing Clays," *Trans.* AIME (1959) 216, 209.

3. Black, H. N., and Hower, W. F.: "Advantageous Use of Potassium Chloride Water for Fracturing Water-Sensitive Formations," API Paper 850-39-F (1965).

4. Simpson, J. P., and Barbee, R. D.: "Corrosivity of Water-Base Completion Fluids," 23rd Annual NACE Conf., Los Angeles, CA (March 3, 1967).

5. Hutchinson, S. O.: "Foam Workovers Cut Costs 50%," *World Oil,* (Nov. 1969).

6. Christensen, R. J.; Connor, R. K.; and Millhone, R. S.: "Applications of Stable Foam in Canada," *Oilweek* (Sept. 20, 1971).

7. Bruist, E. H.: "Better Performance of Gulf Coast Wells," SPE 4777, New Orleans, Feb. 1974.

8. Tuttle, R. N., and Barkman, J. H.: "The Need for Non-Damaging Drilling and Completion Fluids," *JPT* (Nov. 1974) p. 1221.

9. Suman, George O., Jr.: *Sand Control Handbook,* Gulf Publishing Company, Houston (1975).

10. Sparlin, Derry, and Guidry, J. P.: "Study of Filters Used for Filtering Workover Fluids," SPE 7005, Third Annual Formation Damage Symposium (Feb. 1978).

11. Llfrey, W. T.: "Recommended Procedures for Utilizing High Cost Non-Damaging Fluids," SPE 8794, Fourth Annual Formation Damage Symposium (Jan. 1980).

12. Sharp, Keith W.: "Filtration of Oil Field Brines—A Conceptual Overview," SPE 10657, Fifth SPE Symposium on Formation Damage Control (March, 1982).

13. Scheuerman, R. F.: "Guidelines for Using HEC Polymers for Viscosifying Solids Free Completion and Workover Brines," SPE 10666, Fifth SPE Symposium on Formation Damage Control (March, 1982).

14. Morganthaler, L. N.: "Formation Damage Tests of High-Density Brine Completion Fluids," SPE 14831 (Feb. 1986).

15. Houchin, L. R.; Hudson, L. M.; Caothien, S.; Daddazio, G.; and Hashemi, R.: "Reducing Formation Damage Through Two-Stage Polymer Filtration," SPE 15408 (October, 1986).

16. Parks, C. F.; Clark, P. E.; Barkat, Omar; Halvaci, M.: "Characterizing Polymer Solutions by Viscosity and Functional Testing," *Amer. Chem. Soc.* Sept. 1986.

17. Nehmer, W. L., "Viscoelastic Gravel-Pack Carrier Fluid," SPE 17168 (Feb. 1988).

Appendix A

Commercial Completion Fluid Products:

Product Name *Description or Use*

Baroid

ALDACIDE—Paraformaldehyde powder, microbiocide
Aluminum Stearate—Defoamer
BARABUF—Magnesium oxide pH stabilizer
BARACARB—$CaCo_3$, bridging, weighting material
BARACOR A—Corrosion inhibitor, microbiostat
BARA-DEFOAM 1—Surfactant defoamer
BARAVIS—Pure hydroxyethylcellulose
DEXTRID—Filtration control, polysaccharide
INVERMUL—Water-in-oil emulsion mud
Q BROXIN—Thinner, dispersant, ferrochrome lignosulfonate
BARAZAN—Xanthan gum, suspension agent

Dowell Schlumberger

A-181—bisulfite oxygen scavenger
D-47—defoamer
F-68-Eze Flo—non-ionic surfactant foaming agent
G-2—non-ionic surfactant foaming agent
J-158—fast hydrating HEC
J-168—soluble bridging material
J-211—lgs., $CaCO_3$
J-212—HEC, lgs., $CaCO_3$
J-213—HEC, lgs., $CaCO_3$ large particles
J-214—$CaCO_3$
J-215—HEC and $CaCO_3$
J-237, J-275—liquid oil soluble F.L. additive
J-333—BRINE SAVER liquid oil soluble F.L. additive
M-76—Bactericide 400 water soluble, cationic bactericide
M-129—Oxygen scavenger
M-133—Bactericide 300 water soluble, cationic bactericide
M-155—Bactericide 600 water soluble cationic bactericide
S-54—calcium bromide brine
S-55—dry sacked calcium bromide
DOWCIDE G—sodium pentachlorphenate compound bactericide

Imco Services, A Division of Halliburton Co.

DRIL—S—Polymers, biocides and F.L.C. materials for drilling
FOAMBAN—surface active, readily dispersible, liquid defoamer
PRESERVALOID—paraformaldehyde to prevent starch fermentation
SAFE—PAC—selected polymers and other compounds
SAFE-PERFSEAL—lgs., synthetic polymers and sized carbonates for L.C.
SAFE-SEAL—sized carbonates
SAFE-SEAL-X—sized carbonates
SAFE-TROL—filtrate control additives
SAFE-VIS—synthetic polymers for viscosity and F.L.C.

Kelco Oil Field Group Inc.

XANVIS—clarified xanthan gum, suspension agent-low shear rate viscosity
BIOZAN—cement tolerant biopolymer

Magcobar Operations, Oilfield Products Division, Dresser Industries, Inc.

MAGCONAL—aluminum stearate defoamer
CEASCAL—lgs. and $CaCO_3$ F.L.C.
CEASTOP—Polymers, $CaCO_3$, lgs. and complex heavy metals L.C. pill
MIXICAL—sized $CaCO_3$ and surface active agents bridging material
POLYBRINE—polymers and carbonates viscosifier and F.L.C.
IRON CARBONATE—W.M.

Milchem

AEP-132—bactericide
W.O. DEFOAM—alcohol-base defoamer
W.O.20—polymeric viscosifier and F.L. additive
W.O.30—$CaCO_3$ in Fine and Coarse grades
W.O.50—$CaCo_3$

Appendix B

TABLE 8B–1
Fluid Density Conversion Table

Degrees API	Specific Gravity	Density lb/gal	Fluid Gradient psi/ft
60	0.739	6.16	0.320
58	0.747	6.23	0.324
56	0.755	6.30	0.327
54	0.763	6.36	0.330
52	0.771	6.43	0.334
50	0.780	6.51	0.338
48	0.788	6.57	0.341
46	0.797	6.65	0.345
44	0.806	6.72	0.349
42	0.816	6.81	0.354
40	0.825	6.88	0.357
38	0.835	6.96	0.362
36	0.845	7.05	0.366
34	0.855	7.13	0.370
32	0.865	7.21	0.375
30	0.876	7.31	0.380
28	0.887	7.40	0.384
26	0.898	7.49	0.389
24	0.910	7.59	0.394
22	0.922	7.69	0.399
20	0.934	7.79	0.405
18	0.946	7.89	0.410
16	0.959	8.00	0.416
14	0.973	8.11	0.421
12	0.986	8.22	0.427
10	1.00	8.34	0.433
	1.03	8.6	0.447

TABLE 8B–1 (*cont.*)
Fluid Density Conversion Table

Degrees API	Specific Gravity	Density lb/gal	Fluid Gradient psi/ft
	1.08	9.0	0.468
	1.13	9.4	0.488
	1.18	9.8	0.509
	1.20	10.0	0.519
	1.25	10.4	0.540
	1.29	10.8	0.561
	1.34	11.2	0.582
	1.39	11.6	0.603
	1.44	12.0	0.623
	1.49	12.4	0.644
	1.53	12.8	0.665
	1.58	13.2	0.686
	1.63	13.6	0.706
	1.68	14.0	0.727
	1.75	14.4	0.748
	1.77	14.8	0.769
	1.82	15.2	0.790
	1.87	15.6	0.810
	1.92	16.0	0.831
	1.97	16.4	0.852
	2.01	16.8	0.873
	2.06	17.2	0.894
	2.11	17.6	0.914
	2.16	18.0	0.935
	2.21	18.4	0.946
	2.25	18.8	0.977
	2.30	19.2	0.997
	2.35	19.6	1.020

TABLE 8B–2
Mixing Schedule: Calcium Chloride–Calcium Bromide Brine

	Quantities for 1 bbl Brine						
	Pelletized (95%) Calcium Chloride Flake (95%) Calcium Bromide			*14.2 lb/gal Calcium Bromide Brine 11.6 lb/gal Calcium Chloride Brine Pelletized (95%) Calcium Chloride*			
Brine Density lb/gal @ 60°F	*Water bbl*	*$CaCl_2$ lb*	*CaBr lb*	*CaBr Brine bbl*	*$CaCl_2$ Brine bbl*	*$CaCl_2$ Pellets lb*	*Crystallization Temperature °F*
11.8	.8281	189.81	16.28	.0507	.9429	6.1	51
12.0	.8206	184.73	32.57	.1016	.8857	12.1	52
12.2	.8129	179.66	48.86	.1524	.8286	18.2	54
12.4	.8053	174.58	65.15	.2032	.7715	24.2	54
12.6	.7977	169.51	81.43	.2540	.7143	30.3	55
12.8	.7901	164.43	97.71	.3048	.6572	36.4	55
13.0	.7826	159.36	114.00	.3556	.6000	42.4	57
13.2	.7749	154.29	130.28	.4064	.5429	48.5	57
13.4	.7673	149.21	146.57	.4572	.4857	54.6	58
13.6	.7597	144.14	162.86	.5080	.4286	60.6	58
13.8	.7522	139.06	179.14	.5589	.3714	66.7	58
14.0	.7446	133.99	195.43	.6069	.3143	72.8	58
14.2	.7369	128.92	211.71	.6604	.2572	78.8	59
14.4	.7293	123.84	228.00	.7113	.2000	84.9	61
14.6	.7217	118.77	244.28	.7620	.1429	90.9	63
14.8	.7142	113.69	260.57	.8128	.0858	97.0	65
15.0	.7066	108.62	276.86	.8637	.0286	103.0	66

NOTE: Brines in this density range can be mixed in several ways and the formulation listed is not necessarily the preferred formulation.

TABLE 8B–3
Mixing Schedule
Calcium Chloride–Calcium Bromide–Zinc Bromide Brine

	Quantities for 1 bbl Brine			
Brine Density lbs/gal @ 60°F	*56.7% Zinc Bromide, 19.7% Calcium Bromide Brine bbl*	*53% Calcium Bromide Brine bbl*	*95% Calcium Chloride Pellets lb*	*Crystallization Temperature °F*
15.2	.024	.866	103.3	62
15.4	.071	.826	98.6	59
15.6	.119	.783	93.5	59
15.8	.167	.741	88.4	57
16.0	.214	.699	83.4	54
16.2	.263	.656	78.3	52
16.4	.310	.613	73.2	50
16.6	.357	.572	68.2	47
16.8	.429	.508	60.6	40
17.0	.442	.449	90.0	40
17.2	.495	.421	72.2	38
17.4	.561	.380	51.8	38
17.6	.619	.339	40.4	45
17.8	.667	.296	35.3	44
18.0	.714	.254	30.3	43
18.2	.762	.212	25.3	41
18.4	.810	.169	20.2	35
18.6	.857	.127	15.2	28
18.8	.905	.084	10.1	23
19.0	.952	.043	5.1	18
19.2	1.000	—	—	16

NOTE: Brines in this density range can be mixed in several ways and the formulation listed is not necessarily the preferred formulation.

Chapter 9 Through-Tubing Production Logging

Questions for production logs
Basics of through-tubing devices
Application of production logging
Field examples
Location of tubing and casing leaks
Flow behind casing
Fluid entry location
Producing zone evaluation-reservoir management
Water-injection well profiling
Primary cement evaluation
Theory of temperature log interpretation

INTRODUCTION

Through-tubing Production Logging refers to logs run after the production string casing has been cemented and the well placed on production. Measurements are made under dynamic as well as static conditions.

Surface fluid measurements are usually not adequate to determine the efficiency of the downhole production or injection system. In many wells downhole malfunctions, related to mechanical problems or communication problems, may be reducing ultimate recovery in the reservoir. For effective management of the reservoir, measurement of fluid saturations is needed periodically at various points in the reservoir.

Production logs have application in three major areas: diagnosis of mechanical problems, analysis of individual well performance in relation to the reservoir, and perhaps most important from the standpoint of recovery, management of reservoir fluids.

Major questions which can be addressed to Production Logs are:

—Mechanical Condition of the Well.
Are there casing, tubing, or packer leaks?
Is there internal or external corrosion damage?

—Anomalous Fluid Movements Between Zones.
Is there flow behind casing through inadequate primary cementing?
Is flow from inside casing moving into thief zones?

—Evaluation of Completion Efficiency (Producing Well).
Are some zones not contributing?
Are some zones contributing only gas—or only water?
Are contributing zones producing up to the potential shown by other data sources?

—Evaluation of Completion Efficiency (Injection Well).
Where are injected fluids going?
How much into each zone?

—Design and Evaluation of Stimulation Treatment.
Which zones need to be stimulated?
Where did stimulation fluids go?
Did stimulation achieve the desired result?

—Reservoir Management.
What are initial fluid saturations in each zone?
What fluid saturation changes have occurred, due to production—or extraneous fluid movement?
Is the reservoir being depleted in the desired manner?

Confident answers to these questions require careful design and application of Production Logging techniques. Usually more than one logging device or measurement is required. For producing wells, the combination of radioactive devices to evaluate fluid type and saturations behind casing, and flow devices and fluid differentiation devices to evaluate fluids and fluid movements inside casing, can provide most of the clues needed to effectively deplete all hydrocarbon zones penetrated by the well. For flow behind the casing, the combination of temperature and noise measurements usually provides optimum definition. Data from all available sources must be considered to provide most effective diagnosis.

DOWNHOLE MEASUREMENTS

There is no lack of measurements that can be made downhole, each providing clues aiding problem diagnosis. These measurements are outlined as follows.

Temperature is a simple economical measurement and is affected by many factors associated with problems both inside and outside the casing. Measurements can be recorded in various manners:

—Temperature at a particular level,
—Temperature at various points around the casing circumference at a particular level.
—Temperature change with time at a particular level,
—Temperature change with depth, and
—Rate of temperature change with depth.

Temperature inside the tubing or casing is a result of heat flow and/or fluid expansion. Heat is transferred through materials and liquids by conduction influenced by temperature difference. Heat is also transferred by convection due to fluid movement influenced by flow rate and fluid type.

Pressure measurements are simple and can also be recorded in several manners. With the well flowing, pressure change with time at a particular level indicates stability of flow conditions. Pressure change with depth (pressure gradient) indicates **fluid density**. Transient pressure analysis is not considered a production logging technique but obviously can add to the store of clues available for problem diagnosis.

Fluid density can also be measured by a radioactive device through gamma ray absorption. Density can distinguish between the amount of water, oil, and gas at a particular level in the wellbore at a given instant. With low gravity crudes, however, there may not be much difference in density between oil and water. Also, it must be recognized that the relative proportion of oil, water, and gas at a point is often not equivalent to the relative rate of movement of the individual fluids passing that point. Unless average flow velocity is quite high, lighter fluids move faster than heavier fluids.

Fluid velocity within the wellbore is often an interesting measurement related to flow entering or leaving the casing. Velocity can be measured with a spinner—if a meaningful relation between spinner speed and flow velocity can be established. This relation depends on downhole fluid density and viscosity and fluid slippage around the spinner, as well as friction in the spinner. For precision, downhole calibration, sometimes at several depths, is required. For multi-phase flow, particularly in deviated holes, additional problems are created because lighter fluids tend to move up the high side of the pipe. Heavier fluids may move up and also back down in lower segments of the cross-section. Diverters, which force a high percentage of the total fluid through the spinner, minimize these problems. Turbulence due to fluid movement into the wellbore, perpendicular to the average flow, make measurements within a perforated zone suspect.

Fluid velocity can also be measured by injecting a radioactive tracer in the flow stream and recording the time to move a particular distance. Problems involve knowing when the tracer actually passes the detector and that it actually moves with the fluid. However, at low flow rates, tracer measurements of velocity may be better than spinner measurements.

Increased radioactivity at a particular depth compared with a base log indicates movement of radioactive fluid or solids into the zone. The rate of decline of the induced radioactivity is related to the rate of movement of the tracer away from the wellbore. Radioactive cement or sand grains can show the position of the cement or frac sand after a treatment; however, it should be recognized that a small amount of radioactive material near the gamma ray detector may give the same reading as a large amount of material several inches further back in the formation.

Decay rate of thermal neutron bursts generated in the wellbore is affected primarily by the amount of chlorine in the area. Since chlorine is usually associated with formation water, the time required for the thermal neutron population to decay by a certain percentage can be related to water and hydrocarbon saturation in the porous rock near the wellbore. However, for this measurement to be meaningful, other factors must be favorable. Formation porosity and water salinity must be reasonably high, and formation shale content must be reasonably low. Also, fluid saturations in the rock near the wellbore may not be the same as those further away due to invasion.

Pulsed neutron bombardment of elements such

as carbon, oxygen, chlorine, silicon, hydrogen, calcium, sulphur, and iron in the rock near the wellbore permits the measurement of the relative amounts of these elements and calculation of the relative saturation of the fluids with which they are associated.

Many factors enter into calculation of fluid saturation by pulsed neutron techniques; thus these devices are most effective when used to compare changes in fluid saturation over a relatively long period of time due to production, or a short period of time due to injection.

Audible noise level and frequency patterns in the wellbore caused by movement of fluid inside or outside the casing can be used to establish the presence of flow, the path of the flow, what fluid phases are involved, and to a degree, the flow rate.

Sound transmission characteristics from a wellbore transmitter to a nearby receiver provide information as to acoustic coupling between the cement and the casing, and the cement and the formation. Under favorable conditions, these characteristics may be related to the possibility of fluid movement between zones outside the casing.

Electrical properties (i.e., conductivity or dielectric constant) usually differ significantly between hydrocarbons and water. Thus they may be used to determine relative amounts of these fluids at a particular level in the wellbore. These measurements are more definitive at low water percentages, and where the fluids are intimately mixed. Again, the relative amounts of fluids residing at a particular level are not necessarily the same as the relative amounts of the fluids passing that level.

LOGGING DEVICES

Most through-tubing logging devices operate on an electric line and record at the surface. Tools have also been developed to run on a slick wireline and record on bottom-hole charts or magnetic tape. Instrumentation will perform properly at hydrostatic pressures up to 15,000 psi and temperatures up to 350°F. Wellhead pressure control equipment, Figure 9-1, allows safe entry into producing wells with a surface pressure as high as 10,000 psi with a 3/16-in. diameter single conductor cable. With a suitable cable and inhibitor, high percentages of H_2S can be handled with safety.

Most significant through-tubing tools are:

—Temperature devices
—Spinner flowmeters

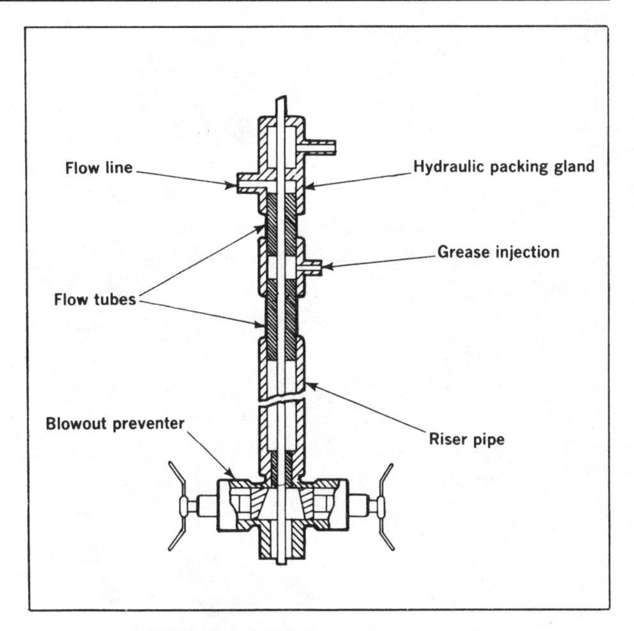

FIG. *9-1—Wellhead pressure control equipment.*

—Gradiomanometer or radioactive density devices
—Radioactive tracer and detector
—Noise device
—Pulsed neutron devices
—Gamma ray-neutron devices
—Bottom-hole pressure device

Invariably production logging devices are measuring parameters that do not give concrete answers to specific questions—but give clues indirectly related to these questions—thus, the problem of the production logger is to put enough clues together from every source available to develop answers having an acceptable confidence level. Experience with specific devices in specific areas is an important factor in effective analysis.

Production logging interpretation many times depends on measurement of small differences; thus, tool operational characteristics and limitations must be known and must be a constant consideration during the logging and the interpretation process. The following brief discussions of typical tools are intended primarily to provide background.

Collar Locator

A primary concern in production logging, as in perforating, is depth control. In perforating a new well, normally a gamma ray or neutron device is run that simultaneously records casing collar locations. The nuclear log is then correlated with logs previously run in open hole to position the collars with respect to the open-hole logs. The casing collars subsequently provide a positive depth reference. Usually a "short" joint is positioned in the casing string somewhere near bottom as a further identification aid. Figure 9-2 shows a magnetic collar locator arranged such that as the unit moves through a collar, the increased thickness of metal disturbs the magnetic field and causes a blip to be recorded at that depth on the log. In production logging operations, collar logs should be recorded on each run or rerun of a log. Each reversal of direction of the logging sonde causes some degree of slippage in the depth odometer. After several reversals, the odometer may be off by 10–20 ft as shown in Figure 9-3.

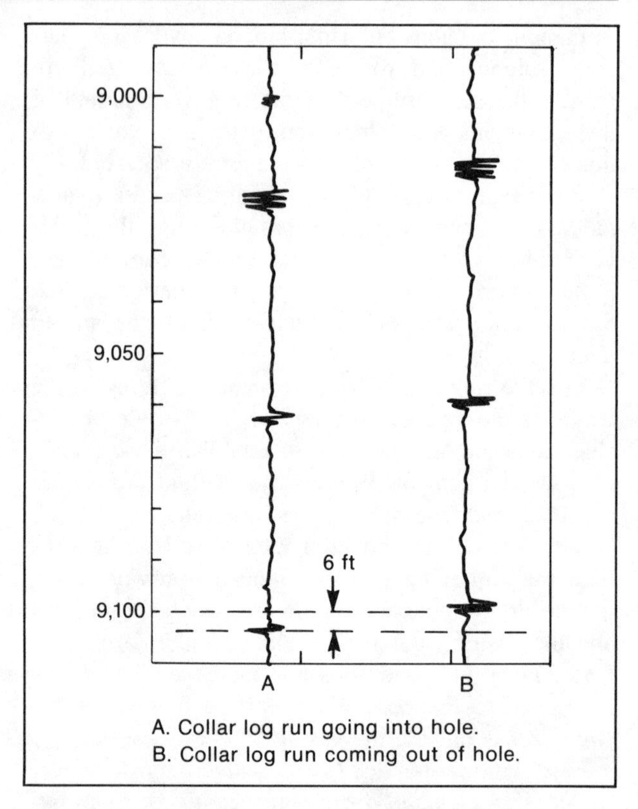

FIG. *9-3—Typical collar record. Permission to publish by the Society of Petroleum Engineers of AIME. Copyright 1983 SPE-AIME.*

Temperature Devices

The **high resolution thermometer** uses an electrical bridge system with an exposed sensing wire as the fourth arm of the bridge, Figure 9-4, to record a temperature profile. Resolution is better than 0.1°F, although actual temperature may be off by 5°F depending on calibration. If desired, differential temperature can be recorded, comparing temperatures at two points a short distance apart and using a more sensitive scale than the absolute temperature curve.

The temperature at a particular level in the wellbore is primarily a result of heat flow. Fluid expansion can be an additional factor in some situations. Heat is transferred through solids and liquids by conduction, or with fluid movement by convection. Rate of heat

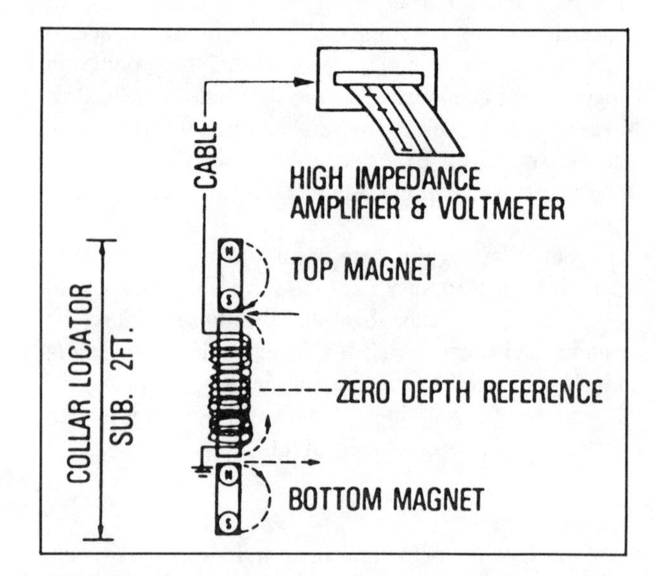

FIG. *9-2—Collar locator section. Permission to publish by the Society of Petroleum Engineers of AIME. Copyright 1983 SPE-AIME.*

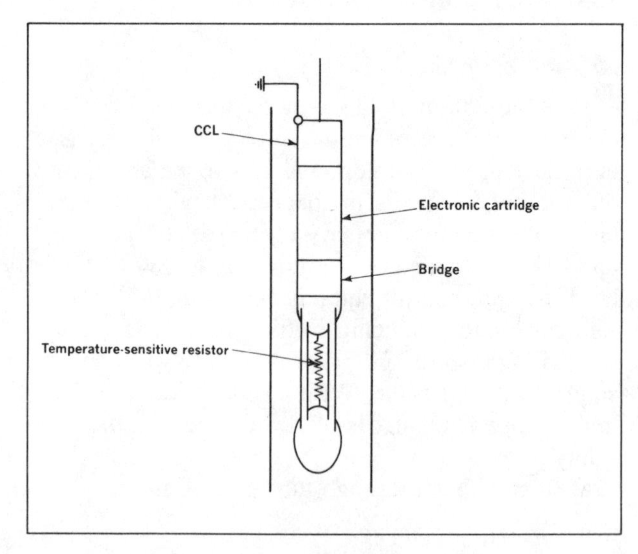

FIG. *9-4—High resolution thermometer.*

transfer through a material at a given temperature difference depends on the thermal diffusivity of the material, which is directly related to the thermal conductivity, and inversely related to the density and specific heat capacity of the material. Table 9-1 shows thermodynamic properties for various materials around the wellbore. Heat transfer by fluid movement depends on the rate of fluid movement and the specific heat capacity of the fluid. Expansion of gas results in significant cooling (1°F per 40 psi pressure reduction with methane at adiabatic conditions), whereas expansion of liquids results in a slight heating effect.

For diagnosis of well problems, the temperature log may be run under either flowing or shut-in conditions. Interpretation is basically a heat transfer problem. With a dynamic situation involving flow in the tubing, annulus, or outside the casing, temperature inside the tubing is primarily affected by convective heat transfer or by fluid expansion effects. With a static situation, temperature in the tubing is primarily affected by conductive heat transfer. Where static conditions have existed for a long period, wellbore temperature reflects a geothermal gradient that varies with location and depth but "typically" is about 1.7°F/100 ft in sand-shale sequences. This gradient is a result of heat flow toward the surface from some point in the earth. Geothermal gradient in a particular area depends on the conductivity of the various formations, and also on geological features such as salt domes (with high conductivity) or perhaps volcanic activity with convective heat transfer. Gradients as high as 4.5 to 5.0°F/100 ft have been recorded. Figure 9-5A and 9-5B show effects of lithology on geothermal gradient. Table 9-2 shows various lithologies in order of decreasing thermal conductivity and thermal diffusivity.

TABLE 9-1
Thermodynamic Properties

	Rock	*Cement*	*Steel*	*Water*
Thermal Conductivity btu/hr/ft °F	1.4–2.0	0.3–0.8	26.0	0.346
Specific Heat btu/lb °F	0.21	0.21	0.12	1.0
Density lb/cu ft	159	109	485	62.4

Courtesy Schlumberger

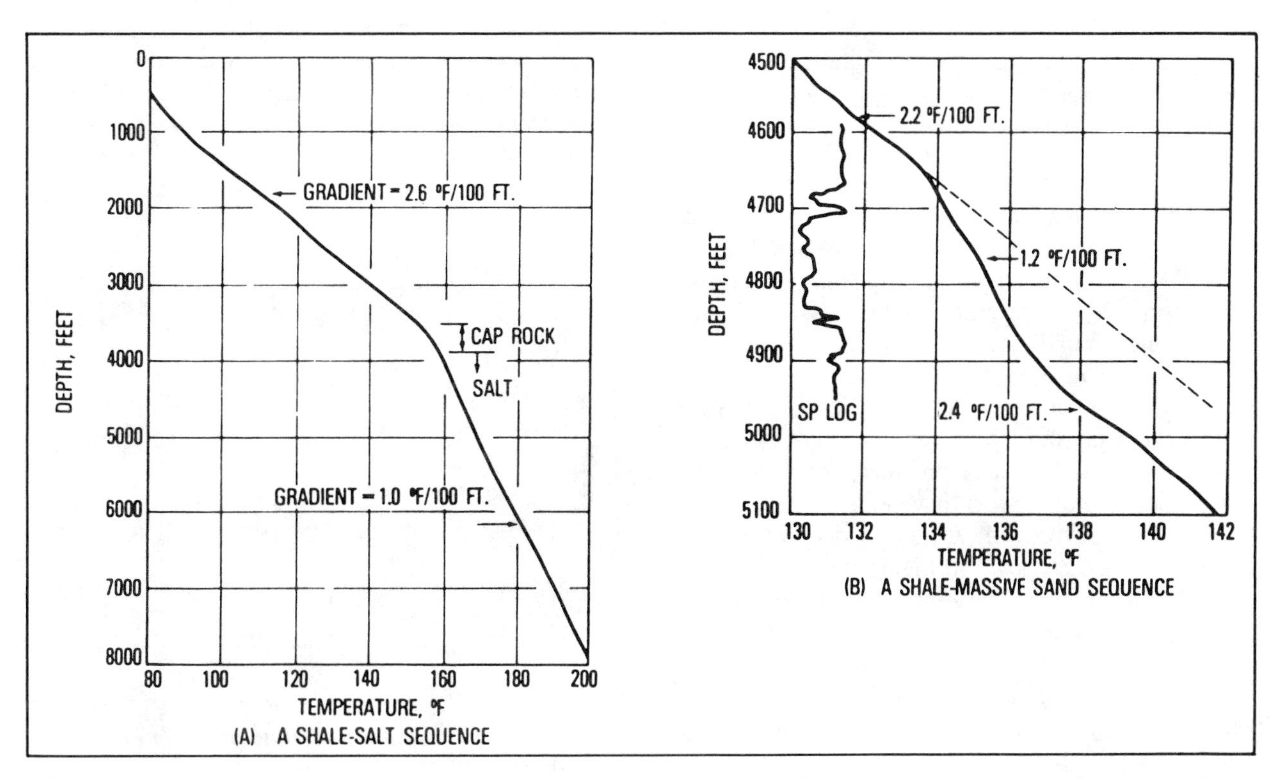

FIG. *9-5—Lithological influence on static temperature gradient. Permission to publish by the Society of Petroleum Engineers of AIME. Copyright 1983 SPE-AIME.*

TABLE 9-2
Lithological Effects on Temperature Logs

Materials in order of	
Decreasing thermal conductivity and increasing static geothermal gradient	*Decreasing thermal diffusivity and increasing lag time*
Quartzite	Quartzite
Salt	Salt
Anhydrite	
	Anhydrite
Dolomite	Dolomite
	Limestone
Limestone	Sandstone
Sandstone	
	Shale
Shale	Gypsum
Gypsum	
Cement	Cement
Water	Gas
Oil	Water
Gas	Oil

With fluid movement in the wellbore convective heat transfer overrides lithological detail. When flow is stopped lithology begins to show up since zones with high thermal diffusivity return to the geothermal temperature more rapidly. The nature of the well completion also influences the temperature profile as it returns to the geothermal. A greater mass of cement in a washout or static fluid in the tubing-casing annulus or behind the casing provides insulation such that temperature of the rock beyond is not affected as much by the convective heat transfer due to the fluid movement. This is shown in Figure 9-6.

The **radial differential temperature device** has the potential of sensing flow through a channel along one side of the casing. It consists of a section that can be anchored by bow springs at a particular depth. Two arms containing temperature sensors are extended to contact the casing. Difference in the output of these sensors as they are rotated around the circumference of the casing indicates fluid movement in a channel.

Thus, the wellbore temperature measurement is affected by many factors needed in well problem diagnosis, but proper interpretation requires a knowledge of many physical factors around the wellbore.

Spinner Flowmeters

Several configurations of spinner flowmeters are available, as shown in Figure 9-7. Basically, they all consist of a propeller mounted on a jewel-bearing supported shaft. Rotation rate and direction are determined magnetically or optically.

Early flowmeters such as the **inflatable packer flowmeter**, Figure 9-7A, divert all flow directly through the spinner. Measurement precision is excellent even at low flow velocities. Restriction created by the packer tool causes a pressure drop that may alter the flow profile and at higher rates may push the tool up the hole. These effects place an upper limit on flow rate of about 1500 bpd in 7-in. casing. Operational complications with the packer flowmeter encouraged development of the non-diverting spinner or **continuous flowmeter**, Figure 9-7B. The continuous flowmeter logs either with or against the flow direction or can make stationary readings. Although there are no upper limits on flow velocity, only a small portion of the flow stream is sampled; thus, calibration problems are magnified. To improve cross-sectional sampling, the **fullbore flowmeter**, Figure 9-7C, utilizes collapsible blades that unfold below the tubing. Ideal calibration curves for the continuous and fullbore flowmeters are shown in Figure 9-8.

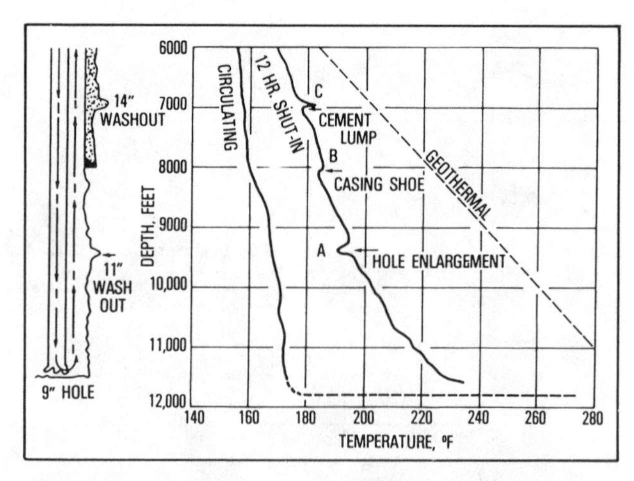

FIG. 9-6—Well completion influences on return to geothermal temperature. Permission to publish by the Society of Petroleum Engineers of AIME. Copyright 1983 SPE-AIME.

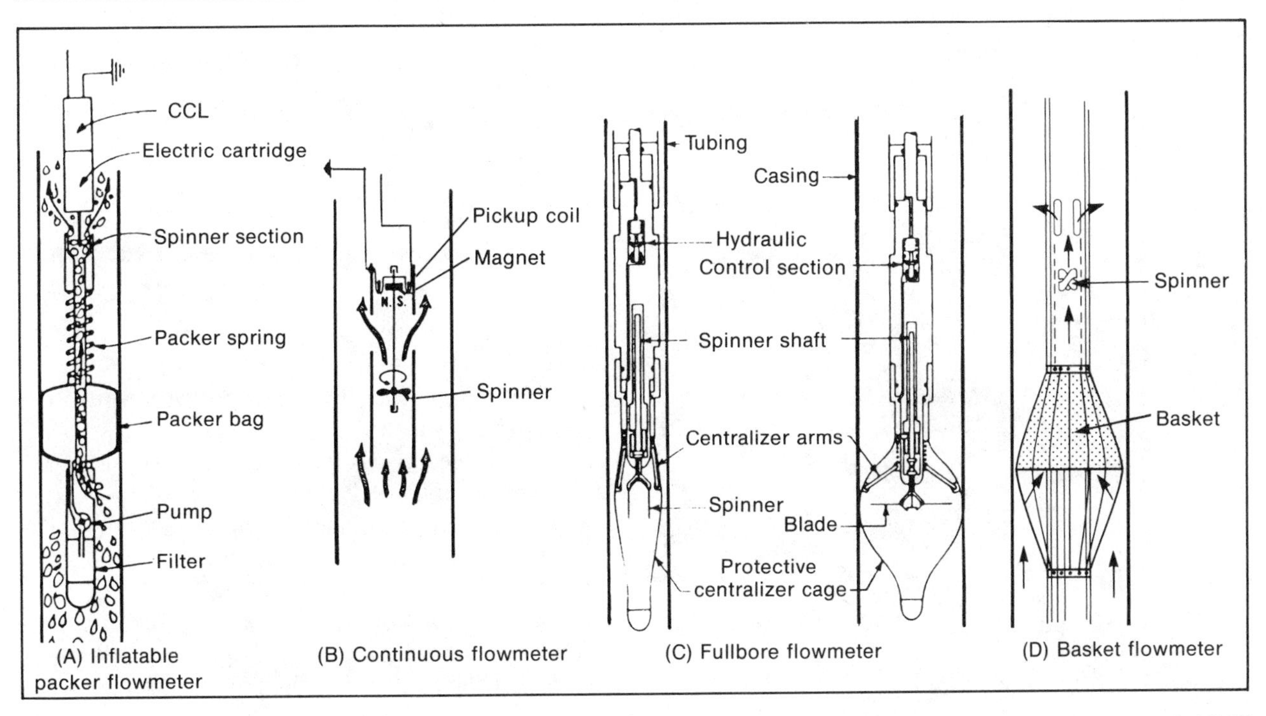

FIG. *9-7—Spinner type flowmeters. Permission to publish by the Society of Petroleum Engineers of AIME. Copyright 1983 SPE-AIME.*

The calibration curve for any spinner flowmeter is non-linear at low fluid velocities. Extrapolations of the curves of Figure 9-8 establish a threshold or frictional cut-off velocity of about 3.5 ft/min with water. In 5 1/2-in. casing this is equivalent to about 110 bpd. The threshold velocity for gas depends significantly on pressure (or density and viscosity) but is much higher than for liquid.

Efforts to improve the ability of spinners to measure multi-phase fluid flow at relatively low flow rates has resulted in the reappearance of the **diverter flowmeter**, Figure 9-7D. With the tool stationary, an inflatable ring can be extended to contact the casing wall to divert most of the flow through the spinner. The upper limit, on flow rate is about 2000 bpd with water.

Downhole Calibration—There is no practical upper limit of flow rate that can be measured with the continuous or fullbore flowmeters. Where precise flow measurements are needed, however, several factors must be considered. The continuous flow profile must be corrected for cable speed to obtain the actual fluid velocity profile. Fluid viscosity has a marked effect on spinner speed; decreasing viscosity (gas compared to oil) increases spinner speed. Thus, the downhole response curve of spinner speed versus fluid velocity must be established for specific well conditions.

Downhole calibration, using the multi-pass technique, consists of recording the tool response in revolutions per second during several runs (both up and down) at known logging speeds. Reading stations are established on either side of perforated zones. Readings within a perforated zone may be affected by local turbulence due to flow from the perforations. One station below the bottom perforations should provide a zero fluid flow measurement.

Ideally a plot of rps versus tool velocity at the various depths should then be two parallel straight lines with a discontinuity between the up and down curves, as shown in Figure 9-9. The midpoint between the discontinuities for Station A (below all perforated zones) corresponds to zero flow. It provides a reference for comparing fluid velocities at other stations. The fluid velocity at Station B is then equal to the difference between the velocity at the midpoint of the Station A runs and the velocity at the midpoint of the Station B runs.

Experience shows that average flow velocity is about 83% of the velocity measured at the center of the flow path. Thus, the flow cross-section correction factor (about 0.83) should be applied to provide higher precision to continuous flow measurements. Inadequate centralization of the tool obviously reduces accuracy.

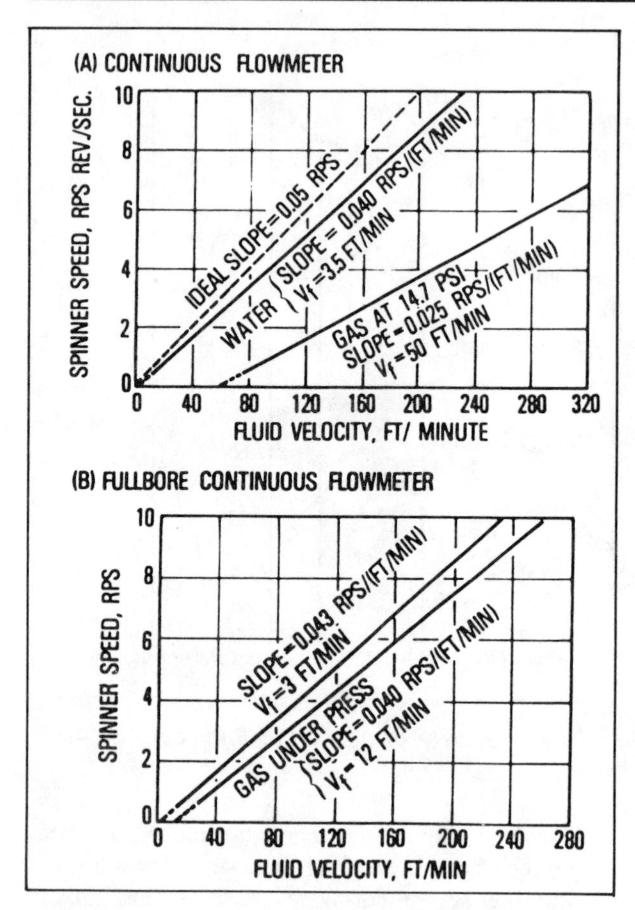

FIG. 9-8—*Spinner flowmeter calibration lines. Permission to publish by the Society of Petroleum Engineers of AIME. Copyright 1983 SPE-AIME.*

Stationary measurements at various levels provide added corroboration.

With two-phase flow the multi-pass technique is sometimes difficult to apply since the response curve becomes nonlinear if the tool logging speed is about the same as the fluid velocity. This problem is minimized if the logging speed exceeds the velocity of the lighter phase, or if the spinner speed is kept above 1.5 rps.

An interpretation method called the two-pass technique is recommended with multi-phase flow. In its simplest form this involves making one pass down and one pass up through the zone. The response curves are then matched in the zone of zero flow below the bottom perforations, as shown in Figure 9-10. The difference (Δrps) between the curves at various levels can then be used to calculate fluid velocity at a particular level, using the equation:

$$\bar{v} = \frac{\Delta rps}{b_{up} + b_{dn}} \times C$$

$\bar{v}$ = average velocity ft/min
C = turbulence correction 0.83 to 0.85
b_{up} = slope of velocity vs. rps response curve based on up-runs
b_{dn} = slope of velocity vs. rps response curve based on downruns

Note: Value for b_{up} or b_{dn}, if not known, requires at least two up and two down passes.

A shift in the center line between the recorded rps curves indicates a change in fluid viscosity (i.e., shift to left, increased viscosity; right, decreased viscosity) as shown in Figure 9-11.

The two-pass technique should reduce interpretation time and permit recognition of relatively small fluid entries.

Measurement Problems—Velocity components perpendicular to the axis of the spinner, turbulence, interfere with spinner operation. Thus, measurements

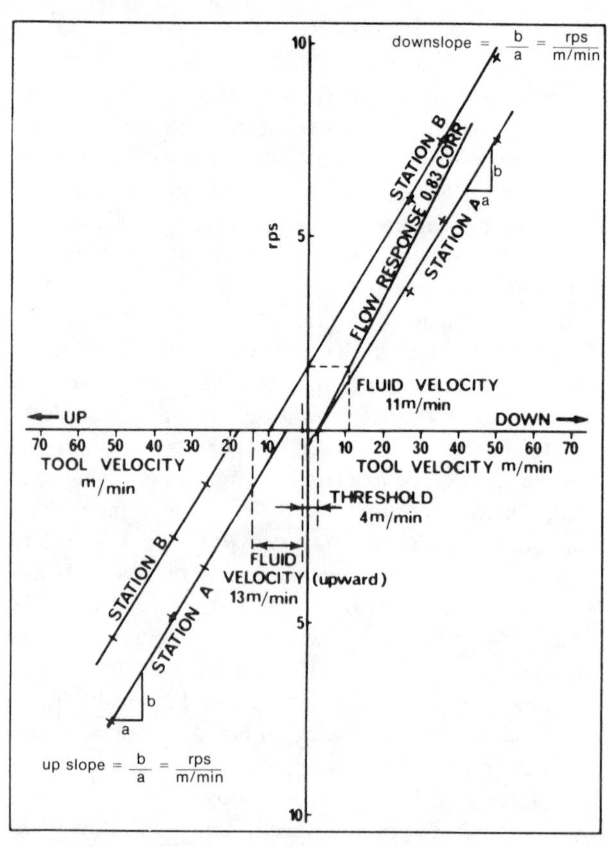

FIG. 9-9—*Multipass calibration technique. Courtesy of Schlumberger.*

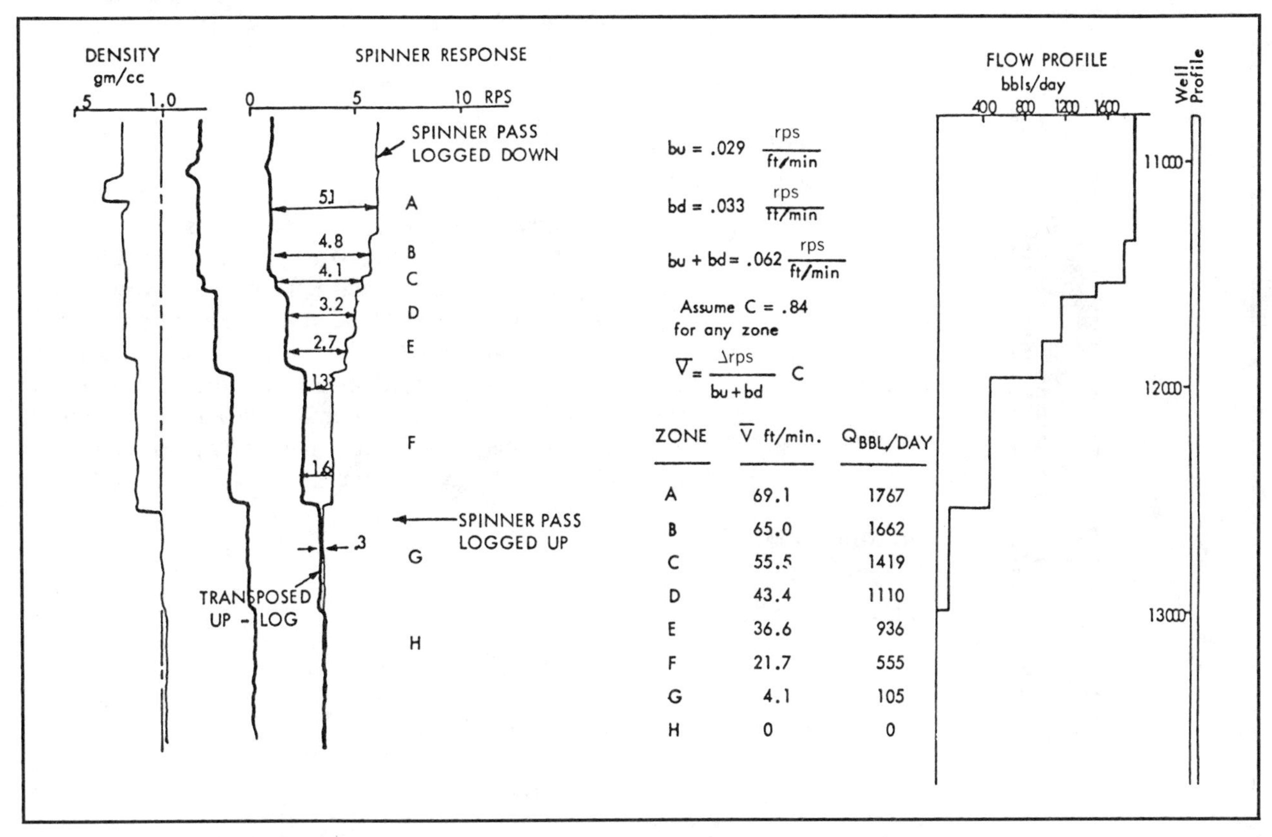

FIG. 9-10—*Two-pass calibration technique for multi-phase flow. Courtesy of Schlumberger.*

within a perforated zone are suspect, and in high rate wells turbulence may persist for a hundred feet above the zone. Multi-phase flow may create intense turbulence because the lighter phases tend to move faster than the heavier phases. The lighter phases flow as discontinuous bubbles or slugs. In moving past a particular depth level this causes an uplift of the heavier phases, then a fall back, creating localized circulation of the heavier fluids. Thus, the "holdup" of the heavier phase (the proportion of that phase at a particular level) may be much higher than the proportion of that fluid effectively moving past the level (the "cut"). "Holdup" doesn't match "cut" until the average flow velocity is about ten times the slip velocity of the fluids. Gas at downhole pressure has a slip velocity of about 40 ft/min through water. At lower pressure this can exceed 100 ft/min. Gas in oil downhole slip velocity is about 20 ft/min, and oil in water is about 10 to 20 ft/min. In deviated holes continuous flowmeter surveys are not meaningful where the lighter phases segregate and move up the top side of the hole. A diverter flowmeter, however, operates in deviated holes with little or no error due to the angle of inclination.

Fluid Density Devices

Borehole fluid density can be measured continuously with depth with the gradiomanometer or with a gamma ray absorption device.

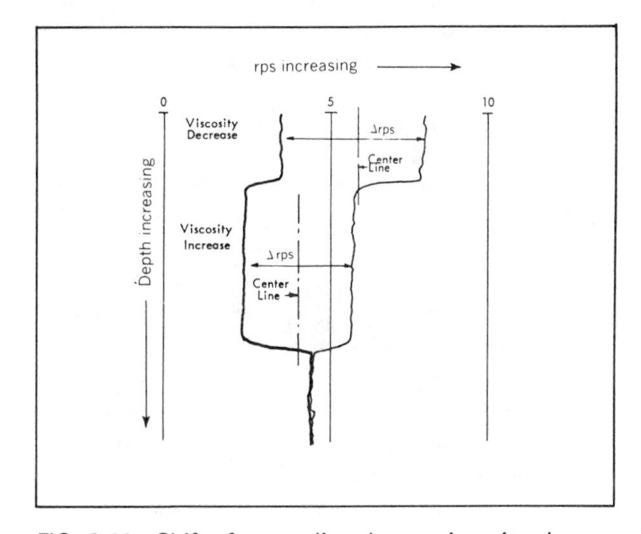

FIG. 9-11—*Shift of center line due to viscosity change. Courtesy of Schlumberger.*

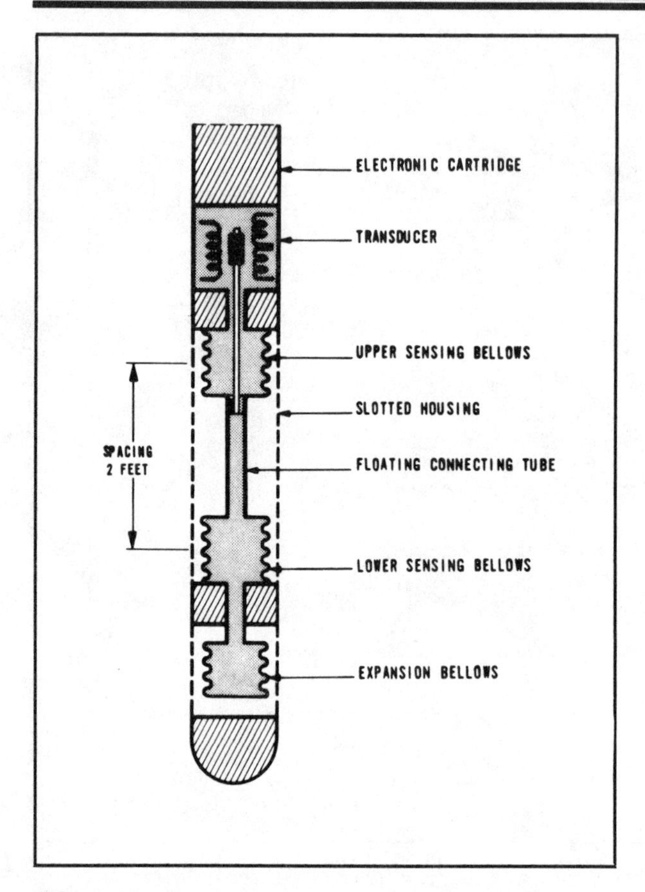

FIG. *9-12—Gradiomanometer. Courtesy of Schlumberger.*

The **gradiomanometer**, Figure 9-12, measures the difference in pressure between two pressure sensors spaced two feet apart vertically (assuming vertical hole). Resolution is about 0.01 gm/cc. The pressure difference is the sum of the hydrostatic head, plus the friction head, plus the difference in kinetic effect between the two bellows. At normal flow velocities, friction is nil, and unless there is a change in flow velocity between the two bellows, kinetic effect is also nil; thus, the pressure difference seen by the gradiomanometer is usually due only to the average fluid density.

The gradiomanometer is most effective for identifying gas entry and locating standing water levels.

Identification of water entry into a well producing oil and water requires that the relation between water holdup and water cut be established. If the downhole density of oil, ρ_o, and water, ρ_w, can be determined from static gradiomanometer measurements and the apparent fluid density, ρ_t, from dynamic gradio measurements at various levels, then water holdup, y_w, can be calculated at various levels from:

$$y_w = \frac{\rho_t - \rho_o}{\rho_w - \rho_o}$$

Slippage velocity, v_s, is the difference between the oil stream velocity, v_o, and the water stream velocity, v_w. It can be determined empirically from Figure 9-13 knowing the difference in density between oil and water and the water holdup. Figure 9-14 relates flow rates of water and oil to water holdup and slippage velocity. With these factors, and the velocity of the total flow stream, q_t, from a flowmeter measurement, the oil rate and water rate moving by a particular level can be determined from:

$$q_o = (1 - y_w)(q_t + v_s A y_w)$$

$$q_w = y_w[q_t - Av_s(1 - y_w)]$$

To get q in bpd:

$A = 1.4(D^2_{\text{csg id}} - D^2_{\text{tool od}})$
v_s = slippage velocity, ft/min
D Diameter, in.

and water cut at that level from:

$$\text{water cut} = \frac{q_w}{q_t}$$

Thus, with oil rate and water rate measurements at

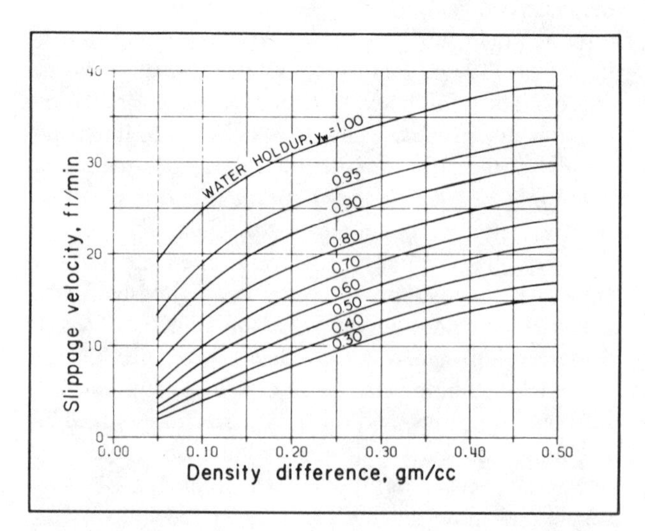

FIG. *9-13—Slippage velocity. Courtesy of Schlumberger.*

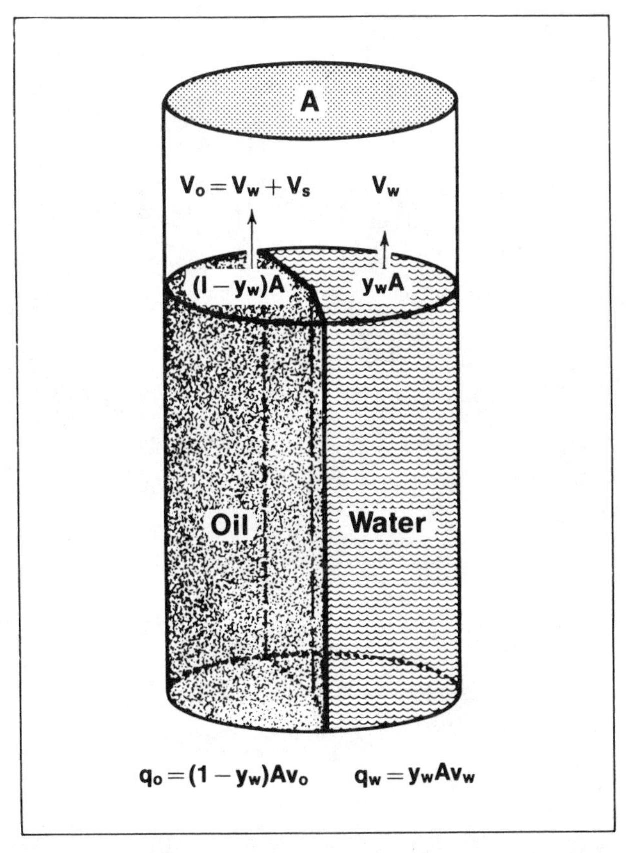

FIG. *9-14—Relationship among flow rate, q; holdup, y; and velocity, v.*

various levels, the zone contributing water can be identified. Obviously, the greater the density difference between the water and oil, the more accurate the resulting identification. With increasing total flow velocity slippage velocity becomes less important. Interpretation of the gradio log is then simpler, since water holdup and water cut are about the same. Oil and water entries can be identified on the basis of density differences read directly from the gradio log; however, a flow measurement is usually required to substantiate flow contribution from each zone.

The **gamma ray density device** consists of a cage open to wellbore fluid. At the base is a gamma ray source, usually cesium-137, and at the top a focused detector measures radiation activity, the logarithm of which is inversely proportional to fluid density. Since this is a statistical measurement, stationary reading improves accuracy.

Radioactive Tracers and Gamma Ray Detectors

These are available in several forms for fluid flow determination. With a properly tailored isotope, the correct tool combination and a planned program, the tracer survey is one of the best available methods for recording fluid movement quantitatively in water injection wells, and particularly for locating flow behind pipe. Tracers are less effective in defining flow from producing wells where multi-phase flow is present, or where surface contamination by tracers produced back to the surface may be a problem.

Tracer selection must consider the radiation intensity, the half-life, the type of fluid in the well, and the bottom-hole temperature.

For water injection well surveys radioiodine in water solution (iodine I-131) is most commonly used since it has a very short half-life (8.1 days) and is miscible in water.

For gas injection well surveys, ethyl iodide (C_2H_5I) or methyl iodide (CH_2I) are commonly used. These are both liquid forms containing the iodine I-131 isotope; thus, both have a short (8.1 day) half-life. Methyl iodide has a relatively low boiling point, 108.5°F.

Gamma-ray detectors in common use are the geiger mueller tube and the scintillation crystal. The scintillation detector is much more sensitive; however, the geiger mueller tube is used in some through-tubing tools since it is more rugged. Resolution power is low

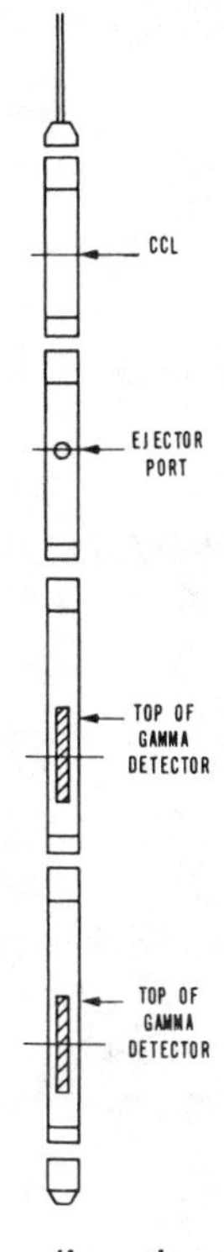

FIG. *9-15—Typical radioactive tracer-detector-tool configuration for water-injection profiling.*

and poor quality correlation logs can be anticipated.

Water Injection Well Profiling—Ejector-type through-tubing tools, Figure 9-15, permit ejecting a small slug of radioactive material, and then measuring travel time of the slug from ejector to detector, or between two detectors. By positioning the tool at several depths and ejecting small slugs, a fluid velocity log can be made. Accuracy is good in the high and medium flow velocity ranges, but at lower rates (i.e., the last 10 to 20% of the flow) the slow movement of the slug makes it difficult to determine the arrival time. However, even at low rate resolution is much better than mechanical spinners.

In open hole a caliper log is needed to determine volume flow rates. The normal variation of flow velocity across the hole diameter underscores the importance of tool centralization and the need for correction of velocity data if actual hole diameter is used to compute velocity rate of flow (with actual hole diameter flow rates will always be too high). Areas of changing hole diameter should be avoided for velocity shots.

With most ejector tools it is possible to selectively eject a relatively large tracer slug. In long zones it is usually advantageous to inject a large slug above the zone and use either of two techniques to generally define flow in the wellbore:

—*Controlled interval*, whereby the detector is positioned at predetermined points and records the slug as it passes.

—*Controlled time*, whereby logging runs are made at definite time intervals to record the position of the slug.

Velocity surveys then pinpoint flow in areas of interest. The rate of disappearance of the tracer into a zone is also indicative of the amount of water going into that zone.

Noise Device

The noise log is a recording at various points in the well of the amplitude of audible sound frequencies generated by moving liquid or gas. A typical logging tool, 1.5 in. diameter and 6 ft long, contains a piezoelectric crystal sound detector section, Figure 9-16. High noise level associated with wireline or tool movement requires that recordings be made with tools stationary. Figure 9-17 is a noise spectrum from a test barrel in which 10 ft^3/hr of air was metered through a porous choke into a column of water. The noise peak at 300–400 hz is due to two-phase flow through the barrel. The 1,000-hz peak is related to single-phase flow through the choke.

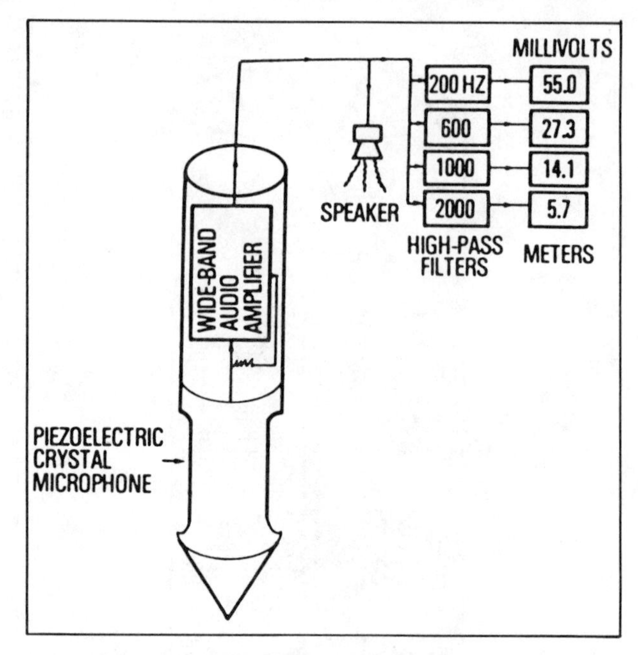

FIG. *9-16—Noise logging sonde. Permission to publish by the Society of Petroleum Engineers of AIME. Copyright 1983 SPE-AIME.*

Usually four tracks are recorded, each showing the average amplitude of all sound frequencies above a cutoff frequency. Cutoff points are 200, 600, 1,000 and 2,000 hz. Thus, the 600-hz curve hears nothing below 600 hz. Figure 9-18 compares the downhole

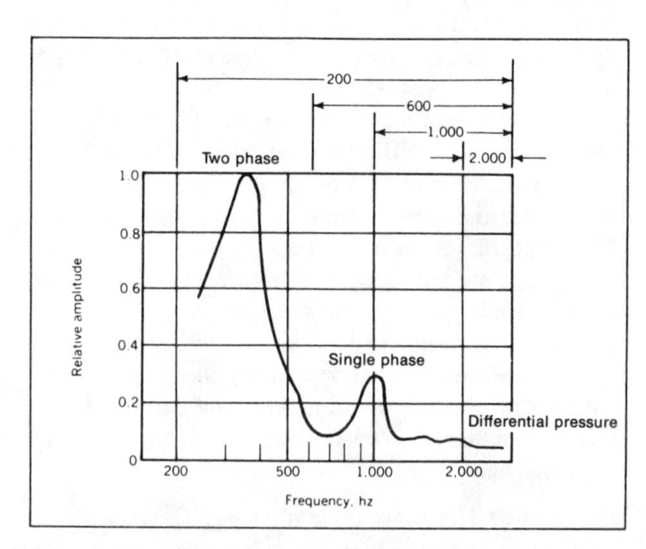

FIG. *9-17—Noise spectrum.[21] Permission to publish by The Society of Petroleum Engineers.*

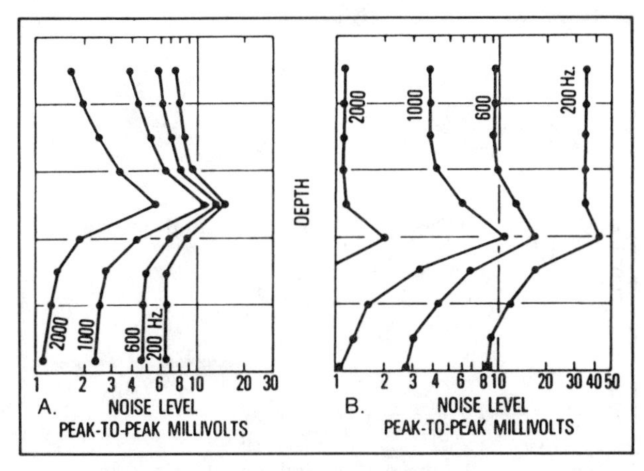

FIG. 9-18—*(A) Frequency character of a single-phase leak, (B) frequency character of gas-into-liquid leak. Permission to publish by the Society of Petroleum Engineers of AIME. Copyright 1983 SPE-AIME.*

noise log recording of a single-phase casing leak with a situation where gas is leaking into liquid.

Raw data adjusted for line attenuation may also be plotted for each depth stop as the energy level between 200–600 hz, 600–1000 hz, 1000–2000 hz, and >2000 hz providing for additional curves.

Interpretation of the noise log is basically as follows: if sound is generated downhole, then something is moving, and the questions become where from or to, what, and how much. Good answers to these questions depend on comparisons with sound patterns developed in laboratory stimulations but the following general points are evident.

1. Gas movement at differential pressure greater than 100 psi creates most of the sonic energy above 1000 hz. At greater differential pressure, frequency peaks move higher.
2. Single phase liquid movement creates more energy below 1000 hz, but again with higher flow rates frequency level of energy peaks move higher.
3. Two phase flow spreads energy over the frequency range, but usually the peak occurs in the 200–600 hz band—particularly with gas moving through a liquid filled channel.
4. Flow in a channel behind the casing creates energy peaks characteristic of restrictions in the channel. Undisturbed flow inside casing or tubing should not create peaks.
5. Attenuation of energy occurs in the medium that carried the sound from source to detector, Figure 9-19. The degree of attenuation is a function of the frequency and the nature of the medium:

—Sound transmission is highly attenuated in gas, twice as much as in water.

—High frequencies are attenuated more than low frequencies.

The noise log, in conjunction with the temperature log, has greatest utility in locating and defining flow behind casing. It has been used to determine relative flow rates from perforated zones, but it is not very quantitative in this respect.

Pulsed Neutron Devices

Pulsed neutron devices have the capability of measuring parameters related to fluid saturation outside the casing during the life of the well. Thus they have

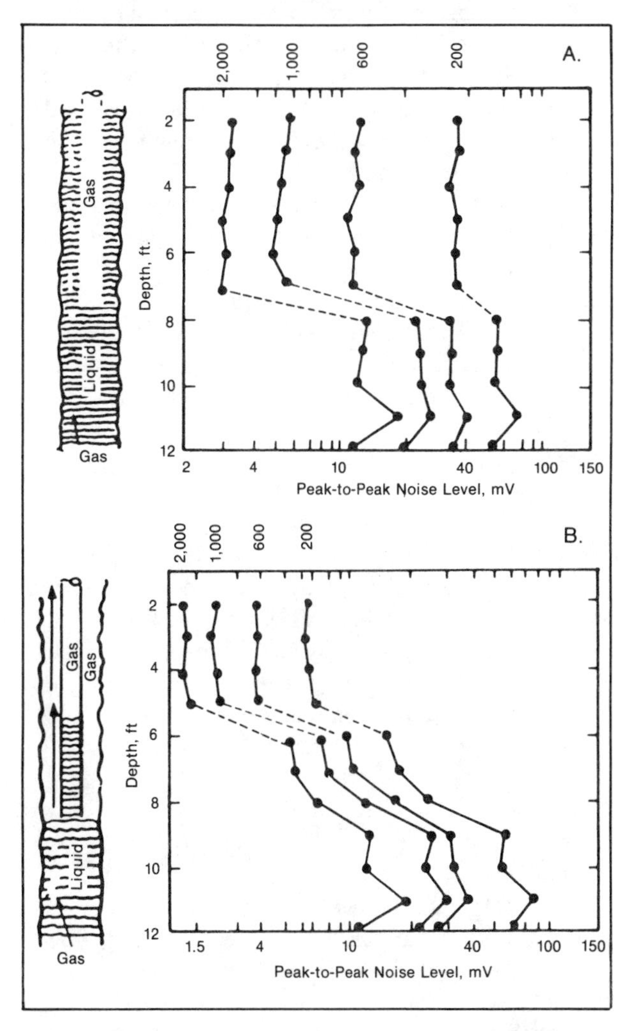

FIG. 9-19—*(A) Change of transmission medium, (B) multiple changes of transmission medium. Courtesy Dresser Atlas.*

the potential to answer very important production logging questions. Two sets of measurements are possible (i.e., fast neutron or neutron capture interactions). *Fast neutrons interactions* (inelastic scattering) require high neutron energy, and the bombarded elements emit gamma rays very shortly after the neutron burst. Gamma ray spectroscopy output includes carbon, calcium, iron, oxygen, sulfur, and silicon. However, to obtain acceptable statistical patterns, the tool must be held stationary at a given level for several minutes.

As the neutrons slow to thermal level, *neutron capture interactions* occur. Calcium, chlorine, iron, hydrogen, and silicon can be detected, along with the thermal neutron macroscopic capture cross section. Continuous logging at normal speeds is possible.

The Schlumberger **gamma ray spectroscopy tool** uses a sodium-iodide detector. The size of the detector is such that in most well situations the tool cannot be used through tubing to log under flowing conditions. The gamma ray spectroscopy method is still somewhat experimental.

The **thermal decay time device** measures only neutron capture actions but is small enough to log through 2 3/8-in. od tubing. The TDT type devices enjoy a longer history of application.

Thermal Decay Time Device—The thermal decay time device contains a downhole generator that periodically emits bursts of high-energy neutrons. These neutrons move through the borehole fluid and into the formation. They are rapidly slowed to a thermal level and captured by various elements. In the process, capture gamma rays are emitted which are detected by a gamma ray detector; thus the device measures the exponential rate of decay of the thermal neutron population after each burst. The time required for the neutron population to decay to 65% of a given value is called thermal decay time, τ. For interpretation purposes the thermal neutron macroscopic capture cross-section, Σ expressed in capture units (cu) is used.

$$\Sigma = \frac{4550}{\tau}$$

Some elements, particularly chlorine, have a very high capacity to capture thermal neutrons. As shown in Figure 9-20, decay rate will be high if chlorine is present. Since chlorine is primarily associated with formation water, porous zones with low decay rates or low capture cross-section should be hydrocarbon zones.

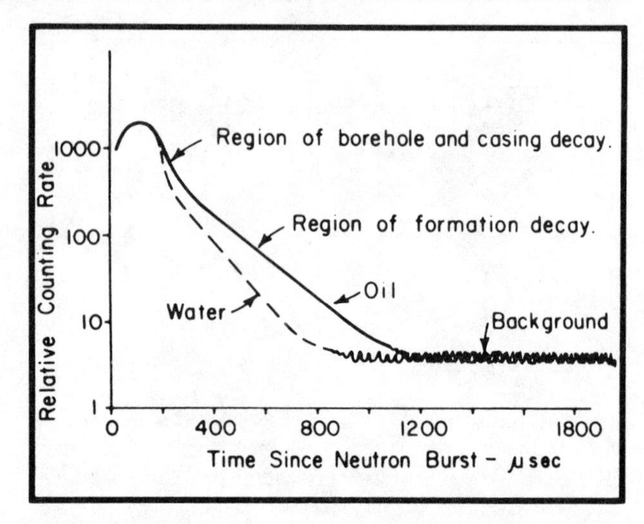

FIG. *9-20—Typical decay curves for oil-bearing and water-bearing formations. Courtesy of Schlumberger.*

Identification of hydrocarbon zones becomes increasingly difficult with lower water salinities, lower porosity, and higher clay content. Experience in an area to establish reasonable capture cross-section values for oil, water, and matrix material can minimize these problems.

Figure 9-21 is a typical thermal decay time log presentation (Schlumberger TDT-M) using a near and far detector system. The "ratio" curve is responsive to formation porosity but is also affected by wellbore geometry.

Interpretation—The capture cross-section of a formation, as recorded on the log, is the sum of the capture cross-sections of each component weighted by their fractional volumes as follows:

$$\Sigma_{\text{formation}} = \underbrace{(1 - V_{sh} - \phi)\Sigma_{ma}}_{\text{matrix}} + \underbrace{V_{sh}\Sigma_{sh}}_{\text{shale}} + \underbrace{\phi(1 - S_w)\Sigma_h}_{\text{hydrocarbon}} + \underbrace{\phi S_w \Sigma_w}_{\text{water}}$$

To solve this equation for water saturation, six parameters are needed:

—ϕ and V_{sh}—at every depth.
—Σ's for matrix, shale, hydrocarbon, and water—usually constant over some depth interval.

Porosity, ϕ, and shale content, V_{sh}, are usually available from other logs—an estimate of ϕ could be

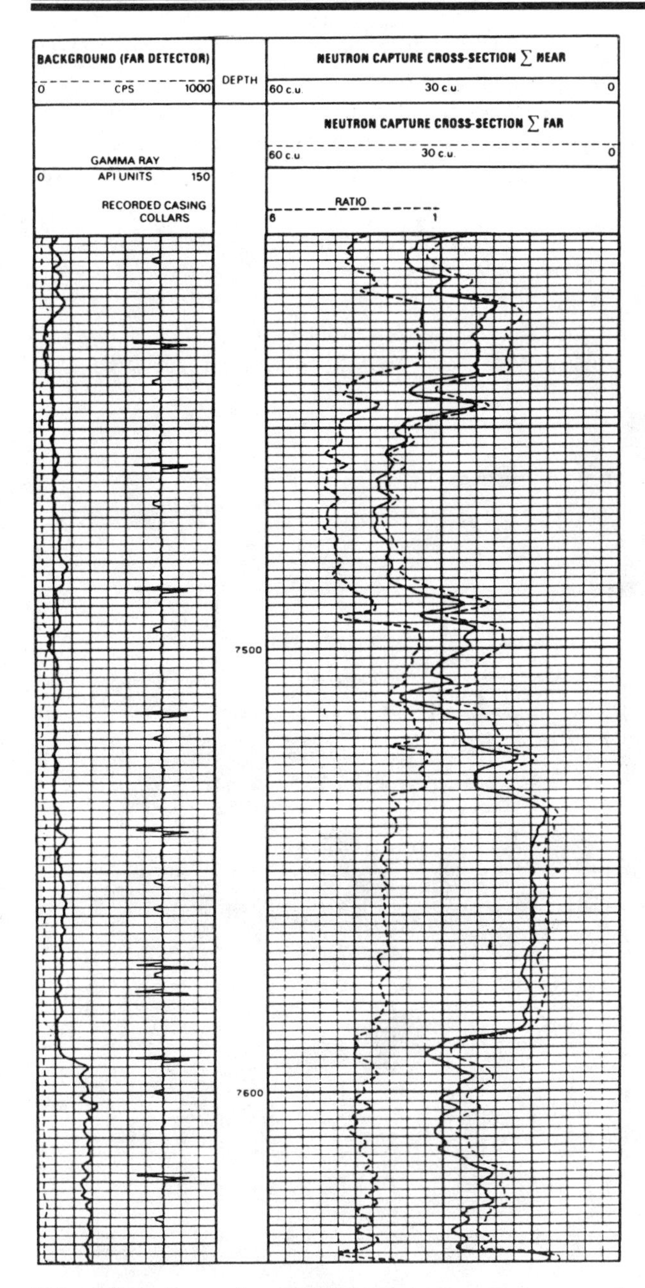

FIG. *9-21—Example of TDT-M field log. Courtesy of Schlumberger.*

obtained from the dual detector TDT. Σ_{ma}, Σ_{sh}, Σ_h, and Σ_w may be approximately known in an area or estimated from log values in nearby zones, or obtained by direct measurement or by various crossplot techniques. Typical Σ values are as follows:

Σ_{matrix}	6–12 cu
Σ_{shale}	30–45 cu
Σ_{water}	60–120 cu (fresh water 22)

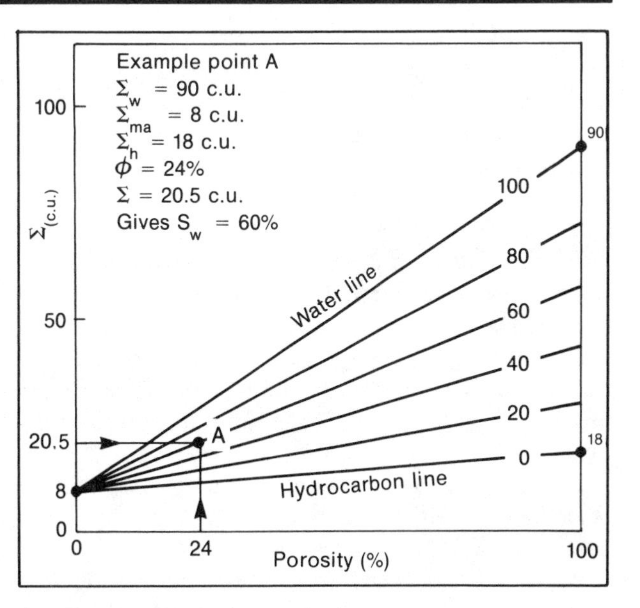

FIG. *9-22—Graphical solution of the TDT saturation equation for clean zones. Courtesy of Schlumberger.*

Σ_{oil}	14–22 cu
Σ_{gas}	2–10 cu

Figure 9-22 shows a graphical solution to the saturation equation assuming a clean formation (i.e., no shale content). Figure 9-23 shows a crossplot technique to determine Σ_{ma}, Σ_w, and Σ_h. The interval logged included a water zone and an oil zone. Logged values of Σ and ϕ at various levels are plotted and are enclosed by the estimated water line and hydrocarbon line. These lines, extended to 0% and 100% porosity, provide estimated values of Σ_{ma} = 7 cu, Σ_h = 16 cu, and Σ_w = 90 cu.

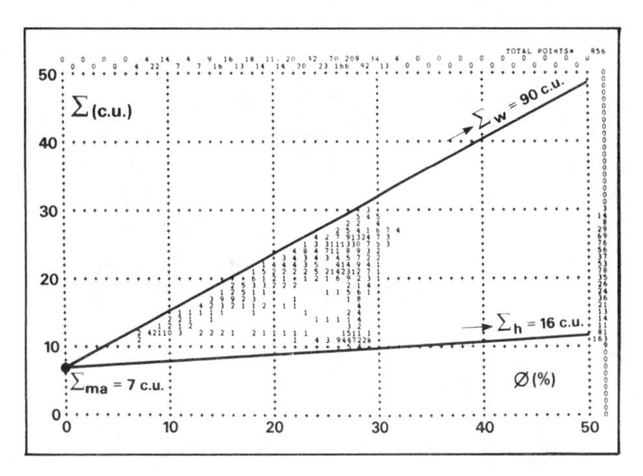

FIG. *9-23—Σ vs. φ crossplot used for determination of parameters. Courtesy of Schlumberger.*

Interpretation Problems—1. Invasion: The TDT device, as with any radioactive device, has a relatively shallow depth of investigation, 10 to 15 in. Thus fluids in the invaded zone influence the capture cross-section measurement. If drilling mud filtrate has displaced hydrocarbon, the altered capture cross-section will increase or decrease calculated S_w depending on the characteristics of the mud filtrate. Mud filtrate invasion behind cemented casing may take several years to disappear. In perforated zones invaded fluid disappears quickly, but the formation can also be reinvaded quickly. The well should be flowing at a rate needed to prevent reinvasion while running the TDT log.

2. Overstated Water Saturation: Matrix rock salt or other mineral with a high capture cross-section will cause high Σ values, therefore high S_w.

Wells that have been acidized will record high Σ values that remain relatively constant until the well begins producing water. This problem can be handled by base log comparison.

The presence of shale in the formation can be a problem because of the difficulty in determining V_{sh}. An error of 10% in V_{sh} will lead to an error of about 20% in S_w.

Time Lapse Technique—Monitoring changes in water saturation during the life of the wells removes many of the interpretation problems since Σ_{sh} and Σ_{ma} remain constant. This method assumes that Σ_h and Σ_w remain constant also. In a water flood, however, Σ_w may well be changed by invading water.

Direct comparison of two TDT log runs in different borehole fluid conditions is difficult because of borehole and neutron diffusion effects. With a dual detector type tool (Schlumberger TDT-M) recording the Σ value with the far detector minimizes the problem of comparison between runs since the far detector is less affected by borehole conditions.

Gamma Ray Spectroscopy Device—The induced gamma-ray spectroscopy method consists of analyzing gamma rays emitted by nuclei excited by particle bombardment. Each element has a characteristic gamma ray emission. Thus the total recorded spectrum can be split into concentrations of specific elements; for example: carbon, oxygen, chlorine, hydrogen, silicon, calcium, iron, and sulfur. The ratio of carbon to oxygen can be a reliable indicator of hydrocarbon saturation irrespective of water salinity.

Measurement Technique—Fast neutrons are emitted with an energy of 14 MeV from a pulsed accelerator source. Gamma rays given off by inelastic scattering are then detected by the tool. To obtain acceptable statistical variations, reading are taken with the tool stationary for about 5 minutes per level. Measurements at one-foot intervals produce a quasi continuous log over the reservoir. Detecting fast neutron interactions outputs C, Ca, Fe, O, S and Si.

Neutron capture interactions can also be detected with the same tool as neutrons slow to thermal levels. Continuous logging at a speed of 600 ft/hr gives acceptable statistical variations. Ca, Cl, Fe, H, and Si content, together with a macroscopic thermal neutron capture cross-section, are obtained.

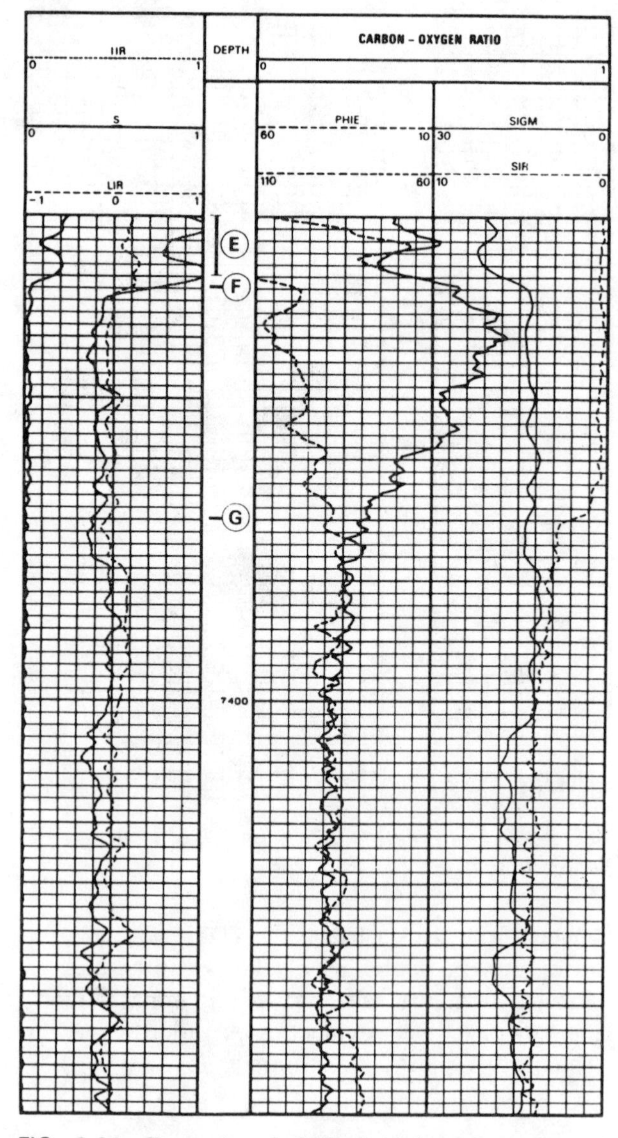

FIG. *9-24—Example of GST field log. Courtesy of Schlumberger.*

The size of the sodium-iodide detector imposes a minimum tool diameter of 3 5/8-in.; thus in many situations the well must be dead at the time of logging. Depth of investigation is quite limited: 5 to 10 in. for inelastic events and 8 to 12 in. for capture events. Invasion of killing fluid is a problem. The contribution of the borehole signal is significant; thus, wellbore fluid must be known. In some cases the wellbore is filled with diesel oil to minimize movement of wellbore water into the formation.

Interpretation—Various ratios may be recorded. With the Schlumberger GST tool the usual ratios are:

—C/O-Carbon Oxygen Ratio (COR)-indicates hydrocarbon saturation independent of salinity—but depends on lithology and porosity.
—Cl/H-Salinity Indicator Ratio (SIR)-varies with salinity but also with fluid porosity.
—H/(Si + Ca)-Porosity Indicator (PIR)-varies with porosity but also with fluid salinity.
—Fe/(Si + Ca)-Iron Indicator (IIR)-considered as a shale indicator but also locates casing seat.
—Si/(Si + Ca)-Lithology Indicator (LIR)-essentially dependent on formation lithology.

Figure 9-24 is a Schlumberger GST log in the Arab formation in the Arabian Gulf area. Best application of the tool currently is in situations where water salinity is so low (less than 40,000 ppm) that the TDT type log cannot be used.

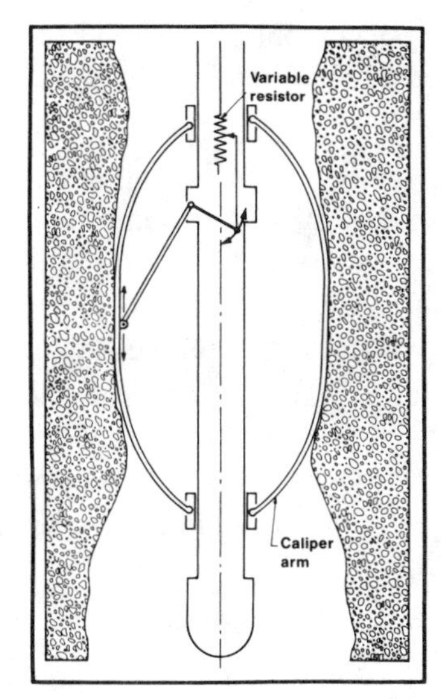

FIG. 9-25—Through-tubing caliper.

Through-Tubing Neutron Log

This can often be used to monitor gas movement into the oil zone. Since gas has a lower hydrogen index on the neutron log, comparison of neutron logs run through tubing during the producing life of the well show a higher count rate in the gas invaded zones. Logs must be run with the well shut in because the through-tubing neutron device is not compensated for borehole fluid movement effects.

Through-Tubing Caliper

The caliper is particularly important for openhole flow profiles, Figure 9-25. Inside casing, damage, scale, or paraffin deposits can seriously affect flow rate logs.

Schlumberger Production Combination Tool

This combines the thermometer, gradiomanometer, continuous or fullbore flowmeter, caliper, manometer, and a collar locator into one device to permit making the five surveys successively on one run in the hole, Figure 9-26. The later PLT tool permits all recording to be made digitally at the same time. The advantage is reduced operating time, reduced time to restabilize the well if several flow rates are used, and increased chance of making several runs before well flowing conditions change in unstable wells.

Table 9-3 shows specifications for each of the individual devices making up the PCT device. They are essentially the same as the corresponding tools previously described.

The manometer is a spiral bourdon tube device. Its resolution is limited by a potentiometer transmitting device which causes pressure changes to appear as discrete steps on the recording. Surface recording of pressure, however, indicates when downhole stabilized conditions have been reached. For precise pressure measurement, an amerada gage can be run also.

APPLICATION OF THROUGH-TUBING PRODUCTION LOGGING TOOLS

Production logging tools are designed to operate downhole under dynamic producing conditions to provide the data necessary to determine the physical condition of the well, to evaluate the performance of the well completion scheme, to diagnose well problems, to evaluate the results of well workover operations, and, in some cases, to provide the best obtainable information concerning basic reservoir parameters.

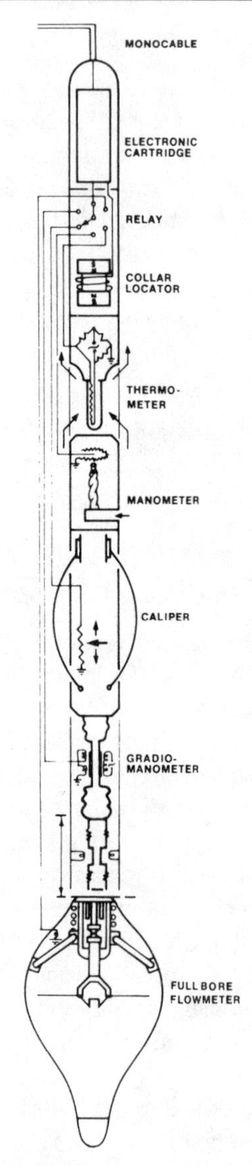

FIG. *9-26—Combination tool including full bore flowmeter, gradiomanometer, caliper, manometer, thermometer, and casing-collar locator. Courtesy of Schlumberger.*

TABLE 9-3

Maximum pressure		*15,000 psi*
Maximum temperature		*350°F*
Sensor	*Accuracy*	*Resolution*
Flowmeter	± 2%	1/4 rev/second
Gradiomanometer	± 3%	.01 gm/cc
Caliper	± 1.4%	.06 inch
Manometer	± 3%	.5% of full scale
Thermometer	± 3%	.04°F

Although costs must be justified in individual cases, it is often advisable to run appropriate production logging tools in key wells early in their life, to confirm that the well is operating as desired, and subsequently, when changes in producing or injection characteristics are noted. When several zones are open to the wellbore, periodic surveys provide the data needed for control of reservoir fluids.

To run downhole surveys, the well hookup must allow movement of logging devices to the desired point in the well. Major application is in flowing oil or gas wells, or in injection wells. However, gas lift wells normally do not present difficult problems, and occasionally tools are operated through the annulus of pumping wells.

Physical Condition of the Well

Early detection of casing or tubing leaks or undesirable fluid movement through channels behind casing should be a constant goal of production personnel. A general program for detecting well problems should include the following points:

—Regular checks of surface pressure on production and surface casing strings.

—Routine observation of producing characteristics to note unusual increases in water or gas production.

—Routine checks of shut-in pressure to note unusual pressure declines.

—In addition to surface checks, temperature surveys made routinely in key wells may be helpful in detection of large leaks.

Once a physical problem is suspected it can usually be confirmed and defined by one or more of several tool combinations.

—Tubing or casing leaks can often be located with the temperature survey run under various flowing and static conditions.

—High volume tubing leaks can sometimes be pinpointed quantitatively with the flowmeter.

—Flow behind pipe may be generally located with the noise log; or with sufficient fluid movement and thermal disturbance, with the thermometer.

—The caliper survey should indicate restricted areas.

Fluid Movement (Single-Phase)

Fluid entry or egress points in single-phase flow can be located by several methods. Flowmeters and temperature surveys have application in either production or injection. Tracers have primary application in injection wells.

Stabilized injection rates corresponding to actual injection conditions are absolutely necessary to accurately define flow. Many examples can be shown on how changing injection rate changes the ratio of fluid accepted by various zones. The same is true of producing rates.

Where injection is related to a stimulation treatment, a static temperature survey, run as formation temperatures are returning toward geothermal, often defines the zones taking stimulation fluids. Use of radioactive tagged solids carried with the stimulation fluid may be helpful.

The following comments summarize the application of various devices in defining single phase fluid movement.

Continuous Flowmeter—For high flow rates (fluid velocity exceeding 40 ft/min) the continuous flowmeter provides an excellent flow profile, is reliable, easy to run, and does not disturb the flow pattern. The tool may be accurately calibrated downhole by recording at two or more known logging speeds. Accuracy also depends on tool centralization, which may be a questionable factor.

Fullbore Flowmeter—For medium to lower flow rates the fullbore flowmeter has much better resolution than the continuous flowmeter. With the fullbore, minimum rates are about 65 bpd in 5 1/2-in. casing compared to 250 bpd with the continuous flowmeter. The tool cannot measure flow within the tubing since blades are folded.

Diverter Flowmeter—The diverter flowmeter is by far the most precise flow measuring device (10 bpd). Its upper limit is about 400 bpd in 5 1/2-in. casing. In addition, it minimizes the problems of multi-phase flow turbulence in vertical and even deviated holes.

Radioactive Tracers—Soluble radioactive tracers can detect fluid movement behind the pipe, as well as measure flow velocity inside the pipe. Tool diameter is such that there is little disturbance of the flow pattern. Flow profiling accuracy suffers at very high flow velocities unless counter spacing is large and at low rates since diffusion makes precise travel time measurements difficult.

Temperature Surveys—Temperature surveys recorded with the well flowing, or during water injection, should reliably show the bottom of the lowest zone producing or taking fluid. In a water injection well, temperature surveys recorded after injection is shut in, while the zones are returning to the geothermal temperature will show which zones have taken fluid. Determination of relative amounts taken by several zones is questionable. A major advantage is that location of injected fluids will be shown whether flow is inside the pipe or outside the pipe through channels.

Radioactive Solids—Tagged frac sand (often 100 mesh) added to an injected stimulation fluid can subsequently be located with a gamma ray device to provide an indication of where the injected fluid went. A basic problem, however, is that a little tagged sand close to the detector looks the same as much more sand a few inches farther away. Tagged solids usually must be confirmed with other information.

Fluid Movement (Multi-phase)

Fluid identification in multi-phase flow is complicated by the variety of possible mixtures of oil, gas and water, and by the effect of relative fluid velocity of these mixtures.

Measurements must be made of parameters that vary sufficiently for reasonable resolution.

- —Density measurements are often of great benefit in defining gas entry into liquids, but are marginal in defining oil-water mixtures.
- —Dielectric constants for hydrocarbons (2–6) are quite different than for water (80). Thus, dielectric constant measurements can define hydrocarbon-water mixtures when oil is the continuous phase.

At low fluid velocities gravity segregation occurs in the casing, and the average in-situ distribution of fluids at a particular point may not coincide with the proportions of fluids being produced at the surface, since the light fluids rise faster.

For locating gas entry in multi-phase flow, the gradiomanometer is perhaps the most useful device. With sufficient flow and sufficient pressure drop, the temperature survey often is also a good choice.

For locating water entry in low rate two-phase (water-oil) production, the packer flowmeter combined with the dielectric constant measurement offers the best definition.

For locating water entry with higher flow rates (where the packer flowmeter cannot be used) the gradiomanometer and fullbore or continuous flowmeter must be used. If the downhole densities of the fluids are known, and if they are sufficiently different, the slippage velocity can be determined. With this value the relative amounts of each fluid in the total flow stream at various levels can be reasonably calculated. With multi-phase flow in deviated hole, conventional flowmeter measurements are meaningless; however, a diverter flowmeter should provide reasonable measurements.

Water movement behind casing can sometimes be detected with a gamma ray device looking for abnormally high radioactivity left by salts or scale from the moving water.

Reservoir Fluid Saturation

Effective depletion of a reservoir and optimum control of reservoir fluids requires knowledge of the changes in fluid saturation that occur with production and injection. These changes can best be determined by making periodic measurements during the life of the well. Looking for changes in saturation, by comparing a current measurement with one made one or two years before, is much more reliable than attempting to make an absolute calculation of current saturations.

Increasing water saturation, with reasonable water salinity and matrix porosity, can be determined with the thermal decay time log. With low water salinity and clean sands the gamma ray spectrometer log, while experimental, may show better definition.

Increasing gas saturation can usually be determined with the through-tubing Neutron device.

Reservoir Parameters—Formation Damage

In longer zones or where several zones are open to the wellbore, the combination of a flow profile, showing the flow rate from each zone, and a wellbore pressure measurement opposite each zone, indicating pressure drawdown, should provide data needed to construct an apparent *kh* or capacity profile for each zone. Comparing production log data with core, or other reservoir data, may indicate need for selective stimulation.

Knowledge that a particular zone is not contributing sufficiently to total flow should be more positive proof of formation damage than transient pressure testing techniques could provide in multi-zone situations.

FIELD EXAMPLES OF PRODUCTION LOGGING TECHNIQUES

The following examples illustrate use of production logging techniques in diagnosing well problems, defining flow in producing or injection situations, and in monitoring reservoir fluid saturations.

Location of Tubing Leaks

A program of measuring pressures at the production casinghead of oil and gas wells is followed by most operators. If pressure on the production string cannot be bled off readily, then a temperature survey can often be used to locate the leak.

In a gas well, the temperature survey should be run with the tubing shut in while gas is bled from the casing. Cooling, caused by gas expansion, should exist at the depth of a small leak. Cooling usually extends upward (assuming most of the cooled stream moves upward), but since the mass rate of flow is usually low, rapidly returns to the normal temperature gradient. Localized cooling from a few degrees to as much as 50°F has been observed.

In an oil well, a temperature survey may not pinpoint the tubing leak with a "cold spot," since there may not be sufficient gas expansion to significantly reduce the temperature. A shut-in temperature survey, however, may show an anomaly from the geothermal temperature due to extraneous flow. At high rates, a flowmeter survey, run under various flow and shut-in conditions, may help pinpoint the leak. Sometimes a fluid density survey, with the tubing shut in at the surface, producing through the tubing leak, may show a gas gradient down to the point of the leak, and an oil gradient below the leak.

Temperature Survey to Locate Tubing Leak in a Gas Well—The Gulf Coast well of Figure 9-27 was

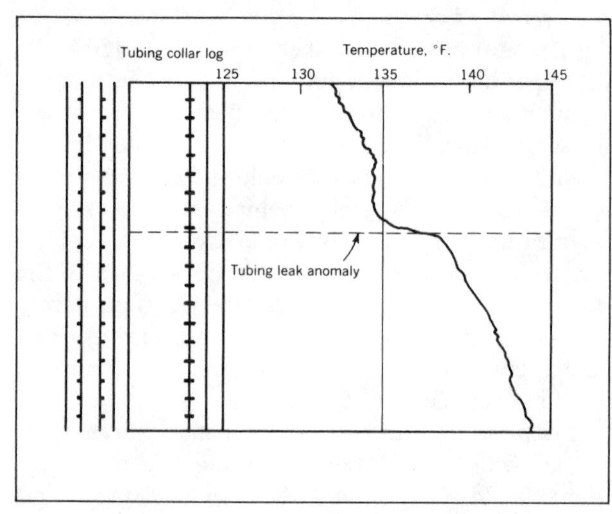

FIG. 9-27—*Gas well with tubing leak. Courtesy of Schlumberger.*

known to have a tubing leak, since casing and tubing pressures equalized after the well was shut in.

The temperature survey was recorded in the tubing with the well shut in. Pressure in the casing annulus was bled to allow gas to escape from the tubing via the leak. Gas expansion, in the casing annulus opposite the leak, created the cold zone identified by the temperature log. Correlation with the CCL log collar indication provides strong evidence of a collar leak, which possibly can be repaired using through-tubing techniques.

Blast Joint Cut-out in a Dual-Tubing-String Oil Well—In a short time, long-string production in the two-string dual completion of Figure 9-28 went from 3,000 bpd clean oil, to 900 bpd oil with 24% water. Short-string production also declined with increased water cut. Excessive sand production began. Communication from other zones was suspected, and the

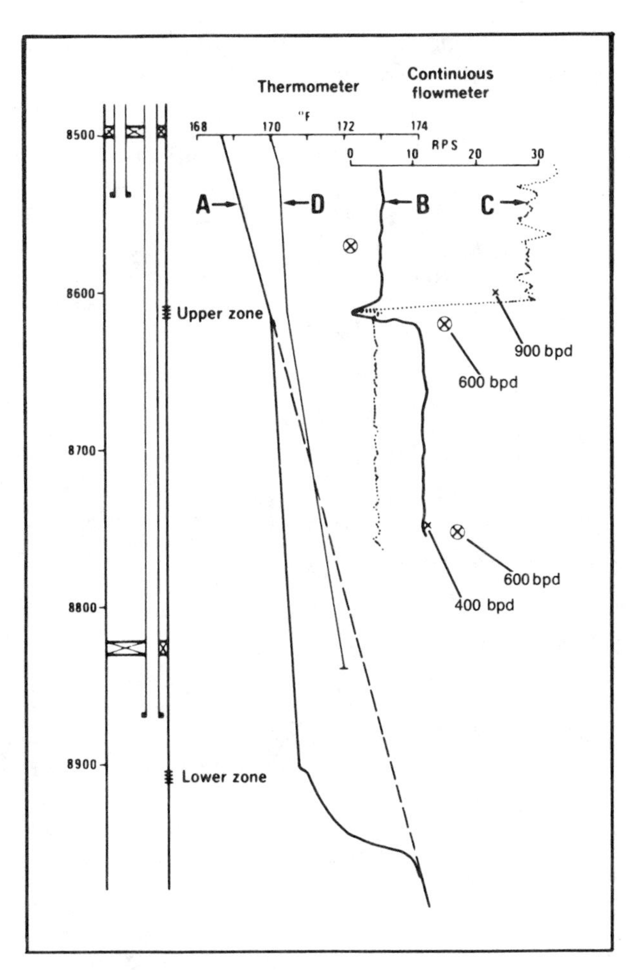

FIG. *9-28—Communication, two-string dual. Courtesy of Schlumberger.*

problem was to locate the source.

After waiting several days for the temperature to stabilize, a shut-in temperature survey was made (Curve A). The abrupt decrease of the temperature gradient below the top perforations and the quick return toward the geothermal gradient below the bottom perforations suggested a strong cross-flow from the upper to the lower sand.

Next, with the well still shut in, a continuous flowmeter was run in the long-string at constant speed through the upper zone (Curve B). At the upper zone, rotation of the spinner decreased to zero, then reversed, and below the upper perforations recorded a sizeable fluid movement downward. Stationary readings ⊗ above and below the upper perforations confirmed 600 bpd downflow.

Third, the continuous flowmeter was run downward through the long-string with the long-string producing to the surface (Curve C). High response above the upper zone shows considerable production to surface, while opposite the upper perforations, fluid velocity again dropped to zero, reversed, and showed downward flow below.

Stationary readings, ×, confirm this. With the long-string producing, downflow rate was reduced to about 400 bpd.

Finally, a tubing plug was set at 8,836 ft, and another shut-in temperature survey run (Curve D). The temperature trend, towards the geothermal gradient between the reservoirs, confirms that the downflow in the long tubing had ceased.

What had happened in this completion seems clear:

—The long-string blast joint cut out opposite the upper zone.

—The upper zone began producing into both strings.

—Very high upper-zone production rates resulted in the higher water percentages and the sand production.

—Some upper-zone water flooded the lower zone through the long-string, and some was produced to the surface through both strings.

—Due to the higher water cut, overall well output declined.

—Communication behind casing, or through the lower packer, was disproved since the tubing plus stopped crossflow.

Subsequently, a tubing patch was placed to repair the blast joint leak. The long-string water production continued for some time, but then gradually declined to zero, while the oil rate returned to normal.

Casing Leak in Pumping Well—In a pumping well, production logging requires that tools be run in the tubing-casing annulus. This can usually be done utilizing an offset wellhead, assuming there is no packer or tubing anchor.

Figure 9-29 shows a combination logging device used for annulus, or below tubing, measurements of fluid density, temperature, or fluid flowrate. The flow device uses the tracer velocity shot technique, and the fluid density analyzer uses a gamma ray device.

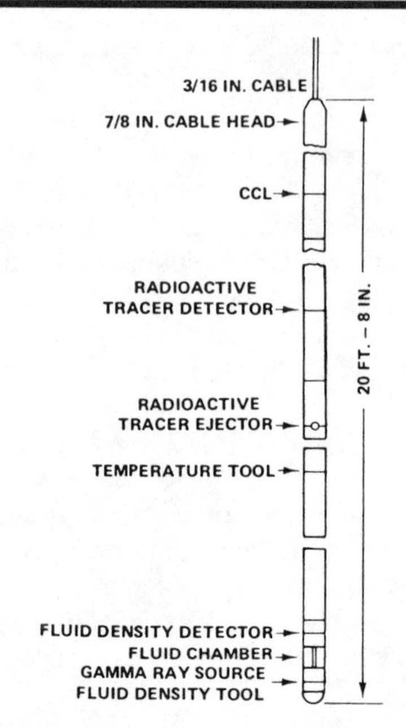

FIG. *9-29—Combination logging device.*[20] *Permission to publish by The Society of Petroleum Engineers.*

The sucker-rod-pumped well, Figure 9-30, was completed with 5 1/2-in. casing and 2 3/8-in. tubing in a water drive reef reservoir. Production dropped from 70 bopd with 40 bwpd, to 186 bwpd with no oil. A casing leak was suspected, but could not be located with a bridge plug and packer technique.

With the pump set at 4,580 ft (perforations at 4,648–4,668 ft and 4,684–4,730 ft), the combination tool recorded the surveys of Figure 9-30. The radioactive tracer fluid velocity survey showed fluid entry at 4,605 ft with 47 bwpd moving upward and 160 bwpd moving downward. The shut-in temperature survey shows water moving downward through a primary cemented channel. After squeeze cementing, production was 110 bopd with 71 bwpd.

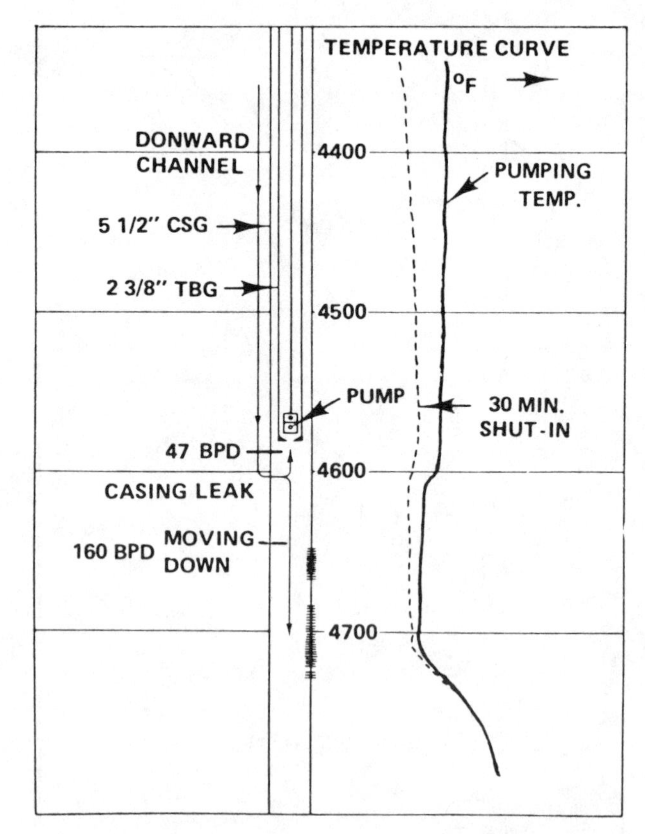

FIG. *9-30—Annular logging below sucker-rod pump.*[20] *Permission to publish by The Society of Petroleum Engineers.*

Flow Behind Casing

Flow behind casing (where no perforations are open) can sometimes be detected with the temperature survey. However, the noise log may be more effective particularly at low flowrates. A gamma ray log run opposite zones where prolonged fluid movement has occurred will sometimes record a greatly increased radiation count. Thus, behind-the-casing water movement, leaving a radioactive crust, may show up dramatically. Experience in an area is needed to verify that this increased radiation occurs with a specific formation water.

Noise Log to Show Fluid Movement Behind Casing—Figure 9-31 shows a 4 1/2-in. casing triple tubingless gas well completion. Before perforating any string, noise log 1 was run in string D, and indicated communication behind casing in the vicinity of 6,650 ft and 6,900 ft.

After a squeeze cement job, string D, loaded with saltwater, was perforated near 7,100 ft, and log 2 was run with the well shut in. Log 2 indicates that squeeze cementing was successful in shutting off the gas chan-

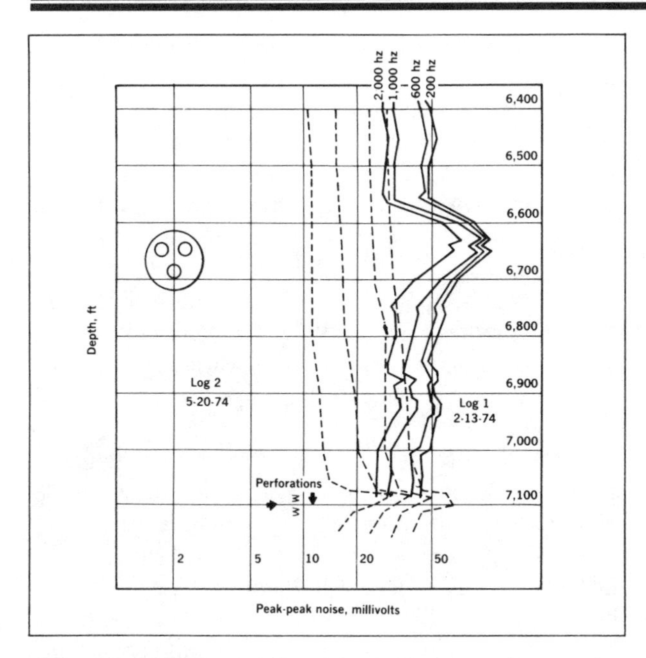

FIG. 9-31—Triple-string tubingless gas well completion in 4 1/2-in. casing.[21] Permission to publish by the Society of Petroleum Engineers.

nel. The low noise peaks at the perforations result from slow downward movement of saltwater into the perforated zone.

Radioactive Tracer to Show Flow Behind the Casing—The oil well of Figure 9-32 is perforated in a 20-ft zone below 9,155 ft, but communication with a gas sand at 9,000 ft is suspected. To check this possibility, the base log was first run using sufficient sensitivity to give reading comparable to a cased hole gamma ray log. An ejector-type tracer tool was then positioned in the perforations, and with the well shut in, a slug of tracer was ejected. The tagged fluid was then pushed into the perforations by surface injection and the interval was relogged. Additional water was then injected to push the tracer into a possible channel. A third logging run confirms that some tagged fluid has moved up outside the pipe to the 9,000-ft level. Experimentation with timing of logging runs and injection pressures may be needed. If pressure differences between zones results in crossflow with the well shut in, the noise log or temperature log may also prove the existence of a channel.

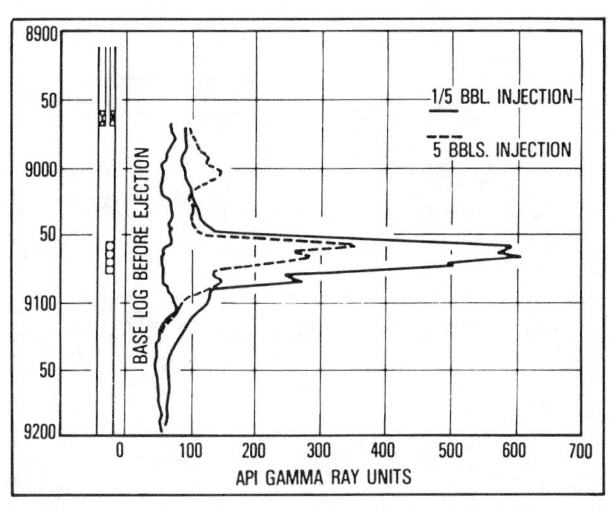

FIG. 9-32—Behind-pipe flow detected by radioactive tracer survey. Permission to publish by the Society of Petroleum Engineers of AIME. Copyright 1983 SPE-AIME.

Gamma Ray to Show Prolonged Water Flow Through Channel—The oil well of Figure 9-33, on initial test, produced 100% water, but after remedial work, made 314 bopd with 1% saltwater. Two years later water increased suddenly to 98%. Since offsetting wells produced no water, a gamma ray and CBL bond log were run to diagnose channeling.

The gamma ray recorded high radioactivity levels in the perforated oil zone, in a water sand 75 ft above the producing zone, and in another water sand 50 ft below the producing sand. High radioactivity was in-

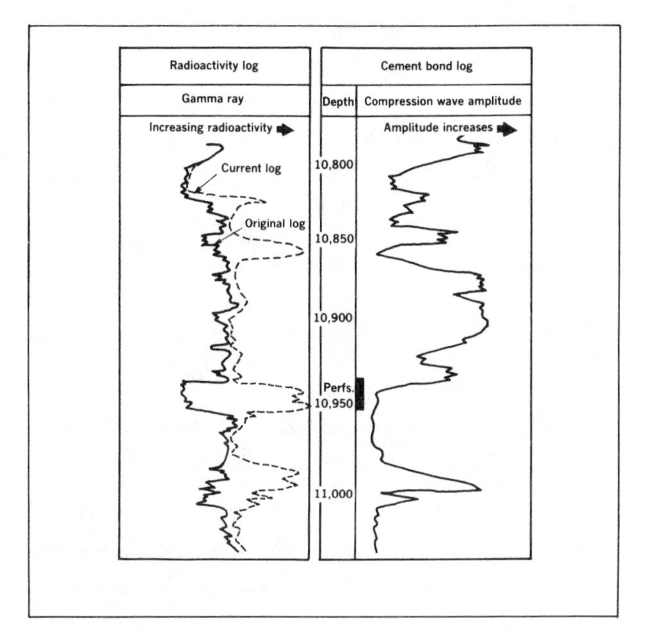

FIG. 9-33—Radioactive crust showing water movement.[17] Permission to publish by The Society of Petroleum Engineers.

dicative of moving water leaving radioactive residue behind. The CBL log showed poor bond above the perforated zone, and only a short section of good bond below the perforations. Combined with the gamma ray indications, the presence of channeling was established and the sands involved were identified.

Producing Zone Evaluation–Reservoir Management

Measurement of fluid type, flow rate, and pressure at various points through the producing zone should provide data needed to deplete the reservoir in an optimum manner.

—In multi-zone producing wells, locating entry of water or gas may permit undesired fluids to be shut off effectively.

—Identification of crossflow between zones may indicate needed revisions in reservoir development schemes.

—Evaluation of the rate of flow from various zones open may indicate need for selective stimulation.

—The combination of flow rate from a particular zone and the well intake pressure opposite that zone should provide reservoir engineering data to prepare a permeability-thickness profile.

—Measurement of fluid saturations and changes in fluid saturations in various zones open to the wellbore, or sealed off behind the casing, should improve ability to optimize reservoir depletion.

—Formation damage effects in multiple zones may be more easily detected by production logging techniques than by transient pressure testing.

—If conditions are favorable, workovers to correct certain situations can be accomplished using through-tubing electric line techniques.

Usually the combination of several logging tools is required to define the flow situation; thus the PCT tool has particular utility.

The following examples show the utility of production logging techniques in many situations. High rate wells are prime candidates—but even lower rate sucker-rod-pumped wells should not be eliminated.

PCT Flowmeter to Define Flow Profile and Crossflow—A high volume (greater than 30,000 bpd) Middle East well was completed in four zones producing water-free oil to the surface with low gor. As a routine observation, the operator wanted to determine the contribution of each zone to the total flow, and to measure crossflow between zones with the well shut in.

Flow Profile—With the well flowing at a rate of 35,000 bpd, the continuous flowmeter was run to define the flow from each of the four zones. Three runs were made downward, and three upward at different logging speeds to record the series of the log up, log down, readings of Figure 9-34.

To permit accurate calculation of volumes, the in-situ calibration of Figure 9-35 was then constructed to account for downhole viscosity variations, and the higher average velocity measured by the centralized flowmeter.

Fluid velocities from the corrected response line of Figure 9-35 converted to bpd (5.7 m/min = 1,000 bpd in 7-in. 26 lb/ft casing) are shown in Table 9-4.

Shut-In Crossflow—Figure 9-36 shows PCT flowmeter, gradiomanometer, and temperature recordings made with well shut in at the surface. Considering flowmeter run No. 6, logged down at 50 m/min, the no-flow response at A is 7.6 rps. The increase at B to 9.8 rps shows upward flow past B entering zone 4.

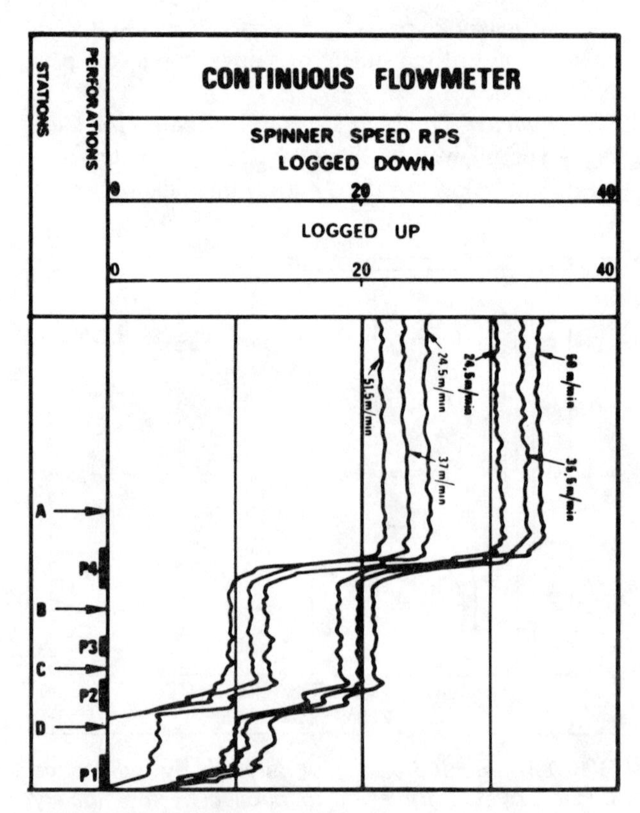

FIG. 9-34—Continuous flowmeter recordings. Courtesy of Schlumberger.

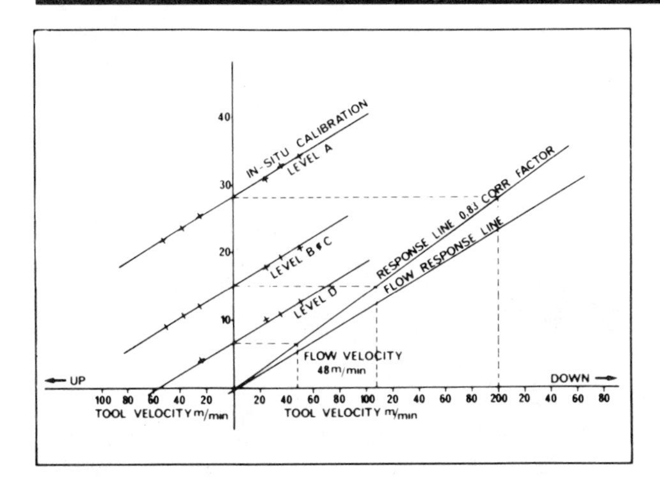

FIG. *9-35—Multipass in-situ calibration. Courtesy of Schlumberger.*

The sharp temperature decrease at interval 4 verifies this and further shows flow entering the lower portion of interval 4. The flow into zone 4 is oil shown by the gradiomanometer, 0.69 gm/cc, and is coming from zone 1 and 2, with perhaps a minor contribution from zone 3.

Using in-situ calibration and correction techniques, Table 9-5 lists actual calculations of shut-in crossflow.

PCT Flowmeter and Gradiomanometer to Determine Water Entry—Water entry into a flowing oil stream can sometimes be located, with the flowmeter to show which zones are producing, and the gradiomanometer to define density of the fluid stream at various points. Downhole density of the oil and water

FIG. *9-36—Recordings made with well shut in at the surface. Courtesy of Schlumberger.*

TABLE 9-4

Station	Fluid velocity m/min	Flow rate bopd	Flow contribution Zone	Flow contribution bopd
A	200	35100	P4	16100
B	108	19000	P3	0
C	108	19000	P2	10600
D	48	8400	P1	8400

TABLE 9-5
Sonic Transit Time (Compressive Wave)

Station	Fluid velocity m/min	Flow rate bopd	Flow contribution Zone	Flow contribution bopd
A	0	0	P4	−1930
B	11	1930	P3	180
C	10	1750	P2	520
D	7	1230	P1	1230

must be known—and must be significantly different. High flow velocities are helpful.

The example of Figure 9-37 is a high rate Middle East well, perforated in three zones shown on the depth track. Above all zones, the total flow rate at the bottom-hole condition was 15,000 bpd. The flowmeter run downward through the three zones shows the percentage of flow contributed by each zone. Note that zone B contributes nothing.

The gradiomanometer measures density varying from 1.08 gm/cc below the zone C to 0.79 above zone A.

Downhole densities were determined to be 0.71 gm/cc for oil, and 1.08 gm/cc for water, permitting the construction of the water holdup chart of Figure 9-38. With the density difference between the water and oil of 1.08 − 0.71 = 0.37 gm/cc slippage velocity is determined from Figure 9-39, using the water holdup values from Figure 9-38, for the actual fluid density at a particular level as measured by the gradiomanometer.

Flow rate of water, q_w, at a particular point is obtained from the relation:

$$q_w = y_w[q_t - 1.4\, v_s(D^2_{csg} - D^2_{tool})(1 - y_w)]$$

q_t = total flow at a particular level, bpd
v_s = slippage velocity, ft/min
y_w = water holdup
D_{csg} = casing id, in.
D_{tool} = tool od, in.

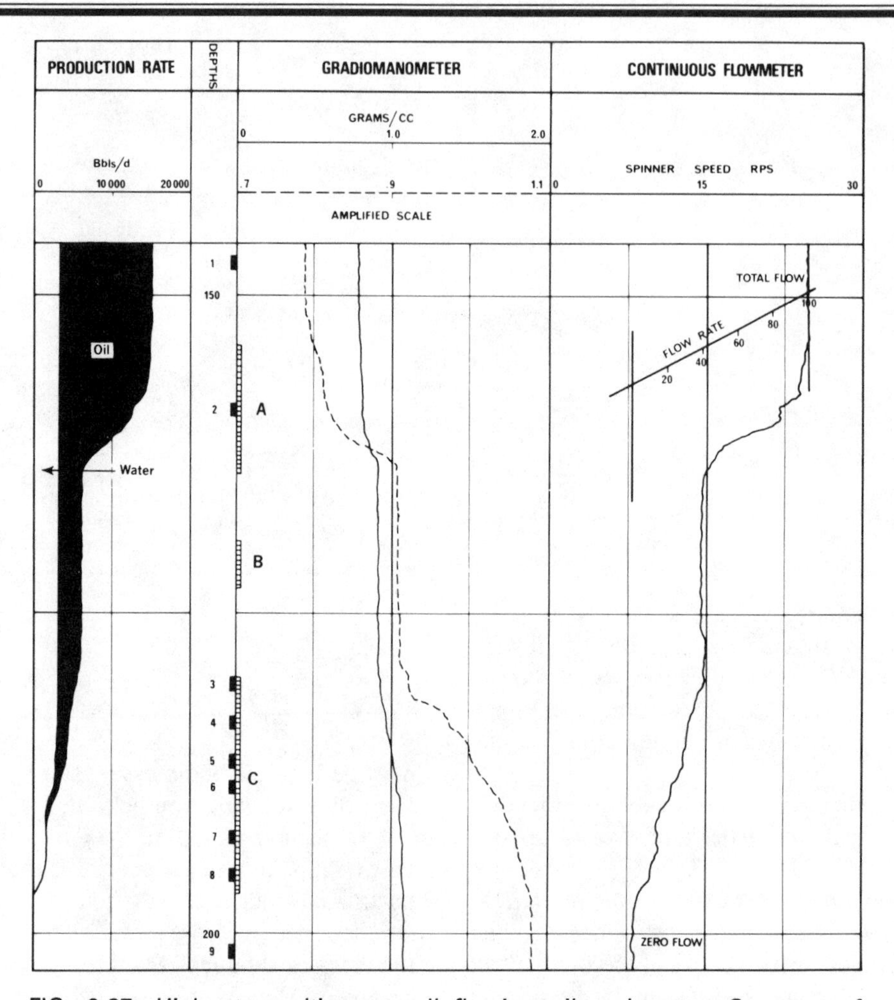

FIG. 9-37—*High-rate multi-zone well flowing oil and water. Courtesy of Schlumberger.*

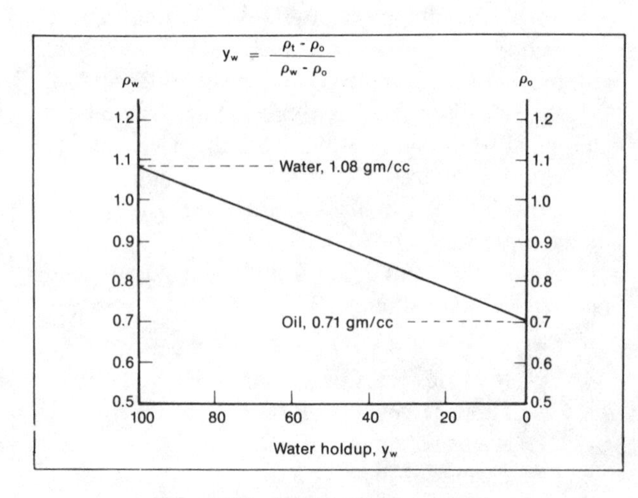

FIG. 9-38—*Water holdup chart.*

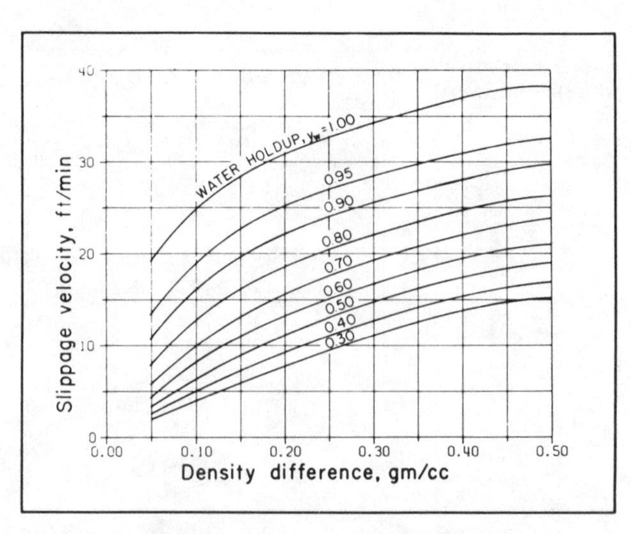

FIG. 9-39—*Slippage velocity. Courtesy of Schlumberger.*

TABLE 9-6

Interval	% Flow	q_t	P_{gm}	y_w	v_s	q_w	q_o
1	100	15,000	0.790	0.23	12	3,350	11,650
2	84	12,600	0.809	0.275	12.5	3,350	9,250
3	41	6,150	0.920	0.570	16	3,320	2,830
4	32	4,800	0.978	0.730	22	3,300	1,500
5	28	4,200	1.010	0.822	25	3,280	920
6	23	3,450	1.030	0.870	26.5	2,860	590
7	13	1,950	1.060	0.955	31	1,800	150
8	10	1,500	1.070	0.980	34	1,450	50
9	0	0	1.080	1.00	—	0	0

Table 9-6 shows complete calculations at each interval.

The production profile, Figure 9-37, shows water entry from the bottom of lower-most set of perforation, with the last point of entry between levels 5 and 6. Most of the oil production is coming from the lower portion of zone A.

Water production could probably be eliminated by setting a plug in the casing between zones B and C (assuming a satisfactory primary cement job in this interval). Oil production lost from the lower zone could probably be recovered by the upstructure wells. Zone B and the top of zone A should be stimulated, perhaps by merely reperforating. By eliminating the water flow, the greater drawdown possibly might permit these zones to contribute to flow even without stimulation.

Periodic Pulsed Neutron Logs to Follow Reservoir Performance—A Middle East well, completed in a higher zone, extended through several lower intervals near the original water-oil contact, providing an observation well for the lower zone.

The TDT logs of Figure 9-40 show four runs made at one-year intervals through three sand zones.

During this 36-month period:

—Water progressively moved through the 5,195–5,225-ft lower sand.

—No flushing of the middle 5,132–5,152-ft zone was indicated.

—First water encroachment of the upper 5,053–5,069-ft zone was shown on the 3rd survey. The 4th survey shows this zone almost completely flushed.

Figure 9-40 summarizes this water movement. It appears that withdrawals from the top zone may be excessive.

PCT for Gas Well Completion Evaluation—Gas well performance can be evaluated with the combination of:

—Flowmeter to measure flow from each zone.

—Manometer to provide bottom-hole flowing pressure, or at least surface indication of stabilized flow.

—Amerada pressure measurement, if needed, for added precision.

—Gradiomanometer and thermometer to substantiate the flowmeter.

Usually gas wells stabilize quickly, so that several flow rates can be obtained, and a conventional 3-point back pressure test can be made to obtain an absolute open-flow potential.

Use of the trigger zone perforating and clean up technique, reduces perforation plugging, permits stepwise evaluation of several zones perforated in consecutive order, and improves the chances of selective stimulation if required.

Figure 9-41 shows a two-stage completion of a North

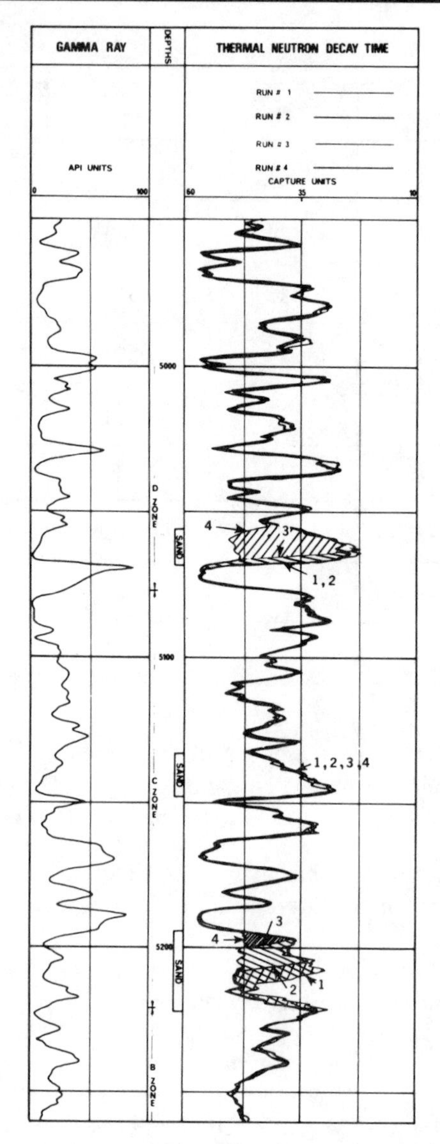

FIG. *9-40—Periodic TDT logs to observe water invasion. Courtesy of Schlumberger.*

Sea gas well. Trigger zone A was perforated first, and the well cleaned up. Zones B and C, relatively low permeability zones, were perforated next.

Figure 9-41A shows surveys, made during the second flow period, of a two-point back pressure test of zones B and C. Flowmeter response corrected by the in-situ calibration technique indicates a flow rate of 1,000,000 m^3/day. The Amerada pressure measurement (not shown) was used to measure downhole pressure. Pressure, gas gravity, and temperature values, provide data to correct downhole gas volumes to standard surface conditions. The first flow period indicated a rate of 240,000 m^3/day.

Figure 9-42 (Completion 1) is a plot of production rate vs. squared drawdown pressure (q vs $P_i^2 - P_{wf}^2$) which, extropolated to P_{wf}^2 = zero shows an AOF of 1,800,000 m^3/day.

Next, zone D was perforated. Figure 9-41B was recorded with the well flowing on the larger of two chokes. Zones A, B, C, and D are open, but the flowmeter is used to look at flow from zone D. With proper corrections and calculations the flowmeter indicates a rate of 1,400,000 m^3/day from zone D.

It also shows that now most of the gas comes from zone D. A smaller choke produced a rate of 500,000 m^3/day from zone D. Figure 9-42 (Completion 2) is the (q vs $p_i^2 - p_{wf}^2$) plot of the zone D two-point back pressure test, indicating an AOF of 8,200,000 m^3/day.

Note that while the thermometer and gradiomanometer are not essential to the analysis they provide backup and lend confidence to the analysis. Also, the gradiomanometer shows that flow from the trigger zone A did not completely remove liquid from the borehole above zone A.

Need for Selective Stimulation—A gas well completed by perforating and acidizing fine-grained dolomitic porosity in the interval 5,902–6,479 ft tested an absolute open flow of 9.5 mmcfd, 8% H_2S, with a wellhead pressure of 1,800 psi.

DST and log porosity-feet data indicated the well should be capable of more. The continuous flowmeter and gradiomanometer were run to define the production profile. Run 1 of Figure 9-43 shows a saltwater column from td to 6,275 ft, with 15 ft of condensate on top. The flowmeter shows 40% of the total production is coming from 6,135–6,180 ft, with no production below 6,275 ft.

After selective acidizing, the well made 16 mmcfd, and logging Run 2 shows that most zones are now contributing to production.

Through Annulus Logging to Define Gas and Water Entry in Pumping Well—The well of Figure 9-44 was completed in open hole, with 7-in. casing set at 4,487 ft and 2 7/8-in. tubing run to td at 4,860 ft. On initial test the well pumped 85 bopd, 340 bwpd, and 40 mcfd.

The combination production log, (tracer, fluid density, and temperature devices) was run to define the producing profile. The tracer runs show the 100% flowrate to be 320 bpd, with 200 bpd coming from

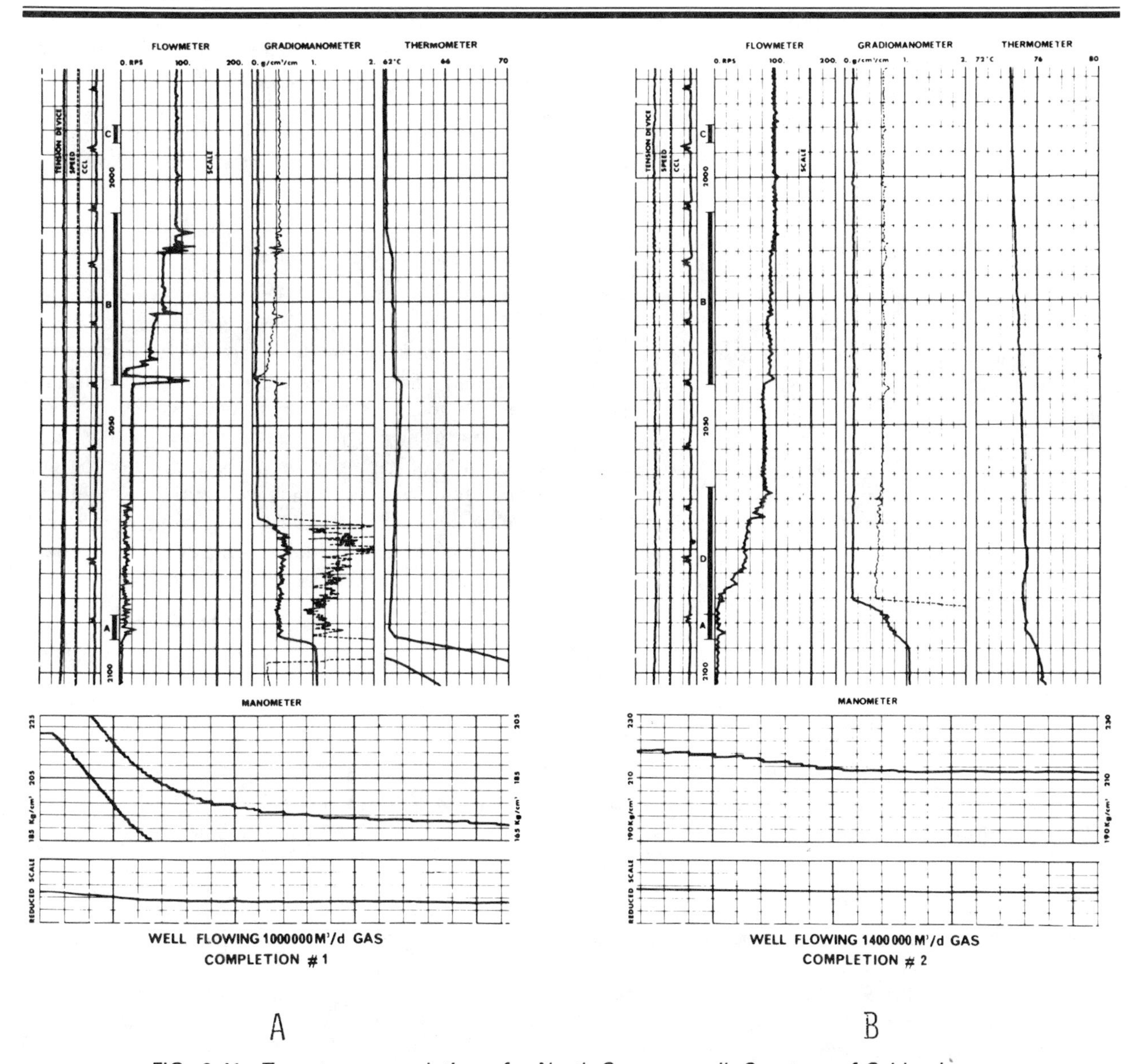

FIG. 9-41—*Two-stage completion of a North Sea gas well. Courtesy of Schlumberger.*

the interval 4,780–4,810 ft. The density device shows this zone to be water productive, and apparently the temperature log indicates this to be cooler waterflood injection water, which has broken through to this well. There is a minor water source (30 bpd) from 4,670–4,695 ft. The fluid density survey shows that all oil and gas in this well is produced from 4,496–4,607 ft.

Expanded Gas Cap Position Located with the Through-Tubing Neutron Log—Two through-tubing GNT-type neutron logs were run in this well, one shortly after completion, the other after one year of production, Figure 9-45.

The logs were made through 3 1/2-in. tubing in 9 5/8-in. casing and were intended to monitor encroachment of the gas cap after continued production from the underlying oil reservoir.

While most of the log sections overlay fairly well, zones A and B in the upper part of the sand body feature significant departures to the right (lower hydrogen index) indicating the presence of free gas.

This information can be used to assess the depletion status of the reservoir and to predict the time at which gas entries into the completed interval are to be anticipated.

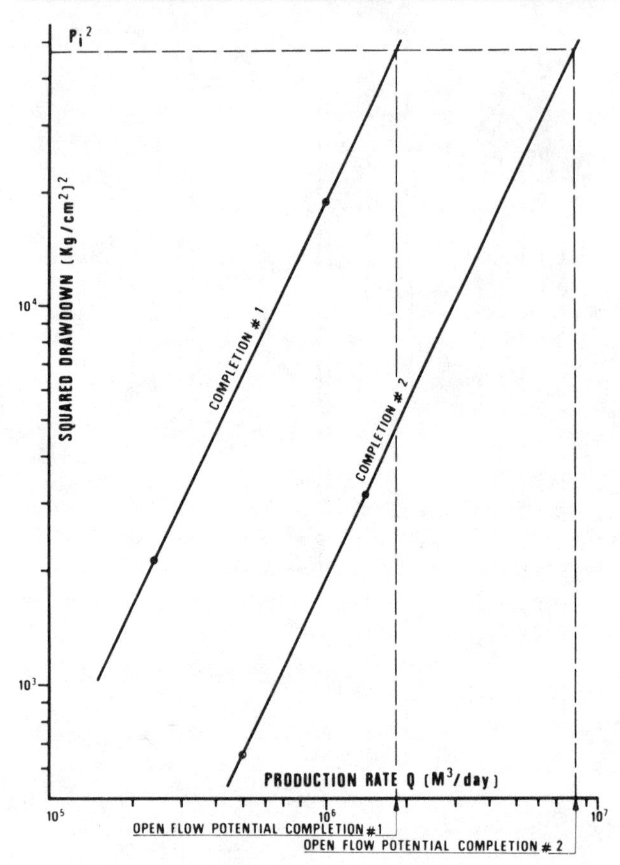

FIG. *9-42—Production rate vs. squared drawdown pressure. Courtesy of Schlumberger.*

Wireline Workover—A Middle East well began producing excessive salt on initial completion (270 lbs/ 1,000 bbl at a flow rate of 2,000 bopd, up to 4,000 lb/1,000 bbl at 5,000 bopd). Water entry from the lower intervals P-1 and P-2 was suspected. The PCT runs of Figure 9-46, made with the well shut in at the surface, indicated downhole crossflow.

In-situ calibration and correction produced the shut-in downhole flow profile of Figure 9-46 showing:

—No flow from P-1.

—170 bpd upflow from P-2. Density (1.13 gm/cc) shows saltwater.

—Of the P-2 flow, 100 bpd enters P-3; 70 bpd enters the bottom of P-4.

—85 bpd of fluid and gas (0.37 gm/cc) downflow from P-7.

—No flow in or out of P-6.

—600 bpd fluid downflow from P-5.

—685 bpd fluid entering top of P-4.

Cement evaluation with the CBL-VDL combination (not shown) indicated at least 6 meters of "good bond" above P-2.

A through-tubing bridge plug, set at 1,070 meters, cut water production such that salt was below limits at 5,000 bopd.

Water Injection Well Profiling

In a water injection well economic success is related to making sure that water enters the desired zones. With high injection pressures, conditions are such that water may move through channels in the primary cement job to undesired zones. Where multiple zones are open, it may be important to know that each zone takes the proper proportion of the total injection volume. Thus, water injection well profiling requires a logging device that looks for fluid movement behind the casing, as well as detailing where, and how much, fluid leaves the casing at various points.

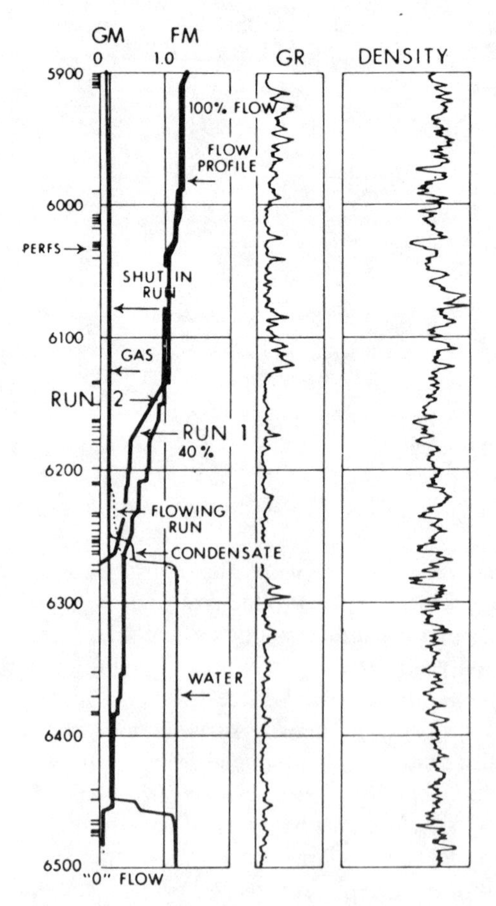

FIG. *9-43—Gas well production profile. Courtesy of Schlumberger.*

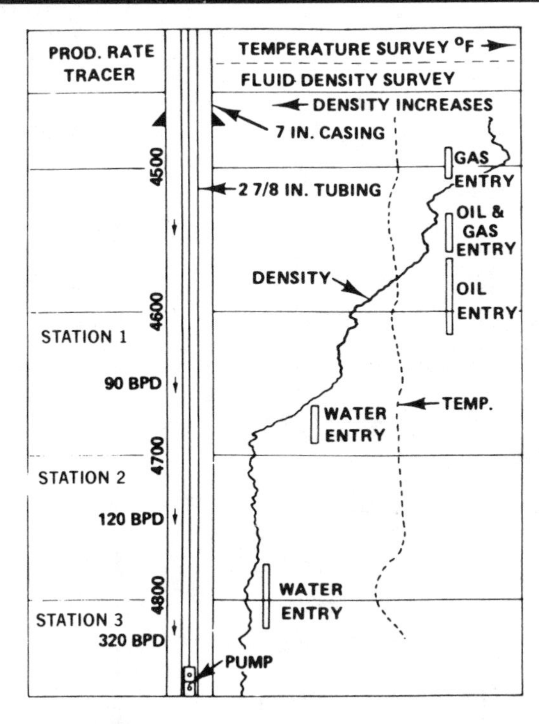

FIG. *9-44—Location of water and gas entry in pumping well.*[20] *Permission to publish by The Society of Petroleum Engineers.*

Usually, the radioactive tracer or the temperature survey is used, with the tracer being more definitive, but the temperature survey more economical.

Radioactive Tracer Survey—Figure 9-47 shows a typical Permian Basin water injection well, completed with 5 1/2-in. casing, 2-in. tubing, and 4 3/4-in. openhole. The problem was to determine the injection profile when injecting water at 800 bpd.

The first step was to establish the 100% velocity measurement and to check for flow up behind casing. An ejector tool was positioned in the casing at 4,895 ft, and a heavy slug of water soluble I-131 tracer was ejected. As shown in Figure 9-47, 10 sec after ejection, the tracer passes the detector. Converting to flow rate inside casing, 10 sec is equivalent to 800 bpd. A short time after the initial kick on the detector, a second kick occurs.

Next, timed logging runs were made through the slug as it traveled to the bottom of the injection interval, as shown in Figure 9-48.

Three zones of interest are noted, B, C, and D, Figure 9-48, where it appears that a portion of the slug entered the formation, since residual radioactivity is indicated on succeeding runs.

To this point, the following information has been established:

—The 100% rate measurement inside casing is 800 bpd.

—There is a channel up behind the casing to about 4,890 ft, (zone A).

—There are three major fluid-loss zones in the borehole: zone B, 4,930–40 ft; zone C, 4,965–75 ft; and zone D, 4,988–5,000 ft. About 75% of the fluid is lost above 4,940 ft; about 10% is lost between 4,965–80 ft; and some water is lost from 4,995–5,000 ft.

The next step was to run several velocity shots to refine the flow pattern. Shots should not be taken in washed out sections of the hole, or where flow velocity is changing rapidly. Figure 9-49 shows 6 velocity shots at various depths, beginning at the bottom. Notice that at low flow rates (shots 2 and 3),

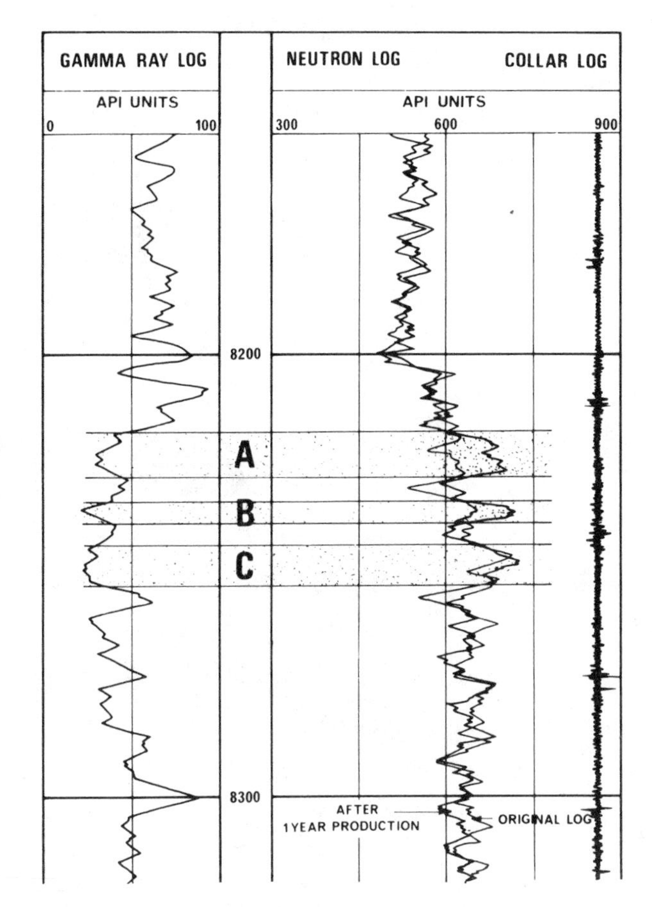

FIG. *9-45—Periodic through-tubing neutron logs to observe gas invasion. Courtesy of Schlumberger.*

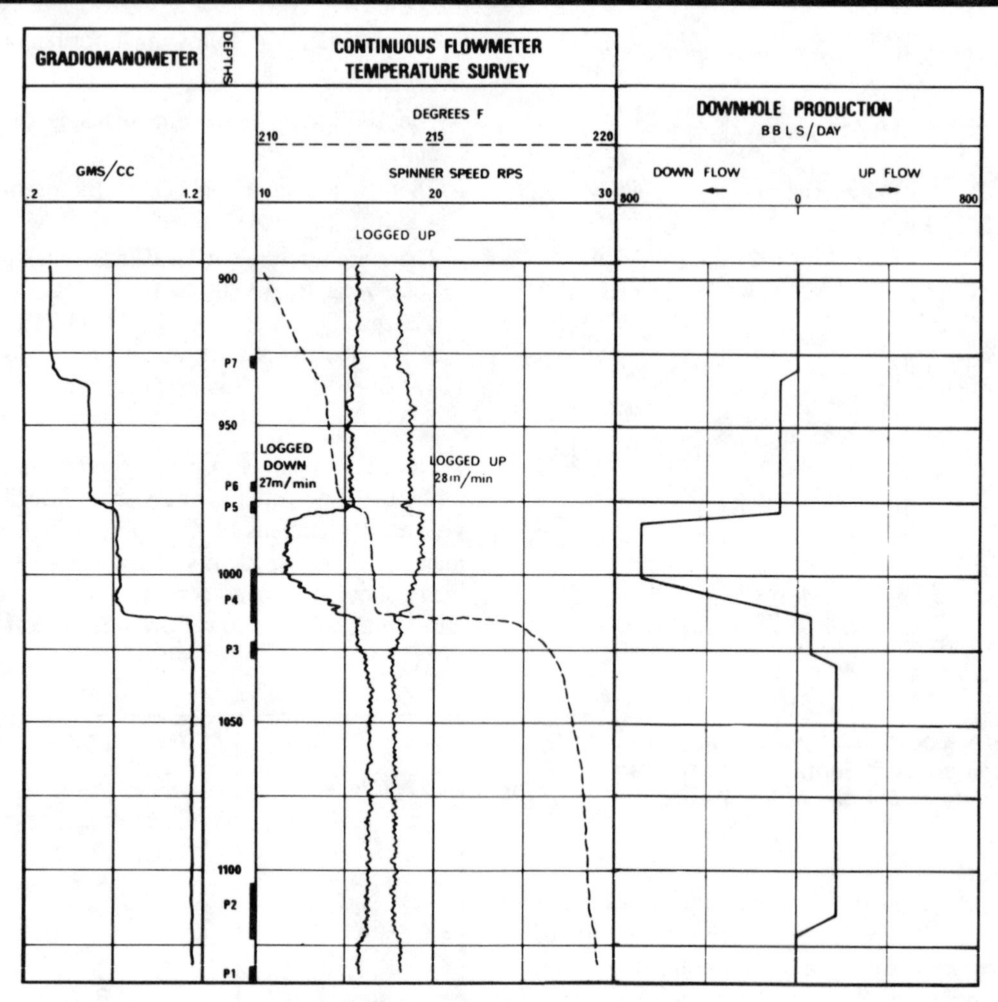

FIG. 9-46—Shut-in survey locating excessive water production and down-hole cross flow. Courtesy of Schlumberger.

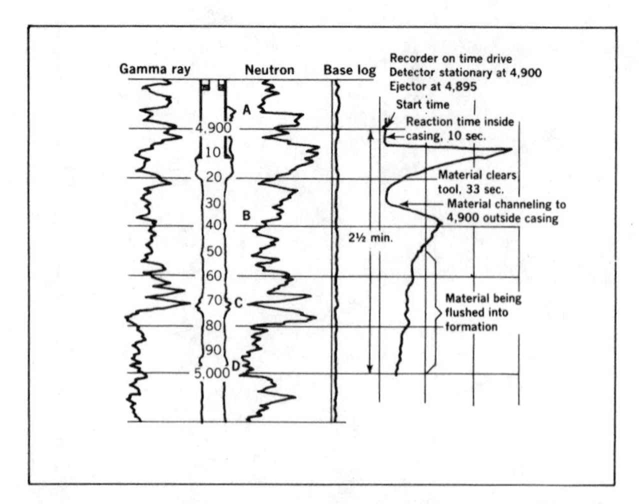

FIG. 9-47—*Typical Permian Basin water-injection well.*[3] *Permission to publish by the API Production Department.*

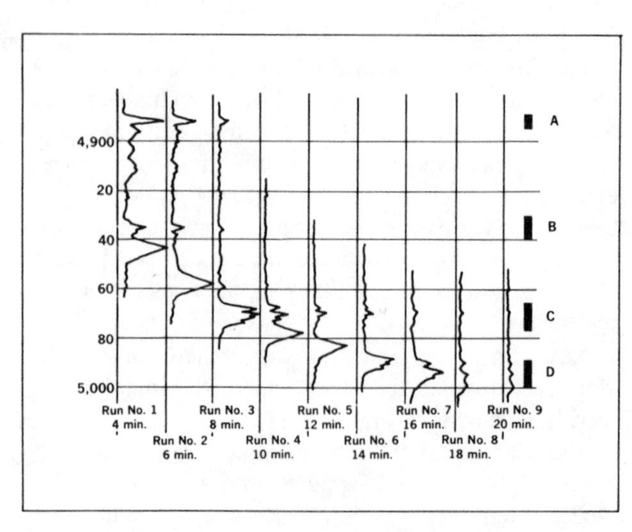

FIG. 9-48—*Timed logging runs.*[3] *Permission to publish by the API Production Department.*

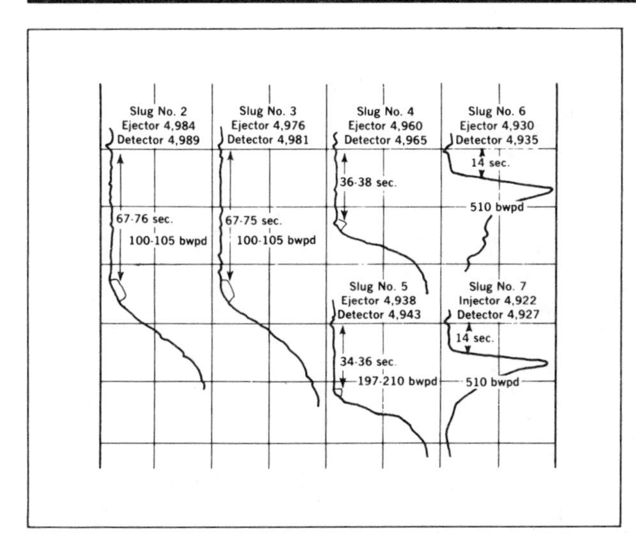

FIG. 9-49—*Detail shots to refine flow pattern.*[3] *Permission to publish by the API Production Department.*

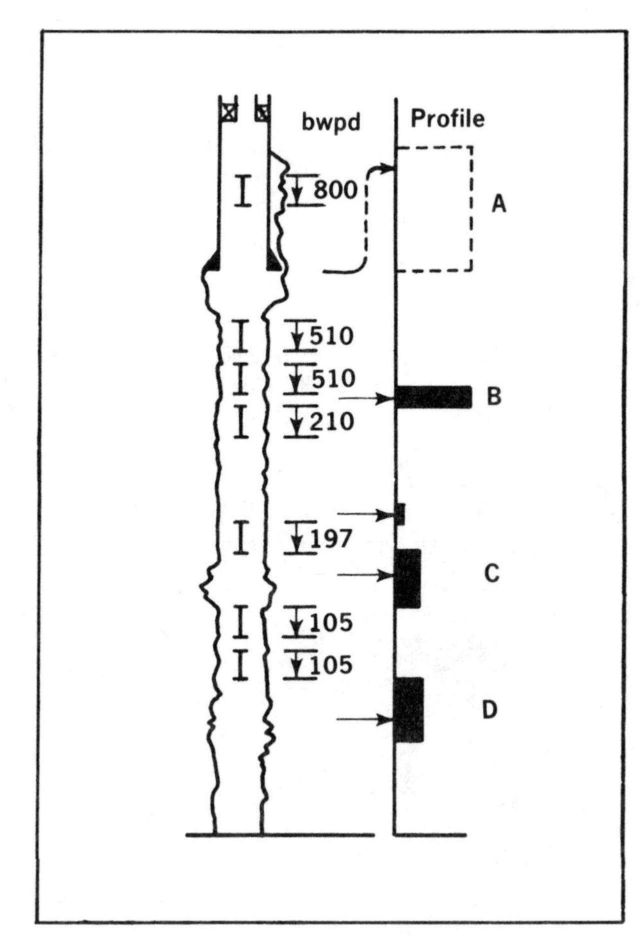

FIG. 9-50—*Final flow profile.*[3] *Permission to publish by the API Production Department.*

accuracy of timing is questionable. The final flow profile was established, as shown in Figure 9-50.

Temperature Survey—Use of temperature logging techniques are often quite helpful in defining *which* zones are taking fluid in a water injection well. *How much* fluid enters each zone in a multi-zone reservoir cannot be effectively defined. Points that favor use of temperature logging include:

—Physical size of the tool is small.
—Cost is relatively low.
—Flow behind casing can be detected.
—Depth of investigation is relatively great.

The basic technique of water injection well logging is shown in Figure 9-51. Water injection rate should be stabilized for 48 hours before running the injection survey. This survey should show the bottom of the injection zone by a sharp return of the temperature curve to the geothermal profile. The top of injected zone is probably not defined; however, the curve should be vertical through the zone taking fluid.

Water injection is then shut in. Formations that did not take water warm up rapidly toward geothermal; whereas, in zones taking water, the cool anomaly persists, to be detected by surveys run periodically after shut-in.

For effective interpretation, several factors influencing heat flow near the wellbore must be considered. The following examples, from Amoco computer model studies of wellbore heat flow in water injection wells,[14] illustrate these factors.

Physical Wellbore Arrangement—Heat flow is primarily by conduction (except in the zone taking water); thus non-moving water, cement sheath, enlarged hole,

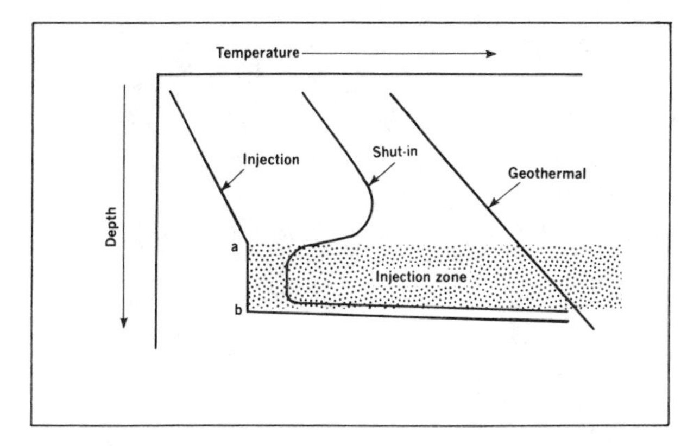

FIG. 9-51—*Basic technique of water injection well logging.*[14] *Permission to publish by The Society of Petroleum Engineers.*

etc., act as insulation to restrict heat flow and affect the recorded temperature curve. Examples in Figures 9-52 through 9-55 show these effects.

Injection Water Temperature—Downhole temperature of the injected water must be 10 to 15°F colder (or warmer) than the geothermal temperature to define the injection zones. The examples of Figures 9-56 and 9-57 show that 5°F difference provides no indication of the injection zone; whereas, 10°F difference provides marginal definition with sufficient shut-in time. Greater temperature differences, Figure 9-58, provide good definition with less shut-in time.

Shut-in Time—Usually 24 to 48 hrs provide optimum definition of zones, Figure 9-57. Experimentation is required, and several runs during the shut-in period aid interpretation.

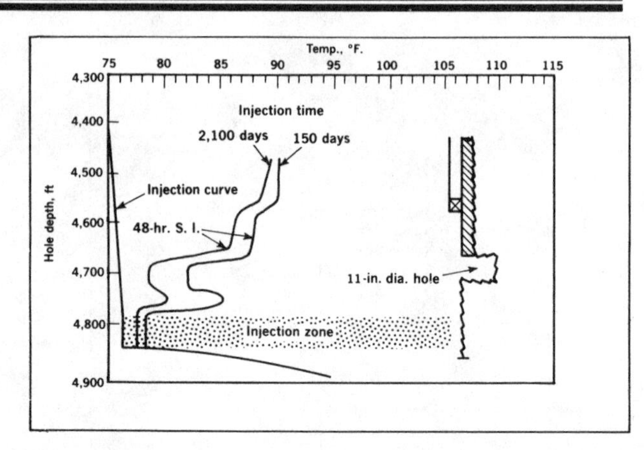

FIG. *9-54—Anomalies due to physical well bore effects.[14] Permission to publish by The Society of Petroleum Engineers.*

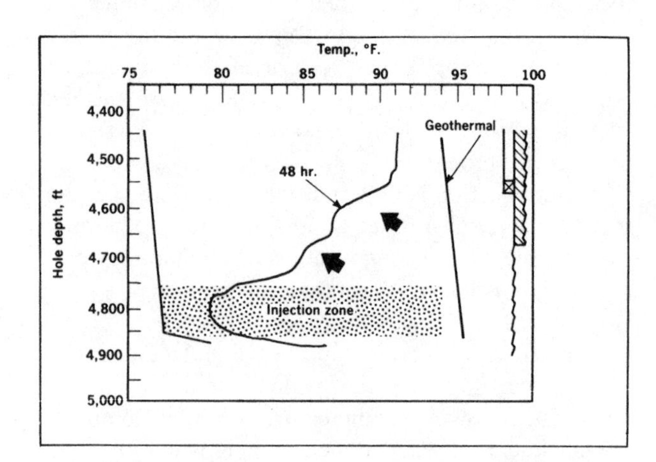

FIG. *9-52—Anomalies due to physical well bore effects.[14] Permissions to publish by The Society of Petroleum Engineers.*

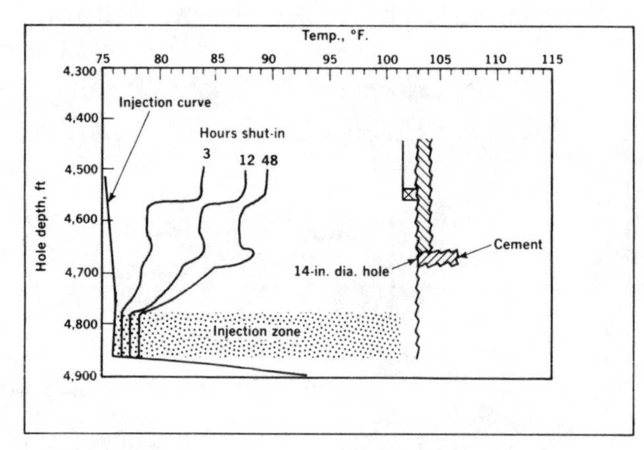

FIG. *9-55—Anomalies due to physical well bore effects. Permission to publish by The Society of Petroleum Engineers.*

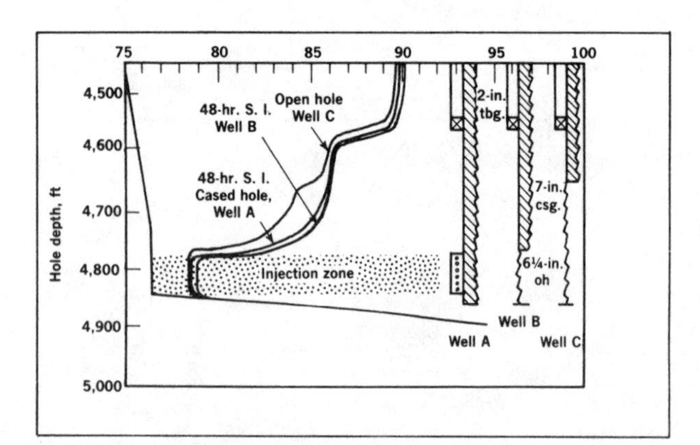

FIG. *9-53—Anomalies due to physical well bore effects.[14] Permission to publish by The Society of Petroleum Engineers.*

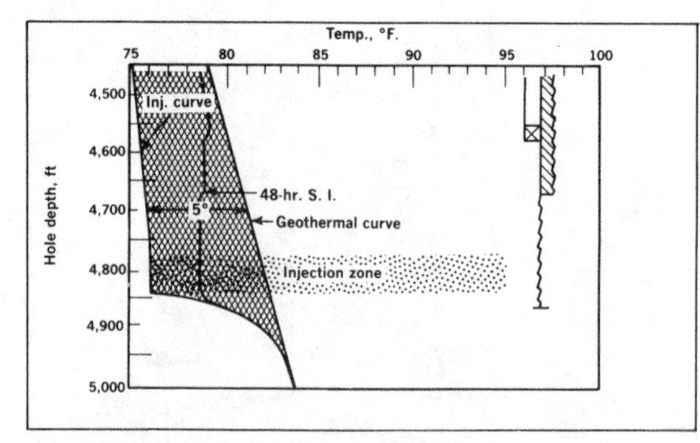

FIG. *9-56—Inadequate temperature difference.[14] Permission to publish by The Society of Petroleum Engineers.*

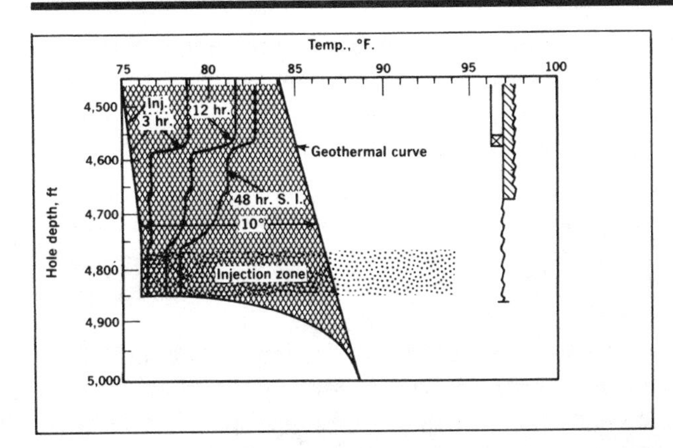

FIG. *9-57—Adequate temperature difference.[14] Permission to publish by The Society of Petroleum Engineers.*

The best definition is obtained by running surveys early in the life of an injection well. Injection over a long period of time tends to wash out zone limits, as shown in Figures 9-59 and 9-60.

For better zone definition in an older well, Figure 9-61, the injection water temperature should be raised (or lowered) 10°F for several hours before shut-in (200–300 bbl of water is sufficient). Shut-in surveys then must be run sooner than normal, since the induced anomaly will dissipate rapidly.

As shown in Figure 9-62, the temperature survey does not tell how much water goes into each zone. A zone taking a small portion of the total injection looks the same as one taking much more.

Quality control is important as usual:

—For routine survey, stabilize temperature and in-

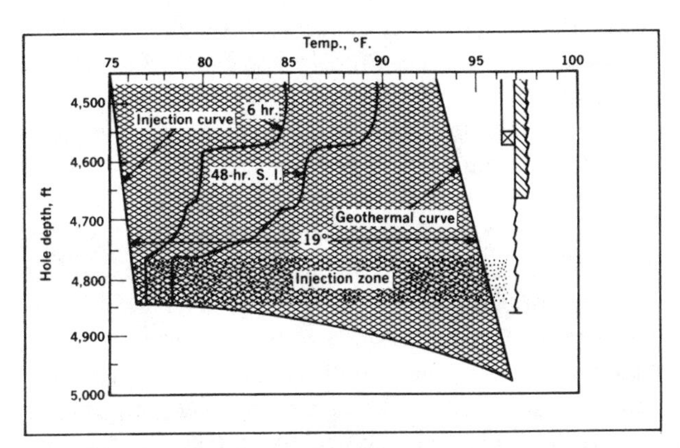

FIG. *9-58—More than adequate temperature difference.[14] Permission to publish by The Society of Petroleum Engineers.*

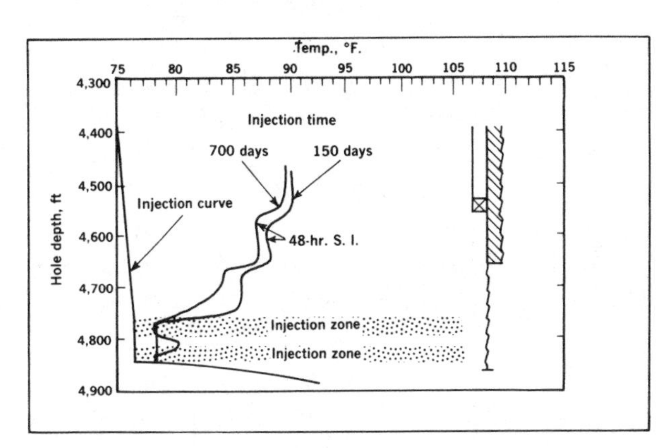

FIG. *9-60—Layered injection effect.[14] Permission to publish by The Society of Petroleum Engineers.*

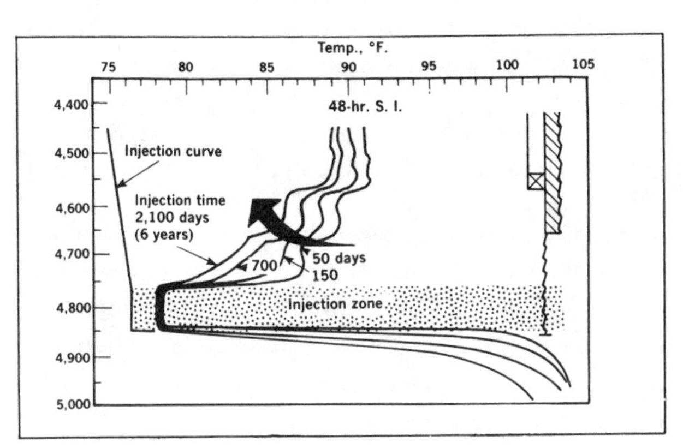

FIG. *9-59—Cumulative injection effect.[14] Permission to publish by The Society of Petroleum Engineers.*

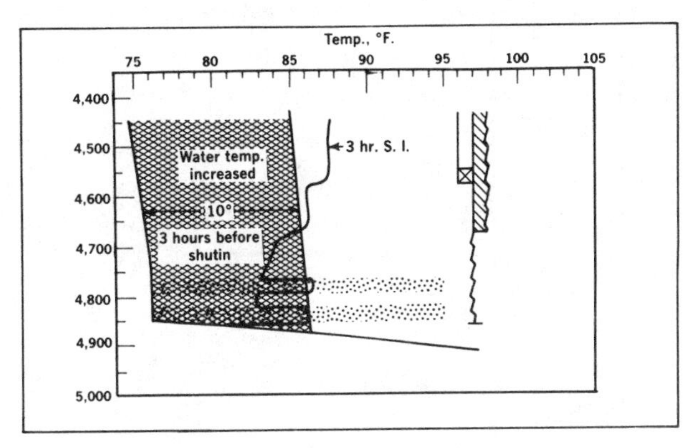

FIG. *9-61—Effect of temperature change just before shut-in.[14] Permission to publish by The Society of Petroleum Engineers.*

jection rate for 48 hrs prior to survey.

—Surface leaks must not be permitted during shut-in survey.

—Log running in the hole, to record undisturbed temperatures. Allow sufficient time between runs for temperature equilibrium to be restored.

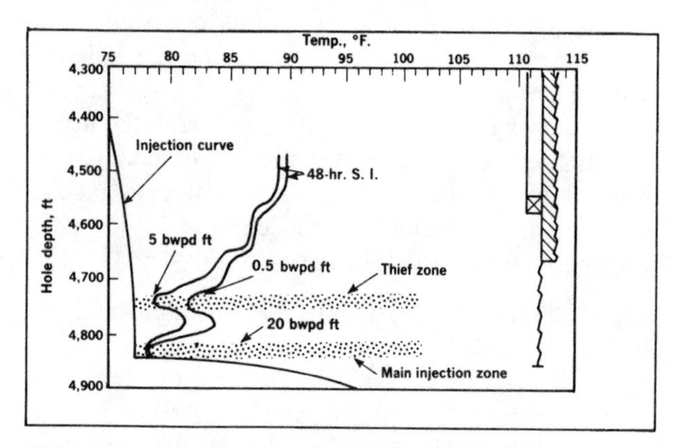

FIG. *9-62—Effect of thief-zone losses.*[14] *Permission to publish by The Society of Petroleum Engineers.*

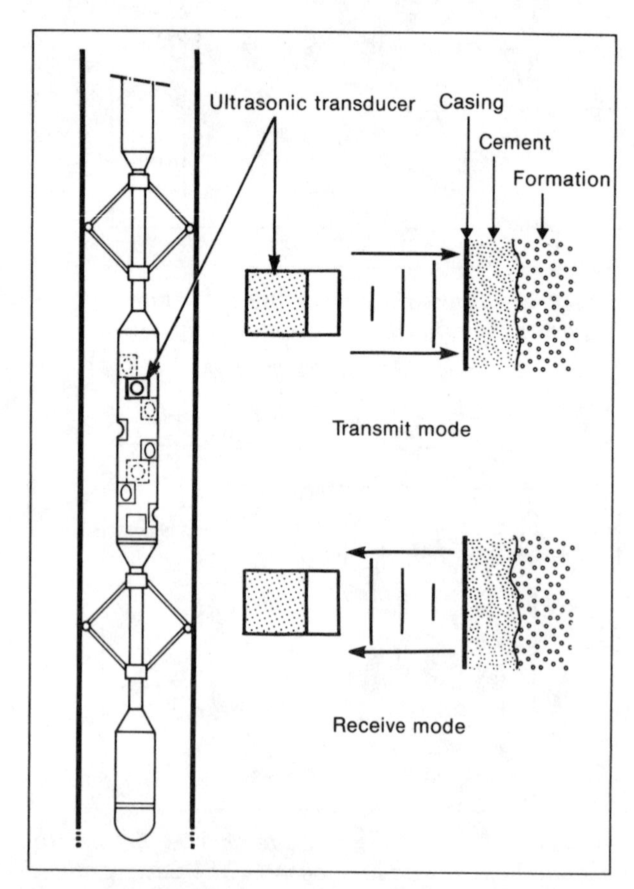

FIG. *9-63—Principle of the Cement Evaluation Tool. Courtesy of Schlumberger.*

—Logging speed should not exceed 20 ft/min. Check temperature response by stopping tool.

PRIMARY CEMENT EVALUATION

Acoustic measurements, correctly interpreted, can be a useful tool in evaluating the success of the primary cementing operation on initial well completion. They may also be applicable if later producing operations indicate a remedial cement job.

Basically, acoustic measurements show the effect of annular cement on sonic energy transmission in the casing and surrounding wellbore area. It should be noted that, even though a high percentage of the annular cross-section is filled with cement, we are not guaranteed a hydraulic seal. Conversely, an annulus only partially filled with cement may present an effective hydraulic seal. The conventional CBL-VDL bond log has an inherent disadvantage in that it records an average condition around the circumference of the casing. Thus a mud channel covering only a small portion of the circumference may not be detected. On the other hand, mud channels are the basic cause of communication between zones.

A recent sonic development to supplement the conventional bond log is the Schlumberger CET (cement evaluation tool), Figure 9-63. It consists of 8 ultra sonic transducers arranged helically over 2 ft. Each transducer in turn transmits a signal, then measures the signal decay rate. Sonic decay rate is related to the compressive strength of the material surrounding the casing, thus mud channels in contact with the casing should be detected.

To look for channels, other logging devices have been used. A focused gamma ray device that can be rotated through 360 degrees to measure the density of the material immediately behind the casing has been used to locate mud channels and, with a zero-phase perforator, to perforate into the channel for squeeze cementing. The radial differential temperature device, which very precisely compares the temperature at various points around the circumference of the casing to the temperature in the center of the hole, has been used for the same purpose. The noise log, sometimes in connection with the temperature log, has been effective in locating and defining actual fluid movement between zones as a cement evaluation procedure.

Principle of the Bond Log Device

Essentially the cement bond device, shown schematically in Figure 9-64, consists of an acoustic section containing a transmitter that generates sonic pulses

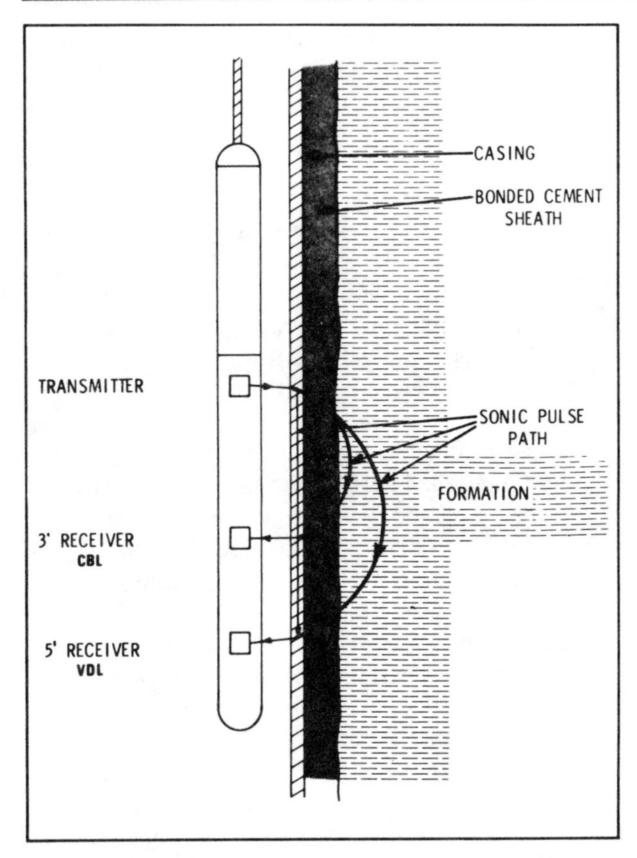

FIG. 9-64—Bond log transmitter-receiver configuration. Courtesy of Schlumberger.

at a cyclic rate high enough to provide a continuous log and a receiver that picks up the reflected wave signal.

An electronic section measures the amplitude of a particular portion of the received signal (CBL), or converts the energy wave to an intensity modulated form (VDL). Both the CBL and VDL are recorded versus depth.

The usual log combination, shown in Figure 9-65, includes a CBL or "pipe amplitude" curve; and a VDL or "full wave train" presentation. Usually a Δt, or transit time curve, is included for quality control, and a CCL or collar locator, and gamma ray for depth control.

As with other sonic tools, the wellbore must be liquid-filled in order to transmit sound energy effectively.

The Pipe Amplitude Log (CBL)—records the first sonic energy to arrive at the receiver, Figure 9-66. This is normally the compression wave traveling along the casing. Transmitter-receiver spacing should be about 3 ft to provide maximum signal level without

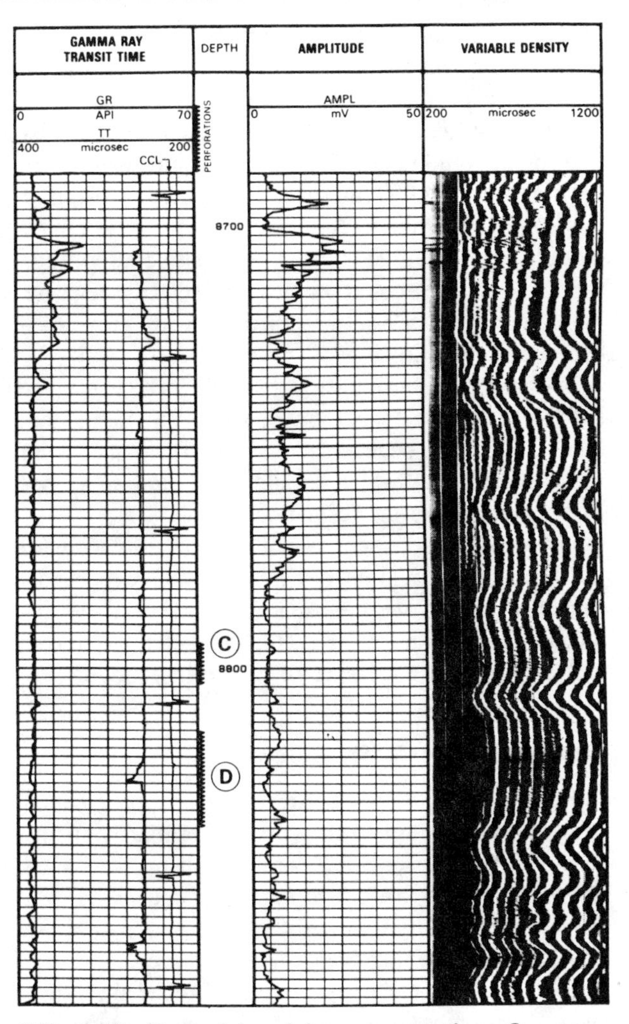

FIG. 9-65—Typical bond log presentation. Courtesy of Schlumberger.

interference from the preceding signal that traveled through the wellbore liquid.

In unsupported casing, amplitude is maximum. In supported casing, (completely surrounded by a well-bonded hard cement sheath at least 3/4-in. thick) am-

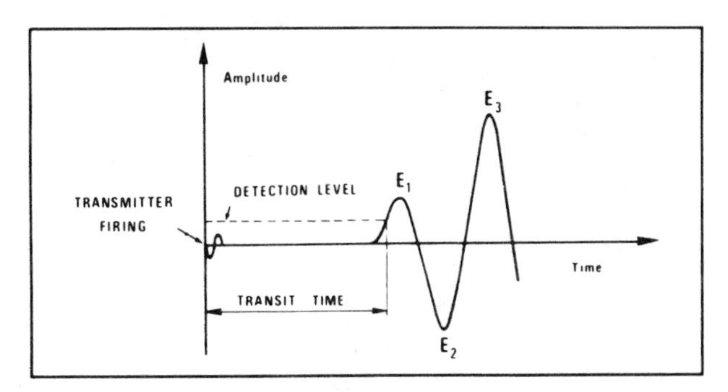

FIG. 9-66—CBL records amplitude of E_1 wave.

plitude is minimum. Thus, with the CBL, comparison of sonic amplitude is the primary measure of cementing success.

Full Wave Train (VDL) Log—is a continuous display of the amplitude of the entire sonic wave form for a period of 1,000 microsec following the firing of the sonic transmitter. Sonic amplitude is intensity modulated, such that increasing positive amplitude produces a darker color; whereas, increasing negative amplitude produces a lighter color, with grey indicating zero amplitude.

Transmitter-receiver spacing for the VDL is usually 5 ft to provide a longer travel path. Thus, the formation signal can be recorded without interference from the casing signal, which should have died away before the arrival of the formation signal.

If the annulus cement is not in firm contact with the formation, sonic energy transmission will be diminished, and a weak formation signal will be detected. Thus, with the VDL, the character of the formation signal is a measure of cementing success.

Mechanics of Sonic Wave Travel

Wave Propagation—Sonic energy is propagated from the transmitter to the receiver primarily in the form of compressional and/or shear disturbances, as shown in Figure 9-67. The compressional wave has the higher velocity, and can be transmitted through any substance having compressibility (i.e., gas, liquid, or solid). The shear wave has a higher amplitude than the compressional wave, but it can only be transmitted through materials having shear strength.

The sonic pulse travels through the wellbore fluid as a compressional wave. As it strikes the casing, it is transformed into both compressional and shear waves. Some of the acoustic energy then travels along the casing to a point near the receiver, and a portion of the pulse travels back to the receiver as a compressional wave.

Arrival Time—There are, of course, many paths for acoustic energy to follow from the transmitter to receiver other than along the casing (i.e., through the borehole fluid, through the annular cement sheath, or through the formation). The signal actually seen by the receiver is the algebraic sum of all the energy reaching the receiver at a particular instant from all paths.

The speed of sonic energy depends on the material it moves through. In effect, the time that a particular signal arrives at the receiver generally indicates the path along which that signal traveled. Ideally the casing signal arrives at one time, and the formation signal arrives at a later time.

Energy waves travel radially outward from the transmitter in all directions. However, with the same path length, energy waves moving along various radial paths through the casing reach the receiver at the same time, and thus reinforce each other. If the transmitter and receiver are not centralized in the casing, path lengths are shorter for energy moving along the low side, as shown in Figure 9-68.

Thus, the casing wave forms arrive out-of-phase, effectively cancelling each other, to provide a low amplitude signal, and a shorter transit time as shown in Figure 9-68. Obviously tool centralization is important, since signal amplitude is a primary interpretation clue.

Table 9-7 shows representative transit times for sonic energy in various media.

Referring to Figure 9-69, normally the first arrival is the compressional wave that traveled along the casing, closely followed by the higher amplitude shear wave, which also traveled along the casing; then the

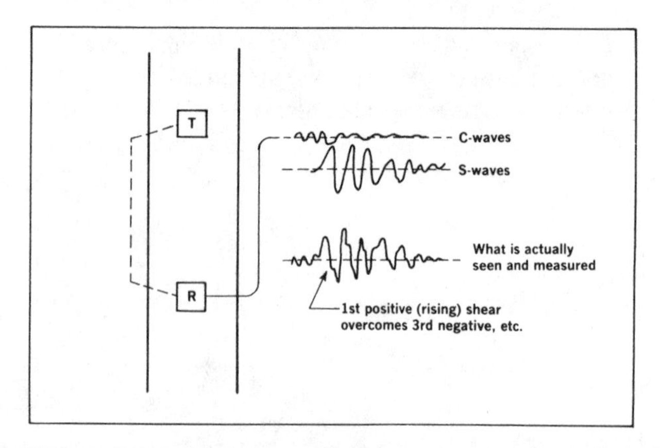

FIG. *9-67—Sonic energy is transmitted as compressional and/or shear disturbances.*

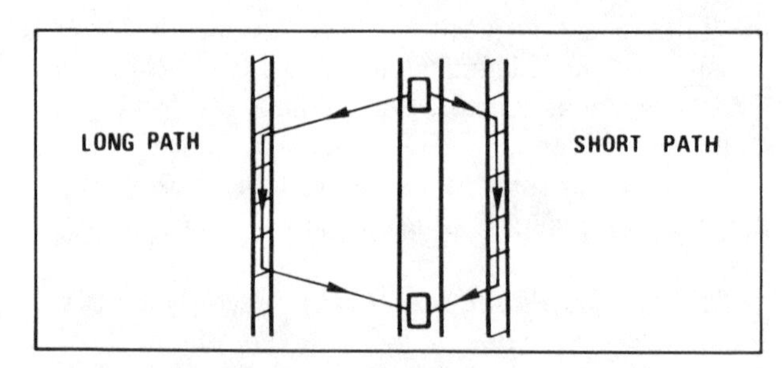

FIG. *9-68—With poor centralization, sonic energy arrives at receiver out of phase.*

TABLE 9-7
Sonic Transit Time (Compressive Wave)

Media	Approximate transit time (microsec/ft)
Steel	57
Mud or water	180–215
Shale	80–160
Sandstone	70–90
Dolomite	45–70
Cement	90–120

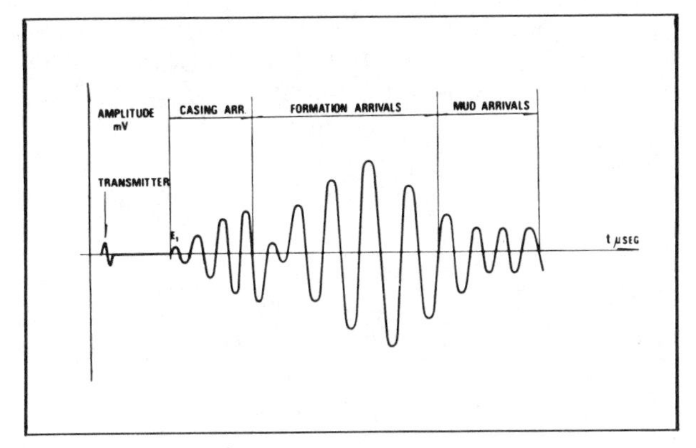

FIG. *9-69—Full sonic wave train used for VDL recording.*

formation arrivals; and finally the mud arrivals. In hard dolomite, which has a transit time about the same as steel, the formation signal may obscure the casing signal.

Arrival time of the E_1 wave at the 3-ft spaced pipe amplitude detector (transit time or Δt) can be estimated from the following relations:

$$1\ 11/16\text{-in. tool: } \Delta t = 173 + 16.9(\text{casing id, in.})$$

$$3\ 5/8\text{-in. tool: } \Delta t = 164 + 16.9(\text{casing id, in.})$$

With well-cemented casing, variations of Δt can occur due to cycle skipping, stretch, or fast formations, Figure 9-70. Cycle skipping results when the E_1 wave is dampened such that the E_1 amplitude is less than the detection level, Figure 9-66. Thus the E_3 wave is detected rather than the E_1 and a longer Δt results. Stretch can increase Δt because a low amplitude wave takes longer to build up to the detection level. Fast-formation can reduce Δt if the formation signal returns before the casing signal. A wiggly Δt above the top of the cement means poor tool centralization and an unusable log.

Attenuation of the Casing Signal

In unbonded casing, little attenuation or dampening of the energy that traveled along the casing occurs, and a very strong casing signal is received, as shown in Figure 9-71. In bonded casing, the cement dampens the energy traveling along the casing, and a low amplitude occurs.

Since the degree of attenuation of the portion of the sonic energy that travels along the casing is our primary measure of cement bond, it is important to know what variables affect this attenuation rate. Pardue et al. of Schlumberger ran a series of laboratory tests to determine the effect of certain variables on attenuation of the casing-borne sonic wave.

Cement Compressive Strength (WOC Time)—The effect of cement setting time for neat cement is shown in Figure 9-72. As compressive strength of the cement increases, attenuation or "cement bond" improves. High strength slurries show greater attenua-

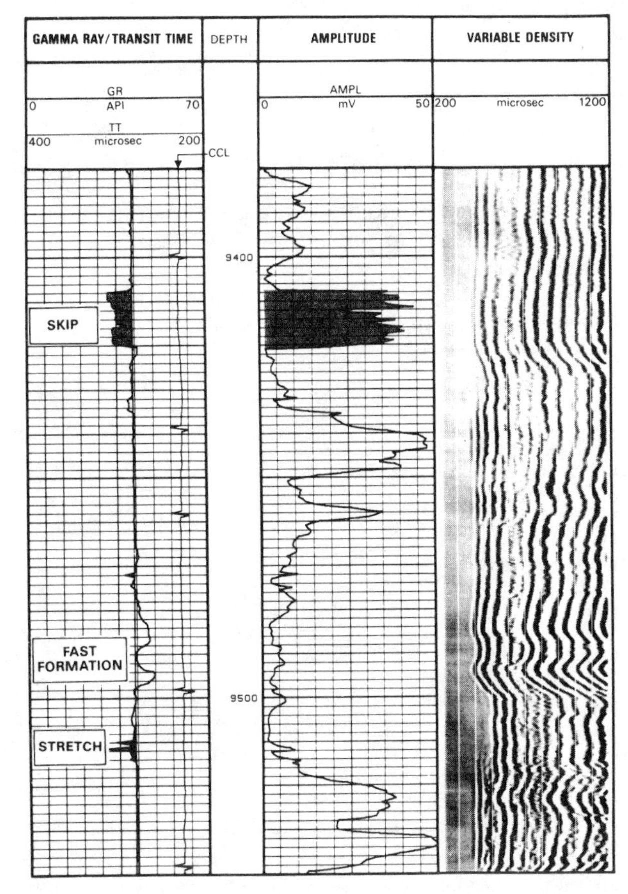

FIG. *9-70—CBL-VDL including examples of stretch, skip, and fast formation arrivals. Courtesy of Schlumberger.*

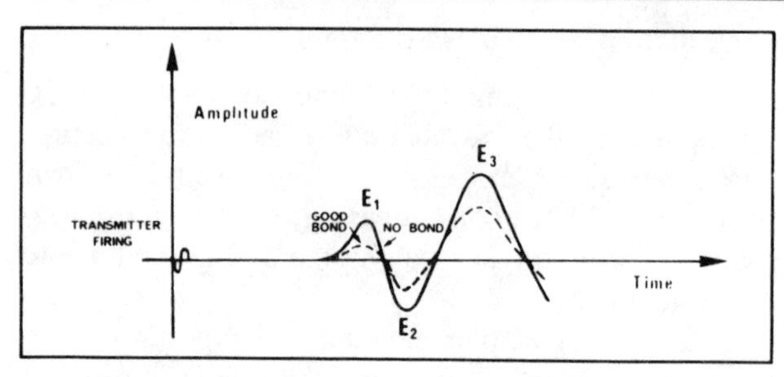

FIG. 9-71—*Attenuation of wave form in cement-bonded casing.*

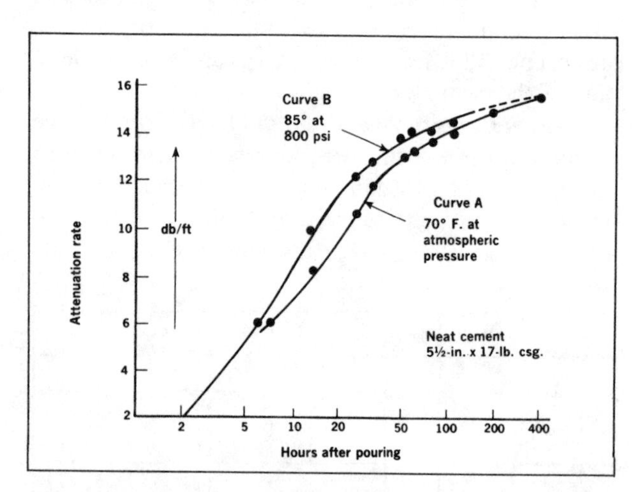

FIG. 9-72—*Typical attenuation rate vs. cement curing time.*[1] *Permission to publish by The Society of Petroleum Engineers.*

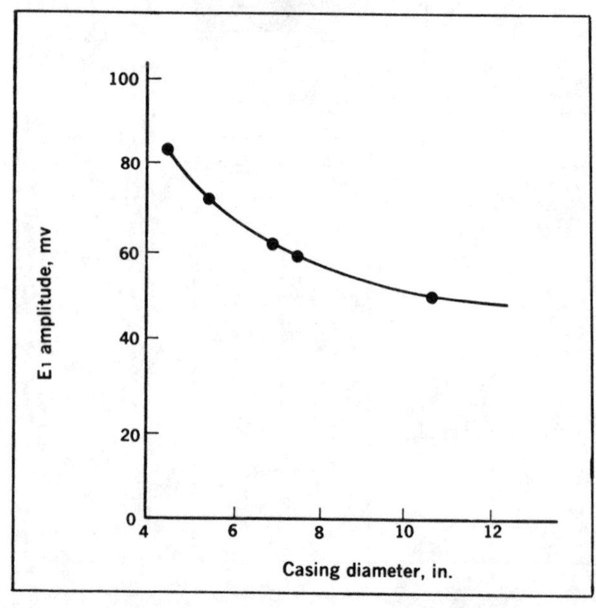

FIG. 9-73—*Effect of casing diameter.*[1] *Permission to publish by The Society of Petroleum Engineers.*

tion under given setting conditions than do lower strength, high water content slurries. Mud contaminated slurries show lower attenuation.

Cement Dimensions—Casing diameter has practically no effect on attenuation rate, although it does affect signal amplitude, as shown in Figure 9-73.

For unbounded casing, wall thickness has little effect on attenuation rate; however, for bonded casing (with at least a 3/4-in. cement sheath) wall thickness does affect attenuation.

Thickness of Cement Sheath—The effect of the thickness of the cement sheath is shown in Figure 9-74. If the cement sheath is thinner than 1/4 wave length (3/4-in.), attenuation drops rapidly unless the cement is securely bonded to a competent formation. Thus, even though a thin cement sheath could provide a hydraulic seal, in soft formation the cement bond log might show a high amplitude indicative of unbonded casing.

Channel—Figure 9-75 shows that the effect of a channel on attenuation is directly proportional to the percent of pipe circumference bonded.

Interpretation of Cement Bond Logs

Interpretation of the cement bond log, both the single curve CBL or the full wave train VDL, must be colored by a knowledge of the many factors that can affect the reading. Also, it should again be noted that there is not necessarily a direct relationship between hydraulic seal and the cement bond log signals.

Uncemented Pipe—Figure 9-76 is a short section of uncemented pipe showing the CBL and VDL presentations along with the CCL, Δt and gamma ray records.

In free pipe, with no coupling to the formation, most

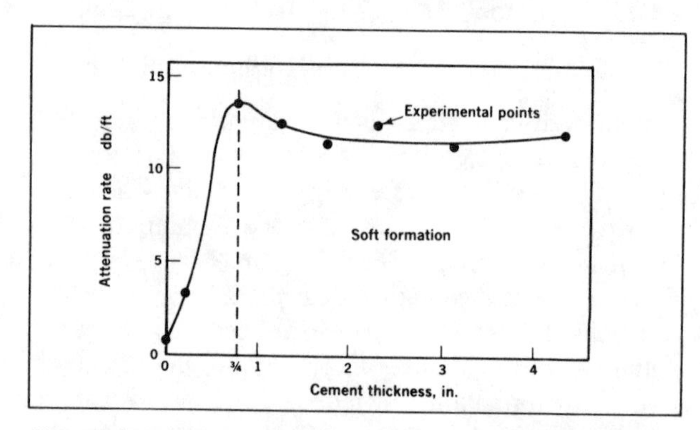

FIG. 9-74—*Attenuation rate vs. cement thickness.*[1] *Permission to publish by The Society of Petroleum Engineers.*

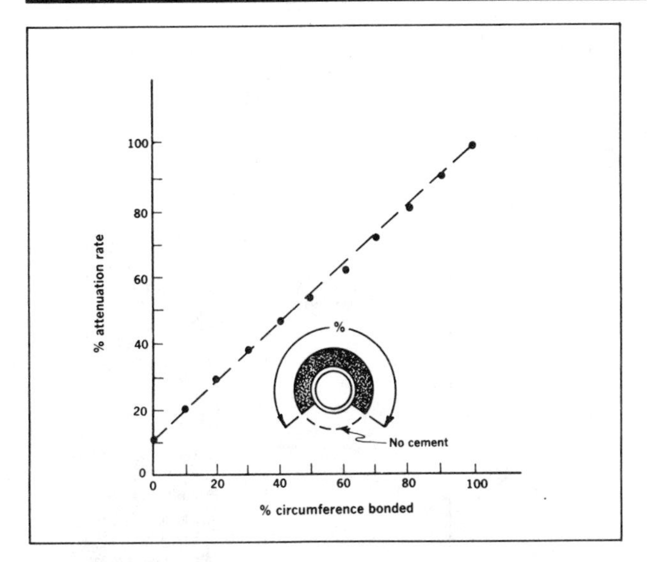

FIG. 9-75—*Percent attenuation rate vs. percent circumference bonded.*[1] *Permission to publish by The Society of Petroleum Engineers.*

of the sonic energy travels through the casing. Thus, casing arrivals on the VDL are strong, with little or no formation signal. Casing collars cause the chevron or double vee patterns on the VDL and the corresponding kicks on the CBL. Note that the CCL records the casing collars higher due to the physical distance between the measuring devices.

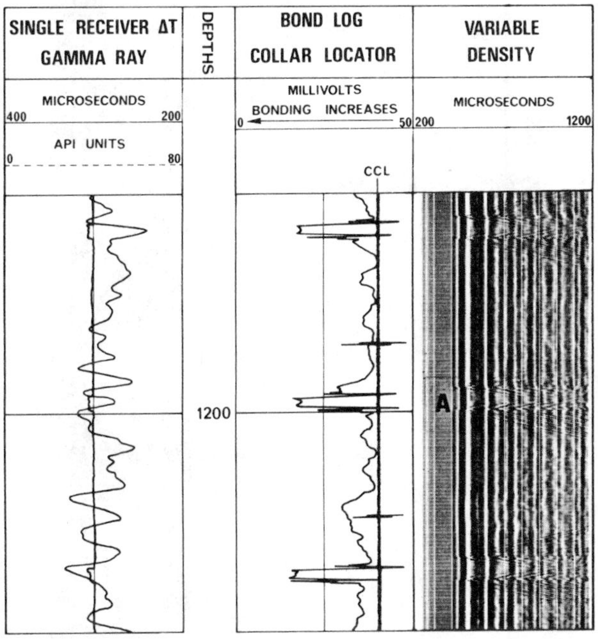

SINGLE RECEIVER ΔT
GAMMA RAY
DEPTHS
BOND LOG
COLLAR LOCATOR
VARIABLE
DENSITY
MICROSECONDS
400 200
API UNITS
0 80
MILLIVOLTS
BONDING INCREASES
0 50
MICROSECONDS
200 1200
CCL
1200
A

FIG. 9-76—*Pipe amplitude (CBL) and full wave train (VDL) in uncemented casing. Courtesy of Schlumberger.*

The lengths of the "kicks" on the CCL, and the distance between the "vees" on the VDL, indicate the transmitter-receiver spacing. The steady Δt shows good centralization. With poor centralization, the Δt should show reduced transit time.

In well-cemented casing, having good cement contact to the casing, and with acoustic coupling to the formation, a major portion of the sonic energy is transmitted from the casing through the cement to the formation. The VDL shows weak casing arrivals, but significant formation signals. Formation signal strength depends on formation characteristics, but can be identified by wiggly lines caused by variation of transit time with depth, which usually correlate with the gamma ray trace. Cement arrivals seldom appear on the VDL due to attenuation.

In well-cemented pipe, the CBL shows low amplitude signals, since sound transmission along the casing is dampened by close contact with the cement.

Figure 9-77 shows a portion of the CBL-VDL log run in 9 5/8-in. casing about 18 hrs after the cement (15.2 lb/gal Class B with 1% bentonite) was in place. The expected cement top was about 480 ft.

On the VDL, below the expected cement top, casing signals fade out, and the chevron casing collars disappear. The formation signals begin to appear, and seem to correlate with the gamma ray.

On the CBL, minimum pipe amplitude values of 6 to 10 mv below 480 ft, indicate the presence of a significant cement sheath. Free pipe further up the hole (not shown) indicated an amplitude of about 50 mv. With good bonding to hard cement, pipe amplitude should be 5 to 10% of free pipe amplitude. Above the expected cement top at 480 ft, some attenuation of the CBL signal occurs, possibly showing cement channeling.

Estimation of cement compressive strength is possible from CBL amplitudes. If these estimates correlate with expected cement compressive strength considering cement composition, setting time, and hole conditions, then confidence in CBL interpretation is increased.

In Figure 9-77, additional waiting on cement time would have produced weaker casing signals and stronger formation signals on the VDL, and lower pipe amplitude on the CBL. Note also that the CBL indications of greater pipe amplitude, below about 550 ft, seem to correlate with shale sections on the gamma ray. Washout of these shales would be expected to

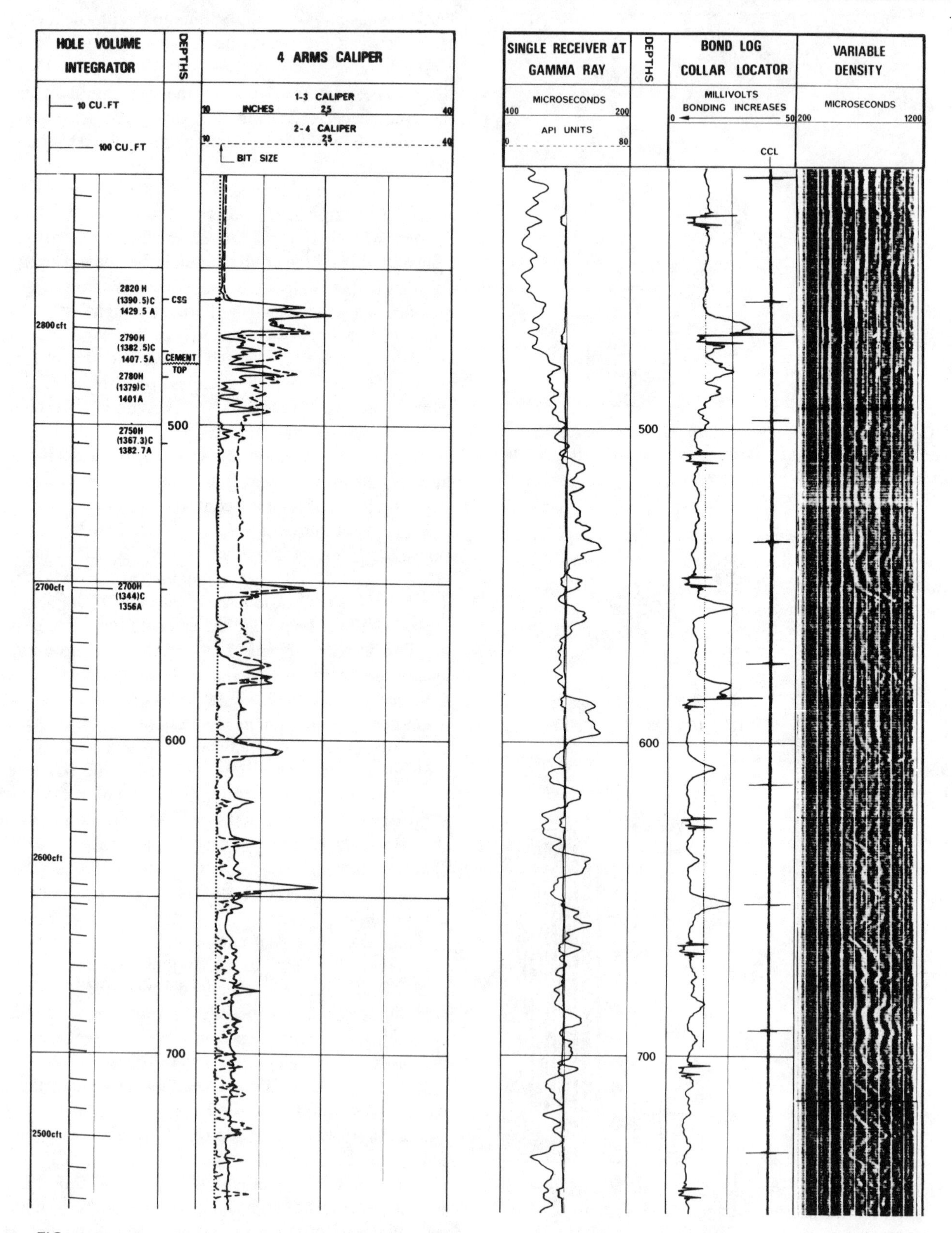

FIG. 9-77—*Correlation of CBL, VDL, GR, CCL, and bore-hole geometry logs. Courtesy of Schlumberger.*

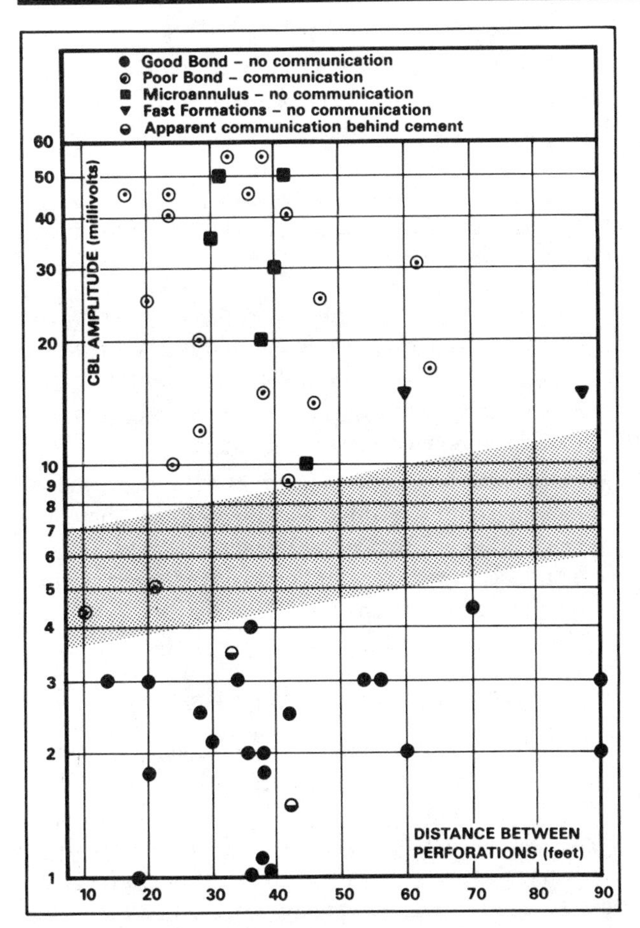

FIG. 9-78—*Local study of CBL amplitude readings and communication test results. Courtesy of Schlumberger.*

increase the difficulty of displacing gelled mud and obtaining good cement coverage completely around the casing.

Figure 9-77 also includes a four-arm caliper device, recording hole diameter in two perpendicular directions. Washouts shown by the caliper do in fact correlate with the shales. Note also that the hole is not round, but elliptically shaped. Thus, the borehole geometry device contributes additional clues that may be helpful for difficult cementing evaluation problems.

Length of "Good Bond" Needed to Prevent Communication—The concept of Bond Index provides a comparative standard for evaluating cementing quality through a particular section using only the CBL recording:

$$\text{Bond Index} = \frac{\text{Attenuation in zone of interest}}{\text{Attenuation in well-cemented section}}$$

Other factors excluded, bond index is proportional to the casing circumference in contact with good cement. Experience shows that a bond index above 0.8 over a 5-ft section of 5 1/2-in. casing, (10 ft in 7-in. casing and 15 ft in 9 5/8-in. casing) usually means no communication along that particular section of casing. Bond index much below 0.8 probably indicates mud channeling or mud contaminated cement.

A Shell study of 28 wells in the Rocky Mountain area in which conclusive communication tests were possible, and in which satisfactory CBL's had been obtained, presented the following conclusions regarding zone isolation as related to CBL amplitude:

1. Communication was likely, if the CBL showed no amplitude less than 10% of unbonded casing amplitude.
2. Communication was unlikely, if the CBL exhibited at least 10 continuous ft, with less than 5% of unbonded casing amplitude.
3. Communication was likely, if the CBL exhibited less than 5 continuous ft, with less than 5% of unbonded casing amplitude.
4. For cases not covered above, the data were too inconclusive to form a confident prediction concerning communication. These cases include:

—CBL amplitude less than 5% of unbonded casing amplitude, thickness between 5 and 10 ft.
—CBL amplitude between 5 and 10% of unbonded casing amplitude, thickness greater than 10 ft.

Confidence level in these conclusions is increased by use of the VDL to substantiate the CBL information on casing arrivals and to provide an indication of acoustic coupling to the formation.

Figure 9-78 summarizes results from communication tests in the southern Arabian Gulf area on some 40 wells. Amplitude reading from the CBL-VDL log is plotted against distance between perforated zones from each test. Six wells showed high amplitude, but no communication. These were probably microannulus situations, but a log run under pressure was not available to confirm microannulus. Two wells showed "good bond" but communicated on test. These probably resulted from a thick filter cake that broke down even though the casing signal was dampened by a cement sheath. During these communication tests 15% HCl acid was routinely used. Communication tests before and after acidizing often differed significantly. Examples were seen where bond log interpretation showed mediocre to good bonding but a small degree of communication was apparent with brine. After acidizing, full communication resulted, showing acid clean out of the existing mud channel. Figure 9-78 confirms that extremely low CBL amplitudes, ideally less than 3mv, must be recorded in order to ensure good isolation, and even then small mud channels are difficult to detect.

Interpretation Problems*—Good Casing Bond—Poor Formation Bond*—The CBL-VDL logs may show good bond between cement and casing (low CBL amplitude) but poor acoustic coupling to formation (no VDL formation signal). This would usually be interpreted as a situation in which a cement sheath surrounds the casing, providing good attenuation of the casing signal, but a thick mud filter cake prevented cement contact and sonic energy transmission to the formation.

Another possibility is that there is no cement behind the casing, but poor centralization of the logging sonde is responsible for the low CBL amplitude. An energy wave traveling along one radial path from the transmitter must go further than a wave from along another radial; thus, they arrive at the receiver out-of-phase and cancel each other. This possibility can be checked by observing the Δt transit time curve in a section where there is obvious free casing. Reduced or varying transit time means poor tool centralization. Steady transit time means CBL indications are reliable. Cycle skipping or stretch in a cemented section means good bond.

Poor Casing Bond—Moderate Formation Bond—The CBL shows high amplitude; the VDL shows a strong casing signal, but also a moderately strong formation signal. This could be interpreted as mud channeling. But, a microannulus is also a likely possibility.

Microannulus means a small clearance (perhaps 0.001 in.) between the casing and cement. This permits the casing to vibrate—or transmit sound energy without significant attenuation. It is due to casing contraction from temperature or pressure changes, perhaps casing surface finish. Significant fluid leakage cannot occur, since pressure drop would have to be unrealistically high to force even a low viscosity fluid along any length of microannulus path. Thus an effective hydraulic seal may be assumed, if we can, in fact, prove that the microannulus condition really exists.

This can be done by rerunning the CBL-VDL with pressure on the casing. With the microannulus, casing expansion against the cement will attenuate the casing signal to produce a low CBL, erase the VDL casing signal, and strengthen the VDL formation signal. With a channel, the casing expansion will not affect the logs. As another test, channeling is usually more localized than is a true microannulus.

Quality Control—Adequate quality control must be achieved for meaningful CBL-VDL interpretation. Quality control depends on:

—*Acoustic properties of fluid in the casing*. The CBL should be run with one single phase fluid—either dead oil or water—filling the logged interval.

—*Repeatability*. Subsequent runs should repeat within 10% of unbonded casing amplitude.

—*Centralization*. Centralization is extremely important in sonic amplitude logging—if adequate repeatability is obtained, satisfactory centralization can be assumed. A wiggly transit time trace in unbonded casing means poor centralization.

Application of Cement Evaluation Logs

Cement evaluation logs require economic justification, as do other logging devices. Many times bond logs are run routinely as a part of completion operations, with justification being that the gamma ray-CCL recordings are required for perforation depth control, and the CBL-VDL curves are recorded at the same time at small additional cost.

The following points should be considered in using cement evaluation logs.

1. Where well conditions are such that the rules of good primary cementing practices can be applied, the cement evaluation logs should not be required.
2. Where conditions make primary cementing difficult, and where experience shows that primary cement success is low, the cement evaluation logs can provide the keys to improve practices.
3. Where fluid movement behind the casing is suspected, the cement evaluation logs may confirm the possibility—and may show the point at which remedial cementing can be effectively applied. The noise log may be even more helpful in this regard.
4. The combination of the CBL-VDL and Δt curves provides much more confident interpretation than the CBL curve alone.
5. The Schlumberger CET or some other channel locating technique is needed in many questionable situations.

REFERENCES

1. Pardue, G. H.; Morris, R. L.; Gollwitzer, L. H.; and Morank, J. H.: "Cement Bond Log--A study of Cement and Casing Variables," *JPT* (May 1963).

2. Harcourt, G.; Walter, T.; and Anderson, T.: "Use of the Micro-Seismogram and Acoustic Cement Bond Log to Evaluate Cementing Techniques," SPE 798 (1964).

3. Johnson and Morris: "Review of Tracer Surveys," API Paper 906-9-E (March, 1964).

4. Wade, R. T.; Cantrell, Rex; Poupon, A.; and Moulin, J.: "Production Logging—The Key to Optimum Well Performance," *JPT* (Feb. 1965).

5. Loeb, J. and Poupon, A.: "Temperature Logs in Production and Injection Wells," Paper presented at the 25th Meeting of the European Association of Exploration Geophysicists in Madrid (May 5-7-9, 1965).

6. Agnew, B. G.: "Evaluation of Fracture Treatments with Temperature Surveys," *JPT* (July 1966) p. 892.

7. Pickett, G. R.: "Prediction of Interzone Fluid Communication Behind Casing by Use of the Cement Bond Log," *Trans.*, SPWLA (May 1966).

8. Bearden, W. G.; Cocanower, R. D.; Currens, Dan; and Dillingham, Mat: "Interpretation of Injectivity Profiles in Irregular Bore Holes," *JPT* (Sept. 1970) p. 1089.

9. Smith, R. C., and Steffensen, R. J.: "Computer Study of Factors Affecting Temperature Profiles in Water Injection Wells," *JPT* (Nov. 1970) p. 1447.

10. Clover, C.; Hoyle, W. R.; and Meunier, D.: "Quantitative Interpretation of TDT Logs," SPE 2658 (Oct. 1969).

11. Schlumberger Well Evaluation Conference, Tripoli, Libya (Feb. 1970).

12. Elkins, L. F.; Skov, A. M.; and Liming, H. F.: "A Practical Approach to Finding and Correcting Perforation Inadequacies," SPE 2998 (Oct. 1970).

13. Connolly, E. T.: "Evaluation of Gas Well Productivity Using Production Logs," Petroleum Society of CIM (May 1970).

14. Smith, R. C., and Steffensen, R. J.: "Interpretation of Temperature Profiles in Water Injection Wells," *JPT* (June 1975) p. 777.

15. Schlumberger Well Evaluation Conference, Lagos, Nigeria (April 1974).

16. Robinson, J. M.; Burton, W. R.; Hamilton, J. M.; Tixier, M. P.: "Improving Producibility with Cased Hole Wireline Technique," Paper 374024, Petroleum Society of CIM (May 1974).

17. Doering, M. A., and Smith, D. P.: "Locating Extraneous Water Sources with the Gamma Ray Log," SPE 5116 (Oct. 1974).

18. Schlumberger Well Evaluation Conference, North Sea (June 1974).

19. Schlumberger Well Evaluation Conference, Iran, (May 1976).

20. Hammack, G. W.; Myers, B. D.; and Barcenas, G. H.: "Production Logging Through the Annulus of Rod-Pumped Wells to Obtain Flow Profiles," SPE 6042 (Oct. 1976).

21. Robinson, W. S.: "Field Results form the Noise Logging Technique," *JPT* (Nov. 1976) p. 1370.

22. Pennebaker, E. S., Jr., and Woody, T. T.: "The Temperature-Sound Log and Borehole Channel Scans for Problem Wells," SPE 6782 (Oct. 1977).

23. Cooke, Claude E., Jr.: "Radial Differential Temperature (RDT) Logging A New Tool for Detecting and Treating Flow Behind Casing," SPE 7558 (Oct. 1978).

24. Schlumberger Well Evaluation Conference, Abu Dhabi (Nov. 1981).

25. McKinley, R. M.: "Production Logging" SPE 10035 (March 1982).

26. Piers, G. E.; Perkins, J.; and Escott, D.: "A New Flowmeter for Production Logging and Well Testing," SPE 16819 (Sept. 1987).

27. Jutten, J. J.: "Studies with Narrow Cement Thicknesses Lead to Improved CBL in Concentric Casing," SPE 16819 (Sept. 1987).

28. Beirute, R. M.; Wilson, M. A.; and Sabins, F. L.: "Attenuation of Casing Cemented with Conventional and Expanding Cements Across Heavy Oil and Sandstone Formations," SPE 18027, Houston (Oct. 1988).

29. Corwith, J. R., and Mengel, F.: "Applications of Pulsed Neutron Capture Logs in the Ekofisk Area Fields," SPE 18147, Houston (Oct. 1988).

Appendix A

Interpretation of the Temperature Log

The following examples[5] show theoretical temperature responses for production or injection of liquids or gases. Geothermal profile is assumed to increase linearly with depth. Oil and water are assumed to be incompressible liquids. Examples showing produced liquid apply equally to oil or water, but not necessarily to mixtures or emulsions of oil and water. Gas is assumed to be methane and, of course, compressible.

Examples apply to dynamic conditions, and it is assumed that conditions have existed sufficiently long so that the temperature profile is no longer a function of time.

Liquid Production—Point Entry

One point P is producing in Figure 9A-1; there is no vertical movement behind the casing, so that the oil entering the casing is at the temperature given by the geothermal profile at the point of entry. Starting from the bottom, the temperature curve follows the geothermal profile up to the point of oil entry. Above the point of oil entry the curve is exponential to an asymptote AA′, which is parallel to the geothermal profile TT′. The horizontal distance ΔT between the geothermal profile and the asymptote is given by the equation: $\Delta T = b\,M/G$, where b is a coefficient that depends essentially on the physical characteristics of the fluid produced and on the thermal conductivity of the formations, G is the geothermal gradient, and M is the mass flow rate. Hence, ΔT is approximately proportional to the weight of the fluid produced per unit time. The size of the casing has practically no effect on the temperature curve. The tangent to the temperature curve is vertical at P; this vertical attitude, away from the geothermal profile, is characteristic of liquid production.

Liquid Production—Dual Point Entry

Two points P, and P are producing oil in Figure 9A-2, with production rates of M_1 and M_2. Starting from the bottom, the temperature curve follows the geothermal profile up to the point of the first entry at P_1. From P_1 to P_2 the temperature reacts exponentially to asymptote A_1A_1' as if it were a single point entry case.

At P_2 the temperature profile moves horizontally toward the geothermal profile TT′; the distance that it moves is a function of the rates of production, M_1 and M_2, from point entries P_1 and P_2. From P_2, up the hole, the temperature moves exponentially to asymptote A_2A_2'. Note that while the tangent to the temperature profile is vertical at P_1, it is less than vertical at P_2. Mathematically, the T's are given by the following equations: $T_1 = b_1M_1/G$ and $T_2 = b_1M_1/G + b_2M_2/G$.

FIG. *9A-1—Liquid production, point entry.*

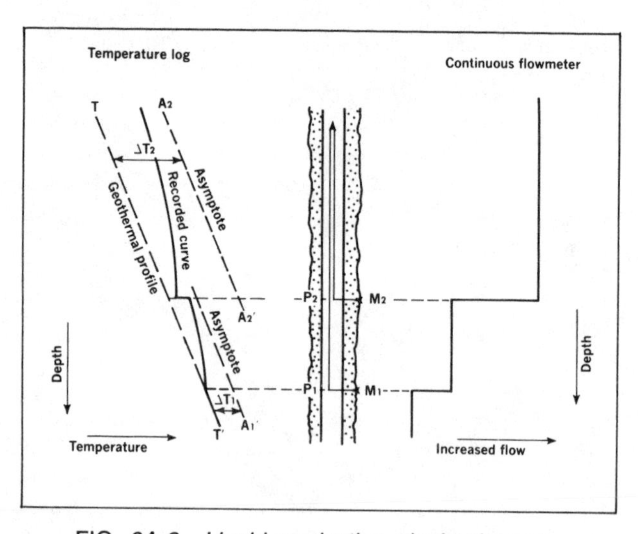

FIG. *9A-2—Liquid production, dual-point entry.*

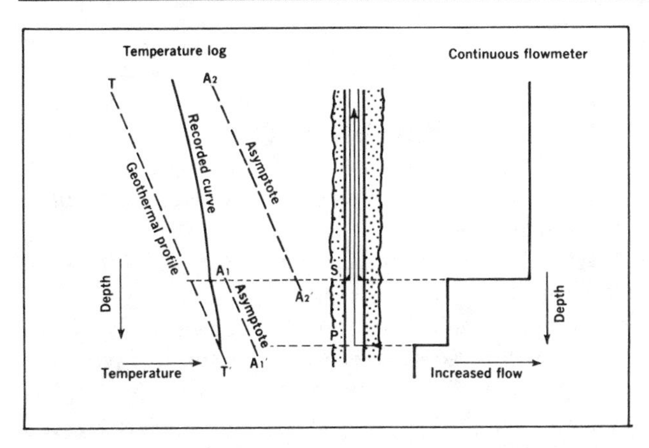

FIG. *9A-3—Liquid production, single-point entry, producing through tubing.*

Liquid Production—Single Point Entry—Producing Through Tubing

One point P is producing oil in Figure 9A-3; there is no vertical movement behind the casing, so that the oil entering the casing is at the temperature given by the geothermal profile at the point of entry. The temperature curve follows the geothermal profile up to the point of entry. Above the point of entry the curve is exponential with an asymptote A_1A_1'. At depth S, all of the flow enters the tubing. Above S the temperature exchange between the flowing liquid and the formation is through natural convection of the oil or water in the annular space. Thus the annulus acts as an insulator and the oil in the tubing cools at a slower rate. Above S the temperature profile reacts exponentially to asymptote A_2A_2'.

Liquid Production—Single Point Entry—Casing Leak

Figure 9A-4 shows response when the logging tools are run through liquid filled tubing, but when all of the upward oil flow is either below the tubing or in the tubing-casing annulus. In each case the casing leak is above the tubing shoe.

After leaving the geothermal profile at P in Figure 9A-4, the temperature curve will react exponentially toward asymptote A_1A_1' until it reaches the level of liquid loss at F. Above F the temperature profile will reach exponentially toward a second asymptote A_2A_2', which is to the left of A_1A_1', if there is a continuing upflow in the annulus past F. Two possibilities are shown in Figures 9A-4A and 9A-4B. If there is no flow up the annulus past point F, the temperature profile will return immediately to the geothermal profile as is shown in Figure 9A-4C.

Liquid Production—Single Point Entry—Flow Upward Behind Casing

Figure 9A-5 shows responses when the oil flow is, first, upward in the casing-formation annulus, and second, into and upward in the casing.

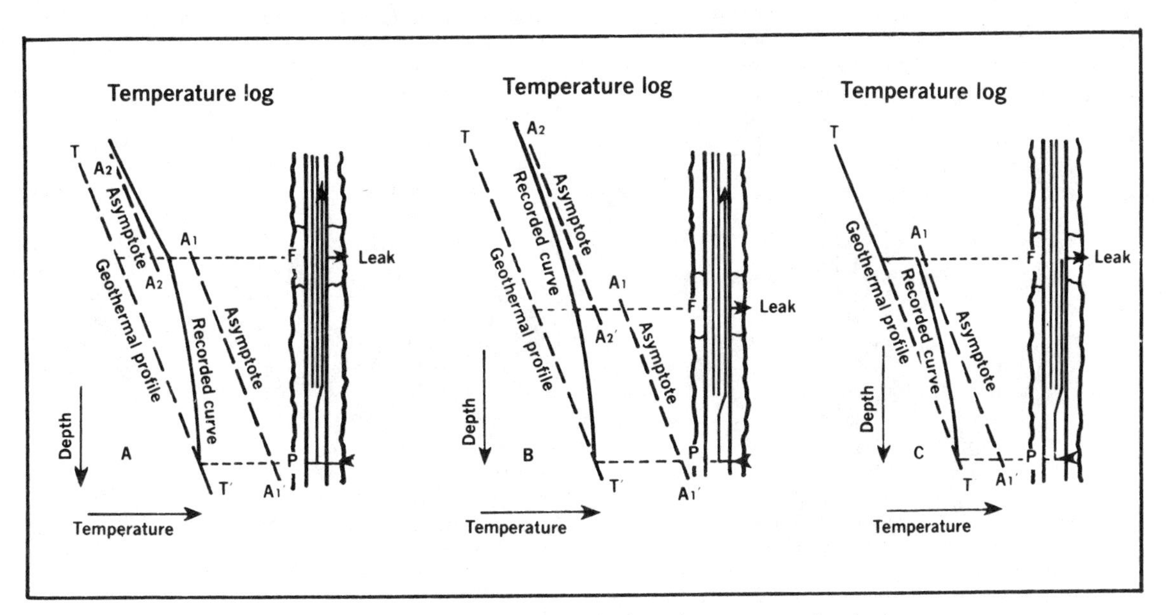

FIG. *9A-4—Liquid production, single-point entry, casing leaks.*

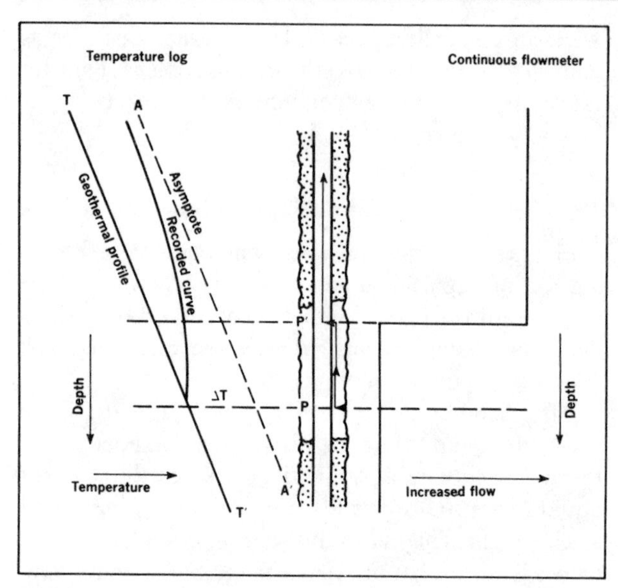

FIG. *9A-5—Liquid production, single-point entry, flow upward behind casing.*

Oil flows upward from P in the casing-formation annulus in Figure 9A-5, and then the flow is into the casing at P′ where it continues its upflow. The temperature profile is practically the same as if the liquid had entered the casing at P.

Liquid Production—Point Entry—Flow Downward Behind Casing

Figure 9A-6 shows responses when the liquid flow is, first, downward in the casing-formation annulus, and, second, into and upward in the casing.

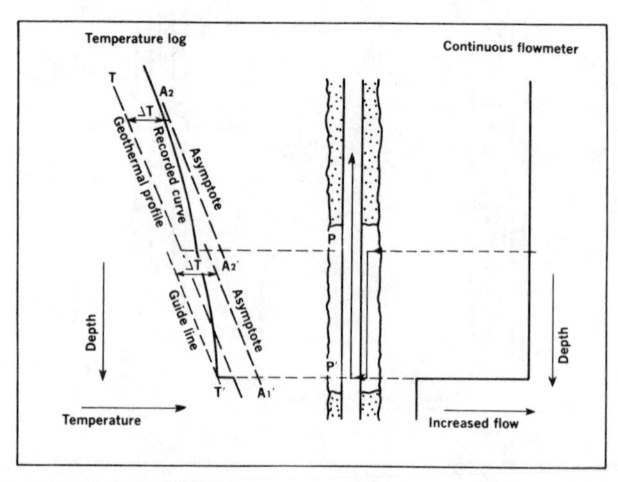

FIG. *9A-6—Liquid production, point entry, flow downward behind casing.*

Oil flows downward from P in the casing-formation annulus in Figure 9A-6, and then the flow is into the casing at P′ where the flow is in an upward direction. Starting up from the bottom, the temperature curve follows the geothermal profile temperatures at P and P′. From P′ to P the temperature curve is exponential to asymptote A_1A_1'.

At P the temperature profile is hotter than the geothermal profile because of the heating that results from the flow of liquid from P to P′ and back to P. Above P the temperature curve is exponential to asymptote A_2A_2'. At P′ the tangent to the temperature curve is vertical, but at P the tangent is less than vertical. ΔT is the same as it would have been without any flow in the casing-formation annulus.

Gas Production—Point Entry

One point is producing methane in Figure 9A-7, and there is no vertical movement behind the casing. Cooling caused by expanding gas causes a shift to the left of both the asymptote AA′ and a parallel guide line LL′. The temperature difference between these parallel lines, ΔT, is computed as was the case of oil: $\Delta T = bM/G$. For the case of methane the temperature differential from the guide line LL′ to the geothermal profile TT′ is about one-third of ΔT.

The pressure drop from the reservoir pressure of the casing pressure is large for low permeabilities and small for high permeabilities. This pressure drop causes cooling at P, to a temperature less than LL′ for low permeabilities, and to a temperature between LL′ and

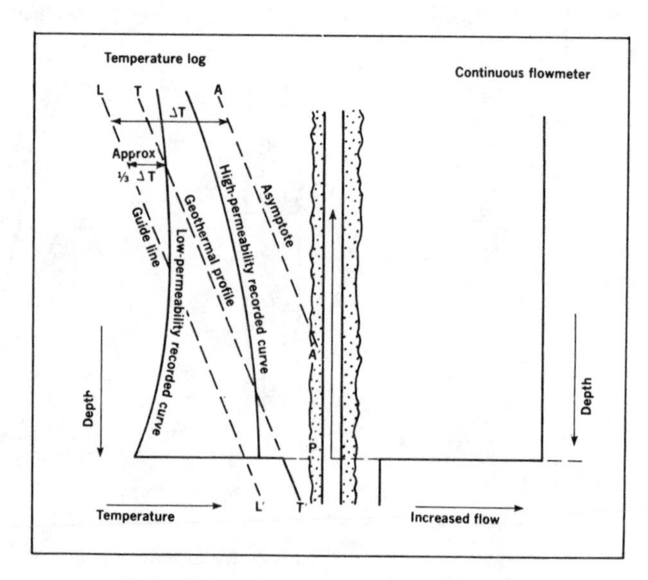

FIG. *9A-7—Gas production, point entry.*

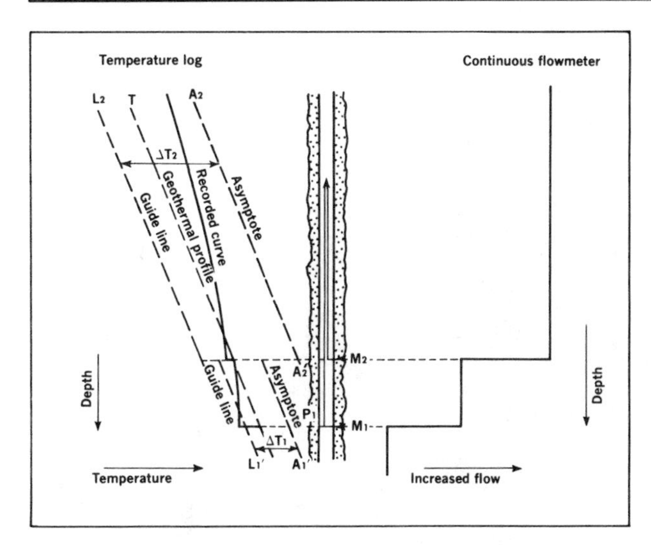

FIG. *9A-8—Gas production, dual-point entry.*

TT′ for high permeabilities. In both cases the temperature curve reacts exponentially to the asymptote AA′. With low permeabilities the temperature curve increases immediately above P as it is heated by the adjacent formation.

Higher up the hole this temperature curve becomes vertical as it crosses LL′, and above LL′ it shows cooling as it approaches the asymptote AA′. With high permeabilities the temperature curve shows a continual cooling from the point of entry.

Gas Production—Dual Point Entry

Two points P_1 and P_2 are producing methane in Figure 9A-8, and there is no vertical movement behind the casing. The flow rate M_1 corresponds to the guide line L_1L_1. The flow rate $M_1 + M_2$ corresponds to the guide line L_2L_2' and the asymptote A_2A_2'. Note that in each case the distance from the guide lines to the geothermal profile is about one-third of their respective ΔT's.

At both P_1 and P_2 the temperature breaks abruptly to a lower value. Above P_1 there is warming because the temperature break at P_1 is to a value less than its guide line L_1L_1'. Above P_2, however, there is no warming because the temperature break at P_2 is to a value between the second guide line L_2L_2' and the geothermal profile TT′.

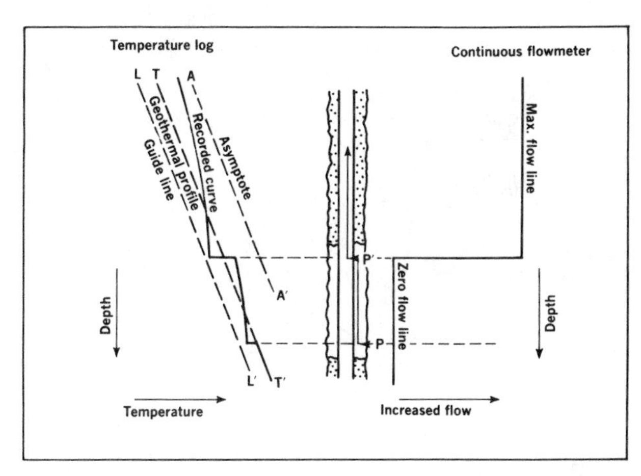

FIG. *9A-9—Gas production, single-point entry, flow upward behind casing.*

Gas Production—Single Point Entry—Flow Upward Behind Casing

Figure 9A-9 shows responses when the gas flow is, first, upward in the casing-formation annulus, and second, into and upward in the casing.

Methane leaves the formation at P in Figure 9A-9, moves upward in the casing-formation annulus, and enters the casing at P′. There is probably a cooling effect due to gas expansion at both P and P′. Both parts of the temperature curve will have the same guide line LL′ and the same asymptote AA′. Starting from the bottom, the temperature curve follows the geothermal profile at P. At P there is an abrupt decrease in temperature. From P to P′ there is exponential cooling to asymptote AA′. Again, at P′ there is an abrupt decrease in temperature. Above P′ there is exponential cooling to asymptote AA′. The abrupt temperature decrease at P and P′ are due to expanding gas; the case shown is typical, but there are an infinite number of possibilities corresponding to the temperature drops at P and P′.

Gas Production—Point Entry—Flow Downward Behind Casing

Figure 9A-10 shows responses when gas flow is, first, downward in the casing-formation annulus, and, second, into and upward in the casing.

Gas leaves the formation at P in Figure 9A-10, moves downward in the casing-formation annulus, and enters the casing at P′. There is probably a cooling effect due to gas expansion at both P and P′. Starting at the bottom, the temperature curve will follow the geothermal profile up to P′. At P′ there will be an abrupt decrease in temperature due to expanding gas. From P′ to P the temperature will react exponentially

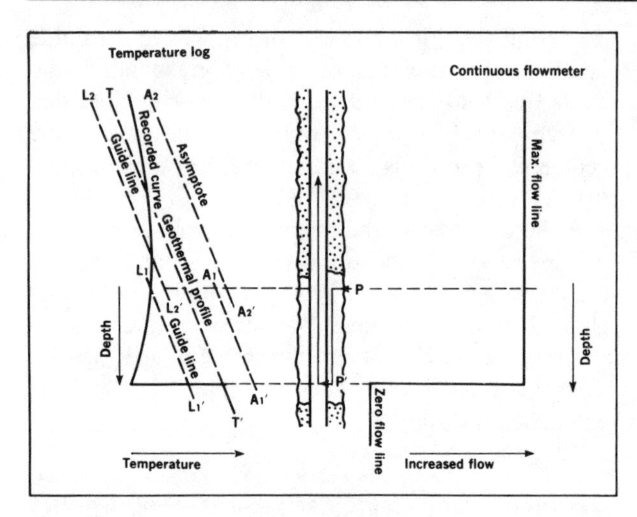

FIG. *9A-10—Gas production, point entry, flow downward behind casing.*

to asymptote A_1A_1'. Above P, without an abrupt change in temperature, the temperature curve will change its slope and react exponentially to asymptote A_2A_2'. Again, this case is typical, but there are an infinite number of possibilities for the temperature drop at P′ and the slope change at P.

Water Injection—Point Exit

In Figure 9A-11 several possible curves are shown which are related to the surface temperature of the injection water. With water injection the temperature profile is, again, exponential. In this case, however, the asymptote is lower than the geothermal profile. The horizontal distance between the geothermal profile TT′ and the asymptote AA′ is given by the equation $\Delta T = -bM/G$.

Starting from the bottom, the temperature curve follows the geothermal profile up to P. At P there is usually an abrupt decrease or increase in temperature; while above P there is an exponential reaction of the temperature curve away from the asymptote AA′. If the temperature curve crosses the geothermal profile, it will do so with a vertical tangent.

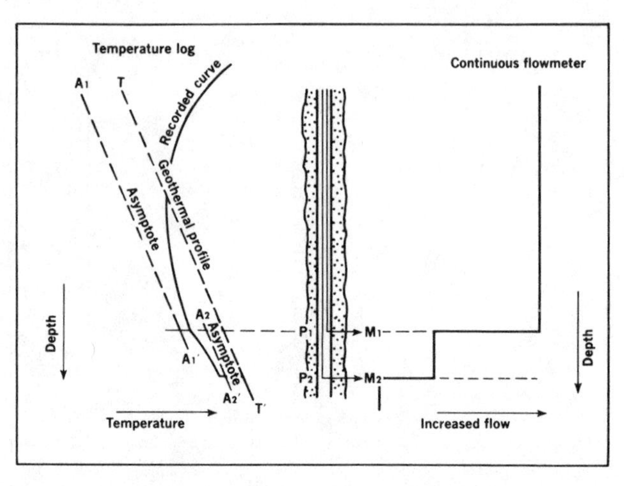

FIG. *9A-12—Water injection, dual-point exit.*

Water Injection—Dual Point Exit

One of the many possible curves is shown in Figure 9A-12. Two points P_1 and P_2 are taking fluid at mass flow rates of M_1 and M_2. Above P_1 the asymptote A_1A_1' corresponds to the total mass flow rate $M_1 + M_2$. Below P_1 the proper asymptote is A_2A_2', which

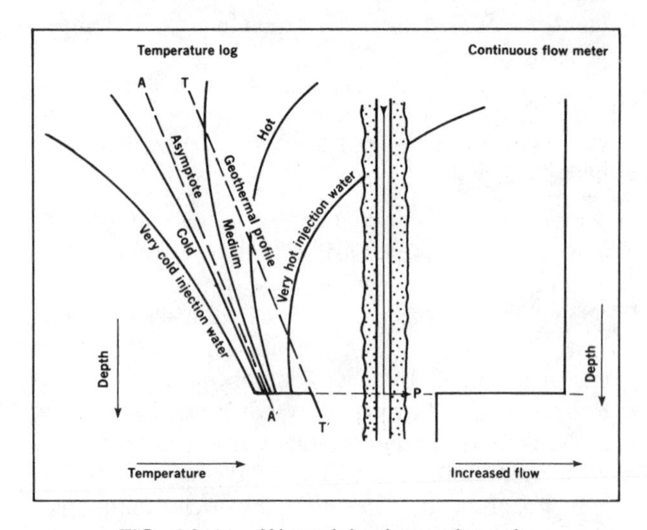

FIG. *9A-11—Water injection, point exit.*

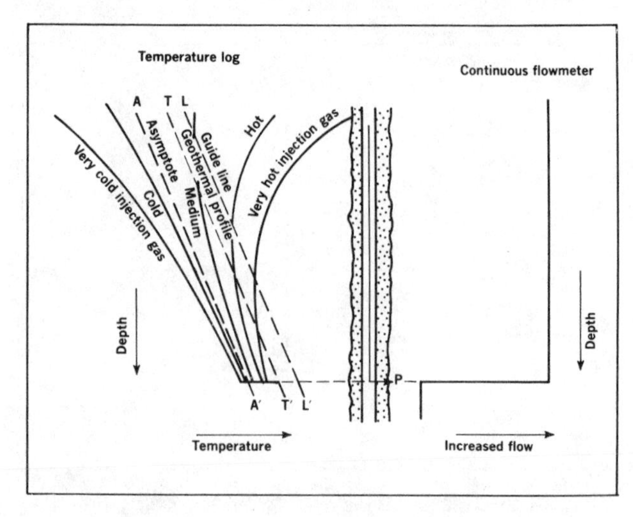

FIG. *9A-13—Gas injection, point exit.*

corresponds to mass flow rate M_2. Starting from the bottom, the temperature curve follows the geothermal profile TT′ to P_2.

At P_2 there is an abrupt temperature decrease to a value between asymptotes A_1A_1' and A_2A_2'. At P_1 there is a change in the slope of the temperature curve, with the new curve being exponential from A_1A_1'. Again, the tangent to the temperature curve is vertical at the point where the temperature curve intersects to geothermal profile TT′.

Gas Injection—Point Exit

Temperature Log—In Figure 9A-13, several possible curves are shown which are related to the surface temperature of the injected methane. With increased depth in gas injection wells there is increased pressure, compression, and heating of the gas. As a result the guide line LL′ and the asymptote AA′ are shifted to the right, toward a higher temperature. The distance between these lines is given by the following equation: $\Delta T = -bM/G$. And, for the case of methane, the distance from the geothermal profile TT′ to the guide line LL′ will be about one-third of ΔT. Starting from the bottom the temperature curve follows the geothermal profile up to P.

At P there is usually an abrupt change in temperature. Above P there is an exponential reaction of the temperature curve away from the asymptote AA′. If the recorded temperature curve crosses the guide line LL′, its tangent at that point will be vertical.

Symbols and Abbreviations, Vol. 1

Reservoir Engineering and Well Testing

A = area
AOF = absolute open flow potential
B = formation volume factor, res bbl/stb
c = compressibility psi^{-1}
DR = damage ratio
E_i = exponential integral
FE = flow efficiency
GOR = gas-oil ratio, cu ft/bbl
h = formation thickness, ft
J or PI = productivity index (stbpd)/psi
J' = modified productivity index for deliverability test
IPR = inflow performance relation
k = permeability, md
K = permeability, darcys
k_f = fracture permeability, md
k_o = permeability to oil, md
k_{rg} = relative permeability to gas, fraction
k_{ro} = relative permeability to oil, fraction
k_{rw} = relative permeability to water, fraction
log = logarithm base 10
ln = logarithm base e (2.7182)
L = length
m = slope of linear portion of transient pressure semilog plot, psi/cycle
n = exponent in productivity index formula or AOF plot
p = pressure, psi
p_c = critical pressure, psia
p_{ff} = final flowing pressure on DST, psi
p_{fhm} = final hydrostatic mud column pressure on DST, psi
p_{ihm} = initial hydrostatic mud column pressure on DST, psi
p_i = initial pressure, psi
p_{wf} = flowing bottomhole pressure, psi
p_{ws} = shut-in bottomhole pressure, psi
$p_{1\,hr}$ = pressure on straight line portion of semilog plot 1 hr after beginning of transient test, psi
p_e = pressure at external radius, psi
$\bar{p}$ = average reservoir pressure, psi
Δp = pressure change, psi
q = flow rate stbd for liquid, Mcfd for gas
q_g = gas flow rate, Mcfd
q_o = oil flow rate, stbd
r = radius, ft
r_e = external radius or drainage radius, ft

r_w = well bore radius, ft
s = van Everdingen-Hurst skin factor
S_g = gas saturation, fraction
S_o = oil saturation, fraction
t = time, hours
t' = time, minutes
t_p = time well on production before shut-in, hours
t'_p = time well on production before shut-in, minutes
t_s = stabilization time, hours
Δt = shut-in time, hours
$\Delta t'$ = shut-in time, minutes
V = volume, bbls
WOR = water-oil ratio, % of water in total flow stream
μ = viscosity, cp
μ_o = viscosity oil, cp
μ_g = viscosity gas, cp
μ_w = viscosity water, cp
ρ = density lb_m/cu ft
ϕ = porosity, fraction
Z = gas deviation factor

Cementing

B_c = Bearden units of consistency—dimensionless
S_s = shear stress, lb/ft^2
S_r = shear rate, sec^{-1}
n' = flow behavior index, power law fluid
K' = consistency index, power law fluid
N = range extension factor for Fann viscometer (usually 1.0)
PV = plastic viscosity, Bingham plastic fluid
YP = yield point, Bingham plastic fluid
Q_b = pumping rate bbl/min
Q_{cf} = pumping rate cu ft/min
D = inside diameter of pipe, in.
D_o = outer pipe id or hole-size, in.
D_i = inner pipe od, in.
N_{Re} = Reynolds Number
f = Fanning friction factor
ρ = density, lb/gal
ΔP_f = frictional pressure drop, psi
V_c = critical velocity, ft/sec
P_h = hydrostatic pressure, psi
H = height of fluid column, ft

Downhole Production Equipment

A_i = cross-sectional area based on inside diameter of tubing, $in.^2$
A_o = cross-sectional area based on outside diameter of tubing, $in.^2$
A_p = cross-sectional area through the packer seal, $in.^2$
A_s = cross-sectional area of steel in tubing, $in.^2$
C = coefficient of expansion of steel per °F
d = inside diameter, in.
D = outside diameter, in.
F = force, lb
I = moment of inertia, $in.^4$

E = modulus of elasticity (30 × 10^6 psi, for steel)
μ = Poisson's Ratio (0.3 for steel)
L = length, in.
P_i = pressure inside tubing at the packer seal, psi
P_o = pressure outside tubing above the packer seal, psi
p_i = pressure inside tubing at surface, psi
p_o = pressure outside tubing at surface, psi
r = radial clearance between concentric tubulars, in.
R = ratio of od to id of tubular
ρ_i = density of fluid inside tubing, lb/cu in.
ρ_o = density of fluid outside tubing, lb/cu in.
δ = pressure drop in tubing due to flow, psi/in.
S_i = stress in inside fiber of tubular, psi
S_o = stress in outside fiber of tubular, psi
Δ = change from initial packer setting conditions
T = temperature, °F
w_s = weight of tubing, lb/in.
w_i = weight of fluid contained inside tubing, lb/in.
w_o = weight of annulus fluid displaced by bulk volume of tubing, lb/in.

Completion Fluids

W = weighting material needed, lb/bbl of initial fluid
ρ_f = fluid density desired, lb/gal
ρ_i = density of available brine, lb/gal
ΔV_i = volume increase, bbl per bbl initial fluid
K, C = constants for weighting material

Production Logging

v_s = slippage velocity—difference in velocity between heavier and lighter fluid
$\bar{v}$ = average flow velocity
C = turbulence correction
b_{up} = slope of flowmeter velocity vs rps response curve (up runs)
b_{dn} = slope of flowmeter velocity vs rps response curve (down runs)
ρ_o = density of oil
ρ_w = density of water
ρ_t = density of composite fluid
y_w = water holdup
q_w = flow rate of water
q_o = flow rate of oil
q_t = flow rate of composite fluid
Δt = difference in time
ΔT = temperature difference, °F
A = cross-sectional flow area
hz = sonic frequency hertz
τ = thermal decay time
Σ = capture cross section
Σ_{ma} = capture cross section, matrix
Σ_{sh} = capture cross section, shale
Σ_h = capture cross section, hydrocarbon
Σ_w = capture cross section, formation water
S_w = water saturation
ϕ = porosity
V_{sh} = shale content in rock

English/Metric Units Standards for Metric Conversion Factors*

The following conversion factors are those published by the American Society for Testing and Materials (ASTM) in E380-76. These same units may be found in literature published by all U.S. Technical Societies, i.e., API Bulletin 2563, American National Standards Institute ANSIZ 210.1, Society of Petroleum Engineers, The Canadian Petroleum Association (CPA) and others.

The metric units and conversion factors adopted by the ASTM are based on the ''International System of Units'' (designated SI for Systeme International d'Unites), fixed by the International Committee for Weights and Measures. This system has been adopted by the International Organization for Standardization in ISO Recommendation R-31.

Conversion factors herein are written as a number equal to or greater than one and less than ten with six or less decimal places. This number is followed by the letter E (for exponent), a plus or minus symbol, and two digits which indicate the power of 10 by which the number must be multiplied to obtain the correct value. For example:

(1) 3.523 907 E−02 is $3.523\ 907 \times 10^{-2}$
or
0.035 239 07

(2) 3.386 389 E+03 is $3.386\ 389 \times 10^{3}$
or
3 386.389

(3) Further examples of conversion are:

To convert from:	To	Multiply by:		
pound-force per square foot	Pa	4.788 026 E+01	means-	1 lbf/ft^2 = 47.880 26 Pa
inch	m	2.540 000 E−02		1 inch = 0.0254 m (exactly)

To convert from	To	Multiply by
	ANGLE	
degree (angle)	radian (rad)	1.745 329 E−02
minute (angle)	radian (rad)	2.908 882 E−04
second (angle)	radian (rad)	4.848 137 E−06
	AREA	
acre (U.S. survey)	meter^2 (m^2)	4.046 873 E+03
ft^2	meter^2 (m^2)	9.290 304 E−02
hectar	meter^2 (m^2)	1.000 000 E+04
in^2	meter^2 (m^2)	6.451 600 E−04
mi^2 (U.S. survey)	meter^2 (m^2)	2.589 988 E+06
yd^2	meter^2 (m^2)	8.361 274 E−01
	CAPACITY (See Volume)	
	DENSITY (See Mass Per Unit Volume)	

*The following conversion factors have been selected from Halliburton Services Technical Data Handbook Section 240

To convert from	To	Multiply by
	ELECTRICITY AND MAGNETISM	
abampere	ampere (A)	1.000 000 E+01
abohm	ohm (Ω)	1.000 000 E−09
abvolt	volt (V)	1.000 000 E−08
ampere hour	coulomb (C)	3.600 000 E+03
ohm centimeter	ohm meter (Ωm)	1.000 000 E−02
statampere	ampere (A)	3.335 640 E−10
statohm	ohm (Ω)	8.987 554 E+11
statvolt	volt (V)	2.997 925 E+02
	ENERGY (includes Work)	
British thermal unit (International Table)	joule (J)	1.055 056 E+03
British thermal unit (mean)	joule (J)	1.055 87 E+03
British thermal unit (thermochemical)	joule (J)	1.054 350 E+03
British thermal unit (39°F)	joule (J)	1.059 67 E+03
British thermal unit (59°F)	joule (J)	1.054 80 E+03
British thermal unit (60°F)	joule (J)	1.054 68 E+03
calorie (International Table)	joule (J)	4.186 800 E+00
calorie (mean)	joule (J)	4.190 02 E+00
calorie (thermochemical)	joule (J)	4.184 000 E+00
calorie (15°C)	joule (J)	4.185 80 E+00
calorie (20°C)	joule (J)	4.181 90 E+00
calorie (kilogram, International Table)	joule (J)	4.186 800 E+03
calorie (kilogram, mean)	joule (J)	4.190 02 E+03
calorie (kilogram, thermochemical)	joule (J)	4.184 000 E+03
erg	joule (J)	1.000 000 E−07
ft·lbf	joule (J)	1.355 818 E+00
ft·poundal	joule (J)	4.214 011 E−02
kilocalorie (International Table)	joule (J)	4.186 800 E+03
kilocalorie (mean)	joule (J)	4.190 02 E+03
kilocalorie (thermochemical)	joule (J)	4.184 000 E+03
kW·h	joule (J)	3.600 000 E+06
therm	joule (J)	1.055 056 E+08
	ENERGY PER UNIT AREA TIME	
Btu (thermochemical)/ft^2·s	watt per $meter^2$ (W/m^2)	1.134 893 E+04
Btu (thermochemical)/ft^2·min	watt per $meter^2$ (W/m^2)	1.891 489 E+02
Btu (thermochemical)/ft^2·h	watt per $meter^2$ (W/m^2)	3.152 481 E+00
Btu (thermochemical)/in^2·s	watt per $meter^2$ (W/m^2)	1.634 246 E+06
cal (thermochemical)/cm^2·min	watt per $meter^2$ (W/m^2)	6.973 333 E+02
	FLOW (See Mass Per Unit Time or Volume Per Unit Time)	
	FORCE	
dyne	newton (N)	1.000 000 E−05
kilogram-force	newton (N)	9.806 650 E+00
ounce-force	newton (N)	2.780 139 E−01
pound-force (lbf)	newton (N)	4.488 222 E+00
poundal	newton (N)	1.382 550 E−01

To convert from	To	Multiply by
	FORCE PER UNIT AREA (See Pressure)	
	HEAT	
Btu (International Table)·ft/h·ft^2·°F (*k*, thermal conductivity)	watt per meter kelvin (W/m·K)	1.730 735 E+00
Btu (International Table)/ft^2	joule per meter^2 (J/m^2)	1.135 653 E+04
cal (thermochemical)/cm·s·°C	watt per meter kelvin (W/m·K)	4.184 000 E+02
cal (thermochemical)/cm^2	joule per meter^2 (J/m^2)	4.184 000 E+04
	LENGTH	
angstrom	meter (m)	1.000 000 E−10
foot	meter (m)	3.048 000 E−01
foot (U.S. survey)	meter (m)	3.048 006 E−01
inch	meter (m)	2.540 000 E−02
micron	meter (m)	1.000 000 E−06
mil	meter (m)	2.540 000 E−05
mile (international nautical)	meter (m)	1.852 000 E+03
mile (U.K. nautical)	meter (m)	1.853 184 E+03
mile (U.S. nautical)	meter (m)	1.852 000 E+03
mile (international)	meter (m)	1.609 344 E+03
mile (statute)	meter (m)	1.609 3 E+03
mile (U.S. survey)	meter (m)	1.609 347 E+03
parsec	meter (m)	3.085 678 E+16
yard	meter (m)	9.144 000 E+01
	MASS	
grain	kilogram (kg)	6.479 891 E−05
gram	kilogram (kg)	1.000 000 E−03
hundred weight (long)	kilogram (kg)	5.080 235 E+01
hundredweight (short)	kilogram (kg)	4.535 924 E+01
ounce (avoirdupois)	kilogram (kg)	2.834 952 E−02
ounce (troy or apothecary)	kilogram (kg)	3.110 348 E−02
pennyweight	kilogram (kg)	1.555 174 E−03
pound (lb avoirdupois)	kilogram (kg)	4.535 924 E−01
pound (troy or apothecary)	kilogram (kg)	3.732 417 E−01
slug	kilogram (kg)	1.459 390 E+01
ton (assay)	kilogram (kg)	2.916 667 E−02
ton (long, 2240 lb)	kilogram (kg)	1.016 047 E+03
ton (metric)	kilogram (kg)	1.000 000 E+03
ton (short, 2000 lb)	kilogram (kg)	9.071 847 E+02
	MASS PER UNIT AREA	
oz/ft^2	kilogram per meter^2 (kg/m^2)	3.051 517 E−01
lb/ft^2	kilogram per meter^2 (kg/m^2)	4.882 428 E+00
	MASS PER UNIT CAPACITY (See Mass Per Unit Volume)	
	MASS PER UNIT TIME (Includes Flow)	
lb/h	kilogram per second (kg/s)	1.259 979 E−04

To convert from	To	Multiply by
lb/min	kilogram per second (kg/s)	7.559 873 E−03
lb/s	kilogram per second (kg/s)	4.535 924 E−01
	MASS PER UNIT VOLUME (Includes Density and Mass Capacity)	
grain (lb avoirdupois/7000)/gal (U.S. liquid)	kilogram per meter^3 (kg/m^3)	1.711 806 E−02
g/cm^3	kilogram per meter^3 (kg/m^3)	1.000 000 E+03
oz (avoirdupois)/gal (U.K. liquid)	kilogram per meter^3 (kg/m^3)	6.236 021 E+00
oz (avoirdupois)/gal (U.S. liquid)	kilogram per meter^3 (kg/m^3)	7.489 152 E+00
oz (avoirdupois)/in^3	kilogram per meter^3 (kg/m^3)	1.729 994 E+03
lb/ft^3	kilogram per meter^3 (kg/m^3)	1.601 846 E+01
lb/in^3	kilogram per meter^3 (kg/m^3)	2.767 990 E+04
lb/gal (U.K. liquid)	kilogram per meter^3 (kg/m^3)	9.977 633 E+01
lb/gal (U.S. liquid)	kilogram per meter^3 (kg/m^3)	1.198 264 E+02
lb/yd^3	kilogram per meter^3 (kg/m^3)	5.932 764 E−01
slug/ft^3	kilogram per meter^3 (kg/m^3)	5.153 788 E+02
	PERMEABILITY	
darcy	μm^2	9.869 233 E−01
millidarcy	μm^2	9.869 233 E−04
	POWER	
Btu (International Table)/h	watt (W)	2.930 711 E−01
Btu (International Table)/s	watt (W)	1.055 056 E+03
Btu (thermochemical)/h	watt (W)	2.928 751 E−01
Btu (thermochemical)/min	watt (W)	1.757 250 E+01
Btu (thermochemical)/s	watt (W)	1.054 350 E+03
cal (thermochemical)/min	watt (W)	6.973 333 E−02
cal (thermochemical)/s	watt (W)	4.184 000 E+00
erg/s	watt (W)	1.000 000 E−07
ft·lbf/h	watt (W)	3.766 161 E−04
ft·lbf/min	watt (W)	2.259 697 E−02
ft·lbf/s	watt (W)	1.355 818 E+00
horsepower (550 ft·lbf/s)	watt (W)	7.456 999 E+02
horsepower (boiler)	watt (W)	9.809 50 E+03
horsepower (electric)	watt (W)	7.460 000 E+02
horsepower (metric)	watt (W)	7.354 99 E+02
horsepower (water)	watt (W)	7.460 43 E+02
horsepower (U.K.)	watt (W)	7.457 0 E+02
kilocalorie (thermochemical)/min	watt (W)	6.973 333 E+01
kilocalorie (thermochemical)/s	watt (W)	4.184 000 E+03
	PRESSURE OR STRESS (Force Per Unit Area)	
atmosphere (standard)	pascal (Pa)	1.013 250 E+05
atmosphere (technical = 1 kgf/cm^2)	pascal (Pa)	9.806 650 E+04
bar	pascal (Pa)	1.000 000 E+05
centimeter of mercury (0°C)	pascal (Pa)	1.333 22 E+03
centimeter of water (4°C)	pascal (Pa)	9.806 38 E+01
dyne/cm^2	pascal (Pa)	1.000 000 E−01
foot of water (39.2°F)	pascal (Pa)	2.988 98 E+03

To convert from	To	Multiply by
gram-force/cm^2	pascal (Pa)	9.806 650 E+01
inch of mercury (32°F)	pascal (Pa)	3.386 38 E+03
inch of mercury (60°F)	pascal (Pa)	3.376 85 E+03
inch of water (39.2°F)	pascal (Pa)	2.490 82 E+02
inch of water (60°F)	pascal (Pa)	2.488 4 E+02
millibar	pascal (Pa)	1.000 000 E+02
millimeter of mercury (0°C)	pascal (Pa)	1.333 22 E+02
poundal/ft^2	pascal (Pa)	1.488 164 E+00
lbf/ft^2	pascal (Pa)	4.788 026 E+01
lbf/in^2 (psi)	pascal (Pa)	6.894 757 E+03
psi	pascal (Pa)	6.894 757 E+03
	STRESS (See Pressure)	
	TEMPERATURE	
degree Celsius	kelvin (K)	$t_{°K} = t_{°C} + 273.15$
degree Fahrenheit	degree Celsius	$t_{°C} = (t_{°F} - 32)/1.8$
degree Fahrenheit	kelvin (K)	$t_{°K} = (t_{°F} + 459.67)/1.8$
degree Rankine	kelvin (K)	$t_{°K} = t_{°R}/1.8$
kelvin	degree Celsius	$t_{°C} = t_{°K} - 273.15$
	VISCOSITY	
Centipoise	pascal second (Pa·s)	1.000 000 E−03
centistokes	$meter^2$ per second (m^2/s)	1.000 000 E−06
ft^2/s	$meter^2$ per second (m^2/s)	9.290 304 E−02
poise	pascal second (Pa·s)	1.000 000 E−01
poundal·s/ft^2	pascal second (Pa·s)	1.488 164 E+00
stokes	$meter^2$ per second (m^2/s)	1.000 000 E−04
	VOLUME (Includes Capacity)	
acre-foot (U.S. survey)	$meter^3$ (m^3)	1.233 489 E+03
barrel (oil, 42 gal)	$meter^3$ (m^3)	1.589 873 E−01
fluid ounce (U.S.)	$meter^3$ (m^3)	2.957 353 E−05
ft^3	$meter^3$ (m^3)	2.831 685 E−02
gallon (Canadian liquid)	$meter^3$ (m^3)	4.546 090 E−03
gallon (U.K. liquid)	$meter^3$ (m^3)	4.546 092 E−03
gallon (U.S. dry)	$meter^3$ (m^3)	4.404 884 E−03
gallon (U.S. liquid)	$meter^3$ (m^3)	3.785 412 E−03
in^3	$meter^3$ (m^3)	1.638 706 E−05
liter	$meter^3$ (m^3)	1.000 000 E−03
ounce (U.K. fluid)	$meter^3$ (m^3)	2.841 307 E−05
ounce (U.S. fluid)	$meter^3$ (m^3)	2.957 353 E−05
pint (U.S. dry)	$meter^3$ (m^3)	5.506 105 E−04
pint (U.S. liquid)	$meter^3$ (m^3)	4.731 765 E−04
quart (U.S. dry)	$meter^3$ (m^3)	1.101 221 E−03
quart (U.S. liquid)	$meter^3$ (m^3)	9.463 529 E−04
ton (register)	$meter^3$ (m^3)	2.831 685 E+00
yd^3	$meter^3$ (m^3)	7.645 549 E−01

To convert from	To	Multiply by
	VOLUME PER UNIT TIME (Includes Flow)	
ft^3/min	$meter^3$ per second (m^3/s)	4.719 474 E−04
gallon (U.S. liquid)/hp·h (SFC, specific fuel consumption)	$meter^3$ per joule (m^3/J)	1.410 089 E−09
in^3/min	$meter^3$ per second (m^3/s)	2.731 177 E−07
yd^3/min	$meter^3$ per second (m^3/s)	1.274 258 E−02
gallon (U.S. liquid) per day	$meter^3$ per second (m^3/s)	4.381 264 E−08
gallon (U.S. liquid) per minute	$meter^3$ per second (m^3/s)	6.309 020 E−05
	WORK (See Energy)